Analog IC Design with Low-Dropout Regulators

About the Author

Gabriel Alfonso Rincón-Mora (B.S., M.S., Ph.D.) was a senior integrated circuit (IC) design team leader for Texas Instruments from 1994 to 2003. He was appointed an adjunct professor at the Georgia Institute of Technology in 1999, where he became a full-time faculty member in 2001 and is now a full professor. Since 2011, Dr. Rincón-Mora has also been a visiting professor at the National Cheng Kung University in Taiwan. He is an IEEE Fellow and an IET Fellow, and the inventor/co-inventor of 38 patents and author/co-author of more than 160 publications. Dr. Rincón-Mora has written 8 books and designed more than 26 commercial power microchips. He received the National Hispanic in Technology Award from the Society of Professional Hispanic Engineers, the Charles E. Perry Visionary Award from Florida International University, a Commendation Certificate from the Lieutenant Governor of California, the IEEE Service Award from IEEE CASS, and the Orgullo Hispano and the Hispanic Heritage awards from Robins Air Force Base. Georgia Tech inducted Dr. Rincón-Mora into the Council of Outstanding Young Engineering Alumni in 2000 and *Hispanic Business* magazine named him one of "The 100 Most Influential Hispanics" in 2000. His current research is on IC systems that draw power from tiny batteries and ambient energy to supply wireless and mobile devices.

Analog IC Design with Low-Dropout Regulators

Gabriel Alfonso Rincón-Mora, Ph.D.
Georgia Institute of Technology
Atlanta, Georgia

Second Edition

McGraw Hill Education

New York Chicago San Francisco
Athens London Madrid
Mexico City Milan New Delhi
Singapore Sydney Toronto

McGraw-Hill Education books are available at special quantity discounts to use as premiums and sales promotions or for use in corporate training programs. To contact a representative, please visit the Contact Us page at www.mhprofessional.com.

Analog IC Design with Low-Dropout Regulators, Second Edition

1 2 3 4 5 6 7 8 9 0 DOC/DOC 1 2 0 9 8 7 6 5 4

ISBN 978-0-07-182663-1
MHID 0-07-182663-7

The pages within this book were printed on acid-free paper.

Sponsoring Editor
Michael McCabe

Editorial Supervisor
Stephen M. Smith

Production Supervisor
Lynn M. Messina

Acquisitions Coordinator
Bridget L. Thoreson

Project Manager
Tania Andrabi,
Cenveo® Publisher Services

Copy Editor
Kevin Campbell

Proofreader
Linda Manis Leggio

Indexer
Robert Swanson

Art Director, Cover
Jeff Weeks

Composition
Cenveo Publisher Services

To my Mami, Papi, and Hermano,
for all I am,
and all I'm not.

And to my beautiful Tai Tai,
for her sweet love,
and unyielding support.

Contents

Preface

My objective with the first edition of this book was to introduce, discuss, and illustrate how to design, simulate, build, test, and evaluate linear low-dropout (LDO) regulator integrated circuits (ICs). The driving inspiration for this effort is the important role that LDO regulator ICs play in modern-day and emerging state-of-the-art applications, as the demand and promise of system-on-chip (SoC) integration continues to drive old markets and create new ones. The fact is, the ubiquity of noisy and unpredictable energy sources and functional loads demands point-of-load (PoL) regulators that draw little to no power, yet generate accurate and fast-responding power supplies. As a result, analog and mixed-signal ICs that traditionally excluded power-conditioning features must now embed system and PoL power supplies, of which linear regulators comprise a large fraction because their switching counterparts alone generate outputs with unacceptably high noise content.

However, a pedagogical presentation of linear regulators is incomplete without the analog IC foundation on which these devices rest. This book therefore presents analog theory, as other books in literature also do, but from an intuitive, design-oriented perspective, one that I have found useful and necessary when designing ICs. The aim is to develop the understanding necessary to predict how semiconductor devices behave individually and collectively in circuits without resorting to cookbook equations, whose underlying premises do not always apply, especially when exploring new technologies. With this kind of insight, reproducing and verifying the equations and theory already found in textbooks is more manageable.

This Second Edition

My goals with this second edition are to expand, improve, and update the presentation so concepts and the state of the art are more clear and insightful. The result of this undertaking is in many ways an entirely new book. For starters, I reorganized the chapters and rewrote the text. I also updated almost all equations and figures, and added examples, review questions, and new sections to nearly every chapter. Plus, the book now includes a chapter on bias currents and reference circuits, on which linear regulators, like most analog systems, rely to wake and function properly.

Chapter 1 from the first edition, for example, splits in two in this rendition, one on power-supply systems and the other one more specifically on linear regulators. Chapter 3 similarly splits in two, one on single-transistor primitives and the other on analog building blocks. Chapters 1–3 now include new sections on bandwidth delay, figures of merit, open design variables, bias points, small signals, absolute and relative accuracies, and subthreshold, weak inversion, MOS capacitors, and channel resistors in metal–oxide–semiconductor field-effect transistors (MOSFETs). New sections in Chaps. 4 and 5 describe two-port models, frequency response, signal flow, signal propagation, gain analysis, direct translations, base degeneration, base/gate-coupled differential pairs, and folded-cascode amplifiers, including discussions on slew rate, power-supply rejection, input-referred offset, and noise.

The presentation on negative feedback in Chap. 6 is entirely new, although some of the basic concepts are not. Here, sections on sensitivity, frequency response, noise, linearity, feedback configurations, and embedded and parallel feedback loops are new. The chapter also describes 13 feedback examples, and offers insight into the design process of negative-feedback loops. Chapter 7 is the new chapter already mentioned, which covers proportional-to-absolute-temperature (PTAT) and complementary-to-absolute-temperature (CTAT) currents, temperature compensation, startup, frequency compensation, and noise suppression in bandgap circuits. And Chaps. 10 and 11 now also describe how to improve power-supply rejection and trim references.

Intended Readers

I wrote the book with the intention of introducing and leading a novice electrical engineer through the entire analog IC design process from the perspective of a linear regulator, which incorporates numerous aspects of the art. The book also aims to update practiced analog designers with little experience in the field of power electronics on linear-regulator concepts and ICs. I think experienced power IC engineers will also find the review of analog and linear-regulator principles refreshing and insightful, and the state of the art in regulator ICs enlightening and stimulating.

Organization

I divided the book into 11 chapters. Chapters 1 and 2 are analogous to the product-definition phase, but with an academic undertone, when a semiconductor company justifies a design effort by defining the role and operational objectives of, in this case, a linear regulator. Before embarking on the design process, however, an inexperienced engineer must first train in the art of analog IC design. The purpose of Chaps. 3–7 is just this, to explain solid-state theory and devices, single-transistor primitives, analog building blocks, negative feedback, and bias circuits. With this background, Chap. 8 returns to linear regulators, and more specifically, to their small-signal response, which corresponds to the second step in the prototype-development cycle. Here, the designer applies the circuit and feedback principles discussed in Chaps. 3–6 to begin designing the system according to the needs described in Chaps. 1–2.

Chapters 9–10 combine the device know-how presented in Chap. 3 with the circuit teachings of Chaps. 4–6 and the compensation strategy devised in Chap. 8 to design the actual regulator IC, first at the component level in Chap. 9 and then at the system level in Chap. 10. The development process culminates in these two chapters because all analog training converges here, with the design of the IC. The last chapter in the book, Chap. 11, is on IC protection and characterization, which constitute the last two steps of the design cycle. As a whole, the book is an example of a top–down–top design because it starts with an abstract view of the system for context, and then dives down to devices for training. The book later ascends slowly through circuits until again reaching the system, but now in the form of a final transistor-level realization.

A novice engineer may use this book to learn about analog IC design by reviewing the entire design process, traversing through all eleven chapters sequentially. They may also seek to enhance their understanding of specific analog-design principles, in which case they might target specific chapters like Chaps. 3–7 on devices, circuits, feedback, and references; Chaps. 1–2 and 8–10 on linear regulator ICs; and Chap. 11 on protection and characterization. Trained analog designers with little regulator experience who do not wish to review basic analog principles but want to design linear regulator ICs may refer to Chaps. 1–2 and 8–11 for regulator-specific issues. Experienced regulator IC designers may also pinpoint specific sections in Chaps. 1–2 and 8–10 to enhance their understanding of the state of the art and in Chaps. 3–7 to review analog IC principles.

To facilitate the process of targeting chapters, sections, and subsections, I divided each chapter into self-inclusive sections. I also partitioned sections into what I thought were relevant and subject-specific subsections, and assigned headings that I thought were insightful. My hope is that, this way, engineers can more easily focus their attention on their particular areas of interest, and use the Contents to jump and navigate to related sections.

The Writing

As a whole, the book presents a fairly comprehensive treatment of analog IC design, from solid-state semiconductor theory, circuit design and analysis of basic analog building blocks, feedback concepts, and bias circuits to frequency response, IC design, and circuit protection and characterization. What sets this work apart from other analog books is the design insight of analog ICs and its application to reference circuits and linear regulators. The tone, format, and thought process presented in the book embodies more than 20 years of my experience in the field, first as an analog IC designer developing commercial power-supply microchips and then as a professor and researcher advancing the state of the art in power- and energy-conditioning ICs.

From industry, I discovered the art of design and the value of product development, which is why the book places emphasis on insight and intuition, system objectives, reliability, and the design process. As an academic, I continue to learn the art of a pedagogical presentation, and the value of technical depth and outside-the-box thinking. What the reader will therefore see in this book is my attempt at drafting a practical, yet academically valuable treatment of analog IC design and linear regulator ICs. I still have much to learn, though, so I hope my devotion to the book and the field at large ultimately wins enough of the reader's favor to pardon deficiencies, inconsistencies, and inaccuracies in this presentation.

Gabriel Alfonso Rincón-Mora, Ph.D.

Analog IC Design with Low-Dropout Regulators

CHAPTER 1
Power-Supply Systems

1.1 Regulators in Power Management

Supplying and conditioning power are the most fundamental functions of an electrical system. A loading application, be it a cellular phone, tablet, or wireless sensor node, cannot sustain itself without energy, and cannot fully perform its functions without a stable supply. The fact is, transformers, generators, batteries, and other off-line supplies incur substantial voltage and current variations with time and over a wide range of operating conditions. They are normally noisy and jittery, not only because of the way they operate but also because high-power switching circuits like central processing units (CPUs) and digital signal processors (DSPs) usually load them. These rapidly changing loads incur transient excursions that cause the supposedly noise-free supply voltage to flicker and exhibit frequency spurs where only a direct-current (dc) component should exist. The role of the voltage regulator is to convert these unpredictable and noisy supplies to stable, constant, accurate, and load-independent voltages, attenuating these ill-fated fluctuations to lower and more acceptable levels.

The regulation function is especially important in high-performance applications where systems are increasingly more integrated and complex. A *system-on-chip* (SoC) solution, for example, incorporates numerous functions, many of which switch simultaneously with the clock, demanding both high power and short response times within short consecutive bursts. Not responding quickly to one of these fast load-current transitions, or *load dumps*, forces storage capacitors to source the full load and subsequently incur considerable transient variations in the supply voltage. The bandwidth performance of the regulator—which refers to its ability to respond quickly—determines the magnitude and extent of these transient fluctuations.

Regulators also protect and buffer integrated circuits (ICs) from exposure to voltages exceeding junction-breakdown levels. The requirement is more stringent and acute in emergent state-of-the-art technologies whose breakdown voltages can be less than 2 V. The growing demand for space-efficient, single-chip solutions, which includes SoC, *system-in-package* (SiP), and *system-on-package* (SoP) implementations, drives process technologies to finer photolithographic and metal-pitch dimensions. Unfortunately, the maximum voltage an IC can sustain before the onset of a breakdown failure drops with decreasing dimensions and pitch because, as the component density increases, isolation barriers drop.

References, like regulators, generate and regulate accurate and stable output voltages that are impervious to variations in the input supply, loading environment, and various operating conditions. Unlike regulators, however, references do not supply substantial steady-state currents, nor do they accommodate widely variable loads. Although a good reference may shunt positive and negative noise currents, its total load-current reach is still relatively low. In practice, references supply up to 1 mA and regulators above that to several amperes.

NOMENCLATURE *To complement and augment the verbal explanations presented in this book, variable names conform to standard small-signal and steady-state naming conventions. Signals incorporating both small-signal and steady-state components use a lowercase name with uppercase subscripts: for example, output voltage* v_{OUT}. *When referring only to the steady-state component, variables use only capital letters, as in* V_{OUT}, *and similarly, when only referring to small-signal values, the entire name, including subscripts, is in lowercase, as in* v_{out}.

As also illustrated by the previous example, the variables adopt functionally insightful and intuitive names. The first letter usually describes the signal type and its dimensional units, such as v for voltage, i for current, A or G for amplifying gain, P for power, and so on. The subscript tends to describe the function or node to which the variable is attached, such as out for the output of the regulator, REG for a regulated parameter, and so on. Additionally, parameters with a zero in the subscript typically refer to variables that describe low-frequency behavior, such as low-frequency voltage gain A_{V0}. *These low-frequency metrics ultimately convey steady-state information about a system.*

1.2 Linear versus Switched Regulators

A voltage regulator is normally a buffered reference: a bias voltage cascaded with a noninverting operational amplifier or Op Amp, as is commonly known, capable of driving high load currents in a shunt-feedback configuration. Bearing in mind the broad range of load currents possible, regulators generally fall under one of two broad classifications:

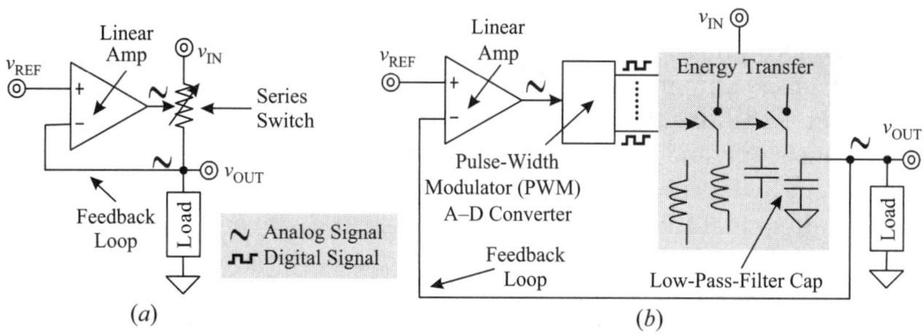

FIGURE 1.1 (*a*) Linear and (*b*) switched regulator circuits.

linear or *switched*. Linear supplies, also called *series* regulators, linearly modulate the conductance of a series pass switch connected between an input supply and the regulated output v_{OUT} to ensure the output voltage is a predetermined ratio of an established reference voltage v_{REF}, as illustrated in Fig. 1.1*a*. The linear feedback amplifier it incorporates compares v_{OUT} against v_{REF} to generate a control signal that keeps v_{OUT} within an acceptably small window of v_{REF}. The term *series* refers to the pass element or switch device that is in series with the unregulated supply and the load. Since the switch's current flow and its control signal are continuous in time, the circuit is linear and analog in nature, and because it can only supply power through a linearly controlled series switch, its output voltage cannot exceed its unregulated input supply: $v_{OUT} < v_{IN}$.

A switched regulator is the counterpart to the linear solution, and because of its switching nature, it can accommodate both alternating-current (ac) and direct-current (dc) input and output voltages, which is why it can support ac–ac, ac–dc, dc–ac, and dc–dc conversion functions. Within the context of ICs, however, dc–dc converters predominate because the ICs derive power from available dc batteries and off-line ac–dc converters, and most loading applications in the IC and outside of it demand dc supplies to operate. Nevertheless, given its conversion capabilities, engineers often call this class of regulator circuits *switched* or *switching converters*.

From a circuit perspective, the driving difference between linear and switched regulators is that the feedback loop of the latter is mixed signal in nature, incorporating both analog and digital components, like Fig. 1.1*b* shows. The basic aim of the switched circuit is to energize inductors and/or capacitors from the supply and drain them into the load in alternate cycles of the switching period, transferring input energy to the output via quasi-lossless energy-storage devices. To control the network, the system feeds back and processes an analog signal before converting it into a *pulse-width modulated* (PWM) digital-pulse train whose on–off states determine the connectivity of the switches in the circuit. From a signal-processing perspective, the function of the switching network is to low-pass-filter the supply-level swings of the digital train down to a millivolt analog signal v_{OUT} whose average remains near v_{REF}.

As Fig. 1.1*b* illustrates, a dc–dc converter system normally incorporates transistors and/or diodes as synchronous or asynchronous switches, inductors and/or capacitors as energy-transfer components, a linear differential amplifier, and an analog–digital converter or pulse-width modulator. As in the linear regulator, the purpose of the amplifier is to compare v_{OUT} against v_{REF} and generate a control signal that keeps v_{OUT} near v_{REF}. The PWM block then converts the amplifier's analog output into a digital stream that determines the connectivity of the network. Many switched-capacitor implementations do not require power inductors, so total on-chip integration is sometimes possible. These integrated inductorless converters, however, typically cannot supply the high current levels that discrete power inductors can without allowing v_{OUT} to droop excessively, which is why they normally satisfy a relatively smaller market niche in low-power applications. Because these inductorless circuits "pump" charge in and out of "flying" capacitors, IC design engineers often refer to them as *charge pumps*.

Switched regulators, unlike their linear counterparts, are capable of generating a wide range of output voltages, including values below and above the input supply. *Buck* converters, for example, generate output voltages that are lower than the input supply: $v_{OUT} < v_{IN}$, while *boost* converters deliver the opposite: $v_{OUT} > v_{IN}$. *Buck–boost* converters, as the name implies, are a combination of both buck and boost circuits, and they are consequently capable of regulating output voltages both above and below the input supply. In spite of the apparent flexibility and advantages of switching supplies, however, linear regulators are also popular in consumer and high-performance electronics, as the next subsection discusses.

1.2.1 Response-Time Tradeoffs

Linear regulators tend to be simpler and faster than switched converters. For one, there are fewer components in the feedback loop of a linear regulator, as Fig. 1.1 indicates, so fewer components delay the feedback signal across the loop. The pulse-width modulator, on the other hand, which converts an analog signal into a pulse train, often requires several blocks, such as a clock, comparators, nonoverlapping digital drivers, and a sawtooth triangular-wave generator. Plus, for a switched regulator in negative feedback to remain stable, its bandwidth must normally fall below its switching frequency f_{SW} by about a decade or so of frequency, further limiting the system's response time to orders of magnitude below the transitional frequency f_T of the transistors available. As a result, dc–dc converters require more time to respond than linear regulators: 2–8 μs versus 0.25–1 μs, where f_{SW} typically falls between 20 kHz and 10 MHz. Although higher switching frequencies can reduce output voltage ripple and/or relax LC filter requirements, raising f_{SW} is often impractical because doing so increases switching power losses in the converter and therefore dissipates more battery energy and reduces operational life.

1.2.2 Noise

Switched regulators are noisier than their linear counterparts are, and Fig. 1.1 illustrates this by the presence of digital signals in the feedback path. Power switches, which are large devices that conduct vast amounts of current, switch at relatively high frequencies, so they require fast and abrupt drive signals that inject noise energy into the substrate and output. What is more, radio-frequency (RF) noise is more prevalent in boost configurations because a diode duty-cycles power abruptly into the load. Start–stop clock operation for on–off sleep-mode transitions further aggravates noise content, adding low- and high-frequency harmonics to the output.

1.2.3 Power-Conversion Efficiency

Switched regulators have one redeeming quality when compared to linear regulators: they are power efficient. The fact is that the voltages (and power) dropped across the power switches in a dc–dc converter are far lower at 10–100 mV, for example, than that of the series pass device of a linear regulator, whose difference between unregulated input v_{IN} and regulated output v_{OUT} can be 0.3–2 V. Allowing the regulator to dissipate more power and therefore deliver a smaller fraction of input power to the output amounts to lower *power-conversion efficiency*, which is an important metric in power-conditioning circuits. Design engineers often define this parameter η_C as the ratio of power delivered to the output P_{OUT} to input power P_{IN}, where the latter accounts for P_{OUT} and consumed losses in the regulator P_{LOSS}:

$$\eta_C \equiv \frac{P_{OUT}}{P_{IN}} = \frac{P_{IN} - P_{LOSS}}{P_{IN}} = \frac{P_{OUT}}{P_{OUT} + P_{LOSS}} \tag{1.1}$$

P_{LOSS} can be so low in switched regulators that they commonly achieve efficiencies between 80 and 95 percent. P_{LOSS} in linear regulators, on the other hand, is often higher because the application typically superimposes what the unregulated supply v_{IN} and regulated output v_{OUT} should be. As a result, the difference between v_{IN} and v_{OUT} and the circuit's quiescent-current flow i_Q limit efficiency to lower levels:

$$\eta_{C(LIN)} = \frac{P_{OUT}}{P_{OUT} + P_{LOSS}} = \frac{i_{LOAD} v_{OUT}}{(i_{LOAD} + i_{GND}) v_{IN}} < \frac{v_{OUT}}{v_{IN}} \tag{1.2}$$

where i_{LOAD} is load current and quiescent current i_Q flows to ground as i_{GND}, not the load. The maximum possible efficiency a series regulator can attain is therefore the ratio of the output and input supply voltages, even when ground current is zero. To comprehend the meaning of this conclusion, consider that the maximum power efficiency a 2.5-V linear regulator can ever achieve when powered from a 5-V supply is 50 percent.

Notwithstanding, power efficiency in a linear regulator rises with lower input–output voltage differentials. For instance, if in the above-stated example the regulator drew power from a 3.3-V supply, the efficiency would have been 76 percent, or from a 2.8-V supply, 89 percent. This trend holds true only if load current i_{LOAD} is considerably greater than quiescent current i_Q, which is typically the case when confronted with a full load, but not when the system idles or sleeps. Consequently, when the voltage drop between the unregulated supply and the output is relatively low, below 0.3 V, for example, designers often prefer linear regulators over their switching counterparts because efficiencies are on par and the circuit is simpler, less expensive, less noisy, and faster. Their only, though significant, drawback is limited power efficiency, and if that is not an issue, or its value is equivalent to that of a switching converter, a linear regulator is often the better choice.

Raising load currents to the point where the system requires heat sinks is detrimental to the application. A heat sink is an additional onboard component that demands area on the *printed-circuit board* (PCB). A common technique used to circumvent this disadvantage is to split the load between several linear regulators sprinkled across the PCB to minimize the power dissipated by each regulating IC, or by using a switched regulator, if the regulated output can tolerate more noise. Another negative side effect of high temperature is higher *metal–oxide–semiconductor* (MOS) switch-on resistance, the results of which are higher conduction losses and lower power-conversion efficiencies. In all, as Table 1.1 summarizes, linear regulators are simpler and faster, and generate lower noise, but their limited efficiency constrains them to lower-noise and lower-power applications. Switched regulators enjoy higher power-conversion efficiencies, but the loads they sustain

Linear Regulators	Switched Regulators
Limited output range: $v_{OUT} < v_{IN}$	√ Flexible: $v_{OUT} \leq$ or $\geq v_{IN}$
√ Low cost: Less PCB and Si area	Expensive: More PCB and Si area
√ Low noise	Switching noise
√ Quick to respond	Slow to respond
Limited efficiency: $\eta_C \leq v_{OUT}/v_{IN}$	√ High efficiency: $\eta_C \approx$ 80%–95%
Low-noise/-power applications	Boosting and high-efficiency systems

Check mark (√) denotes positive attributes (i.e., advantages).

Table 1.1 Comparison of Linear and Switched Regulators

must tolerate higher noise levels, which is why linear regulators in series with switched regulators normally channel power to sensitive high-performance analog subsystems.

1.3 Market Demand

1.3.1 System

Both linear and switched regulators claim their place in today's market. Systems like desktop and laptop microprocessors not only demand substantial amounts of clock-synchronized currents but also low supply voltages. Such systems reap the benefits of power-efficient dc–dc converters. Circuit blocks serving purely analog functions, on the other hand, cannot sustain the noisy supplies that switching regulators generate, so they exploit the low-noise and cost-effective advantages of linear regulators. These analog circuits are inherently more sensitive to noise originated in the supply rails than are digital blocks, which is why they require "cleaner" power supplies.

Today's growing market demand for portable electronics like cellular phones, tablets, laptops, and the like requires the use and coexistence of both linear and switched regulators, since both accuracy and power-conversion efficiency are paramount. In these applications, the integrated power-management circuit drives noise-sensitive circuits from noisy and variable input supply voltages. A dc–dc converter under these conditions steps down the input supply to a lower voltage level, generating in the process a regulated but noisy supply voltage, as v_{NOISY} in Fig. 1.2 depicts. A linear regulator then draws power from this noisy supply to generate a low-noise, ripple-rejected output, like v_{CLEAN}, that can now supply high-performance, noise-sensitive ICs. This way, the voltage across the linear regulator can be low enough to limit its power losses to practical levels, like from 2.4 to 1.8 V in the case of Fig. 1.2. Here, the purpose of the switched supply circuit is to drop as much of the input voltage as possible to save power, since it is more power efficient than the linear regulator. The function of the

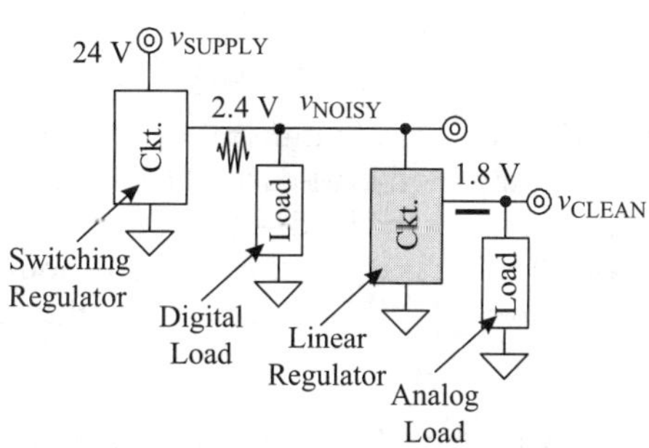

Figure 1.2 Sample low-noise power-management system.

linear regulator is to filter the noise that the dc–dc converter produces and generate the noise-free supply that the system ultimately requires.

Similar operating conditions to those just described arise in many other mixed-signal applications, where active power-supply decoupling is necessary to reduce and suppress noise. Systems demanding high-output voltages from low-input sources, as in the case of single- and dual-battery packs that supply 0.9–1.5 V and 1.8–3 V, for example, require boosting dc–dc converters. As in the previous case, the application may still need series regulators to suppress the switching noise that the switched regulator generates.

1.3.2 Integration

The mobile market's impact on the demands of regulators is profound. Because of high variations in battery voltage, virtually all battery-operated systems require regulators. What is more, most designs find it necessary to include regulators and other power-supply circuits *in situ*, on the same microchip as the system to save PCB real estate and improve performance. This trend is especially prevalent in products that strive to achieve or approach the fundamental limits of integration in the form of SoC, SiP, and SoP solutions. Since volumetric space is scarce, limited energy and power densities are the natural byproducts of such a market, which means surviving practicable lifetime amounts to conditioning power efficiently with low quiescent current.

1.3.3 Operational Life

Current efficiency η_I, which refers to how much of the input current i_{IN} reaches the load, is vital in linear regulators. To be more specific, quiescent current i_Q lost as ground current i_{GND} must be as low as possible during low- and near-zero loading conditions because i_{GND} otherwise constitutes a significant fraction of the battery's total drain current i_{IN}. Fortunately, higher quiescent currents are more acceptable during heavy loading events because their impact on total drain current and, therefore, battery life is miniscule, which is why current efficiency, and not the absolute value of i_{GND}, is important in linear regulators:

$$\eta_I = \frac{i_{OUT}}{i_{IN}} = \frac{i_{LOAD}}{i_{LOAD} + i_{GND}} \tag{1.3}$$

Ultimately, the load alone usually determines the single-charge lifetime of the battery during high load-current conditions and the regulator's ground current when the load is zero or nearly zero.

The *capacity* of a battery is defined in ampere–hours (Ah) and its averaged drain current $i_{DRAIN(AVG)}$ across time sets the operational life of the electronic system:

$$\text{Life [h]} = \frac{\text{Capacity [Ah]}}{i_{DRAIN(AVG)}} = \frac{\text{Capacity [Ah]}}{\text{Avg}\{i_{LOAD} + i_{GND}\}} \tag{1.4}$$

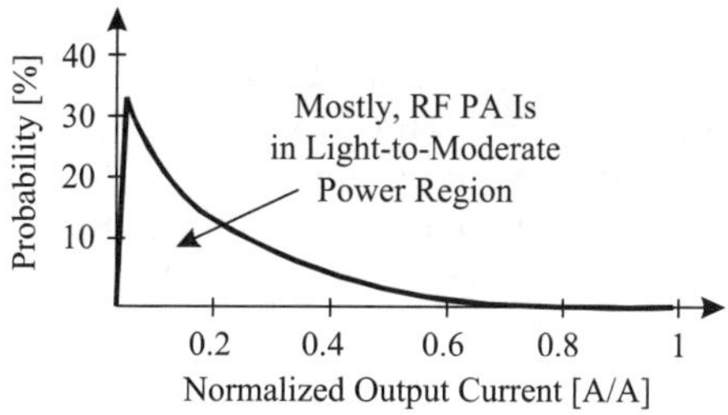

FIGURE 1.3 Probability-density function of the RF power amplifier in a typical portable CDMA handset application.

This relationship, coupled with the fact that the majority of portable devices are idle most of the time, means battery life is a strong function of low load-current conditions, of i_{GND}. A cellular phone, for instance, mostly idles in *alert* mode and consequently requires only a fraction of its peak *talk*-mode power, as depicted in Fig. 1.3 by the *probability-density function* (PDF) of the RF *power amplifier* (PA) of a typical *code-division multiple-access* (CDMA) handset. As shown, the region of highest probability is the zero-to-moderate load-current range, which is where drain current is, for the most part, simply i_{GND}:

$$i_{DRAIN(AVG)} = \int (i_{LOAD} + i_{GND}) \times PDF \times di_{LOAD} \approx i_{GND} \tag{1.5}$$

1.3.4 Supply Headroom

Battery power and state-of-the-art process technologies also imply low-voltage operation. Today's most popular rechargeable (i.e., secondary) battery technologies are *lithium-ion* (Li-Ion), *nickel–cadmium* (NiCd), and *nickel–metal–hydride* (NiMH) batteries, the first of which ranges from 2.7 V when completely discharged to 4.2 V when fully charged and the latter two from 0.9 to 1.7 V under similar states. Microscale fuel cells have even lower voltages, at approximately 0.4–0.7 V per cell. The variable nature of these relatively low-voltage technologies superimposes stringent requirements on the regulator, limiting the amount of supply-voltage *headroom* and *dynamic range* (or *signal-to-noise ratio*, SNR) that circuits have to perform their prescribed functions.

Low-voltage operation is also a consequence of advances in process technologies. The push for higher packing densities forces technologies to improve their photolithographic resolution, fabricating, in the process, nanometer-scale junctions that exhibit inherently lower breakdown voltages. A typical 0.18-μm *complementary* MOS (CMOS) technology, for example, cannot sustain more than roughly 1.8 V. Still more, since financial considerations limit the complexity of the process, that is to say, limit the number of masks used to fabricate the chip and, as a result, the variety of devices available, standard (*vanilla*) CMOS and stripped *bipolar*–CMOS (BiCMOS) process technologies appeal most to semiconductor companies, which translates to less flexibility for the circuit designer.

A low-voltage environment is fundamentally restrictive for an analog IC designer. Many traditional design techniques are prohibitive under these conditions, limiting flexibility and, ultimately, performance. *Cascode* transistors, emitter and source *followers*, and *Darlington*-configured NPN *bipolar-junction transistors* (BJTs), to cite some examples, which are useful for increasing gain, bandwidth, and peak output current, require more supply-voltage headroom than their nonassisted counterparts, and headroom is a precious commodity in battery-operated environments. Because dynamic range suffers at lower supply voltages, low voltages also imply higher accuracy, demanding higher performance from already fundamentally constrained circuits. A 1-percent 1.8-V regulator, for example, must have a total variation of less than 18 mV across load-current, supply-voltage, temperature, process, and noise extremes. These issues typically give rise to more complex and usually more expensive chips, to ICs that require more silicon area and/or more exotic circuits and process technologies. Of course, keeping costs down forces designers to be more resourceful and innovative.

1.4 Power Sources

Fully comprehending the applications in which many, if not most, linear regulators find a home is important. Typical parameters to consider in a power source, for one, from an IC designer's perspective, are capacity, internal resistance, self-discharge, physical size and weight, and if the source is rechargeable, *cycle life*. Cycle life refers to the number of discharge–recharge cycles a battery can endure before its capacity, which refers to its ability to store energy, deteriorates significantly. There are several types of sources, ranging from conventional batteries and emerging fuel cells to atomic batteries and transducers that harvest ambient energy from the surrounding environment. Unfortunately, in spite of considerable advances in all relevant areas of research and development, no single technology can suit all possible applications.

1.4.1 Early Batteries

Most portable electronics today use nickel- or lithium-based chemistries to power their systems. Reusable alkaline and lead-acid batteries are not as suitable for high-performance applications. Alkaline batteries, for example, benefit from long shelf life, but they suffer from short cycle life and low power densities. As a result, their attributes suit best the range of consumer electronics and gadgets that demand reliable

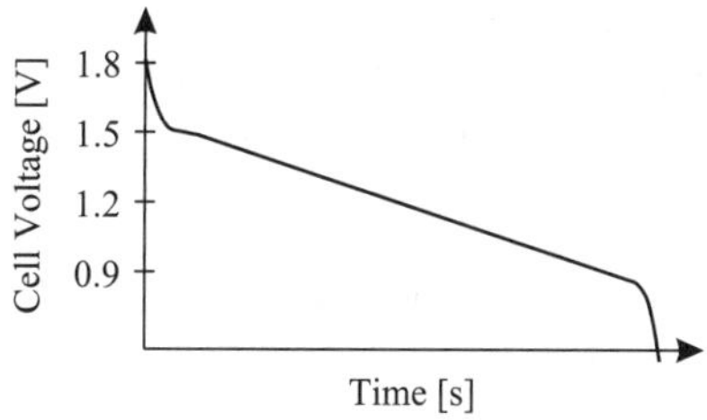

Figure 1.4 Typical NiCd and NiMH discharge profile when loaded with a constant current.

operation on an infrequent basis, like flashlights. Lead-acid batteries are economical and output high power, but they are bulky, which is why they are popular in large-scale applications like the automobile industry, but not in the portable handheld consumer market.

Early cellular phones derived power from nickel-based sources, such as NiCd and NiMH batteries. These nickel-based solutions, however, suffer from a phenomenon known as *cyclic memory*, which refers to how crystalline formation in the battery causes self-discharge rates to rise over time. Fortunately, periodic discharge–charge cycles mitigate these ill-fated effects. Figure 1.4 illustrates the typical discharge profile of these nickel-based devices. Note that, since constant loads discharge them quickly from about 1.8 to 1.5 V and again below 0.9 V, most of their usable energy is in the 0.9–1.5-V range.

NiCd batteries, which contain toxic metals, are the predecessors of the NiMH solution, which is more environmentally friendly and yields slightly higher energy densities and lower memory effects. The advantages of the NiMH chemistry, however, come at the cost of other performance metrics. Figure 1.5, for example, illustrates how NiCd eventually outperforms NiMH in almost every way over their cycle life. Not only is internal equivalent resistance R_{ESR} lower for NiCd batteries but capacity, internal resistance, and self-discharge also remain relatively constant throughout their full cycle life, roughly

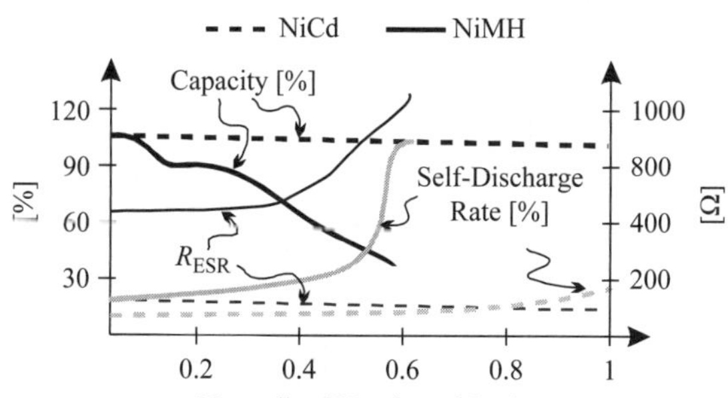

Figure 1.5 Comparative performance of NiCd and NiMH batteries.

1500 cycles' worth. NiMH batteries perform well at first, but quickly degrade within 20 percent of the NiCd's full life after a small number of charge cycles, limiting their life to roughly half the usable life of the NiCd counterpart. NiMH devices, nevertheless, appeal to a sector of the electronics industry that values environmentally friendly features and whose life expectancy is relatively short, on the order of a year, limiting the number of recharge cycles to within the range of NiMH technologies.

1.4.2 Lithium-Ion Batteries

Next in the evolutionary chain of rechargeable energy-storage devices are lithium-ion batteries, or Li Ions for short. To begin, they do not contain toxic metals or the dreaded memory effects from which their nickel-based counterparts suffer. Energy densities are also higher, and capacity and internal resistance are relatively constant over most of their entire cycle life, which extends up to approximately 1000–1500 cycles. Plus, their self-discharge rates are miniscule when compared to nickel-based technologies. Operationally, most of their usable energy falls within the 2.7–4.2-V range, as the discharge profile of Fig. 1.6 illustrates. One drawback to note, however, is that discharging or charging them beyond their minimum- and maximum-rated limits of nearly 2.7 and 4.2 V causes irreversible and sometimes catastrophic effects, which is why the charging circuit for these batteries is often more complex than for other technologies.

Another disadvantage of this technology is its higher cost premium: it is roughly twice the cost of nickel-based technologies. This is not to say manufacturing costs will not continue to fall in the future, because they will, as more and more products incorporate Li Ions and manufacturing companies supply them to reap the benefits of economies of scale. Increased sales volume will also spark innovation and advancements in technology that reduce cost, ease manufacturability, and improve reliability. At a slightly higher price, Li-Ion polymers offer similar performance, but with the ability to conform to thinner and smaller packages, which is important in handheld, wearable, and wireless-sensor applications. In all, high volumes and technical

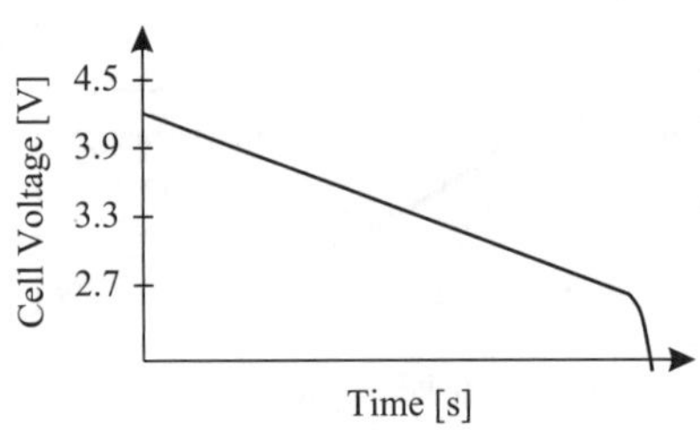

FIGURE 1.6 Typical Li-Ion discharge profile when loaded with a constant current.

superiority ultimately offset cost disadvantages to such an extent that most modern and emerging cellular phones, tablets, laptop PCs, and other portable consumer electronics rely on Li Ions for power.

1.4.3 Fuel Cells

Li Ions are mainstream, but not perfect. They do not store sufficient energy per unit weight or volume to supply the needs of microscale systems that demand moderate power levels, which is why *fuel cells* (FCs), energy harvesters, and nuclear batteries are the subject of much research today. For context, consider the *Ragone* plot that Fig. 1.7 illustrates, which describes the energy-to-power relationships of modern and emerging sources. FCs, for example, like nuclear batteries and energy harvesters, source higher energy at low power levels, whereas Li Ions and *super/ultra capacitors* supply higher energy at higher power levels. In other words, under similar volume constraints, Li Ions and super capacitors outlast FCs and nuclear batteries under high-power conditions, and vice versa.

Fuel cells also require more time to respond to load changes than Li Ions, and with inherently lower voltages at roughly 0.4–0.7 V. Although super capacitors source less energy than Li Ions, their cycle life is considerably higher at approximately 100,000 cycles, which is why researchers developing energy-harvesting wireless microsensors like them. Unfortunately, however, super capacitors also leak considerably more power than Li Ions, so product developers will more than likely continue to steer away from these devices until manufacturers can improve self-discharge performance. Ultimately, none of these sources are ideal. Researchers are therefore looking into ways of

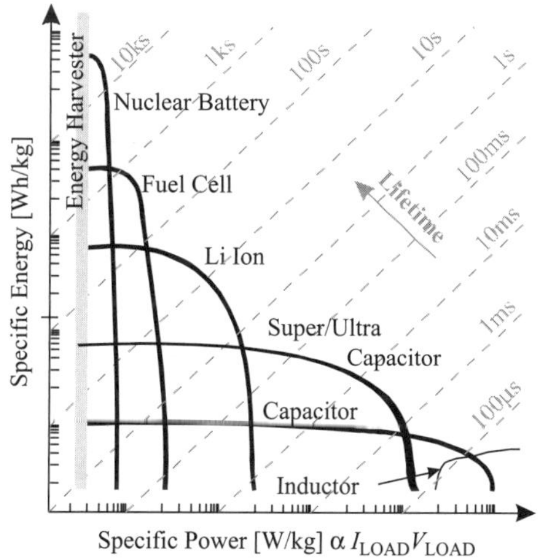

FIGURE 1.7 Ragone plot: Comparative energy–power performance of several energy-storage technologies.

improving each of these technologies separately and, at the same time, leveraging their complementary features in compact hybrid solutions.

1.4.4 Nuclear Batteries

In the battle for maximum energy storage, nuclear batteries overwhelm others at lower power levels, except for energy harvesters, which provide essentially boundless energy. The main drawback with nuclear technology is the radioactive nature of the material used to manufacture it, and associated safety and containment requirements. The secondary, but equally relevant disadvantage is low power. The heat this technology generates, however, can provide long-term fuel for *thermoelectric* generators—for a decade, for instance. Similarly, the electrons emitted by decaying isotopes can establish electric fields across parallel and mechanically compliant piezoelectric plates. The attractive force this field generates bends the *piezoelectric* material to such an extent that the plates touch and release charge on contact. Emitted electrons can also generate electron–hole pairs in PN-junction devices, much like photons do in *photovoltaic solar cells*. Like their solar counterparts, these so-called *beta-voltaic batteries* enjoy the benefits of on-chip integration. In the end, however, safety and the cost of these isotopes keep these batteries from penetrating the marketplace. Nevertheless, because their energy density is unequally high, research certainly continues.

1.4.5 Energy-Harvesting Transducers

Last, but certainly not least in the race for long battery life are energy harvesters. These electric generators extract fuel energy from light, motion, heat, radiation, and other sources from the surrounding environment. Of most commonly available ambient sources, as Table 1.2 tabulates, photovoltaic cells when exposed to solar light offer the

Sources		Transducers	Power Density
Light	Solar	Photovoltaic cells	< 15 mW/cm^2
	Artificial		< 50 μW/cm^2
Motion		Electrostatic	50–100 μW/cm^3
		Electromagnetic	< 1 μW/cm^3
		Piezoelectric	50–300 μW/cm^3
ΔTemperature (10°C)		Thermoelectric piles	5–15 μW/cm^3
Electromagnetic radiation		Transponder: Coupled inductors	mW/cm^3 at mm's μW/cm^3 at cm's

TABLE 1.2 Power Densities from Commonly Available Ambient Energy Sources

highest power level at roughly 10–15 mW/cm^2. Coupling electromagnetic energy can also supply considerable power at maybe 10 mW/cm^3, but only when the receiving coil is within 10–15 mm of a rich radiating source. Piezoelectric and electrostatic transducers extract moderate power densities from motion at 50–300 μW/cm^3. Solar cells when exposed to artificial lighting generate less than 50 μW/cm^3 and temperature gradients across *thermoelectric piles* and vibrating electromagnetic transducers even less at 1–15 μW/cm^3. Needless to say, all these sources are at the forefront of research, and solutions have yet to mature. Notwithstanding, they promise to compete with nuclear batteries in the race for extended, if not perpetual life. Miniaturized energy-harvesting transducers fabricated with *micro-electromechanical systems* (MEMS) and nanometer-scale technologies, like solar cells, electrostatic and piezoelectric transducers, and thermoelectric piles are compatible with ICs and may therefore see the light of day, but now is still too early to tell.

1.5 Computer Simulations

Validating power-supply systems before committing them to fabrication schedules and production is crucial in industry. Therefore, after converging on a circuit topology, IC designers use computer simulations to verify the functionality of their systems and determine their parametric compliance limits. As with all software programs, however, simulation results are as good as the configured schematics and data files used to generate them, which means, as the popular saying goes, "*garbage in, garbage out.*" Designers should therefore simulate only when they think they know what to expect, so they may properly evaluate the output. If the results do not conform to expectations, the simulation setup, the circuit, or both are faulty. As such, the first order of business when debugging a simulation is to ensure operating conditions, models, control statements, and simulation options like steady-state and transient tolerances are all appropriate and correct for the simulation at hand. If the setup is accurate, the circuit might be defective, so reevaluating the circuit *without using a computer* is imperative. Engineers iterate this process of "design" and "validation" until computer results match theoretical predictions, at which point designers are in a better position to draw conclusions, negotiate parametric performance, and improve the system. In other words, people theorize and design what computers can only validate.

After corroborating theory, assessing performance across all possible operating conditions and variations of the fabrication process is critical. Designers generally refer to these extreme conditions as *process–voltage–temperature* (PVT) *corners*, and more generally, as *worst-case corners* because they also include changes in supply, input, and output conditions, such as input noise, load current, output capacitance, and parasitic resistances. Successively repeating simulations of each test circuit with all possible combinations of worst-case models and operating conditions helps designers determine the reliability of the system, or more specifically, how unlikely dies across a silicon wafer and across wafers and fabrication lots are liable to underperform. Weak N-channel and strong P-channel MOS *field-effect transistors* (MOSFETs), for instance, along with nominal NPN BJTs and high resistances and capacitances constitute one of many similar and distinct model sets with which to simulate a circuit. Likewise, high temperature, low input voltage, low output capacitance, high *equivalent series resistance* (ESR) in the output capacitor, and others similarly comprise one of many sets of conditions that a designer simulates with each of the model sets.

Process engineers often guarantee parameters that perform better than what they claim because their primary objective is to increase die yield: to raise the fraction of dies in a wafer that performs as specified, which is what ultimately sets revenues for a semiconductor company. As such, simulations across *quoted* process corners are, on probabilistic terms, unreasonably pessimistic, because an indiscriminate and exhaustive linear combination of systematic and random six-sigma results is unrealistic. The caveat, however, from a designer's point of view, is that an analog circuit has, in theory, an infinite number of operating conditions, and to simulate them all is impossible. Startup, for one, subjects a regulator to an infinite number of bias points that change linearly and continuously with time, so determining the feedback-loop dynamics of the regulator across all these possible operating conditions would extend a product's time to market beyond fiscally sound limits.

Most designers mitigate these technical risks by first using good engineering judgment to define practicable worst-case conditions. Then, when the system is small enough to simulate quickly, within a couple of minutes, for example, they simulate all these cases with all six-sigma process-corner models quoted. Combining the results of these six-sigma simulations *linearly* is a way of offsetting untested bias conditions that are not only impossible to fully enumerate but also difficult to predict. Fast product-development cycles also rely on these extreme corner scenarios to improve the probability of building a sufficiently robust prototype that meets all parametric specifications after only one fabrication cycle, achieving the coveted *first-pass success*.

1.6 Summary

The driving aim of a regulator is to *regulate* its output voltage against all possible variations in operating conditions, from changes in load and supply to temperature. They differ from references in that they supply substantially higher load currents that vary with time. This seemingly insignificant fact adds considerable complexity and challenges to the problem, as follow-up chapters show. Nevertheless, when compared to their switched counterparts, linear regulators are simpler, faster, and less noisy. Their main drawback is limited power-conversion efficiency, which is why designers drop high input voltages from, say, 12 to 1.8 V with switched regulators and cascade linear regulators in series to remove switching noise and further drop the voltage to the desired, more accurate level of 1.6 V in the foregoing example.

Fixing the input and output voltages this way means that power efficiency ultimately hinges on how much of the input current the linear regulator loses to ground. From this perspective, attaining high efficiency is especially challenging in portable applications because mobile devices are idle most of the time, so ground current, no matter how low, can be a substantial fraction of the output current. Similarly, losing leakage power across the input source is problematic, which is one of the reasons lithium-ion batteries are so popular, because they do not suffer significantly from self-discharge. Their performance is also fairly consistent across their recharge cycle life, which is typically long enough at 1000 to 1500 cycles to satisfy the needs of a market that disposes of its products within two to three years of operation, that is, within 730 to 1095 days, or more to the point, within an approximately equal number of recharge cycles.

Needless to say, ensuring that the microelectronic circuit also functions consistently across cycles, operating conditions, and temperatures before committing to production is paramount. Engineers therefore use computer simulations to verify that their designs perform as expected across process and operating corners. Determining what metrics to evaluate in a design is critical in this evaluation process, which is why the next chapter describes how linear regulators operate and the specifications typically used to quantify and qualify how they perform.

1.7 Review Questions

1. Variable v_{out} refers to *small signals only/steady-state signals only/both small and steady-state signals/neither small nor steady-state signals.*

2. What is the most redeeming advantage of switched dc–dc converters over their linear counterparts?

3. Relative to linear regulators, switched regulators are *noisier/slower/more complex/all of these/none of these.*

4. Many microelectronic systems use a *linear/switched* regulator to drop high voltages and cascade a *linear/switched* regulator in series to remove switching noise.

5. Which circuit characteristics determine the power-conversion efficiency of linear regulators?

6. When load currents are nearly zero, the effects of quiescent current on efficiency are *miniscule / profound / irrelevant.*

7. *Lithium-ion / nickel–cadmium / nickel–metal–hydride / nuclear / all of these / none of these* batteries suffer from cyclic-memory effects.

8. Relative to fuel cells of the same size, lithium-ion batteries store *more / less / the same amount of* energy.

9. Relative to super/ultra capacitors of the same size, lithium-ion batteries feature *longer / shorter / the same* cycle life and *higher / lower / the same* discharge rates.

10. Of commonly available ambient energy sources, *vibrations / solar light / artificial light / temperature gradients / radio-frequency radiation* supply(ies) the most power in small-scale applications.

11. Computer simulations are optimal for *design only / validation only / both design and validation.*

12. Combining the results of simulations that use six-sigma models under extreme operating conditions linearly is *realistic / unrealistically optimistic / unrealistically pessimistic.*

13. Design engineers often use six-sigma worst-case corner simulations as a means of projecting reliability performance of a microchip across *a wafer / wafers / fabrication lots / all of these / none of these.*

CHAPTER 2

Linear Regulators

Understanding how linear regulators generally work is important both in the product-definition phase and through the design process. For one, system designers, whose knowledge of the internal workings of an integrated circuit (IC) is often vague, rely on their basic knowledge of linear regulators and how companies qualify and quantify regulators to define, design, and develop power-supply systems. Also, from an IC designer's point of view, priorities vary from one application to the next, and negotiating performance specifications with a customer before committing to a design requires *a priori* knowledge of how regulators *should* work. Even after this, IC engineers couple this basic knowledge of the system with their understanding of what metrics the customer values to determine and weigh design choices at the transistor level. Note that, for IC designers, system designers are, for the most part, their customers.

For these and other reasons, the purpose of this chapter is to familiarize the reader with how linear-regulator systems perform and respond to variations in their supplies, loads, and other working conditions. This chapter therefore begins with a description of the operating regions of a linear regulator. Predicting performance is critical when selecting and designing linear regulators, so quantifying typical performance metrics like accuracy, power-conversion efficiency, load range, and other parameters is next. Discussions on the operating conditions that linear regulators must survive when supplying their loads follow. The ensuing discussion then classifies linear regulators into categories that designers and vendors normally use to qualify linear-regulator systems. Finally, this chapter decomposes the system and identifies the functional blocks whose design implementation subsequent chapters describe.

2.1 Regions of Operation

A linear regulator works properly when it regulates its output voltage v_{OUT} within a small offset of its target V_O'. For this, it compares v_{OUT} against V_O' and modulates the conductance of the output switch S_O that supplies power to v_{OUT} until the error difference between v_{OUT} and

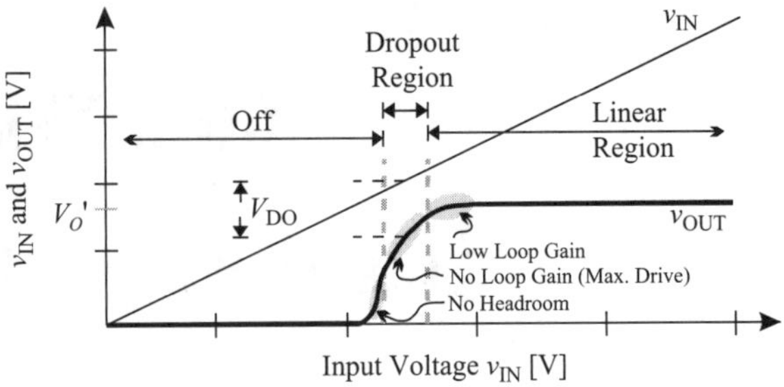

FIGURE 2.1 Typical input–output voltage behavior of a linear regulator.

V_O' is small. In this respect, when the signal path that processes v_{OUT} to control S_O is sensitive and reacts linearly to oppose small fluctuations in v_{OUT}, the system is in the *linear region*. Here, variations in load current and input supply voltage v_{IN} have little impact on v_{OUT}, as Fig. 2.1 illustrates. As a result, because v_{OUT} nears V_O', this region is by and large the preferred mode of operation for a linear regulator.

To remain sensitive to small changes in v_{OUT}, IC engineers normally design analog circuits that ensure transistors in the signal-signal path remain in their high-gain modes. In this respect, v_{IN} limits how much margin, or more specifically, how much dynamic voltage range signals have before they shift away from their preferred states. As such, decreasing v_{IN} in a linear regulator lowers the *dynamic range* with which signals in the feedback loop process v_{OUT} and control S_O. Therefore, when v_{IN} is so low that one or several transistors transition out of their high-gain mode, the loop gain and sensitivity of the system fall, allowing v_{OUT} to deviate from V_O'. This is the point where the system begins to "drop out" of its well-regulated linear region.

When the circuit and, in consequence, S_O are no longer sensitive to v_{OUT}, the system enters the *dropout region*. Here, one or more transistors enter their ohmic or triode regions, where gain is low. As a result, the loop that senses and controls S_O loses gain to the extent it stops adjusting S_O's conductivity in response to variations in v_{OUT}. In other words, S_O behaves like a switch, and v_{OUT} is less than v_{IN} by S_O's ohmic drop:

$$v_{OUT}\big|_{Dropout} = v_{IN} - v_{SW} = v_{IN} - i_{LOAD}R_{SW} \equiv v_{IN} - V_{DO} \tag{2.1}$$

where i_{LOAD} is load current, v_{SW} and R_{SW} are S_O's voltage drop and equivalent resistance, and V_{DO} is the regulator's dropout voltage. Since v_{SW} is nearly constant in this region, v_{OUT} falls with v_{IN}, as Fig. 2.1 shows. In practice, however, R_{SW} rises a little when v_{IN} drops because S_O's driving capability falls with v_{IN}, which means v_{OUT} falls slightly faster than v_{IN}. With further reductions in v_{IN}, one or more transistors eventually lose the supply *headroom* they need to assess and keep

S_O engaged. As a result, S_O loses strength and i_{LOAD} pulls v_{OUT} to ground. Shutting S_O this way marks the onset of the *off region*.

2.2 Performance Metrics

Three general categories aptly describe the operating performance of a linear regulator: *accuracy, power-conversion efficiency,* and operating requirements. Accuracy here refers to the regulator's ability to maintain and regulate its output voltage within a small window of its target across all possible operating conditions. Power-conversion efficiency describes how much of the power drawn from the input supply actually reaches the load. The operating limits of the input voltage, output voltage, output capacitance and associated equivalent series resistance (ESR), and load current are similarly no less important, for they define the environment within which the regulator can maintain accuracy while transferring and supplying power to the load.

2.2.1 Accuracy

Output accuracy, as a whole, incorporates all the adverse effects caused by nonidealities in the fabrication process and the circuit when responding to changes in temperature, input supply, and load current. While process and temperature cause permanent and steady-state offsets in the output, the input and load produce both slow-changing and momentary disruptions, where unfortunately, the latter often overwhelm the former. Understanding how each of these culprits disturbs the output allows an IC engineer to better manage and distribute margins in his or her circuit during the design process. In the end, total accuracy, which represents the combined effects of all variations, is what ultimately specifies a regulator.

Tolerance

Perhaps the most fundamental accuracy error from which a regulator suffers is the initial *offset voltage* V_{OS}^* that fluctuations in the fabrication process induce in an integrated circuit. One culprit in this regard is tolerance in the voltage reference v_{REF} used to establish the output's target voltage of V_O'. To see how this offset arises, consider that linear regulators rely on negative feedback to control the output power transistor S_O that regulates output voltage v_{OUT}. For this, they employ an error amplifier A_{EA} in a noninverting feedback configuration, as Fig. 2.2 shows. Here, v_{OUT} is a β_{FB}-factor translation of feedback signal v_{FB}, and negative feedback ensures that v_{FB} nears v_{REF}, so v_{OUT} is roughly

$$v_{OUT} = \frac{v_{FB}}{\beta_{FB}} \approx \frac{v_{REF}}{\beta_{FB}} \approx v_{REF} A_{CL} \tag{2.2}$$

where A_{CL} is the regulator's closed-loop gain with respect to v_{REF}. Given this characteristic, tolerance in v_{REF} manifests directly in v_{OUT},

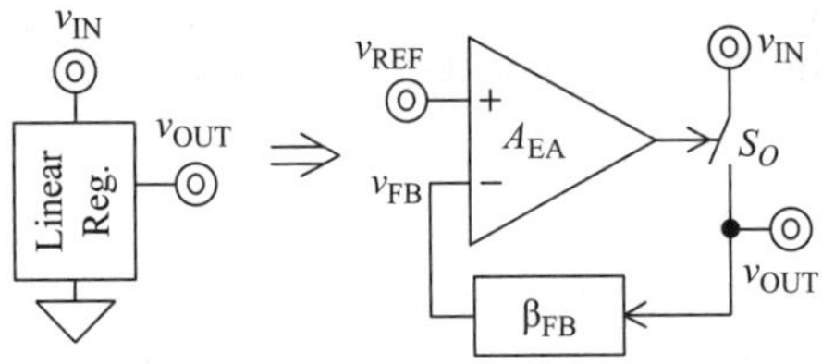

FIGURE 2.2 Linear regulators employ negative feedback to regulate the output v_{OUT}.

as does the input-referred offset $V_{EA.OS}^*$ that mismatches between resistors and transistors in A_{EA}'s input stage produce. Since tolerance variations and mismatches are random phenomena, their effects on v_{OUT} are neither consistent nor monotonic. In fact, probabilistic theory dictates that the combined effects of random components is the root-square sum of its constituent elements, which means

$$V_{OS}^* = \left(\sqrt{(\Delta V_{REF}^*)^2 + (V_{EA.OS}^*)^2}\right) A_{CL} \approx \frac{\sqrt{(\Delta V_{REF}^*)^2 + (V_{EA.OS}^*)^{22}}}{\beta_{FB}} \tag{2.3}$$

where ΔV_{REF}^* is v_{REF}'s steady-state random tolerance about its nominal target.

Gain Error

The sensitivity of the system to fluctuations in v_{OUT} ultimately rests on the gain across the loop that senses v_{OUT} and controls S_O. If an infinite loop gain were possible, v_{FB} would match v_{REF}, but since gains are finite, v_{FB} is not exactly at v_{REF}. In fact, the control loop reacts to ensure that output-derived feedback signal v_{FB} is near v_{REF}, so v_{OUT} is

$$\begin{aligned} v_{OUT} &= \frac{v_{FB}}{\beta_{FB}} = v_{REF}\left(\frac{A_{OL}\beta_{FB}}{1+A_{OL}\beta_{FB}}\right)\left(\frac{1}{\beta_{FB}}\right) \\ &= v_{REF}\left(\frac{A_{OL}}{1+A_{OL}\beta_{FB}}\right) \equiv v_{REF}A_{CL} \end{aligned} \tag{2.4}$$

where A_{OL} or $A_{EA}A_{SW}$ is the forward open-loop gain that A_{EA} and S_O's gain A_{SW} establish and $A_{OL}\beta_{FB}$ is the gain across the loop, which later chapters abbreviate to A_{LG}. Again, if $A_{OL}\beta_{FB}$ were infinite, v_{OUT} would reduce to v_{REF}/β_{FB}, which corresponds to v_{OUT}'s target V_O', but since $A_{OL}\beta_{FB}$ is finite, the circuit introduces a *gain error* ΔV_{GE} equivalent to

$$\begin{aligned} \Delta V_{GE} &= v_{OUT} - V_O' = v_{REF}\left(\frac{A_{OL}}{1+A_{OL}\beta_{FB}} - \frac{1}{\beta_{FB}}\right) \\ &= \frac{v_{REF}}{\beta_{FB}}\left(\frac{-1}{1+A_{OL}\beta_{FB}}\right) \end{aligned} \tag{2.5}$$

Temperature Drift

Because of negative feedback, any variation in the reference v_{REF} propagates to the output of the regulator v_{OUT} through the equivalent closed-loop gain of the system A_{CL}. As a result, a temperature-induced drift in v_{REF} causes variations in v_{OUT}. This drift is in addition to the temperature effects from which the error amplifier A_{EA} in the regulator suffers in the form of a fluctuation in its input-referred offset voltage $V_{EA.OS}^*$, which combines with ΔV_{REF}^* to define overall random offset V_{OS}^*. The general metric that gauges the extent to which temperature changes a variable is *fractional temperature coefficient*, or TC for short. In the case of a regulator, TC_O describes relative or fractional (but not absolute) variations of v_{OUT} in response to changes in junction temperature T_J. In this sense, since V_{OS}^* drifts with temperature and A_{CL} amplifies v_{REF} along with both these drifts, TC_O reduces to the effective fractional variation of the reference:

$$\begin{aligned} TC_O &\equiv \frac{1}{V_{OUT}}\left(\frac{dV_{OUT}}{dT_J}\right) \approx \frac{1}{V_{OUT}}\left(\frac{\Delta V_{TD}}{\Delta T_J}\right) \approx \frac{1}{V_{REF}A_{CL}}\left(\frac{\Delta V_{OS}{}^* A_{CL}}{\Delta T_J}\right) \\ &= \frac{\Delta V_{OS}{}^*}{V_{REF}}\left(\frac{1}{\Delta T_J}\right) \end{aligned} \tag{2.6}$$

where V_{OUT} is v_{OUT}'s nominal steady-state value, V_{OS}^* is the temperature-induced variations of V_{OS}^*, ΔT_J is the corresponding change in junction temperature, and V_{TD} is the resulting temperature-induced drift in v_{OUT}. Not surprisingly, designers design references and regulators with equal care.

Line Regulation

Line regulation is a steady-state metric that specifies how much the output v_{OUT} changes in response to slow, but often pronounced variations in the input supply v_{IN}. In other words, *line* refers to the input supply from which the regulator "hangs" and line regulation to the regulator's steady-state gain from v_{IN} to v_{OUT}, or to be more precise, low-frequency input-supply gain A_{IN0}:

$$A_{IN0} \equiv \frac{\Delta v_{OUT}}{\Delta v_{IN}} \tag{2.7}$$

Unfortunately, since the loop in Fig. 2.2 reacts to oppose variations in v_{OUT} and loop gain is finite, S_O's response does not completely cancel the effects of a changing supply, which means A_{IN0} is not zero. Furthermore, v_{REF} is also sensitive to v_{IN}, so v_{OUT} ultimately displays the line effects of not only the error amplifier and output switch in Fig. 2.2 but also v_{REF}:

$$\begin{aligned} \Delta V_{LN} &= \Delta V_{IN}(A_{IN0.REG} + A_{IN0.REF}A_{CL}) \\ &= \Delta V_{IN}\left[\left.\frac{\Delta V_{OUT}}{\Delta V_{IN}}\right|_{\Delta V_{REF}=0} + \left(\frac{\Delta V_{REF}}{\Delta V_{IN}}\right)A_{CL}\right] \end{aligned} \tag{2.8}$$

where $A_{IN0.REG}$ and $A_{IN0.REF}$ are the low-frequency supply gains of the regulator and reference and A_{CL}, as before, is the regulator's forward closed-loop gain from v_{REF} to v_{OUT}. Since $A_{IN0.REG}$ should not include line-regulation effects in v_{REF}, v_{REF} should be an ideal voltage source when deriving $A_{IN0.REG}$, which is why the preceding expression assumes ΔV_{REF} is zero when deriving $A_{IN0.REG}$.

Power-Supply Rejection

In a typical mixed-signal environment, the input voltage v_{IN} delivers power to several subsystems. Since the underlying aim of the linear regulator is to supply noise-sensitive blocks, circuits that switch often and are less vulnerable to noise usually draw power from v_{IN} directly. This means high-power switching transitions in, for example, digital signal processors (DSPs) and power amplifiers (PAs) produce considerable noise in v_{IN}. What is more, since a switched ac–dc or dc–dc converter is what usually sources v_{IN}, v_{IN} incorporates a systemic ripple of ±5–±25 mV at the switching frequency of the converter, which can be anywhere from 20 kHz to 10 MHz. The problem here is that the linear regulator's input-supply gain A_{IN} relays some, or in some cases, all of this ripple as well as other noise components in v_{IN} to v_{OUT}, to sensitive subsystems like voltage-controlled oscillators (VCOs), data converters, phase-locked loops (PLLs), and others.

Power-supply rejection (PSR) or *power-supply ripple rejection*, as some power-supply designers like to call it, is a parameter that describes how well linear regulators reject noise in v_{IN}. Since saying rejection is good or high means supply gain A_{IN} is low, PSR is the reciprocal of A_{IN}:

$$\text{PSR} \equiv \frac{1}{A_{IN}} = \frac{\Delta v_{IN}}{\Delta v_{OUT}} \tag{2.9}$$

Notice that supply rejection and line regulation both describe how v_{OUT} behaves when confronted with fluctuations in v_{IN}. What differentiates one term from the other is that line regulation refers only to steady-state effects, whereas supply rejection specifies v_{OUT}'s behavior across time, that is to say, across frequency. In other words, PSR at low frequency, which reduces to $1/A_{IN0}$, essentially describes line regulation. In fact, when datasheets specify PSR, they typically include rejection performance at several frequencies, including one at dc that essentially describes the effects of line regulation. Regrettably, loop gain falls with frequency, so the circuit's ability to reject noise degrades at higher frequencies, where noise is worse.

As with line regulation, noise in v_{IN} affects both v_{REF} and the regulator, so PSR in a linear regulator refers to the combined effects of the reference and the regulator circuit:

$$A_{IN} \equiv \frac{1}{\text{PSR}} = \frac{A_{CL}}{\text{PSR}_{REF}} + \frac{1}{\text{PSR}_{REG}} \tag{2.10}$$

where A_{CL}/PSR_{REF} describes how input-induced noise in v_{REF} propagates to v_{OUT} and PSR_{REG} describes how input noise bleeds to v_{OUT} directly through the error amplifier of Fig. 2.2. Normally, PSR_{REF} is considerably higher than PSR_{REG}, so PSR_{REF}'s effects are often insignificant and therefore are normally neglected, but only when justified. PSR ultimately specifies to what extent fluctuations in v_{IN} ripple to v_{OUT} in the form of Δv_{PSR}:

$$\Delta v_{PSR} = \Delta v_{IN} A_{IN} = \frac{\Delta v_{IN}}{PSR} \tag{2.11}$$

Load Regulation

Another manifestation of finite loop gain is v_{OUT}'s response to steady-state variations in load current i_{LOAD}, which is what designers generally call *load regulation*. With finite loop gain, the circuit cannot completely cancel the effects of a changing i_{LOAD}. In other words, the equivalent resistance that the shunt-feedback network in Fig. 2.2 presents to the load is not zero, so load-regulation error ΔV_{LD} is

$$\begin{aligned} \Delta V_{LD} &= \Delta I_{LOAD} R_{CL} = \Delta I_{LOAD}\left(R_{OL} \parallel \frac{R_{OL}}{A_{OL0}\beta_{FB0}}\right) \\ &= \Delta I_{LOAD}\left(\frac{R_{OL}}{1 + A_{OL0}\beta_{FB0}}\right) \end{aligned} \tag{2.12}$$

where R_{CL} is the closed-loop output resistance of the negative-feedback system, R_{OL} is the open-loop counterpart, and $A_{OL0}\beta_{FB0}$ is the low-frequency gain across the feedback loop.

Load-Dump Response

Although linear regulators mostly supply "quiet" and stable subsystems, they are not immune to quick disruptions. On–off transitions, for example, can cause quick and wide variations in i_{LOAD}; so can hopping from one operating mode to another. Unluckily, load regulation cannot capture the full impact of the *load dumps* that these quick transitions spark because the circuit requires time to react in the fashion that load regulation expects. In other words, finite low-frequency loop gain $A_{OL0}\beta_{FB0}$ limits load regulation, and lower high-frequency loop gain further limits load-dump response. The fundamental problem here is that the regulator cannot possibly satisfy the load before it can realize that a mismatch exists between its output current and the load.

To avoid excessive variations in v_{OUT}, engineers enlist the help of a capacitor. The purpose of output capacitor C_O in Fig. 2.3*a* is just this: to cache energy it can use to supply the portion of load current i_{LOAD} that the regulator cannot source and to absorb the fraction of regulator current i_O that i_{LOAD} no longer sinks. C_O, however, is not a perfect solution because its voltage changes when sourcing or sinking current i_C. In addition, C_O's equivalent series resistance R_{ESR} or ESR, which can range between 0.2 and 1 Ω, also drops a voltage when C_O

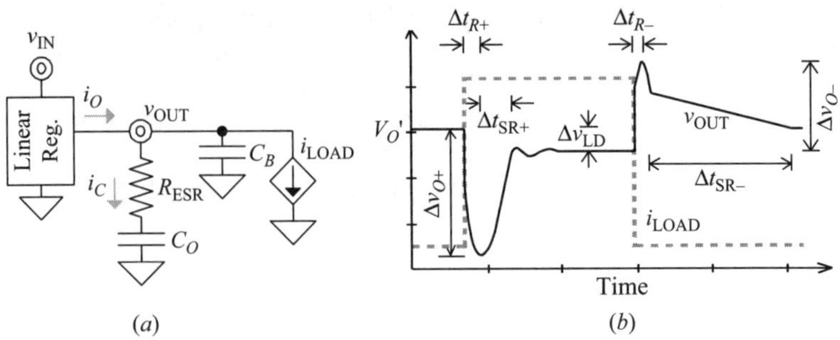

FIGURE 2.3 Load-dump (*a*) circuit and (*b*) its typical time-domain response.

conducts i_C. Because of these drops, power-supply designers often place a more costly low-ESR (i.e., high-frequency) bypass capacitor C_B across the load. In this way, C_B helps supply and sink excess current to mitigate the needs of C_O. Unfortunately, the resulting drops in C_O and R_{ESR} ultimately manifest in v_{OUT}.

The worst-case variation in v_{OUT} occurs when i_{LOAD} suddenly transitions between extreme limits. Because the regulator cannot raise or lower i_O fast enough to match i_{LOAD}, C_O and C_B supply the difference, which means C_O's fraction of the current instantly drops a voltage Δv_{ESR} across R_{ESR} that is equivalent to

$$\Delta v_{ESR} \approx \left(\frac{C_O}{C_O + C_B}\right)\Delta i_{LOAD} R_{ESR} \tag{2.13}$$

where Δi_{LOAD} is i_{LOAD}'s total variation. C_O is normally more than an order of magnitude higher than C_B, which means C_O's impedance $1/sC_O$ is correspondingly lower than C_B's at $1/sC_B$. Therefore, most of Δi_{LOAD} flows through C_O, so Δv_{ESR} can be substantial. This same current also drains C_O until enough response time Δt_R elapses to allow the regulator to respond and adjust i_O, which means C_O slews Δv_C:

$$\Delta v_C \approx \left(\frac{\Delta i_{LOAD}}{C_O + C_B}\right)\Delta t_R \tag{2.14}$$

These two voltages, of course, combine and shift v_{OUT} away from its target of V_O'.

Two distinct mechanisms govern the regulator's response time Δt_R: loop bandwidth and internal slew rate. Bandwidth time t_{BW} refers to how long small-signal variations propagate through the feedback loop to adjust S_O. With one bandwidth-setting pole at f_{-3dB}, the loop delays a signal by roughly 2.3 time constants τ_{RC}, or

$$t_{BW} \approx 2.3\tau_{RC} = \frac{2.3}{2\pi f_{-3dB}} \tag{2.15}$$

which to be more specific, corresponds to how long the output of the system requires to reach 90 percent of its input, as the side note explains. Unfortunately, however, load dumps are not small-signal events, so large capacitors across the loop further delay the response. In the case of linear regulators, S_O is typically a large device because i_{LOAD} can be high, which means the parasitic capacitance at S_O's input control terminal is also correspondingly high. As a result, A_{EA}'s ability to charge and discharge S_O's input capacitance C_{PAR} determines the extent to which S_O's slew-rate limit extends Δt_R:

$$\begin{aligned}\Delta t_R &= t_{BW} + t_{SO.SR} = t_{BW} + C_{PAR}\left(\frac{\Delta v_{PAR}}{i_{O.EA(MAX)}}\right) \\ &= \frac{2.3}{2\pi f_{-3dB}} + C_{PAR}\left(\frac{\Delta i_{LOAD}}{G_{M.SO} I_{SR)}}\right)\end{aligned} \tag{2.16}$$

where $i_{O.EA(MAX)}$ is A_{EA}'s maximum slew-rate-limited output current i_{SR}, $G_{M.SO}$ is S_O's transconductance, and Δv_{PAR} and $\Delta i_{LOAD}/G_{M.SO}$ refer to how much voltage C_{VAR} charges or discharges before S_O can adjust to i_{LOAD}'s new setting.

BANDWIDTH DELAY t_{BW} *Since capacitors sink and supply instantaneous currents, an incoming current i_{IN} that suddenly rises first flows through a capacitor. But because capacitors also block steady-state currents, output capacitor C_O in the equivalent resistor–capacitor model of the linear system illustrated in Fig. 2.4 sinks less and less of i_{IN} as time elapses, so after a while, output resistor R_O absorbs all of i_{IN} in*

$$\frac{v_O}{R_O} = i_{IN} - i_C = i_{IN} - C_O\left(\frac{dv_O}{dt}\right)$$

In other words, the voltage across R_O rises with time until it slowly reaches its final value of $i_{IN}R_O$. And since the derivative of an exponential is also an exponential, output voltage v_O in the preceding expression rises in exponential fashion across time:

$$v_O = i_{IN}R_O\left[1 - \exp\left(\frac{-t}{R_O C_O}\right)\right]$$

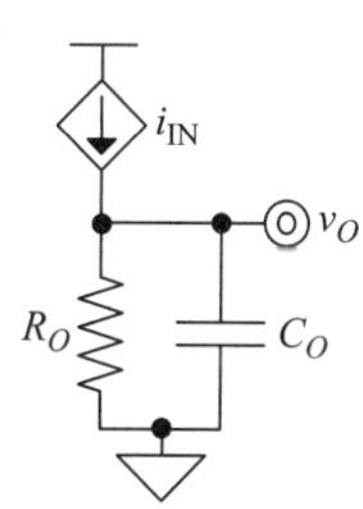

FIGURE 2.4 Equivalent resistor–capacitor model of a bandwidth-limited system.

which means response time t is

$$t = R_O C_O \ln\left(\frac{i_{IN} R_O}{i_{IN} R_O - v_O}\right)$$

So, for v_O to reach 90 percent of its final value, which in this case is $0.9 i_{IN} R_O$, 2.3 time constants of $R_O C_O$ must elapse:

$$t_{90\%} = R_O C_O \ln\left(\frac{i_{IN} R_O}{i_{IN} R_O - 0.9 i_{IN} R_O}\right)$$
$$= R_O C_O \ln(10) = 2.3 R_O C_O \equiv 2.3\tau_{RC} \approx t_{BW}$$

which roughly corresponds to the delay across a linear system with one bandwidth-setting pole.

Since C_O's current in steady state is zero, C_O holds v_{OUT}'s target voltage V_O'. Therefore, in the case of a rising load dump, when i_{LOAD} suddenly rises, C_O's fraction of Δi_{LOAD} discharges C_O and establishes a voltage across R_{ESR} that combined with C_O pulls v_{OUT} below V_O' by Δv_{ESR} and Δv_C, or as Fig. 2.3*b* shows, by Δv_{O+}. v_{OUT} suffers similar effects during a falling load dump, when i_{LOAD} suddenly drops. Here, excess current from the regulator charges C_O through R_{ESR} to raise v_{OUT} above V_O' by Δv_{ESR} and Δv_C, or equivalently, by Δv_{O-}.

The behavior of the output across rising and falling load dumps, however, is usually not the same. The fundamental reason for this asymmetry in response is that regulators need not, and therefore cannot for the most part, sink much current. As a result, after excess regulator current charges C_O above V_O' in response to a falling load dump, only a small current on the order of microamperes is available to discharge C_O back to V_O', so C_O slews for an extended period Δt_{SR-}. After a rising load dump, on the other hand, the regulator can momentarily raise i_O above i_{LOAD} by some considerable margin to more quickly recharge C_O across Δt_{SR+} back to its steady-state value. Plus, A_{EA} drives S_O asymmetrically, so Δt_{R+} also differs from Δt_{R-}.

The driving motivation for engineers to design regulators with this asymmetry is power savings. The fact is, ensuring the regulator can sink as much current as it can supply requires more transistors that demand both additional power and silicon area. Not surprisingly, the speed or bandwidth of the circuit when responding to rising and falling load dumps also differs. This is why response times Δt_{R+} and Δt_{R-} are different in Fig. 2.3.

The last important detail to note is the impact of load regulation on load-dump response. When completely unloaded, v_{OUT} nears its loop-defined target of V_O'. As load regulation already describes, raising i_{LOAD} slowly pulls v_{OUT} below V_O' by ΔV_{LD}. As a result, after a rising load dump, v_{OUT} settles to a steady state that is below its initial value by ΔV_{LD}; similarly, after a falling load dump, v_{OUT} settles to a value that is above its prior state by ΔV_{LD}. In other words, the ultimate

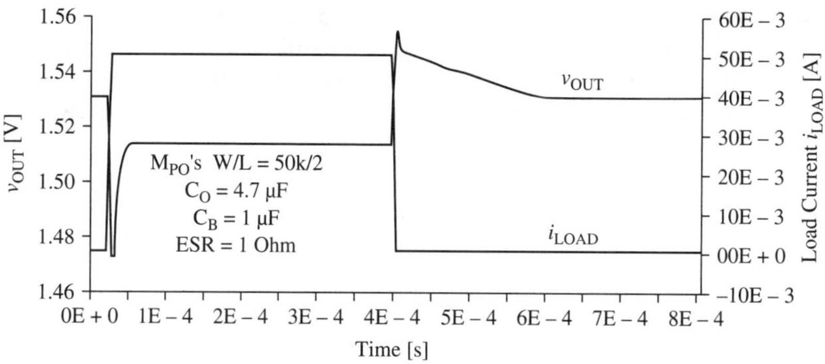

FIGURE 2.5 The load-dump response of a linear regulator.

impact of load dumps on v_{OUT}'s accuracy is the portion of Δv_{O+} and Δv_{O-} that excludes ΔV_{LD}:

$$\Delta v_{DUMP} = \Delta v_O - \Delta V_{LD} = \Delta v_{ESR} + \Delta v_C - \Delta V_{LD} \tag{2.17}$$

Figure 2.5 shows the simulation results of a 1.53-V regulator responding to rising and falling 50-mA load dumps when assisted by a 4.7-μF electrolytic capacitor with 1 Ω of ESR and a 1-μF low-ESR ceramic bypass capacitor. Here, a macro model of error amplifier A_{EA} in Fig. 2.3 drives S_O, which a P-channel metal–oxide–semiconductor (PMOS) pass transistor with a channel width–length ratio of 50 mm to 2 μm implements. The results show that v_{OUT} drops 55 and 40 mV in response to the aforementioned rising and falling load dumps, respectively. Since load regulation already accounts for 20 mV, load dumps effectively degrade accuracy by 35 mV. In other words, load regulation contributes 20 mV of 1.53 V or 1.3-percent variation in v_{OUT} and load-dump transitions another 35 mV or 2.3 percent.

As the example shows, load-dump accuracy is a significant portion of the overall performance of the regulator, which is why power-supply designers often allocate a 3- to 8-percent margin to this phenomenon. Using high-frequency capacitors, which have low ESRs and cost more money, reduces this variation. Shortening the response time of the regulator has similar benefits, but at the expense of more quiescent current, which shortens battery life.

Extending the rise and fall times of load dumps to the extent that transition times exceed the regulator's response time Δt_R, which can be 0.5 to 5 μs, practically negates all adverse load-dump effects in v_{OUT}. And because load dumps are difficult to predict, and overly pessimistic specifications can dissuade a customer from buying a part, many data books extend transitions to several microseconds or even milliseconds, to the point that load-dump error is nearly nil. This type of response is, of course, overly optimistic for many state-of-the-art applications. Microprocessors (μPs), for example, can clock anywhere from several megahertz to gigahertz, so their synchronized transitions can load a regulator in nanoseconds, well before typical regulators can react. As a result, determining and using a realistic load is important in the design process, when negotiating how much power and silicon area to sacrifice for the sake of accuracy.

Example 2.1 Determine the load-regulation and load-dump variations and corresponding accuracies of a 100-mA, 1.5-V regulator with a bandwidth of 500 kHz and an equivalent closed-loop resistance of 300 mΩ when subjected to 1-ns, 0- to 100-mA load dumps. Assume the regulator relies on a 10-μF electrolytic capacitor with an ESR of 200 mΩ and a 1-μF bypass ceramic capacitor for assistance. The output stage of the error amplifier in the regulator is a class-A buffer that, while sourcing plenty of current, can only sink up to 5 μA. Assume the transconductance gain and input capacitance of the power pass device are 300 mA/V and 100 pF, respectively.

Solution

1. $\Delta V_{LD} \approx \Delta I_{LOAD} R_{CL} = 100\,\text{mA} \cdot 300\,\text{m}\Omega = 30\,\text{mV}$

   so load-regulation accuracy is 30 mV/1.5 V, or approximately 2.0 percent.

2. $\Delta v_{DUMP} = \Delta v_{O(MAX)} - \Delta V_{LD} = \Delta v_{ESR} + \Delta v_{C(MAX)} - \Delta V_{LD}$

   where $\Delta v_{ESR} = \left(\frac{\Delta i_{LOAD} C_O}{C_O + C_B}\right) R_{ESR} = \left(\frac{100\,\text{mA} \cdot 10\,\mu\text{F}}{10\,\mu\text{F} + 1\,\mu\text{F}}\right) 0.2\,\Omega = 18\,\text{mV}$

   $$\Delta v_{C(MAX)} = \left(\frac{\Delta i_{LOAD} C_O}{C_O + C_B}\right) \frac{t_{R(MAX)}}{C_O} = \left(\frac{100\,\text{mA}}{10\,\mu\text{F} + 1\,\mu\text{F}}\right) t_{R(MAX)}$$

   and $\Delta t_{R(MAX)} \approx t_{BW} + t_{SO.SR(MAX)} \approx \frac{2.3}{2\pi f_{-3dB}} + \frac{C_O}{i_{O.EA(SINK)}} \left(\frac{\Delta i_{LOAD}}{G_{M.SO}}\right)$

   $$= \frac{2.3}{2\pi(500\,\text{kHz})} + \frac{100\,\text{pF}}{5\,\mu\text{A}} \left(\frac{100\,\text{mA}}{300\,\text{mA/V}}\right) \approx 730\,\text{ns} + 7\,\mu\text{s} \approx 8\,\mu\text{s}$$

   so Δv_{DUMP} is 91 mV and load-dump accuracy after excluding load regulation is 61 mV/1.5 V, or roughly 4.1 percent.

Total Accuracy

When quantifying the accuracy of a regulator with a single specification, the value should account for all possible variations in the output. Therefore, the parameter should comprehend the systemic and random effects of all steady-state errors and transient disruptions. Since systemic offsets are continuous, consistent, and even monotonic at times, their combined effect on the output is roughly the linear sum of the individual components. If, for instance, a 1- to 50-mA rise in load current pulls v_{OUT} 20 mV below its target when supplied from a 5-V source and a 5- to 3-V reduction in the supply pulls v_{OUT} 5 mV when loaded with 1 mA, their combined impact on v_{OUT} can be approximately a 25-mV drop. Notice, however, that systemic offsets have a polarity and will, at times, partially cancel. Understanding how operating conditions change and affect a regulator is important to avoid mistakes. Consider, for example, that the same 49-mA rise in current at 3 V can pull v_{OUT} 20 mV below its nominal value, and a 2-V rise in the input supply at 50 mA can negate that change by 5 mV. To say the worst-case variation in this case is 15 mV is optimistic because a user can start the regulator at 5 V with 1 mA, then raise the load to 50 mA and drop the supply by 2 V and realize that the total spread is actually 25 mV.

Since random offsets are neither consistent nor monotonic, combining them with their systemic counterparts into a single specification is not straightforward. Considering that the effects of tolerance on v_{OUT} in V_{OS}^* are random and those of gain error in ΔV_{GE}, temperature drift in ΔV_{TD}, line regulation in ΔV_{LN}, and load regulation in ΔV_{LD} are systemic, the combined steady-state fractional error is

$$\text{Error}_0 \equiv \Delta E_0 = \frac{\Delta V_{GE} + \Delta V_{TD} + \Delta V_{LN} + \Delta V_{LD} \pm V_{OS}^*}{V_{OUT}} \tag{2.18}$$

where V_{OUT} is v_{OUT}'s nominal steady-state target of v_{REF}/β_{FB}. Unfortunately, time-domain variations in v_{OUT} also degrade accuracy, so including the effects of supply ripple in Δv_{PSR} and load dump in Δv_{DUMP} in the specification is oftentimes necessary:

$$\text{Error} \equiv \Delta E = \Delta E_0 + \frac{\Delta v_{PSR} + \Delta v_{DUMP}}{V_{OUT}} \tag{2.19}$$

although they are sometimes difficult to combine without including redundancies or negating other offsets.

In practice, many product engineers test ICs across wafers and fabrication lots and plot and enclose all measured values of v_{OUT} at extreme temperature, supply, and load settings into one graphical "box." Then they determine the probabilistic variability of the boundary of the box to quote the steady-state error (Error_0) of the device. System designers later superimpose the effects of supply ripple and load dump on this value to estimate the combined effects of all possible variations on v_{OUT}.

In the end, as alluded to in earlier discussions, load-dump effects normally dominate all others, but as with other effects, though for different reasons, overall accuracy specifications often exclude them. The fact is, these variations depend on how fast the load rises or falls: across nanoseconds, microseconds, or milliseconds. In other words, they depend strongly on the application. Package-stress effects and other process-induced random variations in v_{REF} are normally so severe, although not to the same extent as load dumps, that line-regulation and temperature-drift fluctuations in v_{REF} are difficult to discern. As a result, ΔV_{REF}^* in V_{OS}^* usually absorbs all systemic variations in v_{REF}. V_{OS}^* similarly overwhelms process-induced variations in G_E, so accuracy also omits the random component of G_E.

Load-dump and supply-ripple effects in v_{OUT} are normally so problematic in mixed-signal systems that inherent *shot*, *thermal*, and 1/*frequency* or 1/*f noise* in transistors become insignificant. This is not to say, however, that systemic switching noise always overwhelms inherent noise, especially when considering extremely sensitive applications like PLLs and high-resolution data converters. Still, most quoted steady-state errors today typically fall in the range of 1 to 3 percent, and load dumps degrade accuracy by another 1 to 7 percent, depending on the load profile of the application.

Example 2.2 Determine the total accuracy performance of a 2.4-V regulator whose reference voltage is 1.2 V, forward open-loop gain is 60 dB, input supply ripple is ±25 mV at 100 kHz, power-supply rejection is 10 dB at 100 kHz, and output voltage variations with their corresponding three-sigma values in response to load dumps and changes in temperature, load current, and input supply are 140 mV ± 5mV, 3 ± 24 mV, 20 ± 5 mV, and 8 ± 2.5 mV, respectively.

Solution

1. Note that the random component of temperature drift in 3 ± 24 mV overwhelms its systemic counterpart. This is typical in plastic-packaged voltage references because the physical shapes of the fillers used to reduce the *thermal coefficient of expansion* of the package and increase its reliability are random. The odd-shaped fillers therefore exert uneven stresses across the die that cause mismatches between adjacent and otherwise matched transistors.

2. Deriving 2.4 V from a 1.2-V reference requires a feedback factor β_{FB} of 2, so

$$\Delta V_{GE} = \frac{v_{REF}}{\beta_{FB}}\left(\frac{-1}{1+A_{OL}\beta_{FB}}\right) = \frac{1.2\text{ V}}{2}\left(\frac{-1}{1+1000\cdot 2}\right) \approx 0.3\text{ mV}$$

3. At any extreme, only one half cycle of the supply-induced ripple in v_{OUT} degrades accuracy, so

$$\Delta v_{PSR} = \frac{\Delta v_{IN}}{PSR} = \frac{25\text{ mV at }100\text{ kHz}}{\log^{-1}\left\{\frac{10\text{ at }100\text{ kHz}}{20}\right\}} = \frac{25\text{ mV}}{3.16} = 7.9\text{ mV}$$

4. Load-dump variation Δv_O includes load-regulation error ΔV_{LD}, so

$$\Delta v_{DUMP} = \Delta v_O - \Delta V_{LD} = 140\text{ mV} - 20\text{ mV} = 120\text{ mV}$$

5. Note that v_{OUT} measurements incorporate all systemic and random variations in v_{REF}. As a result, V_{OS}^* is effectively the square-root summation of all squared random variations, so

$$V_{OS}^* \approx \sqrt{(\Delta v_{DUMP}^*)^2 + (\Delta V_{TD}^*)^2 + (\Delta V_{LD}^*)^2 + (\Delta V_{LN}^*)^2}$$
$$= \sqrt{(5\text{ mV})^2 + (24\text{ mV})^2 + (5\text{ mV})^2 + (2.5\text{ mV})^2} = 25.1\text{ mV}$$

which means random temperature effects dominate V_{OS}^*.

6. $$\text{Error} = \frac{\Delta V_{GE} + \Delta V_{TD} + \Delta V_{LN} + \Delta V_{LD} + \Delta v_{PSR} + \Delta v_{DUMP} \pm V_{OS}^*}{V_{OUT}}$$

$$\approx \frac{0.3\text{ mV} + 3\text{ mV} + 8\text{ mV} + 20\text{ mV} + 7.9\text{ mV} + 120\text{ mV} \pm 25.1\text{ mV}}{2.4\text{ V}}$$

$$= 6.6\% \pm 1.0\%$$

so the worst-case overall accuracy of the regulator is roughly 7.6 percent, and if supply ripple and load dumps were excluded, 1.3% ± 1.0%, or 2.3 percent worst case.

2.2.2 Power-Conversion Efficiency

The basic functions of a voltage regulator are to condition and transfer power from an input source to an electronic load. Viewed in this way, accuracy specifies how well the regulator conditions power and power-conversion efficiency, or *efficiency*, for short, how much of the energy drawn reaches the load. Ideally, the load receives all the energy the input source provides, which means efficiency is 100 percent. In practice, however, the power pass device S_O and the error amplifier A_{EA} and, in general, the feedback controller dissipate power in the form of heat as P_{LOSS}. As a result, power-conversion efficiency η_C, which again is that fraction of the input power P_{IN} which reaches the load as P_O, is always less than 100 percent:

$$\eta_C \equiv \frac{P_O}{P_{IN}} = \frac{P_{IN} - P_{LOSS}}{P_{IN}} = \frac{P_O}{P_O + P_{LOSS}} < 100\% \tag{2.20}$$

Efficiency is a critical metric because it ultimately determines how much energy the system loses, and in the case of battery-powered devices, it also determines the operational life of the source because it can only store and supply finite amounts of energy. This is worse in the case of microsystems like wireless sensors and biomedical implants because batteries are tiny and their single-charge lifetimes are correspondingly short.

While power consumption in the controller need not change with the load, losses in the switch necessarily do. The fact is, dropping v_{IN} to v_{OUT} while running output current i_O or load current i_{LOAD} through S_O dissipates ohmic power P_{SO}:

$$P_{SO} = i_O(v_{IN} - v_{OUT}) = i_{LOAD}(v_{IN} - v_{OUT}) \tag{2.21}$$

Similarly, but not with respect to i_O or i_{LOAD}, the controller consumes power when drawing quiescent current i_Q from v_{IN} and dumping it to ground as ground current i_{GND}:

$$P_{CTRL} = i_{GND}(v_{IN} - 0) = i_{GND} v_{IN} \tag{2.22}$$

Similar to P_{LOSS} in power-conversion efficiency, i_{GND} is the fraction of input supply current i_{IN} that the system loses to ground. Ideally, the fraction of i_{IN} that reaches the load, which is what *current efficiency* η_I quantifies, is nearly 1, although it never can be in practice:

$$\eta_I \equiv \frac{i_{LOAD}}{i_{IN}} = \frac{i_{LOAD}}{i_{LOAD} + i_{GND}} < 1 \tag{2.23}$$

As a result, conversion efficiency η_C expands and simplifies to

$$\begin{aligned} \eta_C &= \frac{P_O}{P_O + P_{SO} + P_{CTRL}} = \frac{i_{LOAD} v_{OUT}}{i_{LOAD} v_{OUT} + i_{LOAD}(v_{IN} - v_{OUT}) + i_{GND} v_{IN}} \\ &= \frac{i_{LOAD} v_{OUT}}{(i_{LOAD} + i_{GND}) v_{IN}} = \eta_I \left(\frac{v_{OUT}}{v_{IN}} \right) < \frac{v_{OUT}}{v_{IN}} \end{aligned} \tag{2.24}$$

which means v_{IN}, v_{OUT}, and η_I ultimately set η_C. Because applications and power-supply designers often define v_{IN}, v_{OUT}, and i_{LOAD}, η_I via i_{GND} is the only variable that IC engineers control to determine η_C, and all they can hope to achieve with an ideal circuit that loses next to no i_{GND} is v_{OUT}/v_{IN}.

Assigning what v_{IN} and v_{OUT} should be is critical. System designers normally try to bridge as much voltage as possible with switched converters because their efficiencies are both high and weakly dependent on the voltages they drop. The outputs of these converters therefore define low enough v_{IN}'s for linear regulators to drop to v_{OUT}'s and supply their loads. Therefore, IC engineers try to design linear regulators that can survive low input supplies and regulate high output voltages with as little quiescent current as possible.

In this respect, operating in the dropout region (from Fig. 2.1), when v_{IN} is barely above v_{OUT}, is appealing. The problem with this option is that the regulator does not have loop gain with which to regulate its output. This is why operating the regulator just above the dropout region is optimal, because v_{OUT} is both close to v_{IN} for S_O to dissipate little power when conducting i_{LOAD} and low enough to keep the system from entering the dropout region. This means that dropout voltage V_{DO}, which specifies the minimum voltage that S_O can sustain in the linear region, ultimately establishes how high conversion efficiency can be:

$$\eta_C = \eta_I\left(\frac{v_{OUT}}{v_{IN}}\right) < \eta_I\left(\frac{v_{IN}-V_{DO}}{v_{IN}}\right) < 1-\frac{V_{DO}}{v_{IN}} \tag{2.25}$$

assuming η_I is close to ideal at nearly 1.

Not surprisingly, V_{DO} is a critical specification. To quantify this parameter, consider that S_O's terminal voltages in dropout are low and the feedback loop that drives S_O exhausts its dynamic range to the extent that S_O's drive can no longer rise. This means S_O operates like a resistor, so V_{DO} is the ohmic voltage across S_O's equivalent switched-on resistance R_{ON}:

$$V_{DO} = I_{LOAD}R_{ON} \tag{2.26}$$

Note that R_{ON} includes all series parasitic resistances between v_{IN} and v_{OUT}, so metal, contact, bond-wire, and switch resistances all play a role:

$$R_{ON} = R_{METAL} + R_{CONTACT} + R_{BOND\text{-}WIRE} + R_{SWITCH} \tag{2.27}$$

In practice, R_{ON} is not always consistent across dropout, so product engineers define it at the onset of dropout, at the edge of the linear region.Unfortunately, R_{SWITCH} normally rises with reductions in v_{IN}, so R_{ON} is lower when the regulator is closer to the linear region. R_{ON} also climbs with temperature, which means V_{DO} rises. Because of this and all other factors, including today's technological constraints, portable applications normally call for dropout voltages that are on the order of 200–300 mV.

A few idiosyncrasies are worth noting at this point. To start, quiescent current i_Q is normally more than an order of magnitude lower than the regulator's maximum output current, so current efficiency η_I is typically high and nearly 1 when the regulator supplies the full load. When lightly loaded, on the other hand, i_{GND} usually becomes a substantial fraction of the load, so η_I is correspondingly lower. This means power-conversion efficiency is normally high with heavy loads and low otherwise.

For context, consider that many portable devices, for the most part, idle, so average load current is low and, as a result, so is average conversion efficiency. This is why many engineers today ensure i_{GND} drops with the load. Some products, in fact, have a sleep mode in which i_{GND} is just high enough to energize vital blocks in the system, like a crude reference and a slow comparator used to monitor and ensure that the circuit is alert and ready to change states. Ensuring i_{GND} drops this way, however, is no more important than designing for a low dropout voltage V_{DO}, because S_O loses considerable power when supplying the full load.

Example 2.3 Determine the worst- and best-case full- and no-load power-conversion efficiencies of a 0- to 100-mA, 1.2-V linear regulator whose equivalent switch-on resistance and steady-state quiescent current are 500 mΩ and 10 μA, respectively. Assume the regulator draws its power from a 0.9- to 1.6-V NiCd battery.

Solution

1. Worst-case full-load power efficiency occurs when i_{LOAD} and the voltage across S_O are both at their highest levels, so v_{IN} is $v_{IN(MAX)}$, or 1.6 V, and

$$\eta_C = \frac{i_{LOAD(MAX)}v_{OUT}}{(i_{LOAD(MAX)}+i_{GND})v_{IN(MAX)}} = \frac{100\text{ mA}\cdot 1.2\text{ V}}{(100\text{ mA}+10\ \mu\text{A})1.6\text{ V}} \approx \frac{1.2\text{ V}}{1.6\text{ V}} = 75\%$$

2. Best-case full-load power efficiency occurs when i_{LOAD} is at its highest level and the voltage across S_O is at its lowest, so v_{IN} is just above dropout,

where $$V_{DO} = i_{LOAD(MAX)}R_{ON} = 100\text{ mA}\cdot 500\text{ m}\Omega = 50\text{ mV}$$

and $$\eta_C = \frac{i_{LOAD(MAX)}v_{OUT}}{(i_{LOAD(MAX)}+i_{GND})v_{IN(MIN)}} = \frac{i_{LOAD(MAX)}v_{OUT}}{(i_{LOAD(MAX)}+i_{GND})(v_{OUT}+V_{DO})}$$

$$= \frac{100\text{ mA}\cdot 1.2\text{ V}}{(100\text{ mA}+10\ \mu\text{A})(1.2\text{ V}+50\text{ mV})} \approx \frac{1.20\text{ V}}{1.25\text{ V}} = 96\%$$

3. Because i_{LOAD} is zero in the zero-current case, conversion efficiency is also zero, irrespective of other operating conditions:

$$\eta_C = \frac{i_{LOAD(MIN)}v_{OUT}}{(i_{LOAD(MIN)}+i_{GND})v_{IN(MAX)}} = \frac{(0\text{ A})v_{OUT}}{(0\text{ A}+i_{GND})v_{IN(MAX)}} = 0\%$$

which is why ground current alone is so important in portable applications, where load current is almost always near zero and battery life hinges on i_{GND}.

2.2.3 Operating Requirements

The operating range of the input voltage v_{IN}, output voltage v_{OUT}, output capacitor C_O, C_O's equivalent series resistance ESR or R_{ESR}, and output or load current i_O or i_{LOAD} describe the working limits of the regulator. While the application determines the minimum acceptable range, the integrated circuit sets outer boundaries. Generally speaking, source, breakdown, headroom, load, and efficiency considerations limit v_{IN}'s and v_{OUT}'s operating ranges; stability, load-dump, and supply-ripple response limit those of C_O and R_{ESR}; and loading demands, power ratings of the fabricated IC, and efficiency limit those of i_{LOAD}.

Input and Output Voltages

Perhaps the most fundamental requirement to consider when establishing the working range of a linear regulator's terminal voltages is the power source. In the case of a Li-Ion battery, for example, the typical useful range is 2.7–4.2 V. This means that, for maximum battery life, the regulator should draw power from the supply across its entire useful range. For this, design engineers must ensure that nodes and junctions in the circuit do not "break" at 4.2 V, and that transistors have sufficient headroom at 2.7 V for the regulator to supply its load.

With the source specified, the designer must select, from what is available and financially feasible, a process technology with transistors that can withstand the maximum operating voltage. Other considerations also influence the decision, of course, like transistor options, speeds, and leakage currents. In the foregoing Li-Ion example, while the Li Ion limits v_{IN}'s operating range to 4.2 V, the breakdown voltages of transistors in the circuit set the upper boundary to, say, 6 V. The IC engineer must then design a circuit that can operate at the Li Ion's lower limit of 2.7 V. For margin, and to ensure that the system starts properly when first connected to the supply, the engineer designs the regulator to survive even lower voltages. In other words, the regulator's minimum headroom requirement ultimately sets the outer boundary for v_{IN}'s lower limit to, say, 1.8 V.

The process of specifying the value and operating range of the output must first account for the lowest possible headroom limit that the load can survive. In other words, if an analog–digital converter can resolve 12 bits of accuracy only when its supply is above 1.8 V, the linear regulator's output should not fall below 1.8 V. Accuracy and efficiency will then dictate how much higher v_{OUT} can be. If accuracy degrades, or power losses rise above acceptable limits when v_{OUT} climbs above 2 V, to expand on the example, the output should remain between 1.8 and 2 V. Efficiency can similarly prescribe the input when power is supplied from a switching converter, in which case, because the switched regulator is more efficient, the switching network drops as much of the supply voltage as the linear regulator can tolerate. If supplied from a 24-V source, for instance, dropping 22 V across the switched circuit and another 200 mV across the linear regulator reduces losses in the latter while at the same time supplying the load with sufficient headroom to perform as prescribed.

Output Capacitor

Stability is another parameter to consider. The challenge here is that the load usually spans a vast range of currents. Wall-supplied applications 20 to 30 years ago demanded load-current ranges that spanned two to three decades, from maybe 0.001 to 1 A, because wall outlets, when compared to batteries, offer boundless energy. Battery-powered systems today, however, cannot survive sustained amp-level loads, so system engineers shut down subsystems whenever possible. In such cases, load current i_{LOAD} can span five decades of current, if not more, from 1 μA or less to 50–200 mA. Because i_{LOAD} sets the regulator's small-signal output transconductance and resistance, shifting i_{LOAD} this way across so many decades of current changes the feedback dynamics of the system to such an extent that guaranteeing the circuit remains stable is difficult. Datasheets, as a result, constrain output capacitance C_O and equivalent series resistance R_{ESR} to within a specific and often narrow window of acceptable values.

Load dumps play an equally important role in specifying C_O and its R_{ESR}. To see the correlation, consider that most electronic applications nowadays save energy by shutting down subsystems and engaging them on demand. These smart devices monitor and react to their surrounding environment, so waking subsystems in time to address asynchronous events is imperative. Quickly enabling several subsystems this way, however, constitutes a substantial load dump Δi_{LOAD} that can range anywhere from a few milliamps to hundreds of milliamps across a few microseconds or even hundreds of nanoseconds. Because the regulator requires several hundreds of nanoseconds to a few microseconds to react, C_O temporarily supplies most, if not all the dump current. This means Δi_{LOAD} loads C_O across several hundreds of nanoseconds and, as a result, C_O droops and R_{ESR} drops substantial voltages that pull v_{OUT} considerably below its nominal target. Therefore, to keep load-dump response within tolerable limits, datasheets specify that C_O should be high and R_{ESR} low.

Supply ripple can also be problematic. Because the gain across the regulator's feedback loop falls at higher frequencies, the circuit cannot completely negate the effects of a ripple in the supply on the output. The situation is worse at and past the loop's bandwidth, where the loop can no longer help. Unfortunately, cellular phones and other mobile applications feed linear regulators from a master dc–dc converter whose switching frequency nears the bandwidth of the linear regulator. In other words, a converter produces ripples of, say, 10–50 mV at 50 kHz–10 MHz that a 100-kHz–1-MHz linear regulator is barely able to reject. In these cases, C_O is the last line of defense, which is why power-supply engineers specify C_O above a minimum level. Because R_{ESR} limits how much ripple energy C_O can shunt, R_{ESR} should similarly stay below a specified maximum.

Output Current

Although the load normally sets the output's current range, the regulator circuit ultimately imposes outer limits. In particular, the *safe-operating area* (SOA) of the power pass device, which specifies its combined current–voltage–temperature ratings, sets "hard" upper-boundary limits on output current i_O under extreme voltage and temperature settings. Sometimes feedback stability also constrains i_O, because a wider range in i_O has considerable impact on the small-signal loop dynamics of the system. Some product engineers also use efficiency to specify i_O's maximum level as the current that produces 200 mV of dropout. Irrespective of the means, engineers must challenge and fully justify a load-current range before unnecessarily designing for impossible conditions.

2.2.4 Figures of Merit

With so many performance parameters, comparing a linear regulator to the state of the art can be challenging. Thankfully, most applications dictate what metrics are most important, and engineers weigh them accordingly, based on necessity. Engineering managers, researchers, and academics sometimes resort to *figures of merit* (FoM) to evaluate and compare a design to others in literature. The hallmarks of a good linear regulator are higher output current $i_{O(MAX)}$ and worst-case power-supply rejection PSR_{MIN} and lower steady-state errors ΔE_0, ground current i_{GND}, dropout voltage V_{DO}, output capacitance C_O, and response time Δt_R. FoMs combine several or all of these parameters into one metric. Since the absolute value of the resulting number carries little meaning, normalizing to a particular product or the median of several products is helpful. Referencing a regulator that outputs 100 mA with 1-percent steady-state error, 1-μs response time, 20-dB worst-case PSR, 5 μA of ground current, and 200 mV of dropout across 1 μF, for example, can produce a ranking FoM of

$$\text{FoM} \equiv \frac{i_{O(MAX)}\text{PSR}_{MIN}}{\Delta E_0 i_{GND} V_{DO} C_O \Delta t_R}\left(\frac{0.01\cdot 5\ \mu\text{A}\cdot 200\ \text{mV}\cdot 1\ \mu\text{F}\cdot 1\ \mu\text{s}}{100\ \text{mA}\cdot 10}\right)$$

$$= \frac{i_{O(MAX)}\text{PSR}_{MIN}}{\Delta E_0 i_{GND} V_{DO} C_O \Delta t_R (10^{20})} \tag{2.28}$$

Still, practical designs seldom value all parameters equally, or in the same way, so most FoMs of this sort are more academic than practical.

2.3 Operating Environment

The source and load alone do not completely describe the operating environment of the regulator chip. The *bond wires* connecting the die to the *lead frame* and pins of the package, for one, introduce series inductances L_{BW} and resistances R_{BW} to every off-chip node in the system, as Fig. 2.6 illustrates. The printed-circuit board (PCB) also adds parasitic impedances L_{PCB} and R_{PCB} between both the input supply and chip and the chip and load. This is why power-supply designers normally place external bypass capacitors C_{IN} and C_B' near the regulator's input and ground terminals and the load, to shunt the noise that parasitic resistances and inductances produce when conducting current from v_{IN} and to and from v_{LOAD}. In fact, because the purpose of the regulator is to generate a stable supply and the load is sensitive to ripple noise, C_B' is normally a high-frequency capacitor, such as a *multilayer ceramic-chip capacitor*, which is why C_B' has next to no ESR.

In the model shown, the IC refers to the semiconductor die that incorporates the microelectronics used to monitor, regulate, and supply the output v_{OUT}. Along the periphery of the IC are exposed landing pads from which bond wires attach to connect to the lead frame of the package. These *bond pads* exclude the nitride layer that protects other parts of the IC from contaminants in the surrounding environment. The outer extremities of the lead frame are the pins of the packaged IC, or chip, for short, or microchip, as others also call it. In many cases, neither the source nor the load is next to the regulator chip, which is why parasitic PCB impedances often separate them in practice. And since neither the source nor the load is ideal, they also introduce series resistance R_S and parallel impedances R_{LOAD} and C_{LOAD} at their respective terminals.

The purpose of output capacitor C_O is to stabilize the negative feedback loop and suppress load-dump and supply-ripple–induced variations at the output. As a result, C_O is normally on the order of several microfarads. Input filter capacitor C_{IN} is on the same order for similar reasons, to suppress noise at the input v_{IN} in light of other loads and nonidealities in the system. Because of their high values, designers

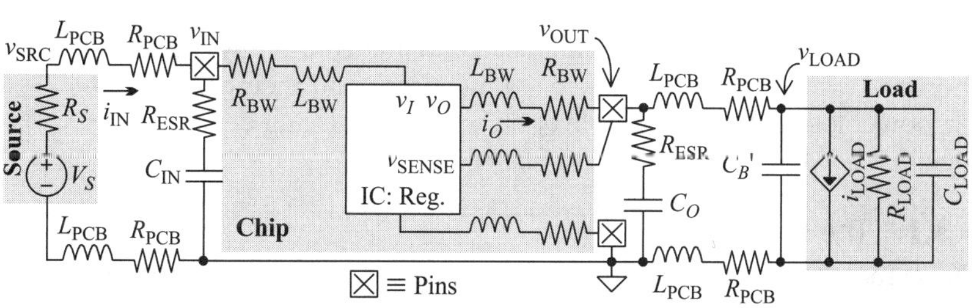

Figure 2.6 Typical operating environment of a linear regulator IC.

often resort to *tantalum capacitors* for C_O and C_{IN}, which typically cost less than other capacitors. Unfortunately, tantalum capacitors trade ESR for capacitance, so R_{ESR} can be as high as several ohms.

Ceramic and multilayer ceramic-chip capacitors generally cost more per farad than tantalum capacitors. On the positive side, however, these capacitors feature low ESR values that enable them to shunt more noise current. As a result, designers tend to place a small ceramic bypass capacitor C_B' across each load. Each bypass capacitor is usually in the nanofarad region, and together they sum to less than 1 μF. Although not often specified or well understood by the user, loading the regulator with more capacitance can compromise the stability of the regulator's negative-feedback loop.

Capacitors also introduce an *equivalent series inductance* (ESL), which is typically less than 5 nH. Luckily, the effects of this parasitic inductance are often negligible in low-power and low-bandwidth applications. A 10-mA, 1-μs rising load dump through a 5-nH inductor, for example, produces a series ESL voltage v_{ESL} of 50 μV:

$$v_{ESL} = L_{ESL}\left(\frac{di_L}{dt_L}\right) \approx L_{ESL}\left(\frac{\Delta i_{LOAD}}{\Delta t_{LD}}\right) \tag{2.29}$$

where Δt_{LD} is the time the load took to climb across the 10-mA dump Δi_{LOAD}. On the other hand, a 100-mA, 50-ns rising load dump across the same 5-nH inductor produces an ESL voltage of 10 mV, which on a 1-V output constitutes a 1-percent variation, and this excludes the effects of load and line regulation, temperature, process, supply ripple, and C_O, R_{ESR}, and C_B on the output.

Generally, all parasitic components in the regulator's power path degrade power-conversion efficiency, steady-state accuracy, load-dump response, or power-supply rejection. Input and output currents i_{IN} and i_O flowing through ground and series parasitic bond-wire and PCB impedances between the source at v_{SRC} and the IC at v_I and between the IC at v_O and the load at v_{LOAD} drop ohmic voltages that consume power. In addition, the dynamic and steady-state voltages across the components between v_O and v_{LOAD} also pull v_{LOAD} below its nominal target. Therefore, using multiple bond wires in parallel between v_{IN} and v_I and between v_O and v_{OUT} appeal to the designer because they introduce lower parasitic resistances and inductances. For similar reasons, placing the regulator close to the source and load and interconnecting them with short and wide traces also reduce voltage drops and their corresponding power losses. Using multiple bond wires and placing devices closer together, however, are not always feasible options.

2.3.1 The Load

The actual "load" is normally difficult to model. In fact, designers are more often than not unaware of what subsystems their regulators will ultimately supply, except in system-on-chip (SoC) and other application-specific cases. Generally speaking, though, the Norton-equivalent circuit that models the load in Fig. 2.6 is sufficient to predict the loading behavior of most microelectronic subsystems. The steady-state value of load current i_{LOAD}, for example, establishes the biasing point about which the circuit regulates v_{OUT}, and the dynamics of i_{LOAD} dictate how fast and to what extent the load disturbs v_{OUT}. Similarly, R_{LOAD} and C_{LOAD} define the impedances that the regulator must supply when responding to small- and large-signal variations in v_{OUT}. Combined, i_{LOAD}, R_{LOAD}, and C_{LOAD} define the bias and loading conditions that the designer must consider to ensure the feedback loop of the regulator remains stable.

In light of the regulator's sensitivity to i_{LOAD}, R_{LOAD}, and C_{LOAD}, not knowing the actual load is a source of anxiety for design engineers. If, for example, a low-power amplifier whose lowest impedance path from its supply to ground is a diode-connected transistor and an active load with a combined small-signal resistance of roughly $1/g_m$ and r_o loads the regulator, R_{LOAD} can be on the order of tens to hundreds of kilo-ohms. Power amplifiers (PAs), on the other hand, deliver substantial supply currents to low-impedance outputs via low-resistance switches. This means the slewing PA presents a sub-kilo-ohm load to the regulator. In the case of digital microprocessors, "pull-down" and "push-up" complimentary metal–oxide–semiconductor (CMOS) transistors in inverters and other gates conduct shoot-through currents simultaneously when transitioning their outputs in response to clocked events. Although these excursions are short, the resistance these switches present from their supply to ground can be tens of ohms. In other words, R_{LOAD} may span a wide range of resistances, from tens of ohms to several hundred kilo-ohms.

To guarantee stable operating conditions, the designer must therefore consider all possible extremes. One such extraordinary instance is when the load is purely resistive, that is, when i_{LOAD} is zero and R_{LOAD} is v_{OUT}/i_{LOAD}. Another exceptional case is when the load is only active, which means R_{LOAD} is infinitely high and i_{LOAD} alone emulates the load. In the case of a 2.5-V, 1- to 50-mA regulator, for example, four possible extremes exist: when R_{LOAD} alone sinks (1) 1 mA at 2.5 V with 2.5 kΩ and (2) 50 mA at 2.5 V with 50 kΩ and when R_{LOAD} is infinitely high and i_{LOAD} alone carries (3) 1 and (4) 50 mA. Considering only a purely resistive load, as in (1) and (2), or only a purely active load, as in (3) and (4), assumes R_{LOAD} is always low or always high, neither extreme of which *alone* is realistic. The problem is that C_O is high, so the small-signal resistance at the output of the regulator establishes what amounts to a low-frequency bandwidth-limiting node that may compromise the feedback dynamics and stability conditions of the feedback loop in the regulator. Therefore, ensuring the regulator remains stable when confronted with all four extreme loading conditions reduces risk.

Fortunately, equivalent load capacitance C_{LOAD} is often negligible when compared to external bypass and output capacitors C_B' and C_O.

This is especially true for standard *commercial-off-the-shelf* (COTS) and moderate- to high-power regulators whose output capacitances are high. SoC realizations, however, do not always enjoy this luxury because C_O is on-chip and therefore considerably lower than its off-chip counterparts. Some SoC implementations, in fact, rely entirely on C_{LOAD} for stability and load-dump response, altogether eliminating the need for C_O, which is equivalent to saying C_{LOAD} *is* C_O. As a result, as with R_{LOAD}, C_{LOAD} in SoC implementations can play a pivotal role in establishing the stability conditions of the regulating loop, which is why designers, when in doubt, must consider all possible extremes of C_{LOAD}. In the presence of substantially higher C_O and C_B' combinations, however, C_{LOAD} is less important.

2.3.2 Point of Regulation

The regulator's negative-feedback loop controls what it monitors, so when possible, regulators *sense* and control the voltage at the load in *point-of-load* (PoL) fashion. When supplying several loads, however, regulating all load points separately with one regulator is not possible, so engineers connect the loads to one centric location that the loop senses and regulates. In the model of Fig. 2.6, output pin v_{OUT} is this *point of regulation*, where the voltage is closest to the regulator's target. Here, v_{SENSE} is a low-current path that monitors and regulates v_{OUT}.

The purpose of the *star* connection that v_{SENSE} and v_{OUT} implement is to incorporate parasitic voltage drops in the power-conduction path into the feedback loop—in other words, to keep v_{OUT} close to its target in spite of drops across R_{BW} and L_{BW}. For this, the sense node must itself be free of voltage drops, which means v_{SENSE} neither sinks nor sources appreciable current. Engineers refer to this star connection, where current- and noncurrent-carrying signal paths belonging to the same node converge at one single point and nowhere else, as a *Kelvin* connection.

In the best of cases, as already mentioned, the regulator senses v_{LOAD} at the point of load. This way, the feedback loop incorporates R_{BW}, L_{BW}, R_{PCB}, and L_{PCB} and keeps their voltage drops from affecting the load. Dedicating a pin for the sensing function is not always feasible, however, which is when engineers adopt the next best approach: to bond-wire the sense pad to the output pin (i.e., v_{OUT}) inside the package, as Fig. 2.6 shows. Doing so keeps drops across R_{BW} and L_{BW} from affecting v_{OUT} and allows v_{LOAD} to suffer the unmitigated effects of R_{PCB} and L_{PCB}. Unfortunately, microsystems nowadays integrate so many functions on-chip that allocating a separate sense pad and connecting multiple bond wires to a single pin is impossible. Under such constraints, designers resort to their last line of defense: to star-connect v_{SENSE} to the output pad v_O. In these cases, the feedback loop compensates for drops inside the IC, but does nothing to correct the series voltage drops that output current i_O produces when flowing through R_{BW}, L_{BW}, R_{PCB}, and L_{PCB} to v_{LOAD}. All this is to say that pushing the point of regulation to the load as much as the application and technology allow is important.

2.3.3 Parasitic Effects

Including parasitic components in the feedback loop does not mean the system is free from their adverse effects. When the regulator enters dropout, for example, the loop gain drops so much that feedback is no longer effective. This means all parasitic series resistances in the power path, whether they are inside the loop or not, drop voltages to the load v_{LOAD}. In quantitative terms, bond-wire resistances between input and output pins v_{IN} and v_{OUT} in Fig. 2.6, which R_{BW}' combines in the simplified model of Fig. 2.7, and between the IC and the load, which R_{PCB}' incorporates, raise the regulator's on-resistance R_{ON}. R_{ON} and load current i_{LOAD} then establish dropout voltage V_{DO} or $i_{LOAD}R_{ON}$.

Unfortunately, resistances outside the loop also drop v_{LOAD} when the regulator is in the linear region. These external resistances R_{EXT} add to the closed-loop resistance of the regulator R_{CL} to raise the circuit's effective output resistance R_O to

$$R_O = R_{CL} + R_{EXT} \tag{2.30}$$

which together with steady-state load-current variations ΔI_{LOAD} establish load-regulation response ΔV_{LD} or $\Delta I_{LOAD}R_O$. (Note that R_{EXT} in the case of Fig. 2.7 reduces to R_{PCB}'.) Parasitic inductances between output capacitor C_O and the load also add to C_O's L_{ESL} to further affect v_{LOAD} when confronted with load dumps. More specifically, the fraction of load-dump current Δi_{LOAD} that C_O supplies through L_{ESL} and L_{PCB}' drops a voltage Δv_{ESL} that degrades load-dump response Δv_{DUMP} by

$$\Delta v_{ESL} = (L_{ESL} + L_{PCB}')\left(\frac{di_C}{dt_L}\right) \approx (L_{ESL} + L_{PCB}')\left(\frac{\Delta i_{LOAD}C_O}{C_O + C_B}\right)\left(\frac{1}{\Delta t_{LD}}\right) \tag{2.31}$$

where C_B combines load and external bypass capacitors C_{LOAD} and C_B' in Fig. 2.6 and Δt_{LD} is the time that load current i_{LOAD} requires to traverse

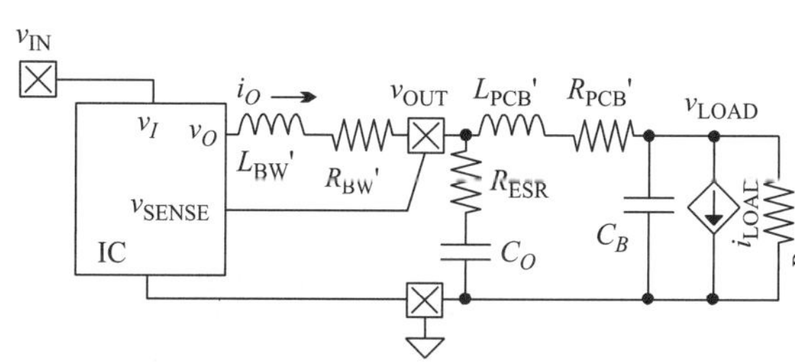

FIGURE 2.7 Simplified model of the regulator's operating environment.

across load dump Δi_{LOAD}. Usually, however, C_O's droop and ESR voltages Δv_C and Δv_{ESR} in Δv_{DUMP} overwhelm Δv_{ESL}.

2.4 Classification

2.4.1 Output Current

There are various types of linear regulators catering to an assortment of different applications. Generally, the most obvious distinction between them is how much current or power they deliver. Low-power regulators, for example, normally supply output currents of less than 1 ampere, which is typical of many portable and battery-powered electronics. High-power regulators source higher currents to satisfy the needs of many automotive, industrial, and similarly demanding applications. For the sake of higher power-conditioning efficiency, however, these high-current devices are quickly losing ground to their switching counterparts. In fact, many systems today use a high-power master switching supply and sprinkle low-power linear regulators across several loads to regulate them in PoL fashion. As a result, many linear regulators find most of their market below 300 mA.

2.4.2 Dropout

Dropout voltage V_{DO} is another critical specification that product engineers and system designers use to classify linear regulators. This voltage is important because power dissipation in linear regulators also hinges on V_{DO}, on how much voltage the regulator drops when operating near the edge of the linear region. In fact, since *low-dropout* (LDO) regulators can dissipate less power than their *high-dropout* (HDO) counterparts, LDOs enjoy more popularity in the marketplace, especially in battery-powered applications where the input supply can be low. Ultimately, the power transistor plays a pivotal role in this regard, because emitter- and source-follower configurations drop considerably higher voltages than their common-emitter and -source counterparts, whose outputs are collectors and drains. Accordingly, linear regulators employing followers, which typically drop over 600 mV, belong to the high-dropout class and circuits with collector and drain outputs, which can drop 300 mV or less, belong to the low-dropout group.

2.4.3 Compensation

Feedback compensation also establishes another important category. For example, regulators that use output capacitor C_O to establish the dominant low-frequency pole of the feedback loop require high output capacitances. Circuits that rely on an internal node for this purpose, on the other hand, limit how high C_O can be because C_O in these cases establishes a parasitic pole in the feedback loop that compromises stability. The motivation for internally compensated regulators is the area savings in the PCB, because lowering C_O eases integration requirements. The drawback to a lower C_O, however, is higher droop variations in response to load dumps and ripples in the supply, which is to say worse load-dump and power-supply-rejection responses Δv_{DUMP} and Δv_{PSR}. Output capacitance requirements in output compensated regulators therefore align well with noise-suppression performance, which is why users normally prefer output-compensated regulators over their internal counterparts, in spite of the dollar and PCB real-estate costs associated with off-chip capacitors.

Example 2.4 In the presence of a 50-ns, 1- to 50-mA load dump, a regulator with a bandwidth of 100 kHz and a 0.47-μF output capacitor allows its output voltage to droop approximately 380 mV before responding and sourcing the full load:

$$\Delta v_C = \left(\frac{\Delta i_{LOAD}}{C_O}\right)\Delta t_R \approx \left(\frac{49\text{ mA}}{0.47\ \mu\text{F}}\right)\left(\frac{2.3}{2\pi \cdot 100\text{ kHz}}\right) \approx 382\text{ mV}$$

The advent of SoC and system-in-package (SiP) integration is slowly changing the trend from output to internally compensated schemes, however. As the IC absorbs more circuits and regulators into its common silicon substrate, external capacitors are increasingly more difficult to accommodate. Output-compensated circuits are therefore appearing under the guise of internally compensated ICs simply because the compensating output capacitor, although still hanging at the output, is *inside* the microchip, either on chip or copackaged. Still, because the effects of pole location on loop dynamics are more profound than capacitor size, linear regulators are output compensated when their dominant low-frequency pole is at the output and internally compensated otherwise.

2.4.4 Categories

Although many parameters describe how well linear regulators perform, power and accuracy are arguably the most important, and they ultimately depend on output current, dropout voltage, and the loop dynamics that compensation strategies produce. As a result, classifying regulators this way, as Fig. 2.8 illustrates, helps users narrow the scope of their search to a more manageable range. Battery-powered systems, for example, rely on low-power devices to extend battery life. Automotive and industrial applications rely on their high-power counterparts to energize power-hungry components under what many describe as extreme conditions. Because of higher efficiency, however, switching converters now fill a wider high-power market segment, replacing high-power linear regulators with a switching converter from which several low-power linear regulators draw power to energize derivative subsystems. Similarly, since internally compensated circuits with no off-chip capacitors cannot supply vast load dumps, they are not as popular in many commercial products today. They are

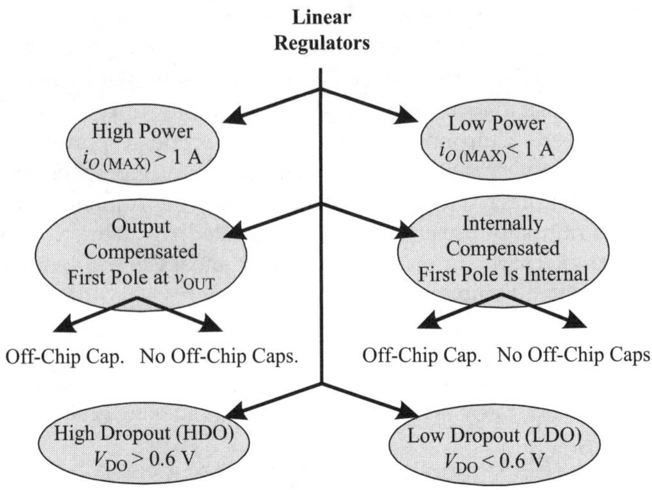

FIGURE 2.8 Classification of linear regulators.

nonetheless desirable because they cost little and occupy minimal space, which is why low-power, *application-specific ICs* (ASICs) that integrate one or several linear regulators on the same silicon die as their loads demand them.

Combining the general characteristics just described with the rising market demand for mobile, battery-powered electronics has created a niche market that demands low power, low dropout, and no off-chip capacitors. Many portable applications, in fact, require small-footprint and extended-life solutions that internally compensated low-dropout and low-power regulators can often satisfy. What is more, because LDO topologies output higher voltages than HDOs when supplied from partially drained batteries, they relax the minimum headroom limits of their loading circuits and raise the dynamic range, and by translation, the *signal-to-noise-ratio* (SNR) with which their loads operate, which is normally the first parameter to suffer when the supply voltage drops. Not surprisingly, users opt for LDO regulators whenever possible.

Example 2.5 If an almost drained lithium-ion battery at 2.7 V powers an LDO regulator with a dropout of 200 mV, which then powers an operational amplifier whose noise floor is 50 mV, the operational amplifier would have to operate under a 2.5-V regulated supply. All signal processing must therefore fall within a 2.45-V window, which constrains SNR to less than 49:

$$\text{SNR} \equiv \frac{\text{Dynamic Range}}{\text{Noise Floor}} = \frac{\text{Maximum Voltage} - \text{Noise Floor}}{\text{Noise Floor}}$$

$$< \frac{2.5\text{ V} - 50\text{ mV}}{50\text{ mV}} = 49$$

Had the regulator been an HDO device with a dropout of 0.7 V, the operational amplifier would have had to work under a 2.0-V regulated supply, which reduces dynamic range to less than 1.95 V and SNR to less than 39:

$$\text{SNR} \equiv \frac{\text{Dynamic Range}}{\text{Noise Floor}} = \frac{\text{Maximum Voltage} - \text{Noise Floor}}{\text{Noise Floor}}$$
$$< \frac{2.0\ \text{V} - 50\ \text{mV}}{50\ \text{mV}} = 39$$

to further restrain the amplifier's ability to process analog signals.

Low dropout is a requirement for many applications today, from automotive and industrial to medical applications. The automotive industry, for instance, exploits the low-dropout characteristics of LDO regulators during cold-crank conditions, when the car-battery voltage is between 5.5 and 6 V and the regulated output must remain around 5 V, which means the loaded dropout voltage must be less than 0.5 V. The increasing demand is especially apparent in mobile battery-operated products, such as cellular phones, tablets, cameras, and laptops, where the voltage of a partially drained battery can be very low. In the case of two alkaline or NiCd cells, for example, whose range is roughly 1.8 to 3.2 V, the maximum output voltage of a 0.6-V HDO regulator over the life span of the battery is less than 1.2 V, which is not sufficiently high to meet the headroom requirements of many high-performance analog circuits. The restriction is even worse and therefore often prohibitive when powered from a single battery cell.

2.5 Block-Level Composition

Decomposing a linear regulator into its constituent components reveals the functional requirements of the system. Generally, a feedback control loop senses and regulates output voltage v_{OUT} about a target that a voltage reference v_{REF} ultimately sets. For this, feedback network β_{FB} in the typical implementation of Fig. 2.9 senses v_{OUT}, and error amplifier A_{EA} compares v_{OUT}'s sampled translation v_{FB} to v_{REF} to generate an error-correcting signal that ensures that pass transistor S_O supplies whatever current the load demands. In essence, the gain across the loop that A_{EA}, S_O, and β_{FB} implement determines how close v_{FB} is to v_{REF}, and β_{FB} translates this to v_{OUT}. As a result, to keep external factors like the load, input supply, and junction temperature from pushing v_{OUT} outside a small window of its target, v_{REF} and β_{FB} must be accurate and the loop gain must be high. The result in the end is a noninverting series–shunt feedback amplifier whose output is v_{OUT} and effective input is steady-state reference v_{REF}.

Housekeeping functions are essential to the overall health of the regulator. Generally, they ensure the circuit operates safely and reliably and interfaces well with other subsystems. Although sometimes

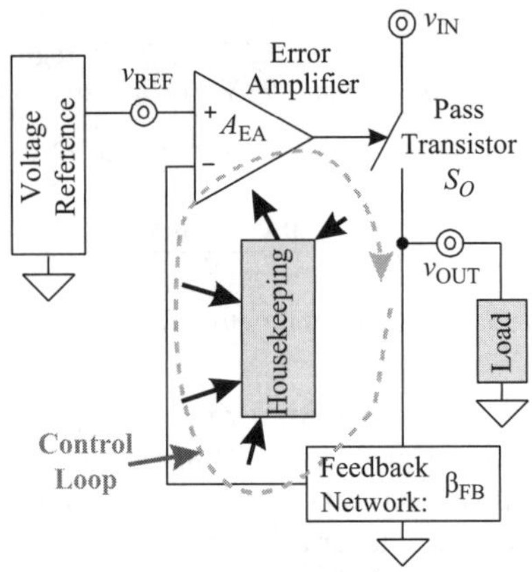

FIGURE 2.9 Typical block-level composition of a linear regulator.

included in the reference block, a current-generator circuit, for example, outputs a reference current that biases and ensures that the error amplifier has enough power to respond quickly to load dumps and other events. Other subblocks similarly protect the IC from exposure to sustained and momentary over-current, over-voltage, over-temperature, short-circuit, and *electrostatic-discharge* (ESD) conditions. Housekeeping circuits can also have application-specific features like enable–disable and soft- or slow-start functions that power-mode and start and restart the system when necessary.

2.6 Summary

The fundamental component that keeps the output of a linear regulator within a small window of its target is the feedback loop. In fact, with enough loop gain, the system can counter most of the effects that variations in the input supply and load produce. When the supply falls below a threshold level, however, loop gain drops to pull the regulator out of its linear region and into dropout, where feedback loses strength and the system, as a result, becomes increasingly sensitive to the supply and load, which is why operating in dropout is generally undesirable. Allowing the supply to stay high, however, raises the power lost across the regulator. So, when considering accuracy and power, operating the system near but not in dropout is optimal, which is why low-dropout regulators are so appealing.

Since accuracy and power are so important, designers specify linear regulators with parameters like tolerance, gain error, temperature drift, line regulation, power-supply rejection, load regulation, load-dump response, power-conversion efficiency, and dropout voltage. The application also superimposes requirements that the regulator must satisfy, such as input and output voltages, output capacitance, and output current. Parasitic resistances and inductances in the power path also affect the load. Therefore, because the feedback loop can counter most of their parasitic effects, designers prefer to regulate the output at the point of load. Dedicating a pin and board space to trace and star-connect a low-current path to the load, however, is not always possible, so negotiating a compromise is part of the design process.

In the end, power and accuracy are what users first assess when searching for a circuit to satisfy their needs. Therefore, product engineers classify linear regulators according to output current, dropout voltage, and frequency compensation. In this regard, although high-power regulators that are accurate are ideal, they are difficult to fully integrate on-chip. Internally compensated low-dropout regulators, on the other hand, which are optimal for portable applications with extended-life requirements, cannot satisfactorily supply fast, wide-ranging load dumps. All this means the *ideal* linear regulator is, at least for now and in the foreseeable future, like the ideal Op Amp: impossible to realize in practice.

The linear regulator decomposes into functional blocks that engineers must ultimately implement with resistors, capacitors, diodes, and transistors. So to build and model these semiconductor devices, designers must first understand how they work individually, which is the focus of Chap. 3. Subsequent chapters describe how to combine and use these devices in negative feedback to realize the functionality and objectives described in this chapter.

2.7 Review Questions

1. Which two parameters ultimately determine the tolerance of a linear regulator at room temperature?

2. Is the regulator more accurate when loop gain is low or high, or is loop gain irrelevant?

3. Which two parameters determine how much the output of the regulator will drift across temperature?

4. Line regulation refers to *steady-state/momentary (i.e., transient)* variations in the supply.

5. Power-supply rejection is *worse/better* when supply gain to the output is high.

6. True or False: The measured load-dump response of a linear regulator includes the effects of load regulation.

7. How much will the output of a 500-kHz linear regulator with a 1-μF, 250-mΩ capacitor at the output droop when confronted with a 10-ns, 5- to 20-mA load dump?

8. When ground current is negligible with respect to load current, what parameters or variables limit the power-conversion efficiency of the regulator?

9. What is the power-conversion efficiency of a linear regulator that steers 20 mA of the 20.4 mA it draws from a 3.6-V battery to a 1.8-V load?

10. What is the power-conversion efficiency of a linear regulator that steers 100 μA of the 140 μA it draws from a 3.6-V battery to a 1.8-V load?

11. What is another parameter used to describe the effects of dropout voltage V_{DO}?

12. True or False: Connecting the sense node to the output bond pad allows the feedback loop to counter the load-regulation effects of parasitic bond-wire and board resistances.

13. True or False: Connecting the sense node to the load allows the feedback loop to counter the effects of parasitic bond-wire and board resistances on dropout performance.

14. Which type of linear regulator best satisfies the needs of miniaturized battery-powered sensors, which demand low power and a small footprint?

15. Which type of linear regulator best satisfies the needs of battery-powered cellular phones that suffer from 50-ns, 1- to 100-mA load dumps, which demand high accuracy and high power efficiency?

16. List the fundamental functional blocks of a linear regulator.

CHAPTER 3

Microelectronic Devices

Integrated circuit (IC) engineers rely on their understanding of how microelectronic components work individually and collectively when they design and build ICs that perform within the context described in Chap. 1 to meet the target specifications outlined in Chap. 2. This chapter is therefore the first of a sequence of five aimed at describing and illustrating how to design analog ICs. The focus in this chapter is devices: their operation, structure, and models. The next couple of chapters use this foundation to build analog blocks with which to design the bias circuits, feedback networks, and linear regulators that subsequent chapters describe. Only with these concepts in hand can IC engineers ultimately design the linear-regulator systems that modern electronic products demand. But again, it all starts with resistors, capacitors, diodes, and transistors.

3.1 Resistors

3.1.1 Theory of Operation

The most basic, yet most powerful component an IC engineer has is a *resistor*. Although not frequently thought of in this way, designers most often use resistors as short-circuit links to conduct electrical signals from one circuit to another. In fact, any material used to connect two points, be it metal, silicided polysilicon, or highly doped silicon, introduces resistance that impedes the flow of charge and dissipates energy in the form of heat. More specifically, a current i_R that flows through a material whose resistance is R_X produces an ohmic voltage v_R across R_X that consumes power P_R:

$$v_R = i_R R_X \tag{3.1}$$

and

$$P_R = i_R v_R \tag{3.2}$$

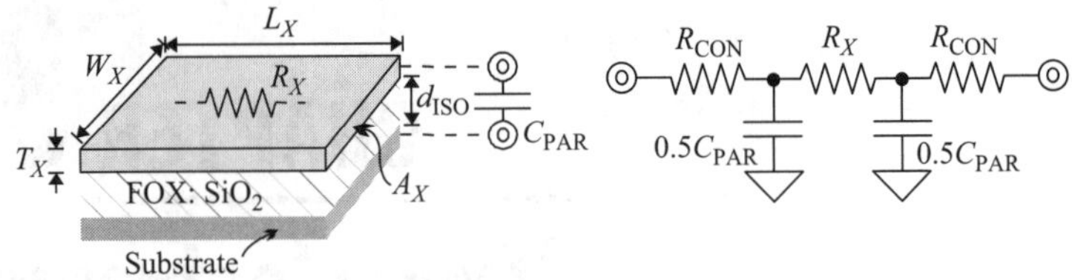

FIGURE 3.1 Three-dimensional view of a resistor strip and its electrical model.

Since every current-carrying link in a circuit drops voltage and consumes power, designers use metal or heavily doped polysilicon materials to interconnect electrical points in a circuit. The resistances of the electronic components these links connect, which range from kilo- to mega-ohms, normally overwhelm the effects of the milli- to subkilo-ohm resistances that these materials introduce.

In a more traditional sense, resistors convert currents into voltages, and when used for this purpose, resistance is no longer parasitic but intentional. In these cases, tens to hundreds of kilo-ohms and even mega-ohms are desirable. With this in mind, extending the length L_X of an unsilicided polysilicon or lightly doped silicon strip of material, like the one in Fig. 3.1, lengthens the path through which charges traverse to raise the total resistance across the structure. Similarly, decreasing the cross-sectional area A_X through which charges travel impedes flow, so lowering the width W_X and thickness T_X of the material also raises R_X:

$$R_X = \frac{\rho_X L_X}{A_X} = \frac{\rho_X L_X}{W_X T_X} \tag{3.3}$$

where ρ_X is the inherent *resistivity* of the material. Interestingly, higher temperatures induce atoms to vibrate more and electrons to accelerate and collide, further impeding the flow of charge. This means the resistance of a strip normally climbs with temperature, which is a more insightful way of saying resistor strips typically exhibit positive temperature coefficients.

3.1.2 Parasitic Components

The intrinsic advantage of integrated-circuit technologies is their ability to build numerous electronic components on a common, relatively low-impedance silicon substrate. To isolate individual resistors, process engineers insert lateral and vertical dielectric media between adjacent devices and surrounding layers. Figure 3.1, for example, illustrates how silicon dioxide SiO_2, more commonly known as *field oxide* or FOX, for short, isolates a metallic or polysilicon resistor strip from the substrate, which in the case of P-doped substrates is also ground or the most negative supply. Unfortunately, the resistor and substrate are, effectively, parallel conducting plates that establish a parasitic capacitance C_{PAR} between them. In the case of diffusion resistors, the dielectric is the depletion layer established by reverse-biasing the PN junction between the resistor and substrate or well. This parasitic capacitance rises with the overlapping surface area of the "plates" or junction, that is, with the width W_X and length L_X of the strip, and falls with longer plate-separation distances d_{ISO} or depletion widths, the principles of which the sections on capacitors and diodes later explain.

Designers link these resistors to the circuit via metallic contacts. Although these contacts offer little impedance, they nonetheless introduce series parasitic resistances R_{CON} to each terminal of the resistor, as Fig. 3.1 shows. Here, the model splits the effects of C_{PAR}, which appears along the length of R_X, between the terminals of the resistor. Distributing the effects of C_{PAR} equally between small fractions of R_X more accurately models the effects of C_{PAR} on R_X, except doing so increases the number of equations a computer must solve to simulate the effects of R_X on a circuit. Extending simulation time this way for the sake of added accuracy is, in many cases, difficult to justify because the improvement is incremental and oftentimes inconsequential. Ultimately, parasitic resistors drop voltages and dissipate power, and parasitic capacitors shunt high-frequency energy away from propagating signals.

3.1.3 Layout

Since only two-dimensional geometrical properties like width and length and material choice are open design variables, as the side note explains, IC designers define electronic components by means of top *layout* views. In the case of a resistor, the length–width dimensions of the strip of material that the top view of Fig. 3.2 illustrates defines its resistance. Extending L_X and shortening W_X, for example, raises R_X. Here, the ends are wider than the strip to not only accommodate a metallic contact but also reduce the parasitic resistance of the contact. This layout strategy normally results in the classical "doggy-bone" structure shown.

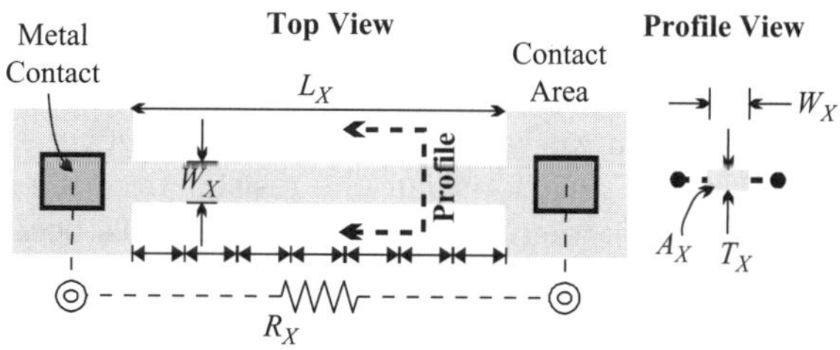

FIGURE 3.2 Top layout view of a resistor strip.

Open Design Variables *The interdependent physical properties of the materials used to fabricate components in and on a silicon substrate and their resulting manufacturing requirements superimpose electrical and geometrical constraints on the layers used to build the devices. Doping concentration, which deposition and diffusion manufacturing processes define and constrain, is one such property, which is why process engineers adjust and fix doping concentrations to particular levels. These same processes also limit how deep materials grow on and diffuse into the silicon substrate. Process designers optimally define depths by, for example, adjusting the deposition rate of polysilicon or the annealing time and temperature of diffusing dopant atoms. Doping concentration and material depths therefore fall out of a circuit designer's jurisdiction, which means only planar geometrical parameters, like width and length, and choice and sequence of materials are open design variables to IC design engineers. As a result, designers define components with top, two-dimensional* layout *views.*

Because material resistivity ρ_X and thickness T_X are fixed process-specific parameters, designers think of ρ_X/T_X as a constant and call it *sheet resistance* R_S:

$$R_S = \frac{\rho_X}{T_X} \tag{3.4}$$

Interestingly, since designers adjust width W_X and length L_X and resistance is higher for longer and narrower structures, R_X is equivalent to the product of R_S and the number of squares of material N_X between the terminals of the strip:

$$R_X = \left(\frac{L_X}{W_X}\right) R_S = N_X R_S \tag{3.5}$$

where one square is $W_X \times W_X$ and N_X is the number of W_X segments in L_X. In other words, R_S is the resistance of one square of material, irrespective of the area that W_X and L_X define. When neglecting the resistance of contact areas, R_X in Fig. 3.2, to cite an example, is roughly the resistance of seven squares, or more succinctly, $7R_S$. Not surprisingly, designers also use $R_\square$ in $\Omega/\square$ to represent R_S and classify resistor materials by R_S, not by ρ_X.

A good resistor is one whose resistance is easy to define and capacitance to substrate is negligibly low. The resistance of the contact areas, whose value is more difficult to predict, should be negligibly low, that is, large areas with no more than one square's worth in length. The dielectric used to isolate the resistor from the substrate should be thick with a high dielectric constant. Plus, because long and thin strips can reach or surpass the length of a chip, designers prefer "snaking" high-value resistors into the *serpentine* structure that Fig. 3.3 illustrates. Note, however, that charges traverse through the least resistive portions of the corners of the serpentine, so corner

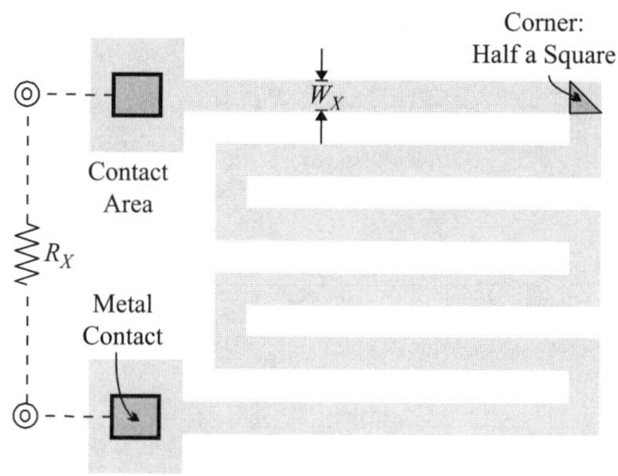

FIGURE 3.3 Top layout view of a serpentine resistor.

resistances are not exactly the equivalent of one square. Engineers often approximate these resistances to half a square's worth.

Example 3.1 Two thousand squares of a 1-μm thick 500-Ω/□ unsilicided polysilicon strip over field oxide offers 500 kΩ of resistance and minimal parasitic capacitance.

3.1.4 Absolute and Relative Accuracy

Imperfections in the manufacturing process cause doping concentrations, depths, widths, and lengths to vary across chips and fabrication lots to the extent that absolute tolerance is on the order of ±20 to ±50 percent. Temperature exacerbates this variation by another 10 to 50 percent. Luckily, these parameters affect all nearby resistors in the same way, so resistances track reasonably well. In other words, while absolute tolerance can be high, relative tolerance can be considerably lower at ±0.5 to ±5 percent. IC engineers therefore rely on the *matching* performance of similarly built devices for precision, not on their absolute values.

From a designer's perspective, improving the matching performance of resistors amounts to reducing the fractional variations of width W_X and length L_X, which means wide and long strips match better than narrow and short pieces. The resistance of strips with widths that are three to five times wider than the process allows, for example, can match within ±1 percent because their respective width variations, which can be ±100 nm, constitute a smaller fraction of their total widths, which can be 3 to 10 μm. Because corner resistances do not always scale with length, designers typically steer away from serpentine structures when matching resistors and, instead, connect similarly sized strips with metallic links, which is what Fig. 3.4 does with R_A, whose resistance is equivalent to two strips of R_B.

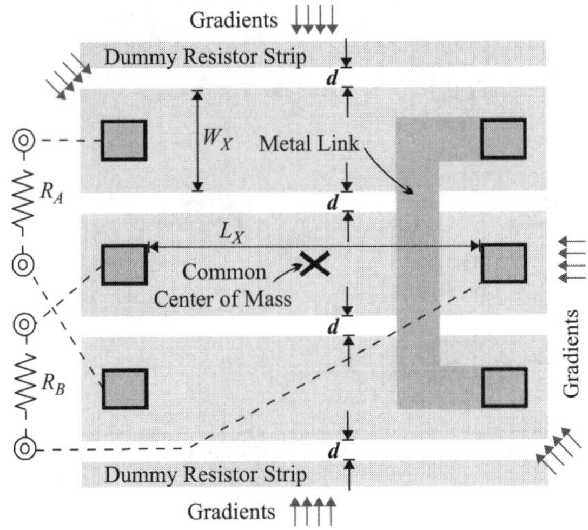

FIGURE 3.4 Top layout view of matched resistors with interdigitated segments and a common center of mass.

In the case of polysilicon resistors, whose width an *etching* step in the fabrication process defines, the width variation depends on the uniformity of the etching rate across its length, which as it happens depends on what lies next to each strip. Establishing similar peripheral conditions around all strips by placing adjacent resistors built of the same material at equidistant intervals, like Fig. 3.4 shows, helps keep etching rates consistent across the array. To ensure that the outer strips of the structure have similar etching conditions, designers similarly place thin dummy devices on their outer edges. However, maintaining similar etching conditions for the inner side of a corner in a serpentine structure is virtually impossible, which is another reason designs seldom rely on serpentine structures for accuracy.

There are other ways of improving matching performance between resistors. *Interdigitating* their constituent segments is one, which is what Fig. 3.4 implements in inserting R_B between two equally sized segments of R_A. This essentially spreads the linear effects of process-induced gradients across the die among the resistors. Better still is to establish a common center of mass, which Fig. 3.4 also achieves, because *common-centroid* configurations average the effects of all linear two-dimensional gradients. Because nonlinear effects exacerbate across large surface areas, reducing the size of the array also improves matching. In the case of equally sized devices, for example, splitting the array equally into four and *cross-coupling* the sections, like Fig. 3.5 illustrates, is a popular method of implementing a common-centroid configuration that reduces the nonlinear effects that large spreads otherwise worsen. Even more, increasing the number of interdigitated

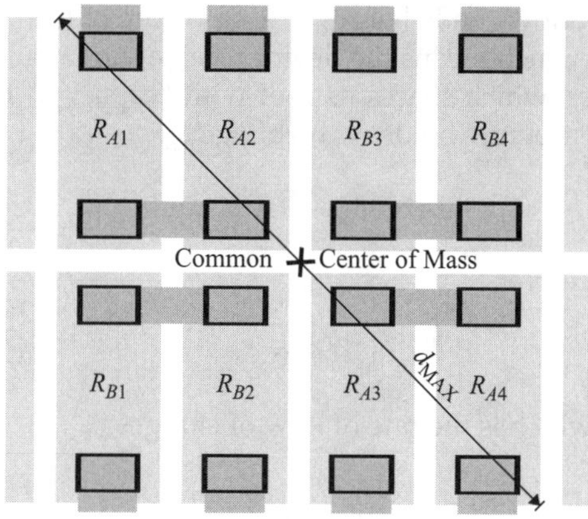

FIGURE 3.5 Top layout view of a cross-coupled pair of resistors.

resistor segments increases the statistical granularity and resolution of the layout, further improving matching performance.

3.2 Capacitors

3.2.1 Theory of Operation

To understand how capacitors work, consider what happens when a charged battery and a pair of parallel conducting plates separated by a nonconducting medium connect. Before the connection, the plates hold no energy, so there is no voltage. Afterward, the battery voltage v_{BAT} establishes an electric field ε_{FLD} between the battery and the top plate as shown in Fig. 3.6 that induces positive charges to flow toward the plate. Because no metallic link connects the two plates, charges accumulate on the top plate and therefore establish an electric field and a voltage v_C between the plates that, in turn, induces negative charges to flow from ground to the bottom plate. The net results are

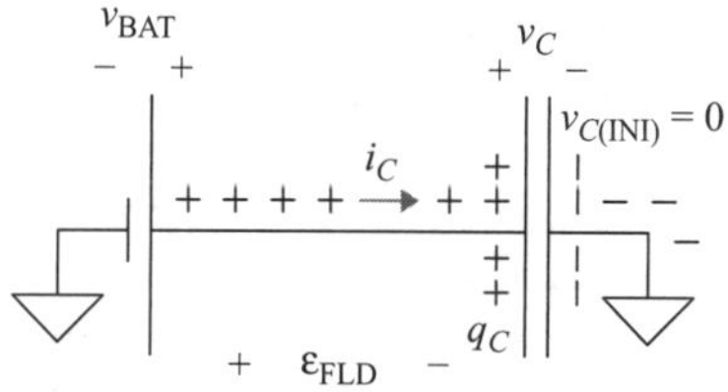

FIGURE 3.6 The effects of connecting a charged battery and a pair of parallel conducting plates separated by a nonconducting medium.

that current i_C flows and charges q_C accumulate at the boundaries of the structure, which means the device receives and holds energy.

Capacitance, which shares its root with "capacity," is how much charge q_C the structure holds with voltage v_C:

$$C_P = \frac{q_C}{v_C} \tag{3.6}$$

Charge is therefore

$$q_C = C_P v_C \tag{3.7}$$

and current, which is the rate of flow of charges, is

$$i_C \equiv \frac{dq_C}{dt} = C_P\left(\frac{dv_C}{dt}\right) \tag{3.8}$$

Not surprisingly, the structure holds more charge when the overlapping area of the plates A_P is larger. Similarly, less separation between the plates intensifies the electric field and attracts more charges. As a result, C_P rises with higher plate width W_P and length L_P and lower plate distance d_P:

$$C_P = \left(\frac{A_P}{d_P}\right)\varepsilon_P = \left(\frac{W_P L_P}{d_P}\right)k_P\varepsilon_0 = W_P L_P C_\varepsilon'' \tag{3.9}$$

Here, *permittivity* ε_P is a measure of how many charges the sandwiched dielectric "permits" across its opposing plates, k_P is how many times higher ε_P is than in vacuum ε_0, and C_ε'' is the resulting capacitance per unit area of the structure. Because v_C rises with i_C over time, the structure's effective impedance is

$$Z_C \equiv \frac{dv_C}{di_C} = \frac{dt}{C_P} \rightarrow \frac{1}{sC_P} \tag{3.10}$$

which also rises with time, or equivalently, falls with frequency, where *s* in Laplace notation represents frequency. This means higher frequency components of a propagating signal tend to shunt or couple through a capacitor more easily.

3.2.2 Parasitic Components

Suspending and isolating a parallel-plate structure from substrate with a thick dielectric like the field oxide (FOX) in Fig. 3.7 creates a parasitic capacitor C_{PAR} between the bottom plate and substrate. Here, the separating medium is, by design, substantially thicker at d_{ISO} than the separation distance of the plates d_P, so C_{PAR} can be an order of magnitude lower than the intended capacitance C_P of the parallel-plate structure. Still, C_{PAR} shunts some high-frequency energy away from propagating signals, so its effects are undesirable. Fortunately, C_{PAR}

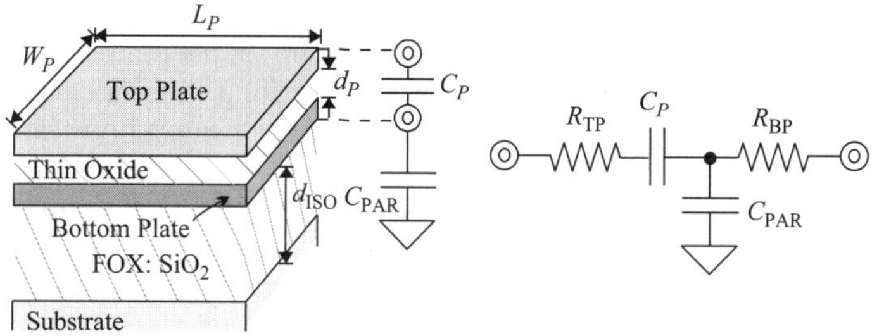

FIGURE 3.7 Three-dimensional view of a parallel-plate capacitor and its electrical model.

usually hangs on the bottom plate only, so designers normally connect this plate to a low-impedance point whenever possible.

Notice the plates are also resistor strips, so they drop voltage and dissipate power. To mitigate these undesirable effects, engineers typically implement the plates with highly conductive materials like aluminum, silicided polysilicon, or highly doped silicon. The material and its metallic contact nevertheless add resistance to both the top and bottom plates of C_P, which is what the two equivalent series resistors (ESRs) R_{TP} and R_{BP} in Fig. 3.7 model show.

3.2.3 Layout

As with resistors and other on-chip devices, while process engineers set material depths that ensure reliable performance, IC designers control their top-level geometries. In the case of capacitors, design engineers define the widths and lengths of plates they choose from a list of available layers. For optimal results, surface areas should be large enough to ensure the series resistances across the plates are low, and parallel-plate capacitance dominates over fringing edge effects. This is why capacitors normally conform to the rectangular or square-like structures of Fig. 3.8.

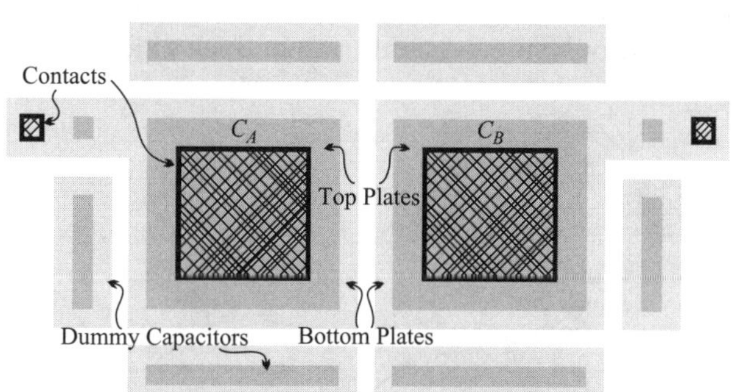

FIGURE 3.8 Top layout view of matched capacitors with peripheral dummy devices.

The top plate incorporates a metallic contact that is immediately above and within the confines of the periphery of the plate. Unfortunately, the bottom plate hides underneath, so for a vertical contact to land on the bottom plate, the bottom plate must extend outside the structure. Plus, since aligning edges perfectly is impossible and the uncertainty of two edges is worse than that of one, the bottom plate overlaps the top plate. This means the top plate, whose edges now define how much the plates overlap, sets the capacitance of the structure, and only inconsistencies along its edges dictate how much it varies across samples.

3.2.4 Absolute and Relative Accuracy

Tolerance ultimately depends on the homogeneity of the sandwiched dielectric across wafers and fabrication lots. Typically, the absolute tolerance of on-chip parallel-plate capacitors is around ±20 percent. As a result, as with resistors, designers prefer to rely on matching performance whenever possible. In this light, since the dielectric of nearby structures is nearly the same, the uniformity of the edges of the top plate ultimately dictates how well two capacitors match. Edge variations along the peripheries cause more mismatches in smaller plates than in larger ones. Interestingly, because etching profiles and lateral diffusion distances also depend on what lies next to the edges, peripheral strips of material along all sides of the structure reduce these proximity errors. This is why capacitors with large surface areas and similar peripheral conditions can match within ±0.5–±1 percent.

The best way of ensuring that all edges have similar and consistent conditions is to place capacitors at the same distance from the matched devices and along all sides of the structure. Since not all capacitors may be necessary, using small dummy devices, as Fig. 3.8 illustrates, produces similar benefits. Note, however, that a dummy top plate alone is not necessarily as high above the substrate as one that sandwiches a dielectric with a corresponding bottom plate. The top plate of the poly–poly parallel-plate capacitor in Fig. 3.9, for example, is above thin and thick oxides and a bottom plate while a strip of plain polysilicon is only above thick oxide. This vertical

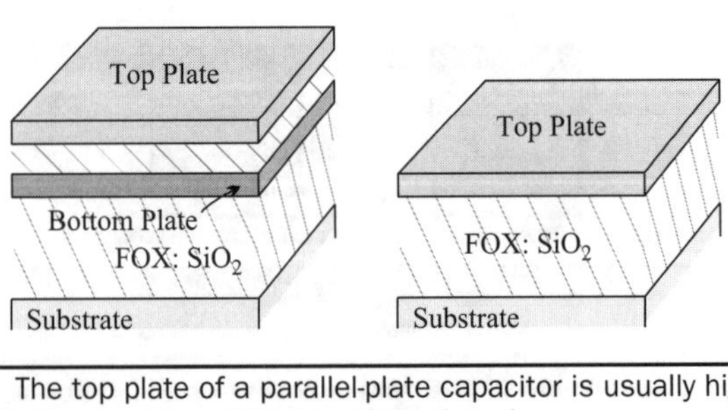

FIGURE 3.9 The top plate of a parallel-plate capacitor is usually higher above the substrate than a piece of the top plate alone is.

misalignment explains why capacitors surrounded by complete dummy devices tend to outmatch those with dummy layers.

Note that leaving these dummy devices disconnected is acceptable, but connecting them to low-impedance nodes helps shunt away stray noise energy. Also notice that the overhanging terminals of the bottom plates in Fig. 3.8 break the continuity of some of the dummy devices. If these discontinuities in the devices are small and similar to those of its matching counterparts, errors can remain minimal. Finally, as with resistors, common-centroid and cross-coupling strategies average and cancel two-dimensional gradients across the die and reduce spreads, which cause second-order mismatches. Partitioning larger capacitors into smaller segments and interdigitating them also improves matching performance.

3.3 PN-Junction Diodes

3.3.1 Theory of Operation

Concentration gradients of *charge carriers* across the metallurgical junction of two oppositely doped pieces of silicon allows current to flow only in one direction. Loosely bound valence *electrons* from donor dopants in the N-type material of Fig. 3.10*a*, to be more precise, *diffuse* from their highly concentrated home sites to regions of lower concentration in the P side. Similarly, voids of electrons or *holes* in the valence band of the P side diffuse to the N side. These migrating

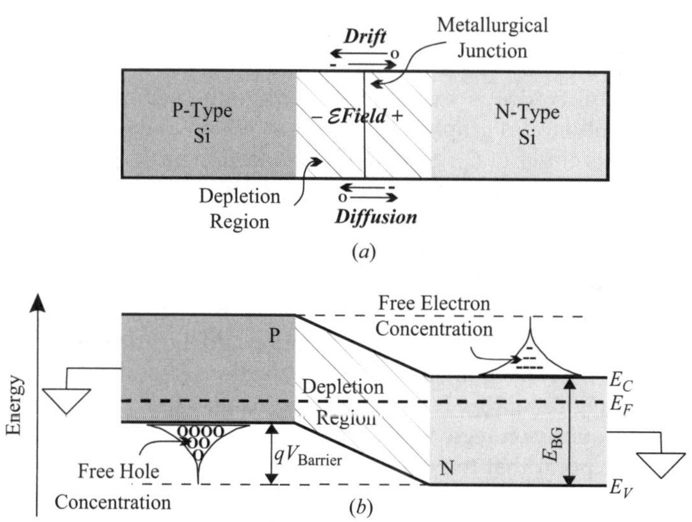

FIGURE 3.10 (*a*) Physical-profile view and (*b*) energy-band diagram of a PN-junction diode.

charge carriers not only leave their otherwise neutral home sites charged or *ionized* but also *deplete* them to establish a region about the junction that is free of charge carriers. The resulting electric field induces charge carriers to *drift* in the opposite direction, so drift and diffusion currents in the disconnected device, which cannot conduct current without a power source, balance and cancel in equilibrium. Also, since highly doped materials are more difficult to deplete, depletion widths are shorter across highly doped junctions than in lightly doped junctions.

The energy-band diagram in Fig. 3.10*b* is useful in describing the operation of the device because it conveys important information about its physical properties and electrical state. The top line denotes the onset of the *conduction band* E_C, the energy level above which electrons are free to roam as charge carriers, and the bottom line denotes the onset of the *valence band* E_V, the energy level below which electrons are strongly bound to the atom. Just as electrons above E_C are charge carriers, holes below E_V, given their complementary features, are also charge carriers.

The energy separation between the two bands is constant for a given material and is commonly referred to as the *bandgap energy* E_{BG}. *Fermi level* E_F, which lies between E_C and E_V, is an abstract variable that represents the energy level where a 50 percent probability of finding a charge carrier exists. This probability drops exponentially for increasing and decreasing energy levels. This is not to say, however, that a charge carrier exists at or around E_F, because none can reside in the band gap, only outside the bands. This does mean, however, that the presence of charge carriers above E_C and below E_V falls exponentially at energy levels that are further away from E_F. Consequently, if E_F is closer to E_C, the probability of finding an electron charge carrier above E_C is greater than finding a hole below E_V, and the probability drops exponentially at higher energy levels, as Fig. 3.10*b* illustrates graphically for the N region. Similarly, E_F in the P material is closer to E_V, and hole concentration peaks just below E_V, and drops exponentially at lower energy levels.

At equilibrium, when there is no external source, E_F is flat, and there is no net current flow across the junction. The conduction and valence bands through the mostly P- and N-type bulk materials are therefore flat. However, since the number of charge carriers start dropping at the edge of the depletion region and the material becomes less doped past this point and into the junction, E_V at the edge of the depletion region on the P side falls with respect to E_F as it traverses across the depletion region. Because E_{BG} is constant in silicon, E_C also falls to the point that the material is neither P nor N type, but what engineers call *intrinsic* or neutral silicon, which means E_F is halfway between E_C and E_V. Since electron concentration in the N side of the metallurgical junction gradually increases until it reaches the N-type bulk, E_C and E_V continue to drop with respect to E_F across the junction, the net result of which is the band bending shown in Fig. 3.10*b*. This bending represents the potential energy barrier that a charge carrier must overcome to diffuse across the junction, more simply described as the *built-in barrier voltage* V_{Barrier}.

Forward and Reverse Bias

Applying a voltage across the junction shifts its barrier potential and forces it out of equilibrium. A positive voltage v_D on the P side, for example, as Fig. 3.11*a* depicts, reduces the barrier and therefore exposes an exponentially rising number of charge carriers over the barrier threshold, allowing them to diffuse to the oppositely doped side, and in the process induce a current flow that rises exponentially with v_D. With a reverse voltage v_R, on the other hand, as Fig. 3.11*b* shows, the barrier increases, impeding the flow of diffusion current. The end result is a device that, like Fig. 3.11*c* models, conducts (1) a current i_D into the P side of a PN junction that rises exponentially with linear increases in *forward-bias* voltage v_D, (2) zero current with zero volts, and (3) little to no current when v_D falls below zero:

$$i_D = I_S\left[\exp\left(\frac{v_D}{V_t}\right) - 1\right] \tag{3.11}$$

where I_S is the *reverse-saturation current* for the junction and V_t the *thermal voltage*, the former of which is on the order of femtoamps and the latter about 25.6 mV at room temperature. Since high temperatures aid the diffusion process, the *diffusion coefficient* and *mobility* components of I_S rise, and diode voltage v_D, as a result, falls with climbing temperatures. Said differently, the diode is "stronger" at higher temperatures, and v_D therefore has a negative temperature coefficient.

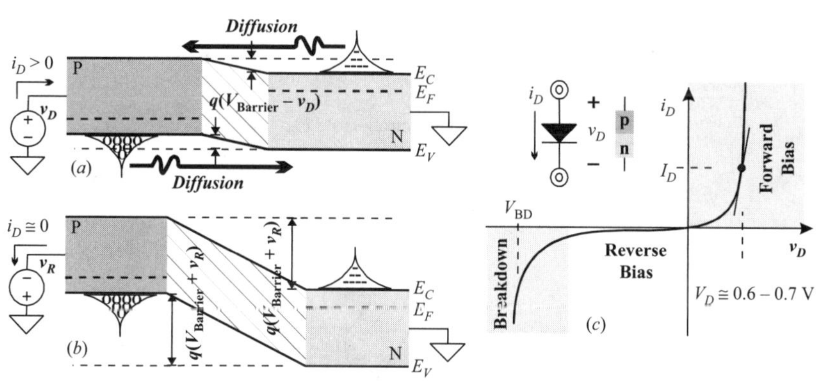

FIGURE 3.11 Energy-band diagrams of a PN junction under (*a*) forward- and (*b*) reverse-bias conditions and (*c*) its current across voltage.

Breakdown

When applying a sufficiently high reverse voltage across the junction, the device "breaks down" and conducts current in the negative direction. Under these conditions, past its *breakdown voltage* V_{BD}, one or a combination of two breakdown mechanisms occurs. In a highly doped PN junction, for example, where the region near the metallurgical junction is relatively difficult to deplete of carriers, the depletion width is narrow and the resulting electric field is high. A large number of electrons therefore gain enough energy to *tunnel* through and across the short region, as Fig. 3.12 shows.

Across lightly doped junctions, the depletion width is longer, so the number of electrons that tunnel successfully through the region diminishes. Raising reverse-bias voltage v_R further, however, increases the electric field to such an extent that roaming electrons accelerate and crash into atoms with such force that bounded electrons break free. This means energized electrons, as Fig. 3.12 exemplifies, generates *electron–hole pairs* (EHPs) through a process that scientists call *impact ionization*. Since one roaming electron frees another, and the other another, the number of free charge carriers increases geometrically in *avalanche* fashion. Negative current therefore rises substantially when reverse voltages fall below the breakdown level, below V_{BD} in Fig. 3.11*c*.

Most moderately doped PN junctions exhibit a combination of both tunneling and avalanche effects with breakdown voltages of 6–8 V. Higher doped devices, which have narrower depletion widths, have lower breakdown voltages and produce more tunneling than avalanche currents, and vice versa. Irrespective of the mechanism, however, engineers refer to junction diodes optimized for breakdown conduction as *Zener* diodes, where Zener voltages correspond to breakdown voltages. Unfortunately, subjecting a diode to extreme

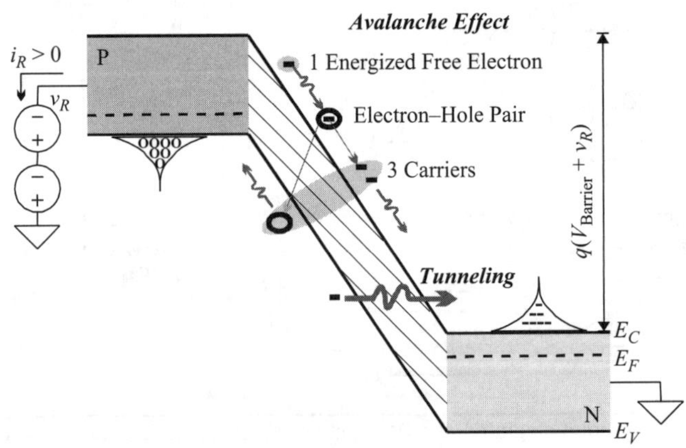

FIGURE 3.12 A PN-junction diode in breakdown.

and sustained high-voltage conditions in either reverse (i.e., $V_R >> V_{BD}$) or forward (i.e., $V_D >> 0.6–0.7$ V) bias induces so much current that the material and the package can heat and strain beyond reliable limits. Pushing the device beyond its safe-operating area (SOA) in this way can cause irreversible damage.

Symbol

Perhaps the most fundamentally useful aspect of a PN junction is that, for the most part, it conducts current in one direction only. For that reason, the head of the symbol that represents it in Fig. 3.11*c* is an arrow that points in the direction of forward-bias current i_D: from *anode* to *cathode*. Because current cannot flow in the opposite direction when reverse voltages are less than a few volts, the head of the arrow incorporates a "current-blocking" bar or line. This is not to say, however, that breakdown is a parasitic region. Most chips today, in fact, rely heavily on breakdown characteristics to clamp and protect integrated circuits from short-circuit, electrostatic-discharge, and other destructive conditions. Engineers generally refer to devices that conduct current this way—in one direction and not in the other—as *diodes*.

3.3.2 Parasitic Components

The bulk P- and N-type materials of a reverse-biased PN junction are a pair of parallel conductive plates that sandwich a nonconducting medium. As such, they exhibit the characteristics of a parallel-plate capacitor, where the depletion capacitance C_{DEP} that results rises with increasing cross-sectional areas A_D and decreasing depletion widths W_{DEP}. Here, W_{DEP} drops with higher doping concentrations because highly doped materials are more difficult to deplete of charge carriers. Also, because higher reverse-bias voltages v_R attract more carriers away from their home sites near the junction, W_{DEP} rises with v_R, or said differently, it falls with increasing v_D:

$$C_{DEP} = \frac{A_D C_{J0}''}{\sqrt{1 - \frac{v_D}{V_{BI}}}} \propto \frac{A_D}{W_{DEP}} \tag{3.12}$$

where V_{BI} is the built-in barrier potential, which is roughly 0.6 V, and C_{J0}'' is the zero-bias capacitance per unit area when v_D is zero. Both C_{J0}'' and V_{BI} depend on doping densities, especially that of the less doped side because it normally dominates the overall width of the depletion region. Note that the peripheral sidewall of the device in the silicon also contributes capacitance.

In standard P-substrate complimentary metal–oxide–semiconductor (CMOS) process technologies, engineers can immerse either P^+ diffusions in N wells above a substrate, as Fig. 3.13*a* illustrates, or N-well or N^+ regions directly in the substrate to create PN junctions. Of these, N well and N^+ are less practical because their P sides connect directly

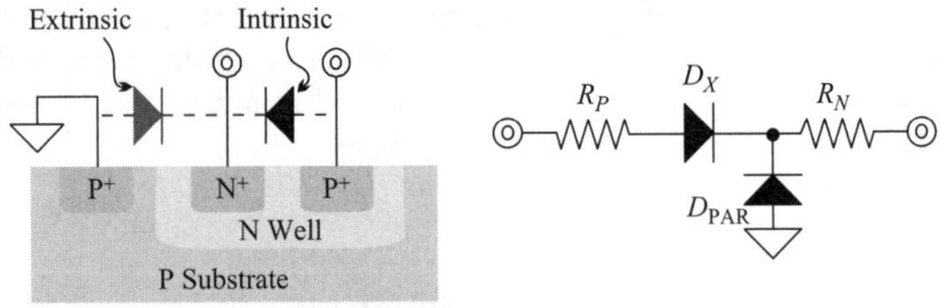

FIGURE 3.13 Physical profile view of an isolated p+ diffusion in an n well PN-junction diode and its equivalent electrical model.

to the substrate. This is challenging in two respects. One is, engineers isolate components across a chip by reverse-biasing them against the substrate, which means they connect P-type substrates to the most negative potential available. As a result, N-well and N^+ diodes in standard *junction-isolated* ICs cannot connect their P sides to anything but the most negative potential. The other challenge with *substrate diodes* is that, in drawing substrate current, substrate diodes generate noise energy that can permeate through the substrate to affect other circuits in the system.

The drawback to *isolated diodes* is that they incorporate a parasitic PN junction. In the case of the P^+ in an N well diode in Fig. 3.13, for example, a parasitic N well in substrate diode D_{PAR} hangs off the N-well terminal. This *extrinsic* diode is normally in its reverse-bias mode, which is not to say it is not parasitic to the structure. Plus, since the diode is a series combination of oppositely doped silicon pieces, each piece and its contact introduces series parasitic resistances R_P and R_N that drop voltage and dissipate power.

3.3.3 Layout and Matching

As before, process engineers specify diffusion depths and IC designers define top two-dimensional geometries. Accordingly, embedding a P^+ area inside an N well, like the top layout view of Fig. 3.14

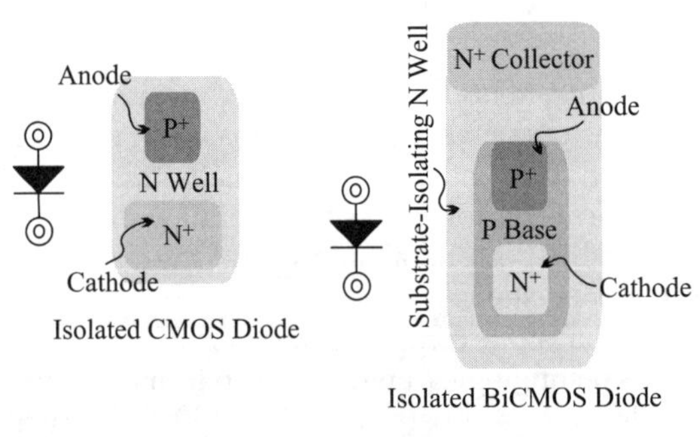

FIGURE 3.14 Top layout views of (*a*) CMOS P^+ in N well and (*b*) bipolar/BiCMOS N^+ in P-base diodes.

illustrates, implements the isolated CMOS diode already discussed and shown in Fig. 3.13. Here, the area and depth of the P^+ region determines the area of the vertical and sidewall components of the intrinsic diode. Because bipolar and bipolar–CMOS or BiCMOS technologies include a P-base region that is deeper than N^+ diffusion and shallower than the N well, an N-well tub isolates an N^+ in a P base junction from the P substrate, as Fig. 3.14 shows. In this case, the area and depth of the N^+ diffusion sets the area of the intrinsic N^+–P base junction, and the area and depth of the P base sets the area of the parasitic P base–N well counterpart.

As with resistors and capacitors, large surface areas, close arrays, similarly oriented devices, common-centroid structures, interdigitation, and cross-coupling improve the matching performance of diodes. Unlike polysilicon resistors and capacitors, however, diodes do not suffer from etching errors because the diffused regions that comprise them are in the silicon. Consequently, adding dummy devices or diffusion strips around the periphery of an array of diodes may not improve matching performance to the same extent that dummy polysilicon strips around polysilicon devices do. This is not to say, however, that doping concentrations around surrounding regions do not affect the out-diffusion process. Adding equidistant dummy diffusion strips improves the uniformity of the out-diffused areas and therefore reduces the likelihood of peripheral asymmetries across a matched array of diodes.

Bias Point and Small Signals in Analog Circuits *While resistors respond to voltage or current stimulation in a linear fashion, diodes and transistors do not. This presents a challenge when amplifying analog signals because nonlinear circuits will not boost voltages or currents evenly across signal strength. Circuit designers reduce this* distortion *by operating nonlinear devices about a* bias point. *This way, biased at steady-state input S_I and output S_O in Fig. 3.15, small input variations Δs_I produce close to linear translations in the output Δs_O that a straight line can predict reasonably well. In other words, variations are so small that the response of a nonlinear circuit approaches that of a*

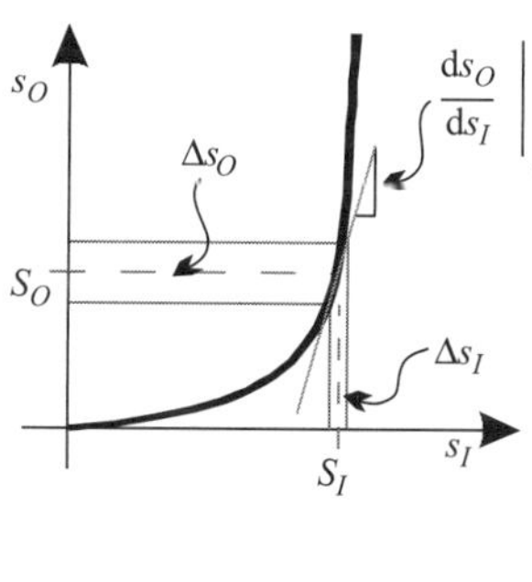

Figure 3.15 Small-signal approximation of a nonlinear response about a bias point (S_I, S_O).

straight line. The best-fitting slope for this line is that of the nonlinear response at the bias point, which corresponds to the response's first derivative ds_I/ds_O when evaluated at S_I.

With this information in hand, engineers first design and characterize the steady-state response of a circuit using the nonlinear large-signal characteristics of its constituent components. After determining the bias point and its associated large-signal limits, designers create an equivalent model with which to approximate the small-signal response of the system about its bias point. The resulting model replaces all exponential, quadratic, and other nonlinear effects with linear approximations, reducing diodes and transistors, for example, to equivalent conductances and resistances, and when the effects manifest in other terminals, transconductances and transresistances. Since current variations in supply and bias voltages produce no small-signal changes in voltage, they appear as short circuits to ground in small-signal models. In the same vein, because currents through current sources do not change, current sources disappear and appear as open circuits in small-signal models.

In this context, analog designers typically operate transistors in the mode of operation that achieves the highest gain possible because, with higher gain, feedback circuits are more accurate and suppress noise better. Plus, with higher gain, buffers exhibit lower output resistances and drive higher currents. High output resistances are similarly attractive for circuits that generate current. As a result, small-signal circuits typically model transistors in their high-gain regions: bipolar-junction transistors (BJTs) in forward active and light saturation and metal–oxide–semiconductor field-effect transistors (MOSFETs) and junction field-effect transistors (JFETs) in saturation.

3.3.4 Small-Signal Model

Since small-signal variations in the diode voltage v_D produce close to linear translations in the current i_D, a straight line can predict the current–voltage small-signal response of a diode reasonably well, as the side note explains. When biased at a steady-state voltage V_D, as Fig. 3.16 shows, the best-fitting slope is that of i_D's exponential

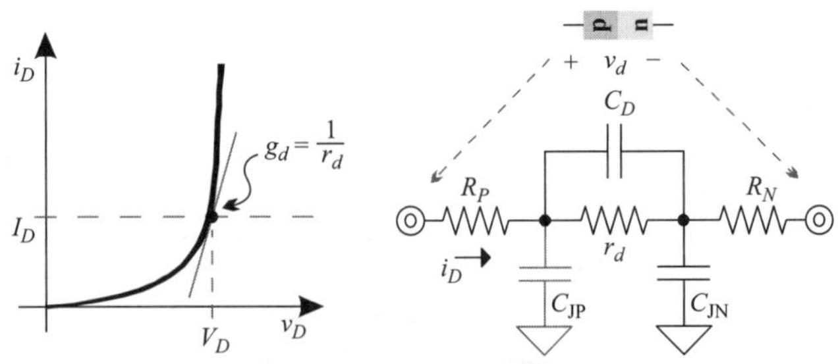

FIGURE 3.16 Large-signal response and small-signal model of an isolated diode.

response at V_D, which from calculus, reduces to i_D's first derivative when evaluated at V_D:

$$i_d = \Delta v_D\left(\frac{di_D}{dv_D}\right) = v_d\left(\frac{I_S}{V_t}\exp\frac{V_D}{V_t}\right) \approx v_d\left(\frac{I_D}{V_t}\right) \equiv v_d g_d \equiv \frac{v_d}{r_d} \qquad (3.13)$$

where i_d and v_d refer to small-signal variations in i_D and v_D, V_t is the thermal voltage, and g_d is to the slope of the line that best describes their response. Because the first derivative of an exponential includes the original exponential, di_D/dv_D in the preceding expression includes the parameters that describe forward-bias steady-state current I_D, so g_d is nearly I_D/V_t.

Interestingly, i_d through a diode produces linear effects in v_d much like a current through a resistor would in the voltage across the resistor. Therefore, for simplicity and insight, engineers often describe conductance g_d in terms of its equivalent resistance r_d, which is why small-signal models typically use a resistance r_d to relate i_d and v_d. Models also include the bulk resistances of each contact and silicon strip in the junction R_P and R_N. Plus, in the case of isolated diodes, as Fig. 3.16 shows, a parasitic capacitor models the effects of the reverse-biased PN junction created by one of the terminals and the substrate or isolating well in which the diode sits. The N-well terminal of P^+ in N well junctions above P substrates, for example, includes a parasitic junction capacitor C_{JN} and, similarly, the P-base terminal of N^+ in P base junctions inside N wells includes C_{JP}. Since reverse-biased junctions do not conduct appreciable currents, no other component is necessary in the model, and since the substrate or isolating well usually connects to a supply voltage or ground, C_J connects to small-signal ground.

The model also incorporates an equivalent capacitance C_D across the junction. Since forward-bias conditions do not eliminate the depletion region, C_D includes C_{DEP}, whose depletion width falls and capacitance therefore rises with increasing forward-bias voltages. Because charge carriers require *forward transit time* τ_F to diffuse across the depletion region, diode voltage v_D establishes a diffusion current i_D that also temporarily deposits charge q_C that is equivalent to $i_D\tau_F$ into the depletion region, as Fig. 3.17 illustrates. This means the

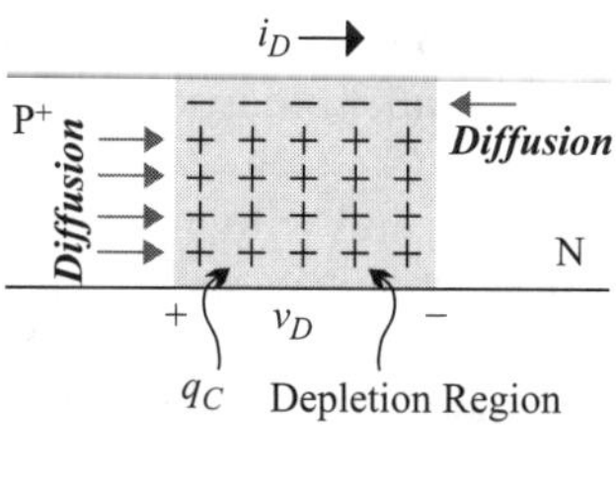

FIGURE 3.17 Charge in the depletion region of a forward-biased PN junction.

region, at any point in time, holds charge with a forward-bias voltage like a capacitor would, except in this case i_D rises exponentially with respect to v_D. However, when biasing the junction and subjecting it to small-signal variations, i_D rises linearly with v_D, so small current variations di_D deposit dq_C into a region whose small-signal voltage dv_D holds dq_C with an effective *diffusion capacitance* C_{DIF}:

$$C_{DIF} = \frac{dq_C}{dv_D} = \left(\frac{dq_C}{dt}\right)\tau_F\left(\frac{1}{dv_D}\right) = \left(\frac{di_D}{dv_D}\right)\tau_F = g_d\tau_F \tag{3.14}$$

where g_d is the small-signal conductance of the diode.

As a result, diode capacitance C_D includes both depletion and diffusion capacitances C_{DEP} and C_{DIF}:

$$C_D = C_{DIF} + C_{DEP} \tag{3.15}$$

However, because forward-bias current deposits considerably more charge into the region than C_{DEP} can normally hold, C_{DIF} overwhelms C_{DEP} in the forward-bias region. In contrast, since i_D is nearly zero in reverse bias, C_{DEP}'s charge overwhelms that of C_{DIF}. In other words, C_D reduces to C_{DIF} when forward biased and to C_{DEP} otherwise. What is more, when conceptualizing circuits, the effects of R_P, R_N, and C_J are so small with respect to r_d and C_D that designers often neglect the former in hand calculations and include them only in nominal and worst-case computer simulations.

3.4 Bipolar-Junction Transistors

3.4.1 Theory of Operation

Forward Active

A *bipolar-junction transistor* (BJT) is, in physical terms, two back-to-back PN-junction diodes whose intermediate oppositely doped piece of silicon is short. Engineers call this middle region, which in the case of the NPN BJT in Fig. 3.18 is a short piece of P-type silicon, the *base* because this region in the first prototyped implementation of the BJT was the physical base of the device. Here, as in diodes, charge carriers diffuse across the N^+–base junction when forward biased with a base voltage v_B. Because doping concentration in the N^+ material is considerably higher than in the P base, diffusing electrons from the N^+ side outnumber diffusing holes by about 50 to 100 times. The *injection efficiency* of the N^+ side is so high, in fact, that engineers refer to this terminal of the BJT as the *emitter* because it emits the dominant charge carriers of the device.

Under these conditions, and with a voltage v_C that reverse biases the base–N^- junction (i.e., when v_C is greater than v_B), electrons diffusing into the base have a tendency to migrate to the N^- side. Because the width of the base is short, the likelihood that diffusing electrons

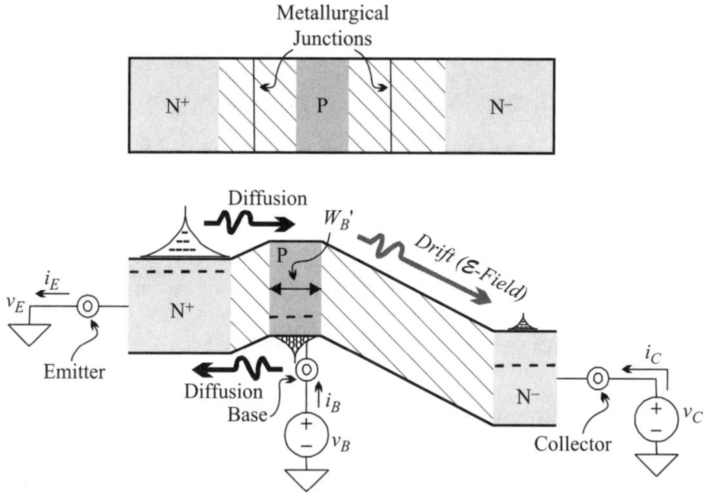

FIGURE 3.18 Physical profile view and energy-band diagram of an NPN BJT in the forward-active region.

will recombine in the base before they reach the edge of the depletion region that butts against the N⁻ region is low. With such a high *transport factor*, the electric field that v_C establishes sweeps diffusing electrons across the ensuing depletion region and into the N⁻ material. Because the N⁻ material collects most of the electrons emitted by the N⁺ region, engineers call this terminal the *collector*.

Currents Since the base is a rich source of holes, the base sources the holes that diffuse to the emitter, thereby establishing a current into the base that i_B in Fig. 3.19 symbolizes. Diffusing electrons, however, outnumber diffusing holes by a factor of 50–100 to 1, so electrons flowing across the base to the collector constitute a current into the

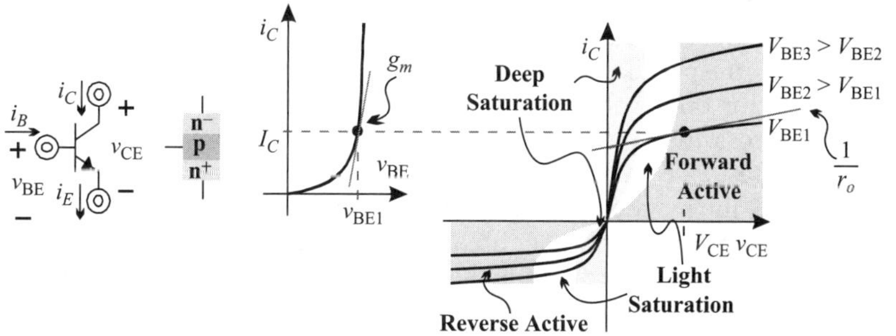

FIGURE 3.19 Symbol and current–voltage curves for an NPN BJT.

collector i_C that is 50 to 100, or β_{F0}, times higher than i_B. As a result, i_C and i_B together flow out of the emitter as i_E:

$$i_E = i_C + i_B = \beta_{F0} i_B + i_B = i_B(\beta_{F0} + 1) \tag{3.16}$$

where β_{F0} refers to the low-frequency current gain of the transistor in the *forward-active* region.

Because i_C is basically the forward-bias current across the base–emitter junction, i_C rises exponentially with base–emitter voltage v_{BE}, as Fig. 3.19 shows:

$$i_C \approx I_S\left[\exp\left(\frac{v_{BE}}{V_t}\right) - 1\right]\left(1 + \frac{v_{CE}}{V_A}\right) \approx I_S\left(\exp\frac{v_{BE}}{V_t}\right)\left(1 + \frac{v_{CE}}{V_A}\right) \tag{3.17}$$

Raising collector–emitter voltage v_{CE}, however, further reverse biases the collector–base junction, in the process extending the depletion region and shortening the transistor's effective *base width* W_B'. With a shorter base, more of the diffusing electrons across the base reach the collector, so i_C rises linearly, albeit slightly, with v_{CE} in the forward-active region of Fig. 3.19. The *base-width modulation* parameter or, as engineers often call it, *Early voltage* V_A describes to what extent i_C rises with v_{CE} in this mode of operation. V_A for conventional transistors is normally between 50 and 100 V.

Symbol The arrow in the symbol of Fig. 3.19 is always at the emitter and in the direction of current flow. PNP devices, like their NPN counterparts, operate the same way, except the diffusing carriers across the base are holes instead of electrons. This means emitters in PNP BJTs supply holes into the base and current therefore flows into the emitter and out of the base and collector terminals. It also means the arrow at the emitter points into the base, unlike in the NPN of Fig. 3.19, which points away from the base.

Optimal BJT A transistor is basically a resistor whose resistance changes with signal variations across a third terminal. In the case of the BJT, the base is this third terminal, so current through the base is by and large parasitic. As such, low-frequency current gain β_{F0} should be high at 50–100, so the doping concentration of the emitter should be correspondingly higher than that of the base. Less doping in the collector, on the other hand, not only diminishes reverse-active currents across the junction but also extends the depletion region and decreases, as a result, the strength of the ensuing electric field, which raises the junction's breakdown voltage. In other words, less-doped collectors exhibit higher breakdown voltages. The drawback to lower doping concentration is that the depletion region, and as a result, the base width and i_C are more sensitive to v_{CE}. All this explains why emitters are typically highly doped, bases are moderately doped, and collectors are lightly doped to the extent β_{F0} and V_A balance at roughly 50 A/A and 50 V.

The BJT has four regions of operation. From an analog perspective, the most appealing mode is the one just described, when forward biasing the base–emitter junction and reverse-biasing the base–collector junction. This way, small variations in collector current produce large variations in collector–emitter voltages, which translates to high current-to-voltage gains. Process design engineers therefore optimize these devices to operate in this region, in forward-active mode.

Reverse Active

Reversing the biasing conditions of the emitter and collector with respect to the base produces similar current-flow dynamics as in forward active, but with less attractive results. As in forward active, forward-biasing the base–collector junction induces carrier diffusion from collector to base, and reverse-biasing the base–emitter junction sweeps the diffused minority carriers in the base into the emitter. Unlike forward active, however, the magnitude of i_C and the resulting reverse-current gain β_{R0} in *reverse-active* mode are lower in Fig. 3.19 because the collector is less doped and therefore emits fewer free charge carriers than an optimized emitter can.

Saturation

Reducing collector voltage v_C to the point where the base–collector junction forward biases induces current flow from base to collector. This current steers current away from the intrinsic base, reducing how much base current the BJT amplifies by β_{F0}. This effective reduction in current gain is the reason engineers say a BJT begins to *saturate* when the base–collector junction forward-biases. Note that forward-biasing the base–emitter junction in reverse mode produces the same effect, but on the lower left quadrant of Fig. 3.19's i_C–v_{CE} graph.

Forward-bias voltages below 0.3–0.4 V, however, induce so little current flow across the junction that current gain hardly drops. Many designers therefore allow their BJTs to operate in *light saturation* (from Fig. 3.19), especially in low-voltage applications when v_{CE} is low. Forward-biasing the junction beyond that point, however, steals appreciable current from the intrinsic base to the extent current gain becomes impractically low. Keeping v_{CE} above *minimum collector–emitter voltage* $V_{CE(MIN)}$, which in practice can be 0.2–0.3 V, ensures that a transistor is not in *deep saturation*.

Cut Off and Temperature Effects

Reverse biasing both junctions shuts both diodes, so no current flows through any terminal, which means the transistor is in *cut off*. Also, as with stand-alone diodes, a higher temperature aids the diffusion

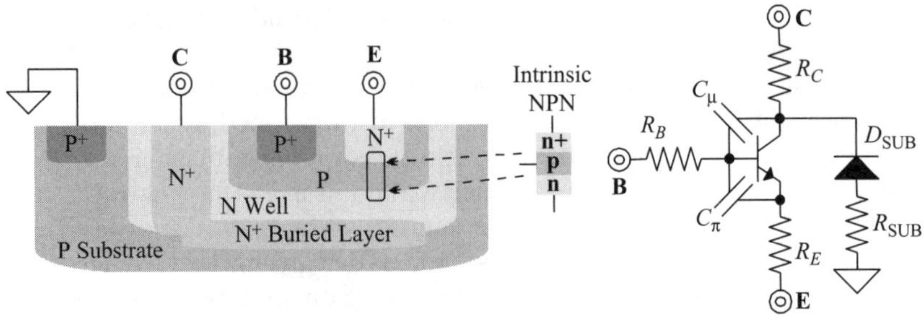

FIGURE 3.20 Vertical NPN BJT in a P substrate and its equivalent electrical model.

process, so more charges flow to the collector under similar base–emitter voltages. This means BJTs at higher temperatures conduct more current and are therefore "stronger" than at lower temperatures.

3.4.2 Vertical BJTs

The emitter–base–collector regions of a *vertical BJT*, as the name implies and Fig. 3.20 illustrates, stack above one another below the surface of the silicon die. The pair of junctions that sandwich the middle base constitute the intrinsic device. The bulk silicon strips that lead to the intrinsic BJT and their corresponding contacts introduce series parasitic resistances R_E, R_B, and R_C. Because the collector is deep inside the silicon and connecting to it from the surface introduces considerable resistance, process engineers often include a highly doped vertical collector plug and a deep buried layer in the manufacturing process flow, reducing R_C to 300–500 Ω. Since there is less material between the extrinsic base and emitter and their corresponding intrinsic terminals, base resistance R_B is normally lower at 200–300 Ω, and emitter resistance R_E is even lower at 100 Ω or less. To these, as the electrical model of Fig. 3.20 shows, base–emitter and base–collector junctions add capacitances C_π and C_μ. Plus, the isolating N well –P substrate junction adds a parasitic substrate diode D_{SUB} with series resistance R_{SUB}.

Layout Embedding an N^+ emitter in the top layout view of Fig. 3.21*a* in a P-base region within a larger N well is the recipe for building the vertical NPN BJT of Fig. 3.20. In this structure, increasing the surface area of the emitter raises the current that flows through the base–emitter junction and the collector that the junction supplies. Since the intrinsic base is beneath the silicon surface under the emitter, the P-base region extends to one of its sides to allow a metallic contact and a low-resistance P^+ diffusion to establish a connection. Similarly, a metallic contact and a deep low-resistance N^+ plug connect to an extension of the N well. Because the N well is large and

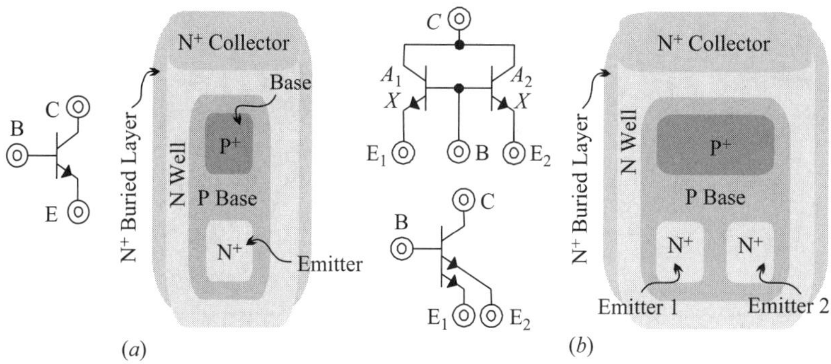

FIGURE 3.21 Top layout views of vertical NPN BJTs.

moderately doped and, thus offers considerable resistance, the highly doped buried layer featured in Fig. 3.20 overlaps the entire N well to reduce the resistance from the intrinsic collector to the extrinsic terminal. Removing this buried layer increases the collector's series resistance R_C, which in turn raises the minimum collector–emitter voltage $V_{CE(MIN)}$ that the device can handle before succumbing to the effects of deep saturation.

Matching Extending the P base and N well to accommodate another N^+ emitter, like Fig. 3.21*b* shows, creates another vertical NPN BJT. The advantage of this setup is the close proximity between the two devices, which enhances their matching performance. The drawback is that the collector and base regions are common to both transistors, so the layout is only useful in circuits where two BJTs share base and collector connections. To illustrate the characteristics of this type of layout on a schematic, designers often add one emitter leg to the BJT symbol. As with resistors, capacitors, and diodes, large emitter areas, close arrays, similarly oriented devices, common-centroid structures, interdigitation, cross-coupling, and dummy devices around the periphery of the array improve matching performance.

3.4.3 Lateral BJTs

Butting two P^+ regions against one another without actually touching inside an N well, as depicted in Fig. 3.22, is another way of building a BJT. In this case, the proximity of the P^+ areas determines how many of the diffusing charge carriers cross the N-well base without recombining, the transport factor of which affects current gain β_0. Since the doping concentration of the N well is normally lower than that of a P base, base-width modulation effects in lateral devices can be worse than in vertical structures with Early voltages V_A as low as 15 V. On the positive side, the device is symmetrical, so either P^+ diffusion

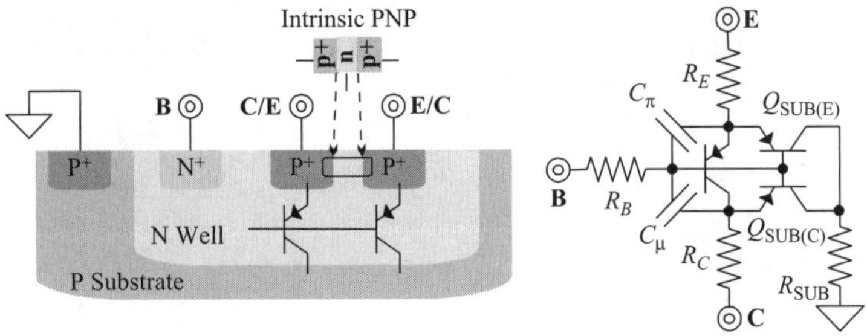

FIGURE 3.22 Lateral PNP BJT in a P substrate and its equivalent electrical model.

can be the emitter. Perhaps the most redeeming feature, however, is that mainstream low-cost CMOS technologies have sufficient diffusions to build these lateral structures.

Unfortunately, each P^+ area, the N well, and the P substrate constitute parasitic vertical PNP BJTs that conduct substrate current to generate noise in the substrate. In fact, since the emitter–base junction is common to the intrinsic device and the substrate BJT attached to the emitter $Q_{SUB(E)}$, that vertical BJT is also active. Fortunately, the N well can be so deep with respect to the P^+ emitter and the resulting base width W_B' so long that current gain β_0 can be low, but not always because lightly doped N wells can deplete easily and, consequently, shorten the effective base region. If available, inserting an N^+ buried layer below and across the N well adds majority carriers with which diffusing holes can recombine to reduce the current gain of these parasitic devices.

Since PNP emitters supply diffusing holes to the base, current flows into the emitter and the emitter arrow in the symbol of Fig. 3.22 points to the base. Both intrinsic emitter and collector terminals have a parasitic substrate BJT whose common collector connects to ground via equivalent substrate resistance R_{SUB}. Plus, as with any BJT, C_π and C_μ model the base–emitter and base–collector junction capacitances. Contacts and diffusion regions also add resistances, so resistors R_E, R_B, and R_C are in series with the emitter, base, and collector terminals of the intrinsic BJT. However, R_C in lateral BJTs is lower than in vertical devices because less material connects the intrinsic collector to the extrinsic terminal. R_B in lateral transistors is similarly higher than in vertical BJTs because the doping concentration of the N-well base in a CMOS process is normally lower than that of a P-base region in a bipolar or BiCMOS process.

Layout Unlike in vertical structures, carriers traverse horizontally across the base of lateral BJTs. This means that a concentric P^+ ring, like Fig. 3.23*a* shows is more likely to collect more lateral holes than

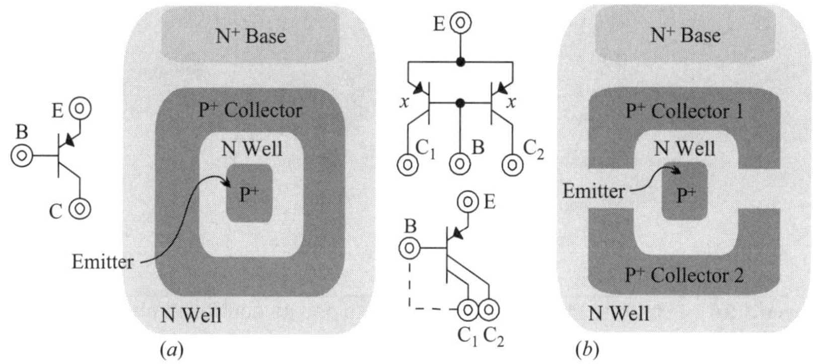

FIGURE 3.23 Top layout views of (*a*) a typical and (*b*) two matched lateral PNP BJTs.

the nonconcentric counterpart. In fact, because these BJTs are peripheral devices, several smaller emitter dots generate more collected holes than one larger emitter. This higher transport factor is the reason why designers build high-power lateral BJTs out of many smaller *emitter-dot* structures in parallel.

Matching Splitting the collector ring in two, as Fig. 3.23*b* shows, is a way of building two even smaller BJTs that are so close and symmetrical that their collector currents match well. In this case, each collector terminal collects roughly half of the diffusing holes that emanate from the emitter. Although sharing the emitter and base terminals in this way can be constricting, they implement the connections that current-mirror circuits require, which are essential analog building blocks, as follow-up chapters later demonstrate. To show they derive multiple collectors from one emitter–base junction, IC designers often add one collector leg to the original BJT symbol, as the figure shows. As before, large emitter areas, close arrays, similarly oriented devices, common-centroid structures, interdigitation, cross-coupling, and dummy devices around the periphery of the array improve matching performance.

3.4.4 Substrate BJTs

Although the vertical BJTs present in the lateral structure are often parasitic, they can at times be useful. Immersing P^+ in an N well above the substrate, as illustrated in Fig. 3.24, is the blueprint for this standalone substrate BJT. Although base-width modulation can be significant, current flows through the substrate, and the collector cannot connect anywhere but to the substrate. Still, using this vertical device to build PNP emitter followers (from Chap. 4) in a low-cost CMOS technology can bring considerable value to the circuit. Plus doping

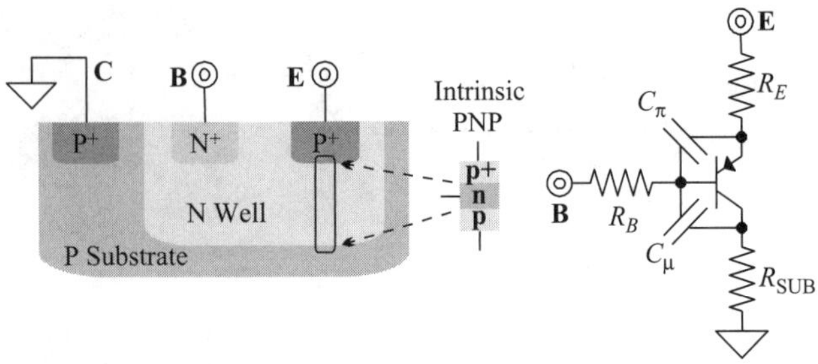

FIGURE 3.24 Substrate PNP BJT in a P substrate and its equivalent electrical model.

concentration in the N well can be so light that current gain β_0 can rise above 50 A/A.

Because running current across the substrate generates noise for the rest of the chip, designers typically add a concentric P^+ ring connected to ground or the most negative potential around the N well to collect as much substrate current as possible near the emanating base. This way, less collector current traverses through the rest of the die. As with the others, the electrical model of this structure includes series terminal resistances R_E, R_B, and R_C and junction capacitances C_π and C_μ. However, since no other PN junctions exist here, there are no other parasitic diodes or transistors to include in the model.

3.4.5 Small-Signal Model

Because high gain and high resistance are useful attributes in analog applications, as the side note on small signals explains, designers typically model the small-signal response of BJTs in the forward-active and light-saturation regions. Here, small changes in base–emitter voltage v_{BE} produce considerable variations in collector current i_C, which as Fig. 3.19 illustrates, ultimately cause substantial variations in collector–emitter voltage v_{CE}. Similarly, although not to the same extent, changes in v_{CE} also alter i_C. Plus, because BJTs conduct base current i_B, varying v_{BE} also changes i_B. In other words, as Fig. 3.25*a* shows, small signal v_{be} produces i_b and one component of i_c, and v_{ce} alters the other part of i_c.

Since linear approximations predict the response of small signals reasonably well, the effect of v_{be} on i_c is the linear translation that is predicted by the slope g_m in Fig. 3.19 and the first partial derivative of i_C with respect to v_{BE} at bias current I_C:

$$i_c\Big|_{v_{ce}=0} \approx v_{be}\left(\frac{\partial i_C}{\partial v_{BE}}\right) \equiv v_{be}g_m \approx v_{be}\left(\frac{I_C}{V_t}\right) \tag{3.18}$$

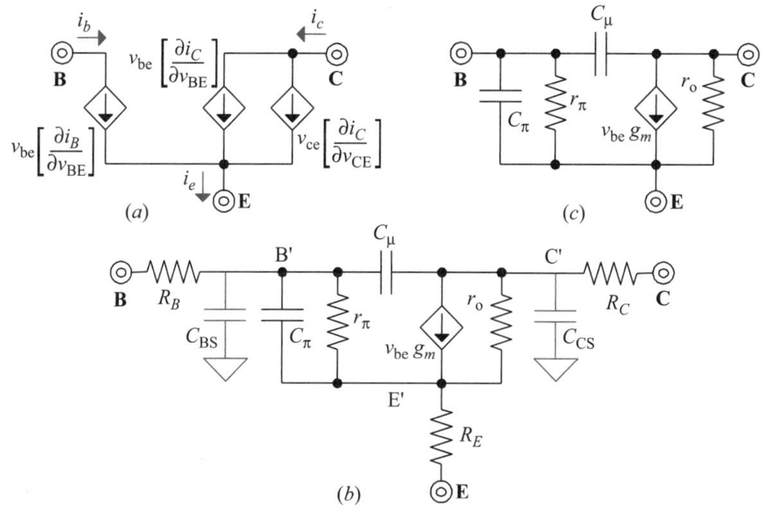

FIGURE 3.25 (a) Small-signal currents and (b) complete and (c) simplified small-signal models of a BJT.

where g_m is the BJT's effective transconductance. Similarly, the slope g_o and first partial derivative of i_C with respect to v_{CE} at I_C forecast the impact of v_{ce} on i_c:

$$i_c\Big|_{v_{be}=0} \approx v_{ce}\left(\frac{\partial i_C}{\partial v_{CE}}\right) \equiv v_{ce}g_o \approx v_{ce}\left(\frac{I_C}{V_A}\right) \equiv \frac{v_{ce}}{r_o} \tag{3.19}$$

where g_o is the effective output conductance of the device. Because this linear variation in current results from a change in the voltage across its own two terminals v_{ce}, output resistance $1/g_o$ or r_o in Fig. 3.25*b* models this behavior well. Since i_C is a current-gain translation of i_B, i_c is also a β_0 translation of i_b and therefore a β_0 and g_m translation of v_{be}:

$$i_b \approx v_{be}\left(\frac{\partial i_B}{\partial v_{BE}}\right) = \frac{v_{be}}{\beta_0}\left(\frac{\partial i_C}{\partial v_{BE}}\right) = v_{be}\left(\frac{g_m}{\beta_0}\right) \equiv \frac{v_{be}}{r_\pi} \tag{3.20}$$

Again, since the variation in i_b is the result of a voltage change across its own two terminals, resistor r_π in Fig. 3.25*b*, whose equivalent resistance is g_m/β_0, models the effects of v_{be} on i_b. In fact, r_o and r_π are so much higher than terminal series resistances R_E, R_B, and R_C in the complete model of Fig. 3.25*b* that designers often neglect the latter in the simplified model of Fig. 3.25*c*.

A small-signal model of the BJT would not be complete without base–emitter and base–collector capacitances C_π and C_μ from Figs. 3.20, 3.22, and 3.24. Since the base–emitter junction is forward-biased in forward-active operation and light saturation, diffusing carriers keep considerably more charge in the depletion region than diode depletion capacitance C_{DEP} can. As a result, diffusion diode capacitance C_{DIF} normally overwhelms C_{DEP} to reduce C_π to

$$C_\pi \equiv C_{BE} = C_{DIF} + C_{DEP}\Big|_{q_{DIF} >> q_{DEP}} \approx C_{DIF} = \frac{dq_{DIF}}{dv_{BE}} = \frac{\partial i_C \tau_F}{\partial v_{BE}} = g_m \tau_F \tag{3.21}$$

where τ_F is the forward-transit time across the base–emitter junction. On the other hand, because the base–collector diode hardly diffuses current, C_{DIF} holds substantially less charge than C_{DEP} across this junction, so C_μ is

$$C_\mu \equiv C_{BC} = C_{DIF} + C_{DEP}\Big|_{q_{DIF} << q_{DEP}} \approx C_{DEP} = \frac{A_{BC} C_{JBC0}''}{\sqrt{1 - \dfrac{v_{BC}}{V_{BI}}}} \tag{3.22}$$

where A_{BC} is the area, V_{BI} the built-in barrier potential, and C_{JBC0}'' the zero-bias capacitance per unit area of the base–collector junction. Vertical and lateral BJTs also include parasitic collector–substrate and base–substrate junctions, but their respective depletion capacitances C_{CS} and C_{BS} are normally so much smaller than other C_π's already present that simplified models often exclude them.

As operating frequency increases, C_π and C_μ steer more input current away from r_π to reduce v_{be}. In other words, C_π and C_μ reduce the effective gain of the BJT. In fact, *transitional frequency* f_T is the frequency beyond which the transistor can no longer amplify small signals well, that is, when short-circuit output current i_c falls below input current i_b. During a short-circuit condition, when v_{ce} is zero, r_o conducts no current and r_π, C_π, and C_μ in the simplified model of Fig. 3.25c are in parallel. Therefore, i_c reduces to $v_{be} g_m$; i_b drops v_{be} across r_π, C_π, and C_μ; and f_T is when the ratio of i_c to i_b is 1:

$$\begin{aligned} \frac{i_c}{i_b}\Big|_{v_{ce}=0} &= \frac{v_{be} g_m}{i_b}\Big|_{v_{ce}=0} = i_b\left[r_\pi \,\|\, \frac{1}{(C_\pi + C_\mu)s}\right]\left(\frac{g_m}{i_b}\right) \\ &= \left[\frac{r_\pi}{1 + r_\pi (C_\pi + C_\mu)s}\right] g_m \Big|_{f_T \approx \frac{g_m}{2\pi(C_\pi + C_\mu)}} \equiv 1 \end{aligned} \tag{3.23}$$

where s refers to frequency in the Laplace domain. This means smaller geometries (i.e., lower capacitances) and higher bias currents (i.e., higher transconductances) raise f_T to expand the useful range of BJTs in analog applications.

3.5 Metal-Oxide-Semiconductor Field-Effect Transistors

3.5.1 Theory of Operation

Metal–oxide–semiconductor field-effect transistors (MOSFETs) differ from BJTs in that current flow is mostly the result of an electric field, not

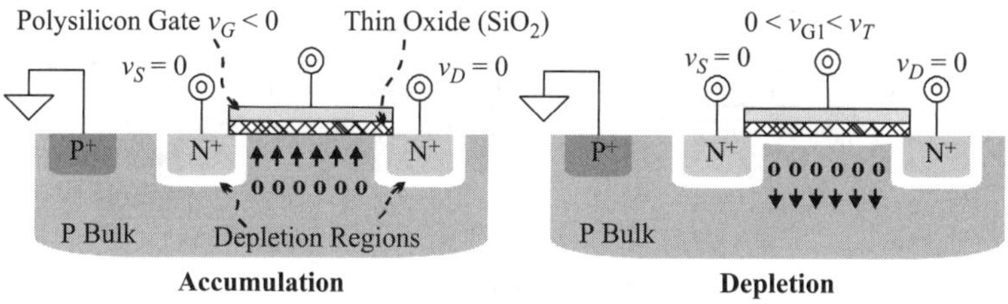

FIGURE 3.26 MOSFET in accumulation and depletion.

diffusion. MOSFETs are, in essence, two-terminal resistors with a conductance-controlling third terminal called the *gate*. To be more precise, the voltage applied to the parallel-plate *metal–oxide–semiconductor* (MOS) capacitor that separates the gate from the resistive channel determines how many charge carriers reach the channel, so gate voltage controls the channel's carrier concentration and, as a result, its resistance. This means that a voltage across the channel establishes an ohmic current. Interestingly, the voltage across the channel can saturate, in which case the ensuing current is less sensitive to the terminal voltages of the transistor. Although modern-day transistors of this sort implement gates with polysilicon instead of metal, they conventionally retain the name MOSFET.

From a physical perspective, as Fig. 3.26 illustrates, a strip of polysilicon and the silicon substrate or bulk material above which the strip sits sandwich a thin layer of silicon dioxide SiO_2. Highly but oppositely doped diffusion regions lie below the extreme edges of the polysilicon gate. In the case of the figure, the bulk material is P type and the diffused terminals are N type. Further away, a metallic contact connects to the bulk via a highly and similarly doped region to ensure that the bulk never forward-biases any PN junction. This means depletion regions surround the oppositely doped terminals of the structure. Note that, because fields are relative, the ensuing discussions of the N-channel MOSFET in Figs. 3.26–3.29 refer all potentials to the *source* terminal v_S, which is at ground, so gate voltage v_G is equivalent to v_{GS} and *drain* voltage v_D to v_{DS}.

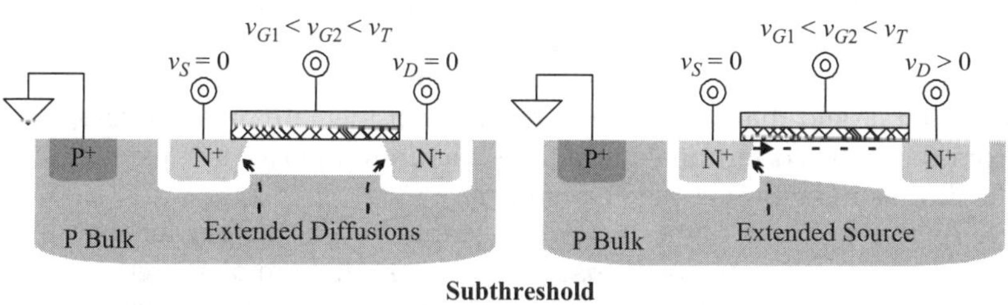

FIGURE 3.27 N-channel MOSFET in subthreshold.

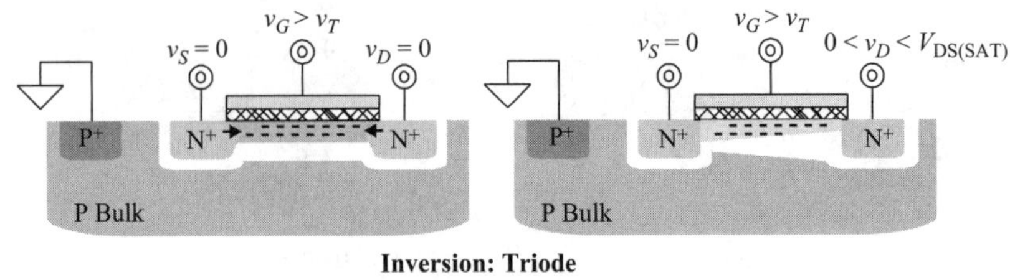

Figure 3.28 Inverted N-channel MOSFET in triode.

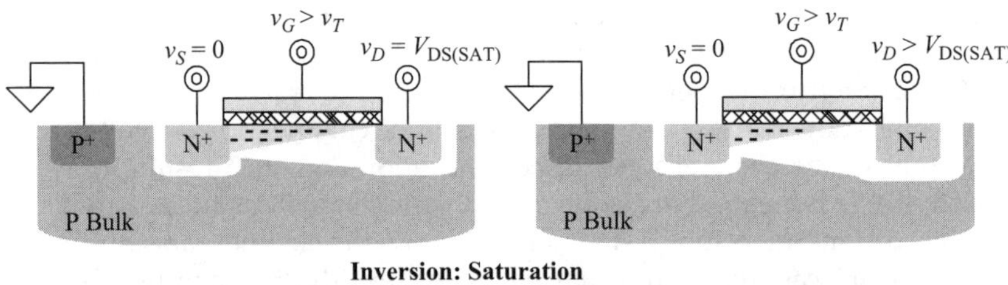

Figure 3.29 Inverted N-channel MOSFET in saturation.

Cut Off

Applying a negative voltage to the polysilicon gate of Fig. 3.26 attracts holes from the bulk material to the surface. Since the region is already P type, drifting holes "accumulate" near the surface. Applying a small positive voltage to the gate, on the other hand, repels holes from the surface and "depletes" the region near the surface of majority carriers. Either way, whether in *accumulation* or *depletion*, no mechanism at this point induces charge flow between the two diffused terminals, so the device is in *cut off*.

Subthreshold In depletion, the structure includes two dielectrics between the gate and bulk material: the oxide layer and the depletion region. Therefore, raising the gate voltage across the two series capacitors increases the intermediate voltage near the surface below the gate. This surface potential is the same across most of the gate, except near the highly doped regions whose terminal voltages are different. The lateral field that results near the edges of the gate pulls charges from the diffused terminals into the structure to, in essence, extend the diffusion regions, as Fig. 3.27 shows.

As in a diode, a higher surface potential lowers the barrier that charges must overcome to diffuse. With a lateral field in place, reducing the barrier frees an exponentially increasing number of charge carriers near the edge of the extended diffusion region that a positive external voltage v_D across the terminals of the device can attract and collect. In other words, a current i_D flows across the surface that rises with geometrical width W and falls with increasing length L and climbs exponentially with gate voltage v_G:

$$i_{D(DIF)} = \left(\frac{W}{L}\right) I_{ST} e^{(v_{GS} - v_{TN})/nV_t} \left(1 - e^{-v_{DS}/V_t}\right)\Bigg|_{v_{DS} > 3V_t} \approx \left(\frac{W}{L}\right) I_{ST} e^{(v_{GS} - v_{TN})/nV_t} \quad (3.24)$$

where W/L is the width–length aspect ratio of the device, I_{ST} is the subthreshold saturation current, v_{TN} is the threshold voltage, n is a process constant between 1.5 and 3, and V_t is the thermal voltage. At first, v_D raises i_D in exponential fashion, but when v_D surpasses $3V_t$, v_D's impact on i_D diminishes to the extent that i_D, for all practical purposes, *saturates*.

Inversion

Raising gate voltage v_G beyond this point and above threshold voltage v_{TN} extends the highly doped diffusion regions to such an extent that, with the aid of a higher surface potential, charge carriers in Fig. 3.28 migrate across and meet in the middle. These carriers form a channel that connects the source and drain terminals of the device. Since the surface underneath the gate is bulk material, positive gate voltages above a P-type bulk ultimately "invert" the region near the surface to N type, placing the device in what engineers call *inversion*. In this case, v_{TN} is the threshold voltage above which the channel becomes N type. Interestingly, carriers require less energy to cross and meet when diffusion regions are closer together. This means v_{TN} is slightly lower for shorter channel lengths and higher otherwise.

Triode The channel is basically a resistor whose resistance $R_{CHANNEL}$ rises with length L and falls with increasing width W and the field strength that *gate-oxide capacitance* C_{OX} and gate voltage v_G establish. As a result, a voltage across the channel v_D induces an ohmic current i_D that is equal to

$$\begin{aligned} i_{D(TRI)} &= \frac{v_{DS}}{R_{CHANNEL}} = v_{DS}\left(\frac{W}{L}\right) K_N'[(v_{GS} - v_{TN}) - 0.5v_{DS}]\Big|_{v_{DS} < V_{DS(SAT)}} \\ &\approx \left(\frac{W}{L}\right) K_N'(v_{GS} - v_{TN})v_{DS} \end{aligned} \quad (3.25)$$

where K_N' is the transconductance parameter of electrons, which is the product of electron mobility μ_N and oxide capacitance per unit area C_{OX}''. Elevating the channel voltage v_D, as Fig. 3.28 shows, attracts some electrons away from the channel to reduce the depth of the channel near the v_D terminal. This raises $R_{CHANNEL}$ and lowers i_D, but still, the $0.5v_{DS}^2$ correction in i_D disappears when v_{DS} is small. Engineers call this ohmic mode of operation the *triode* or *nonsaturated* region of the MOSFET.

Saturation Raising v_D continues to lighten the channel near the drain of charge carriers until the channel finally "pinches," at which point, as Fig. 3.29 illustrates, charge is zero near the drain. This happens when v_D cancels the voltage v_{OX} across the gate–oxide–semiconductor capacitor C_{OX} that attracted charge q_C in the first place. To pinch the channel this way, where q_C at the v_D terminal is zero

$$q_C = C_{OX}v_{OX} = C_{OX}[(v_G - v_D) - v_{TN}]\Big|_{v_D = v_G - v_{TN}} = 0 \tag{3.26}$$

requires v_D to rise to $v_G - v_{TN}$. More generally, because all fields are relative to v_S, pinching the channel requires v_{DS} to climb to $v_{GS} - v_{TN}$, which is what engineers normally call *saturation voltage* $V_{DS(SAT)}$:

$$V_{DS(SAT)} \equiv V_{DS}\Big|_{Pinch} = v_{GS} - v_{TN} \tag{3.27}$$

Ultimately, the device drops v_D across the entire length of the structure. So, since a pinched channel can only drop $V_{DS(SAT)}$, raising v_D above $V_{DS(SAT)}$ shifts the pinch-off point farther away from the drain, as Fig. 3.29 also shows. Because the voltage across the now shorter channel remains constant at $V_{DS(SAT)}$, i_D "saturates" and is no longer as sensitive to v_D:

$$\begin{aligned} i_{D(SAT)} = i_D\Big|_{v_{DS} \geq v_{DS(SAT)}} &= \frac{V_{DS(SAT)}}{R_{CHANNEL}} \\ &\approx V_{DS(SAT)}\left(\frac{W}{L}\right)K'_N[(v_{GS} - v_{TN}) - 0.5V_{DS(SAT)}](1 + \lambda_N v_{DS}) \\ &\approx 0.5\left(\frac{W}{L}\right)K'_N(v_{GS} - v_{TN})^2(1 + \lambda_N v_{DS}) \end{aligned} \tag{3.28}$$

But because the length of the channel decreases slightly with a higher v_D, $R_{CHANNEL}$ also drops and i_D rises with higher drain voltages. Channel-length modulation parameter λ_N models the extent that *channel-length modulation* affects i_D. Realizing $V_{DS(SAT)}$ is $v_{GS} - v_{TN}$ and $v_{GS} - v_{TN}$ in *saturation* is

$$V_{DS(SAT)} \equiv v_{GS} - v_{TN} \approx \sqrt{\frac{2i_{D(SAT)}}{\left(\frac{W}{L}\right)K'_N}} \tag{3.29}$$

helps analog designers determine $V_{DS(SAT)}$ for various combinations of currents and width–length ratios.

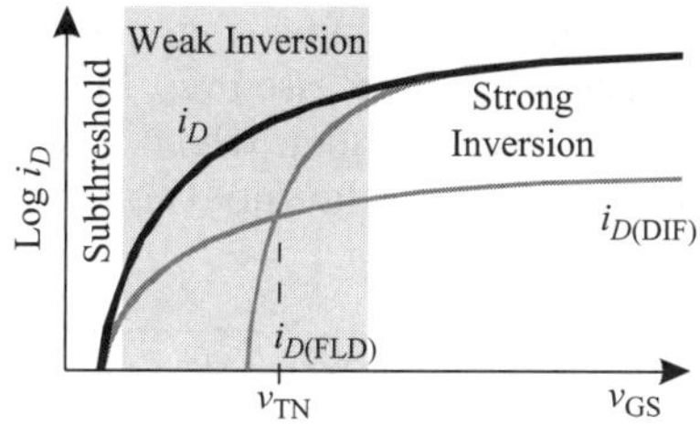

FIGURE 3.30 Drain current across subthreshold and weak- and strong-inversion regions.

Weak Inversion Inverting the channel and impressing a voltage that induces current to flow does not mean subthreshold diffusion current disappears. In fact, drain current carries both diffusion and drift components $i_{D(DIF)}$ and $i_{D(FLD)}$:

$$i_D = i_{D(DIF)} + i_{D(FLD)} \tag{3.30}$$

where $i_{D(FLD)}$ is $i_{D(TRI)}$ in triode and $i_{D(SAT)}$ in saturation. In subthreshold, before the channel inverts, $R_{CHANNEL}$ is so high that $i_{D(DIF)}$ overwhelms $i_{D(FLD)}$, which means i_D reduces to $i_{D(DIF)}$, as the logarithmic graph of i_D in Fig. 3.30 illustrates. As gate voltage v_G rises toward threshold v_{TN}, $i_{D(FLD)}$ rises to the point where both $i_{D(DIF)}$ and $i_{D(FLD)}$ become appreciable components of i_D, characteristics of which correspond to *weak inversion*. Threshold v_{TN}, in fact, is the point where $i_{D(FLD)}$ equals $i_{D(DIF)}$. Raising v_G above these levels, past weak inversion, lowers $R_{CHANNEL}$ so much that $i_{D(FLD)}$ dominates, which marks the onset of *strong inversion*.

Operating in strong inversion is popular because higher currents charge parasitic and loading capacitances faster. As a result, process engineers create computer models that emulate this region of operation reasonably well. The drawback to strong inversion, however, is power, so microwatt applications like biomedical implants and miniaturized sensors often operate transistors in subthreshold. This push for low power motivates engineers to characterize and model the subthreshold region better. However, creating a model that emulates all regions, including the weak-inversion transition, without overtaxing the simulating platform is an arduous undertaking. To appreciate this, consider the computational time required to solve all the relationships that describe each MOSFET in a system that includes thousands of transistors when simulating across time, temperature, process, and a number of other operating conditions.

Body Effect

Discussions thus far assumed the bulks of the N-channel MOSFETs illustrated in Figs. 3.26–3.29 are at the same potential as the diffusion terminal with the lowest voltage, which is not always the case. Here, elevating the bulk-terminal voltage v_B imposes a field that attracts more electrons into the channel, the results of which are a drop in $R_{CHANNEL}$ and a consequential rise in i_D. In other words, the bulk voltage alters the surface potential, which is why engineers model this *bulk effect* with a shift in the threshold voltage. This means that a rising v_B reduces v_{TN} and vice versa:

$$v_{TN} = V_{TN0} + \gamma\left(\sqrt{2\varphi - v_{BS}} - \sqrt{2\varphi}\right) \tag{3.31}$$

where v_{TN} is the threshold voltage of an N-channel MOSFET, V_{TN0} is the zero-bias threshold when v_B and v_S are equal and v_{BS} is therefore zero, γ is the bulk-effect parameter, and 2φ is a process-dependent constant that is normally about 0.6 V.

Current–Voltage Behavior

Engineers call the terminal that supplies charge carriers to the channel the *source* and the terminal that receives and releases them the *drain*. Because the conducting medium lies immediately underneath a gate–oxide–semiconductor capacitor, the symbol for a MOSFET is a capacitor where one plate is the gate and the other, from which source and drain terminals originate, is the channel (see Fig. 3.31). Because the source supplies the charges that establish drain current i_D, the symbol highlights the location of the source with an arrow that points in the direction of current flow. In the case of the N-channel MOSFET of Figs. 3.26–3.29, for example, the arrow in Fig. 3.31 points away from the channel.

Since the gate of a MOSFET is a capacitor, no static current flows into the gate. When strongly inverted, source and drain current i_D rises quadratically with gate–source voltage v_{GS}. In saturation, i_D rises slightly with drain–source voltage v_{DS}, so small changes in i_D correspond to large variations in v_{DS}, which indicates high gain. When v_{DS} falls below saturation voltage $V_{DS(SAT)}$, the transistor transitions to the

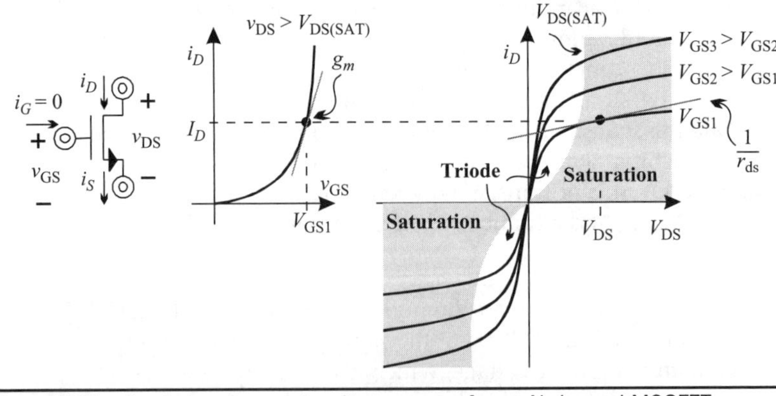

FIGURE 3.31 Symbol and current–voltage curves for an N-channel MOSFET.

ohmic or triode region, where i_D falls linearly with v_{DS}. Because the physical structure of most MOSFETs is symmetrical, engineers can swap source and drain terminals without altering performance. For example, raising v_G above v_D by v_{TN} inverts the same channel that v_{GS} does, and raising v_S above v_D by $V_{DS(SAT)}$ pinches it like v_{DS} does. As a result, the i_D–v_{DS} characteristics in this region, when v_{GS} and v_{DS} are both negative in Fig. 3.31, are symmetrical to those of its counterpart, when v_{GS} and v_{DS} are both positive. Since engineers normally optimize the physical structure of BJTs to operate in the forward-active region, vertical BJTs, unlike MOSFETs, do not enjoy symmetrical performance.

3.5.2 Parasitic Capacitances

The MOS transistor is ultimately a field-effect device because current flows mostly in response to gate–source and drain–source fields. In fact, as the structures of Figs. 3.26–3.29 illustrate, the gate overlaps the source, channel, and drain, so parallel-plate oxide capacitors exist across the length of the gate that increase with channel width W and oxide capacitance per unit area C_{OX}''. The main component is from gate to body or channel, but its value and eventual effects depend on v_{GS} and v_{DS} because the nature and length of the conducting channel changes with v_{GS} and v_{DS}. In cut off, for example, no channel exists, so small overlap capacitances $WL_{OV}C_{OX}''$ across the width of the gate appear between gate and source and gate and drain terminals, as C_{GS} and C_{GD} in Fig. 3.32 show and where L_{OV} is overlap length. The oxide between the gate and bulk material also forms a capacitor, except the depletion region decreases this capacitance to such a degree that its effects are minimal, and considering that the device is in cut off, less relevant in analog applications.

In triode, when the channel connects the source and drain terminals and the length of the channel is practically the physical distance L between the source and drain diffusions, source and drain terminals share gate–channel capacitance WLC_{OX}''. Accordingly, C_{GS} and C_{GD} both include half the channel capacitance $0.5WLC_{OX}''$ plus their corresponding overlap portions $WL_{OL}C_{OX}''$. In saturation, when the

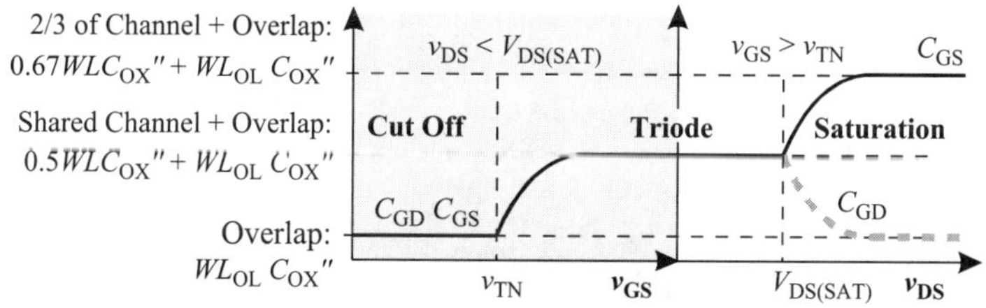

FIGURE 3.32 Gate–source and gate–drain capacitances across MOSFET operating regions.

channel pinches and shortens, the source alone assumes the burden of the channel. Since the channel is shorter, however, C_{GS} includes what amounts to two-thirds of the triode channel capacitance $0.67WLC_{OX}''$ as well as its overlap component $WL_{OL}C_{OX}''$. C_{GD} in saturation, on the other hand, reduces to its overlapped fraction $WL_{OL}C_{OX}''$.

Reverse-biased junctions attached to the drain and source introduce parasitic capacitors in the form of depletion capacitances C_{SB} and C_{DB}:

$$C_{SB} = \frac{A_{SB}A_{JSB0}''}{\sqrt{1+\frac{v_{SB}}{V_{BI}}}} \tag{3.32}$$

and

$$C_{DB} = \frac{A_{DB}C_{JDB0}''}{\sqrt{1+\frac{v_{DB}}{V_{BI}}}} \tag{3.33}$$

Here, A_{SB} and A_{DB} are source and drain surface areas and C_{JSB0}'' and C_{JDB0}'' are zero-bias source–bulk and drain–bulk capacitances per unit area. As with any PN junction, the peripheral sidewall of the diffused source and drain regions contributes additional depletion capacitance.

3.5.3 P-Channel MOSFETs

The P-channel MOSFET is the complement of the N-channel device in every respect. In a PMOS transistor, for example, as shown in Fig. 3.33, a thin oxide strip sits above an N-type bulk material with highly doped P-type diffusion regions underneath the edges of the strip. Raising the gate voltage v_G above the source voltage v_S attracts electrons to the surface, and lowering v_G below v_S repels electrons from the surface, to place the transistor in accumulation and depletion, respectively. Lowering v_G further reduces the surface potential, which allows holes to diffuse across the device. Lowering v_G below a threshold attracts holes into the structure to invert and form a channel that connects the highly doped terminals. Lowering drain voltage v_D therefore collects holes flowing across the channel, and if v_D is sufficiently low with respect to v_S, the channel pinches and current i_D saturates.

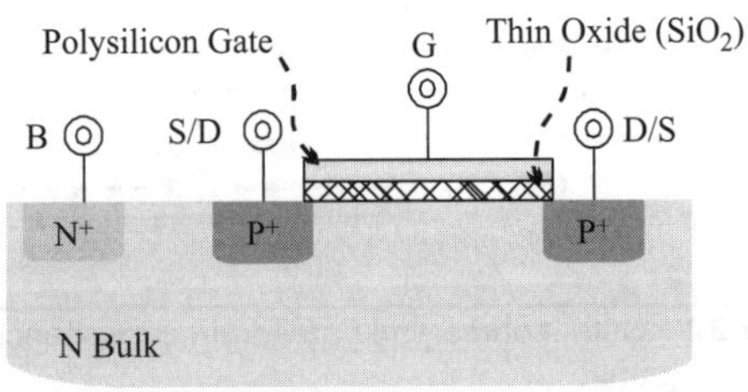

FIGURE 3.33 P-channel MOSFET.

The relationships for i_D are therefore the same as those for the N-channel counterpart, except a positive source–gate v_{SG} must rise above P-type threshold $|v_{TP}|$ to invert the channel, and a positive source–drain voltage v_{SD} must rise above P-type saturation voltage $V_{SD(SAT)}$ to pinch the channel. Conventionally, v_{TP} is a negative number, so absolute signs enclose v_{TP} here for consistency. The result is that replacing v_{GS}, v_{DS}, v_{BS}, K_N', V_{TN0}, and v_{TN} with v_{SG}, v_{SD}, v_{SB}, K_P', $|V_{TP0}|$, and $|v_{TP}|$ in the N-channel equations describes the behavior of the P-channel counterpart.

3.5.4 Transistor Variations

Since the bulk terminal of an N-channel transistor in a P substrate is the substrate, the bulk cannot connect to any point other than the most negative potential. This is why *substrate MOSFETs* are, in practice, three-terminal devices, as Fig. 3.34 illustrates. Immersing the N-channel structure into a P well that lies in an N substrate is another way of implementing an N-channel MOS (NMOS) transistor. Because the well is available, *welled MOSFETs* usually have a fourth terminal called the *bulk* or *body*.

Interestingly, the zero-bias threshold voltage of *native NMOSFETs* is near zero. Digital designers, however, dislike the leakage currents these devices conduct, so process engineers include a threshold-adjust implant step in the fabrication process that shifts V_{T0} from near zero to 0.3–0.6 V. These adjusted devices are what engineers call *enhancement MOSFETs*. Blocking this implant requires an additional step in the process flow, so companies avoid them when possible. When accessible, though, engineers often include a dashed line between the source and drain terminals to indicate that a zero gate–source voltage already inverts the channel slightly. Process engineers sometimes adjust V_{T0} so a zero gate–source voltage inverts the channel to such an extent that current flow is appreciable. As such, the symbol of these *depletion MOSFETs* includes a solid line that represents a strongly inverted channel between the source and drain terminals.

Replacing the thin oxide between the gate and channel with a thicker oxide layer reduces the field across the oxide and increases

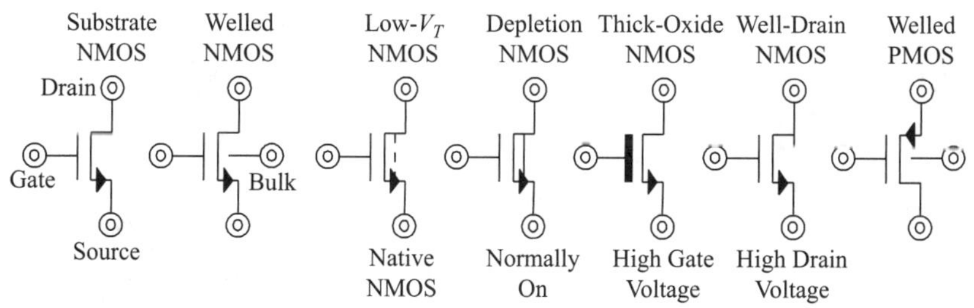

Figure 3.34 Typical symbols for MOSFETs.

the breakdown voltage of the gate. Some engineers model this variation in the symbol with a thicker gate plate. Similarly, a lower doped drain terminal extends the depletion region around the drain and therefore reduces the field and raises the breakdown voltage of the drain. Some engineers indicate this alteration to the device by including a marking near the drain. An N-channel transistor whose drain is an N well, for example, might include a double drain line to indicate the drain region is large and modified, and in this case, lightly doped.

The symbol for a PMOS transistor mirrors the N counterpart. As such, because the source supplies holes into the channel, the terminal with an arrow pointed toward the channel is the source. As with the N-channel devices, welled transistors include a bulk terminal and substrate devices do not. Unlike the N channel, however, the threshold voltage of native P-channel transistors is normally high. In other words, gate voltage v_G must fall well below that of the source to invert and form a P-type channel. As a result, process engineers normally lower the thresholds of P-channel devices, rather than raise them, as they do with their N-type counterparts.

Substrate MOSFETs

Many process technologies today build semiconductors on P-type silicon. In these cases, connecting the P substrate to the most negative potential reverse-biases all derivative PN junctions so that N-channel transistors can sit directly in the substrate, as Fig. 3.35 shows. Connecting an N substrate to the most positive potential similarly isolates P-channel MOSFETs from their substrate. Irrespective of the type, the diffusion terminals of *substrate MOSFETs* form reverse-biased PN junctions with the substrate. Intrinsic source and drain terminals in the electrical model of Fig. 3.35 therefore include parasitic diodes D_S and D_D that connect through substrate resistance R_{SUB} to ground or a power supply. In addition, source and drain contact and diffusion regions and the polysilicon gate introduce series resistances R_S, R_D, and R_G to the intrinsic terminals. As mentioned earlier, the gate also

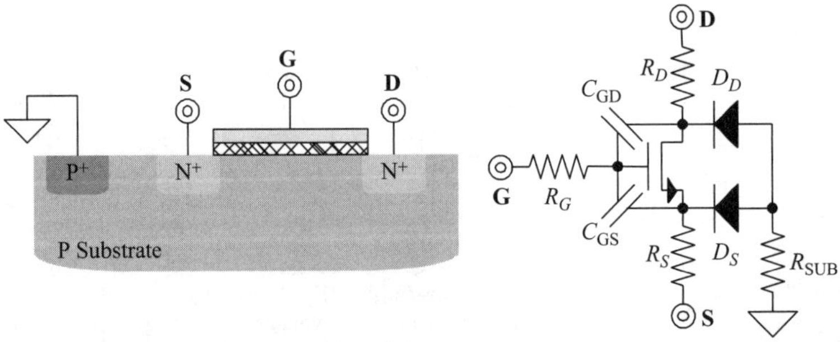

FIGURE 3.35 Substrate N-channel MOSFET and its equivalent electrical model.

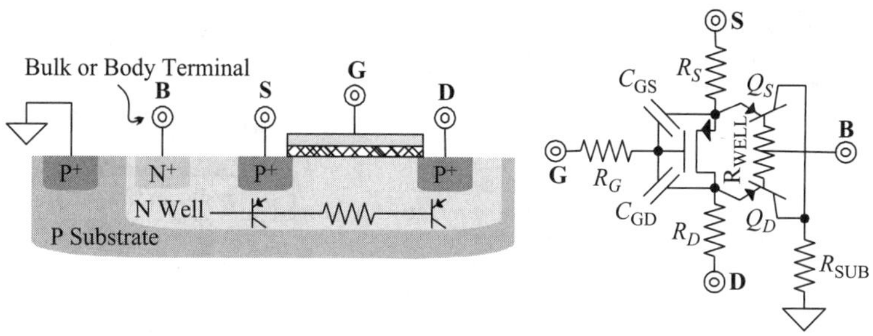

FIGURE 3.36 Welled P-channel MOSFET and its equivalent electrical model.

includes gate–source and gate–drain capacitances C_{GS} and C_{GD} at the source and drain terminals.

Welled MOSFETs

To build P-channel transistors in a P substrate, process engineers embed a PMOS device inside the isolating N well of Fig. 3.36. Similarly, a P well isolates an N-channel MOSFET from an N substrate. Irrespective of the type, the substrate reverse-biases all wells, and the engineer is free to connect the body or bulk of the welled device to any point in the circuit. Like substrate transistors, the electrical model of a welled MOSFET includes parasitic series terminal resistances R_G, R_S, and R_D and gate capacitances C_{GS} and C_{GD}. Unlike their substrate counterparts, however, source and drain diffusions in the well inside the substrate form vertical substrate BJTs Q_S and Q_D. Since the well is typically lightly doped, base widths can be short, so these BJTs can have high current gains with considerable base-width modulation. Luckily, designers can connect their body terminals to a potential that keeps them from forward-biasing their emitter–base junctions. Still, well resistance R_{WELL} can be high, so parasitic currents in the well of high-power devices can still activate Q_S and inject noise current into the substrate.

3.5.5 Layout and Matching

The top layout views of MOSFETs typically display highly doped rectangular and parallel diffusions separated by a polysilicon gate strip, as shown in Fig. 3.37. In the case of substrate devices, the bulk is also the substrate, and its contact is nearby to collect stray carriers in the substrate and to keep the potential of the transistor's body steady near its corresponding supply. Welled transistors lie within a well whose contact is similarly close to the MOSFET. These transistors can share the well with other devices, but only if all body terminals share the same connection in the circuit. Since gates do not carry static currents, designers sometimes interconnect gates by stretching, snaking, and sharing one polysilicon strip between two or more transistors.

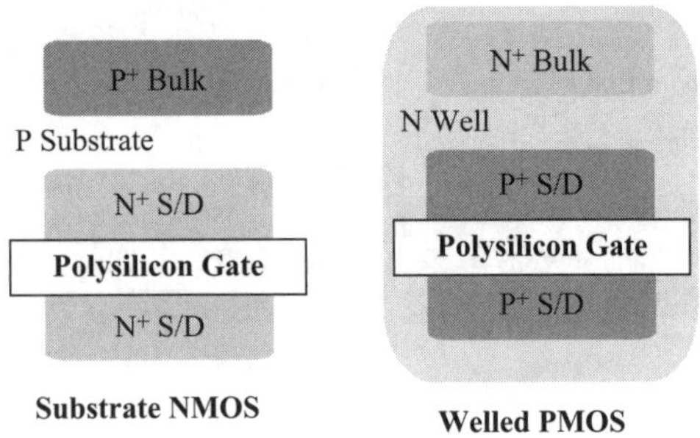

FIGURE 3.37 Top layout views of typical substrate and welled MOSFETs.

Plus, when tied to the same potential, bulk diffusions and contacts can often butt against those of the source to connect them.

As with other microelectronic devices, large gate areas, close arrays, similarly oriented devices, common-centroid structures, interdigitation of gates, cross-coupling, and dummy devices around the periphery of an array improve matching performance. To save space, smaller dummy polysilicon strips can replace dummy transistors. Without specifying otherwise, however, polysilicon strips are normally resistors, so they sit above the thick field oxide FOX that Fig. 3.38*a* shows. The problem with this is that the peripheral boundaries of MOSFET gates are considerably closer to the surface of the semiconductor than those of resistor strips, so etching errors do not abate much. Placing the strips on thin-oxide regions, as in Fig. 3.38*b*, fixes

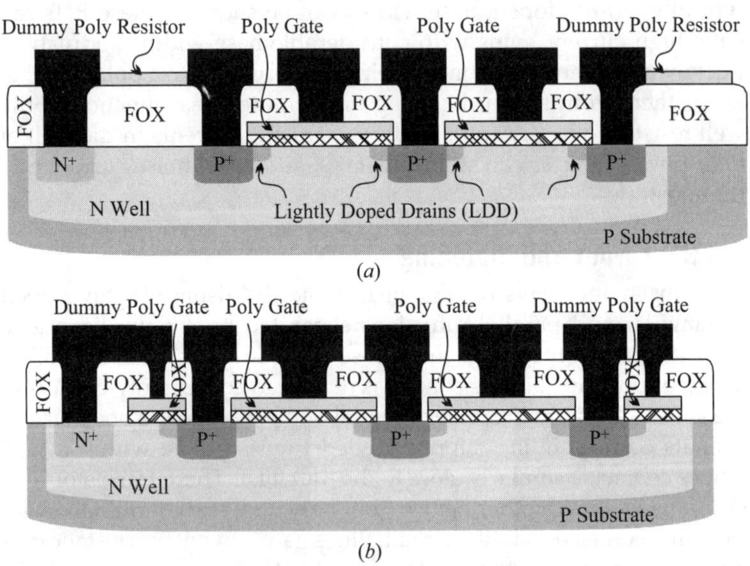

FIGURE 3.38 Welled PMOSFETs surrounded by dummy polysilicon (*a*) resistors and (*b*) gates.

the problem. For this, designers implement them as "active gates," but prescribing this through software tools is not always straightforward, so resorting to dummy MOSFETs is oftentimes a safer choice.

When disconnected, stray signals can couple noise into the substrate through these dummy capacitors. Even worse, noise can inadvertently invert a channel under the surface of the semiconductor that is immediately underneath the strips and load nearby MOSFETs with parasitic capacitance and noise. For these reasons, supply or reference voltages should, if possible, bias these dummy resistors, gates, and transistors so they remain off. This way, they can also shunt and steer stray noise away from nearby MOSFETs. Dummy gates above an N well, for example, should connect to a positive voltage so that, rather than inverting a channel, they will accumulate electrons near the surface of the semiconductor.

State-of-the-art process technologies nowadays use MOS oxide layers on the order of 125 Å or less. With these shorter oxides, the electric field near the drain strengthens to the point that one or two volts at the drain can propel *hot carriers* to not only loosen and energize other carriers but also cross the oxide and reach the gate. Extending the drain with the *lightly doped drain* (LDD) regions in Fig. 3.38 reduces the charge carrier concentration near the edges of the gates, so *impact ionization* is less likely.

3.5.6 Small-Signal Model

Because high gain and high resistance are useful attributes in analog applications, as the side note on small signals explains, designers typically model the small-signal response of MOSFETs in saturation. Here, small changes in gate–source voltage v_{GS} produce considerable variations in drain current i_D, which as Fig. 3.31 illustrates, ultimately cause substantial variations in drain–source voltage v_{DS}. Although not to the same extent, changes in bulk–source voltage v_{BS} shift v_{TN}, which in turn alters i_D. Similarly, variations in v_{DS} also affect i_D. As a result, Fig. 3.39*a* models the small-signal components in i_d that small signals v_{gs}, v_{bs}, and v_{ds} produce.

Since linear approximations predict the response of small signals reasonably well, the effect of v_{gs} on i_d is the linear translation that the slope g_m in Fig. 3.31 and first partial derivative of i_D with respect to v_{GS} at bias current I_D predict:

$$i_d\Big|_{v_{ds}=0,\,v_{bs}=0} \approx v_{gs}\left(\frac{\partial i_D}{\partial v_{GS}}\right) \equiv v_{gs}g_m \approx v_{gs}\sqrt{2I_D K_N'\left(\frac{W}{L}\right)} \tag{3.34}$$

where g_m is the MOSFET's effective transconductance. Similarly, the slope g_{ds} and first partial derivative of i_D with respect to v_{DS} at I_D forecast the impact of v_{ds} on i_d:

$$i_d\Big|_{v_{gs}=0,\,v_{bs}=0} \approx v_{ds}\left(\frac{\partial i_D}{\partial v_{DS}}\right) \equiv v_{ds}g_{ds} \approx v_{ds}(\lambda_N I_D) \equiv \frac{v_{ds}}{r_{ds}} \tag{3.35}$$

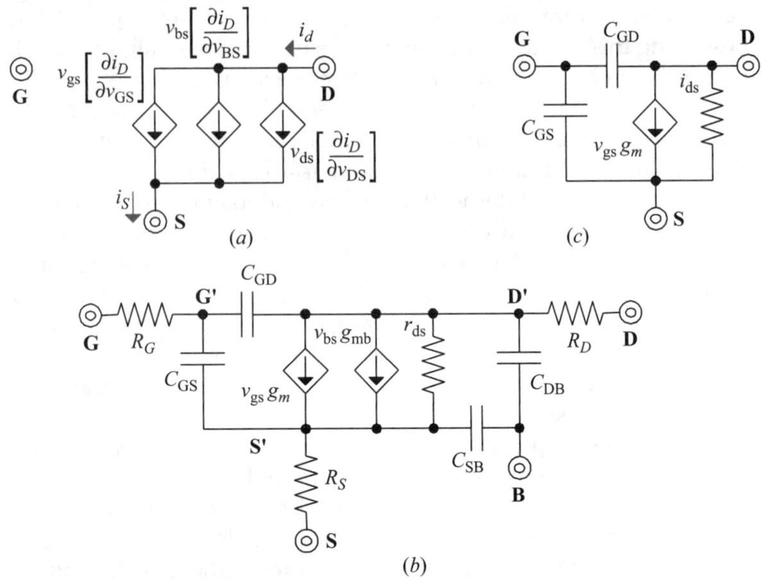

FIGURE 3.39 (*a*) Small-signal currents and (*b*) complete and (*c*) simplified small-signal models of a MOSFET.

where g_{ds} is the effective output conductance of the device. Because this linear variation in current results from a change in the voltage across its own two terminals v_{ds}, output resistance $1/g_{ds}$ or r_{ds} in Fig. 3.39*b* models this behavior well. Lastly, the first partial derivative of i_D with respect to v_{BS} at I_D describes how v_{bs} affects i_d, and to follow the model, v_{bs} shifts threshold voltage v_{TN}, which in turn changes i_d much like v_{gs} does, but with a lower gain:

$$\begin{aligned} i_d\Big|_{v_{ds}=0,\, v_{gs}=0} &\approx v_{bs}\left(\frac{\partial i_D}{\partial v_{BS}}\right) = v_{bs}\left(\frac{\partial i_D}{\partial v_{TN}}\right)\left(\frac{dv_{TN}}{dv_{BS}}\right) = v_{bs}\left(-\frac{\partial i_D}{\partial v_{GS}}\right)\left(\frac{dv_{TN}}{dv_{BS}}\right) \\ &\approx -v_{bs}g_m\left(\frac{-\gamma}{2\sqrt{2\varphi - V_{BS}}}\right) = v_{bs}\left(\frac{\gamma g_m}{2\sqrt{2\varphi - V_{BS}}}\right) \equiv v_{bs}g_{mb} \end{aligned} \tag{3.36}$$

where V_{BS} is the steady-state voltage across the bulk and source terminals and g_{mb} the bulk's effective transconductance, which is normally five to ten times smaller than g_m.

Since there is no gate current and r_{ds} is normally high between hundreds of kilo-ohms and mega-ohms, the simplified model of Fig. 3.39*c* neglects series terminal series resistances R_S, R_D, and R_G, which are between 1 and 100 Ω. Also, because g_m can be an order of magnitude higher than g_{mb}, the model excludes g_{mb}. This is not always a justifiable assumption, however, except when source and bulk terminals connect to the same potential, in which case v_{bs} is zero and its impact on i_d is nil.

A small-signal model of the MOSFET would not be complete without gate–source and gate–drain capacitances C_{GS} and C_{GD}. At high frequency, C_{GS} and C_{GD} steer input current away from v_{gs}, so v_{gs} falls, and so does its impact on i_d. In other words, C_{GS} and C_{GD} reduce the effective gain of the MOSFET. Transitional frequency f_T is therefore the frequency beyond which the transistor can no longer amplify small signals well, that is, when short-circuit output current i_d falls below input current i_g, which flows into the gate through C_{GS} and C_{GD}. During a short-circuit condition, when v_{ds} is zero, r_{ds} conducts no current and C_{GS} and C_{GD} in the simplified model of Fig. 3.39*c* are in parallel. As such, i_d reduces to $v_{gs} g_m$, i_g drops v_{gs} across C_{GS} and C_{GD}, and f_T is when the ratio of i_d to i_g is 1:

$$\frac{i_d}{i_g} = \left.\frac{v_{gs} g_m}{i_g}\right|_{v_{ds}=0} = \left.\left[\frac{i_g}{(C_{GS}+C_{GD})s}\right]\left(\frac{g_m}{i_g}\right)\right|_{f_T=\frac{g_m}{2\pi(C_{GS}+C_{DG})}} \equiv 1 \qquad (3.37)$$

where s refers to frequency in the Laplace domain. This means smaller geometries (i.e., lower capacitances) and higher bias currents (i.e., higher transconductances) raise f_T to expand the useful range of MOSFETs in analog applications. Trench-isolated technologies usually enjoy higher f_T's because they isolate bulk regions from other devices in the same substrate by inserting deep oxide trenches that are so wide that capacitances are substantially lower than in junction-isolated transistors. Unfortunately, inserting trenches into the process flow raises cost, so junction isolation remains popular in commercial applications.

3.5.7 MOS Capacitor

Because the oxide layer in a MOSFET is so thin, designers often resort to a MOSFET to implement a capacitor. As Fig. 3.40 shows, connecting the drain, source, and bulk terminals of a MOSFET together and impressing a voltage v_C across the oxide that surpasses the transistor's threshold voltage establishes a channel against which the gate

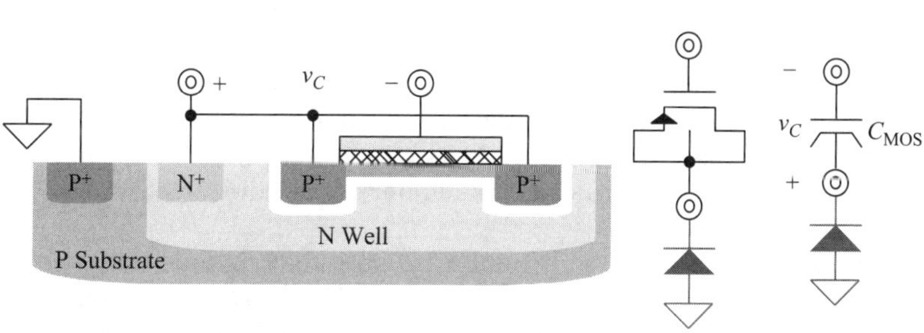

Figure 3.40 Gate–channel P-channel MOS capacitor.

forms a parallel-plate capacitor. The MOS capacitance C_{MOS} that results is the combined overlap and channel capacitance between the gate and source, channel, and drain regions:

$$C_{MOS} = WLC_{OX}'' + 2WL_{OL}C_{OX}'' \tag{3.38}$$

C_{MOS}, however, varies when v_C drops below the threshold voltage, in which case two dielectrics separate the plates, so C_{MOS} is correspondingly lower. The terminal orientation of the capacitor and type of transistor used to build it also affects C_{MOS}. Because of these voltage effects, some designers use this structure to implement variable capacitors, otherwise known as *CMOS varactors*. Note, by the way, that a substrate diode hangs off one of its terminals, so that, too, affects performance by adding parasitic capacitance.

3.5.8 Channel Resistor

Bias-current generators and voltage-reference circuits in nanowatt systems normally require mega- and giga-ohm resistors. Here, too, designers often turn to MOSFETs because inverted channels can be long, narrow, and shallow. Since the mobility of holes is lower than that of electrons, engineers typically favor P-channel transistors over their N-type counterparts for this purpose. In fact, long-length and minimum-width P-channel MOSFETs can establish mega-ohm resistances in strong inversion and giga-ohm resistances in subthreshold and weak inversion. Although less popular, shorter-channel transistors in strong inversion can also implement kilo-ohm resistors.

3.6 Junction Field-Effect Transistors

3.6.1 Theory of Operation

Junction field-effect transistors (JFETs) are what some engineers like to call *pinched resistors* because they are diffused resistors that pinch, and when pinched, their currents saturate. A third terminal called the *gate* controls the depth of the resistor and a voltage across the *source* and *drain* terminals establishes a voltage across the channel that induces current flow. JFETs and MOSFETs are therefore similar in many respects. The fundamental difference between them is how channel resistance changes and pinches. Like in a MOSFET, all fields are relative, so the following discussions refer all potentials to the source terminal v_S, which means gate voltage v_G is equivalent to v_{GS} and drain voltage v_D to v_{DS}.

Physically, a JFET is a diffusion resistor that reverse-biased PN junctions squeeze into a narrower channel. In Fig. 3.41, for example, diffusing a highly doped P^+ gate region into an N well inside a P substrate defines an N-well conductive medium below the silicon surface.

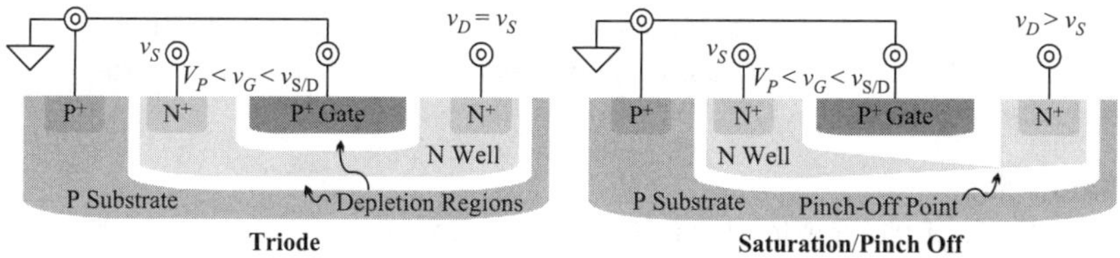

FIGURE 3.41 N-channel JFET in triode and saturation.

Reverse-biasing the PN junctions not only shuts diffusion currents but also compresses the N-type medium into a thinner, more resistive channel.

Cut Off

Because the JFET is a diffused resistor, no fields are necessary to establish a conductive channel. The JFET is therefore normally on, which is equivalent to a depletion-mode NMOS transistor whose threshold voltage is negative. So to eliminate the channel of an N-channel JFET, gate voltage v_G must fall below source voltage v_S to such an extent that the depletion regions *punch through* and the entire channel pinches. This pinching threshold point is V_P, where *pinch-off voltage* V_P is a negative number. However, applying this much negative voltage is usually impracticable, which is why JFETs are seldom in *cut off*.

Triode

Keeping v_G above V_P maintains a conducting channel across the terminals of the device, and as long as v_G is at or below source and drain voltages v_S and v_D, the junctions do not forward-bias against the channel. In this way, channel resistance $R_{CHANNEL}$ increases with geometrical length L and decreases with higher doping concentration, carrier mobility, width W, and depth, where gate voltage v_G controls depth. So like in a MOSFET, a voltage v_D across $R_{CHANNEL}$ establishes a drain current i_D that is equal to

$$i_{D(TRI)} = \frac{v_{DS}}{R_{CHANNEL}} = v_{DS}K_{JN}\left(\frac{W}{L}\right)[(v_{GS} - V_{PN}) - 0.5v_{DS}]\Bigg|_{v_{DS} < V_{DS(SAT)}} \approx v_{DS}K_{JN}\left(\frac{W}{L}\right)(v_{GS} - V_{PN}) \quad (3.39)$$

where effective transconductance parameter K_{JN} depends on doping density and mobility μ_N and v_{DS} is below a saturation threshold $V_{DS(SAT)}$. Raising v_D here extends the depletion region near the drain, so the channel narrows near that junction and $R_{CHANNEL}$ rises and i_D falls as a result. This $0.5{v_{DS}}^2$ correction in i_D, however, diminishes when v_{DS} is low. And as with most diffusion resistors, higher temperatures raise R_{CHANNEL}, and as a result, reduce i_D. Nevertheless, since i_D rises linearly with v_D in ohmic fashion, engineers call this mode of operation *ohmic* or *triode*.

Saturation

Elevating v_D further extends the depletion region near the drain to such an extent that the channel pinches. The point where the channel pinches has no charge, which means v_D cancels the equivalent voltage v_{DEP} across the depletion capacitor C_{DEP} that kept charge q_C in the first place. To pinch the channel this way, where q_C at the v_D terminal is zero

$$q_C = C_{\text{DEP}} v_{\text{DEP}} = C_{\text{DEP}}\left[(v_G - v_D) - V_{\text{PN}}\right]\Big|_{v_D = v_G - V_{\text{PN}}} = 0 \tag{3.40}$$

requires v_D to rise to $v_G - V_{\text{PN}}$, or more generally, because all fields are relative to v_S, calls for v_{DS} to climb to $v_{\text{GS}} - V_{\text{PN}}$, which is *saturation voltage* $V_{\text{DS(SAT)}}$:

$$V_{\text{DS(SAT)}} \equiv V_{\text{DS}}\Big|_{\text{Pinch}} = v_{\text{GS}} - V_{\text{PN}} \tag{3.41}$$

Ultimately, the device drops v_D across the entire length of the structure. So since a pinched channel can only drop $V_{\text{DS(SAT)}}$, raising v_D above $V_{\text{DS(SAT)}}$ shifts the pinch-off point further away from the drain, as Fig. 3.41 also shows. Because the voltage across the now shorter channel remains constant at $V_{\text{DS(SAT)}}$, i_D "saturates" and is no longer as sensitive to v_D:

$$i_{D(\text{SAT})}\Big|_{v_{\text{DS}} > V_{\text{DS(SAT)}}} = \frac{V_{\text{DS(SAT)}}}{R_{\text{CHANNEL}}} \approx 0.5 K_{\text{JN}}\left(\frac{W}{L}\right)(v_{\text{GS}} - V_{\text{PN}})^2(1 + \lambda_N v_{\text{DS}}) \tag{3.42}$$

But because the length of the channel decreases slightly with a higher v_D, R_{CHANNEL} also drops and i_D rises with higher drain voltages. Channel-length modulation parameter λ_N models the extent that channel-length modulation affects i_D. Not surprisingly, engineers call this mode of operation *saturation* or *pinch off*.

3.6.2 P-Channel JFETs

The P-channel JFET is the complement of the N-channel device in every respect. In a P-channel transistor, for example, like Fig. 3.42

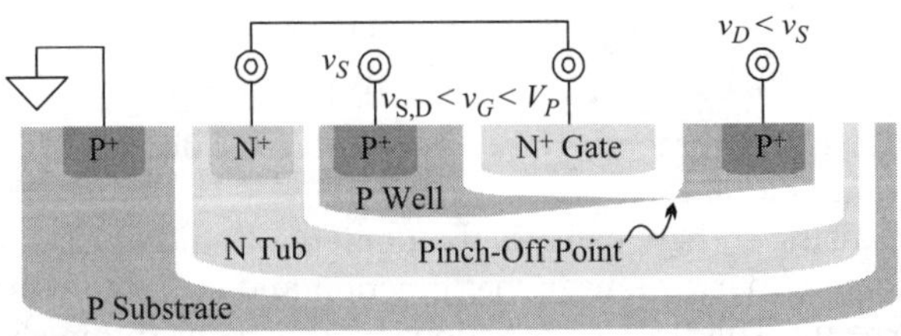

FIGURE 3.42 P-channel JFET.

shows, N-type diffusions squeeze a P well whose terminals ultimately connect to highly doped P⁺ diffusions. Like before, a reverse-biasing gate voltage v_G compresses the channel and a reverse-biasing drain voltage v_D pinches it to saturate the channel current i_D. The relationships for i_D are therefore the same, except v_{SG}, v_{SD}, K_{JP}, and V_{PP} replace v_{GS}, v_{DS}, K_{JN}, and V_{PN} in the N-channel equations.

3.6.3 Large-Signal Model

The JFET is basically a resistor whose charge concentration depends on the depletion region that the gate establishes. In other words, the field across a depletion capacitance controls resistance. So like the MOSFET, the schematic symbol of a JFET, as Fig. 3.43 shows, is a capacitor with one plate as its gate and the other as the channel, from which source and drain terminals originate. Because the gate and channel form a PN junction, the gate includes an arrow that points toward the N-type region: to the channel in the N-channel case and away from the channel in the P-channel counterpart. To reverse-bias the drain in an N-channel JFET, the drain voltage is high, so current flows into the drain. In contrast, to do the same in the P-channel transistor, the drain voltage is low, so current flows out of the drain.

As with all devices, metallic contacts and diffusion regions introduce series resistances R_S, R_D, and R_G to the source, drain, and gate terminals. Because reverse-bias junctions separate the gate and channel, parasitic depletion capacitors C_{GS} and C_{GD} appear between the gate and source and gate and drain. As reverse-biased diodes, these capacitances rise with the surface area of the junctions and fall with increasing reverse-bias voltages. Gate diffusion in a resistive well inside a substrate also forms a vertical BJT. Fortunately, for the JFET to work as described, both junctions should reverse-bias, which means the BJT never activates. Also, for reasons that will become apparent in the layout subsection, the gate of the N-channel JFET in

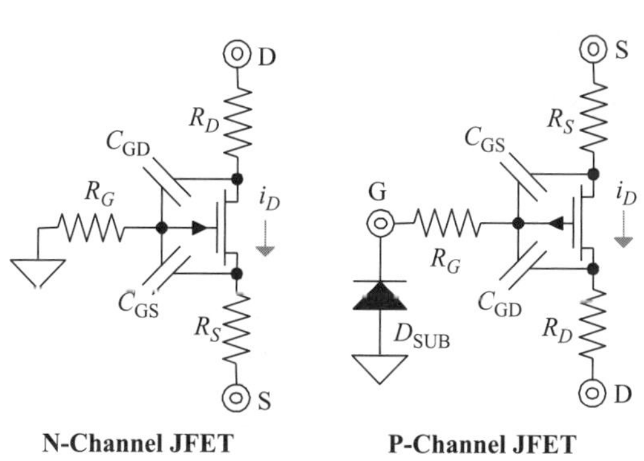

FIGURE 3.43 Electrical models of typical N- and P-channel JFETs.

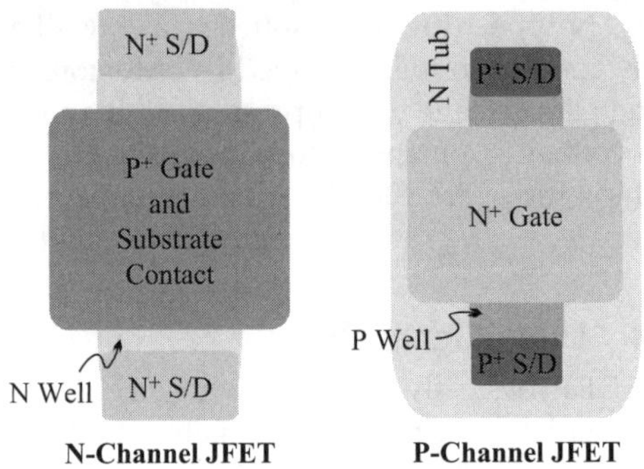

FIGURE 3.44 Top layout views of typical JFETs.

Fig. 3.41 connects to the P substrate, so the gate is automatically at the most negative potential. Lastly, because the P-well resistor in the P-channel JFET of Fig. 3.42 is within an N tub inside a P substrate, the N-tub gate includes a parasitic substrate diode D_{SUB}.

3.6.4 Layout and Matching

Because JFETs are diffused resistors, the N- and P-channel JFETs in Fig. 3.44 include rectangular N- and P-well resistor strips. Similarly, but highly doped diffusion regions connect the extreme ends of the strips to define their corresponding source and drain terminals. Oppositely doped gate regions in the middle then squeeze the resistor strips into narrower channels. These gates must extend beyond the boundaries of the resistor strips to ensure that the depth of the channel narrows across the entire length of the strips. Overlapping the gate regions this way connects the gate of the N-channel transistor in Fig. 3.44 to the substrate and that of the P-type counterpart to the N tub, which is why P-type gates normally short to P substrate and N-type gates to N tubs. As with other microelectronic devices, large resistor areas, close arrays, similarly oriented devices, common-centroid structures, interdigitation, cross-coupling, and dummy devices around the periphery of an array improve matching performance.

3.6.5 Small-Signal Model

Designers typically model the small-signal response of JFETs in saturation because high gain and high resistance are useful attributes in analog applications, as the side note on small signals explains. Since both the JFET and MOSFET are drift devices that, when saturated, respond to gate–source voltage v_{GS} in square-law fashion and drain–source voltage v_{DS} in a linear manner, but not by much, small changes in v_{GS} produce considerable variations in drain current i_D, which in turn cause substantial variations in v_{DS}. As with the MOSFET, variations in v_{DS} also affect i_D, but not to the same degree. So since gate voltage v_G reverse-biases all derivative junctions, the

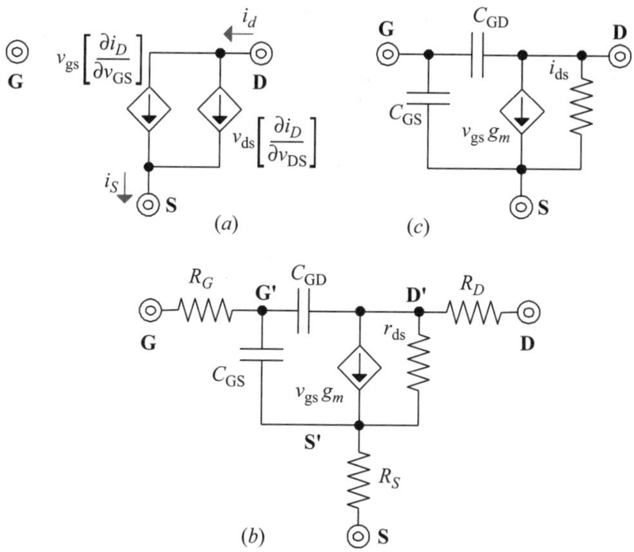

FIGURE 3.45 (*a*) Small-signal currents and (*b*) complete and (*c*) simplified small-signal models of a JFET.

small-signal model in Fig. 3.45*a* excludes gate current and includes components in i_d that small signals v_{gs} and v_{ds} produce.

Since linear approximations predict the response of small signals reasonably well, the effect of v_{gs} on i_d is the linear translation that the first partial derivative of i_D with respect to v_{GS} at bias current I_D predicts:

$$i_d\big|_{v_{ds}=0} \approx v_{gs}\left(\frac{\partial i_D}{\partial v_{GS}}\right) \equiv v_{gs} g_m \approx v_{gs}\sqrt{2 I_D K_{JN}\left(\frac{W}{L}\right)} \tag{3.43}$$

where g_m is the JFET's effective transconductance. Similarly, the first partial derivative of i_D with respect to v_{DS} at I_D forecasts the impact of v_{ds} on i_d:

$$i_d\big|_{v_{gs}=0} \approx v_{ds}\left(\frac{\partial i_D}{\partial v_{DS}}\right) \equiv v_{ds} g_{ds} \approx v_{ds}(\lambda_N I_D) \equiv \frac{v_{ds}}{r_{ds}} \tag{3.44}$$

where g_{ds} is the effective output conductance of the device. Because this linear variation in current results from a change in the voltage across its own two terminals v_{ds}, output resistance $1/g_{ds}$ or r_{ds} in Fig. 3.45*b* models this behavior well. Note that, since both JFET and MOSFET are square-law devices that similarly suffer from channel-length modulation effects, their small-signal parameters g_m and r_{ds} are practically identical.

Since there is no gate current and r_{ds} is normally high between hundreds of kilo-ohms and mega-ohms, the simplified model of Fig. 3.45*c* neglects series terminal series resistances R_S, R_D, and R_G, which are between 1 and 100 Ω. Gate–source and gate–drain capacitances C_{GS} and C_{GD} steer input current away from v_{gs} at higher frequencies, so v_{gs} falls, and so does its impact on i_d and the gain it produces. The frequency f_T beyond which the transistor can no longer amplify small signals well is when short-circuit output current i_d falls below the current that flows into the gate through C_{GS} and C_{GD}. During a short-circuit condition, when v_{ds} is zero, r_{ds} conducts no current and C_{GS} and C_{GD} in the simplified model of Fig. 3.45*c* are in parallel, so i_d reduces to $v_{gs}g_m$, input current i_g drops v_{gs} across C_{GS} and C_{GD}, and f_T, which is when the ratio of i_d to i_g is 1, is the same as in the MOSFET:

$$\frac{i_d}{i_g} = \frac{v_{gs}g_m}{i_g}\bigg|_{v_{ds}=0} = \left[\frac{i_g}{(C_{GS}+C_{GD})s}\right]\left(\frac{g_m}{i_g}\right)\bigg|_{f_T=\frac{g_m}{2\pi(C_{GS}+C_{DG})}} \equiv 1 \tag{3.45}$$

3.6.6 Relative Performance

The JFET is a drift-based device, and its current–voltage relationships therefore mimic those of the MOSFET. The driving difference between the two, from an operational standpoint, is that the JFET normally conducts current with a zero gate–source voltage and the MOSFET normally does not. The JFET also has a small gate current, the reverse-saturation current of its gate-to-channel PN junctions, which is nonexistent in the MOSFET. Because channel dimensions depend on the out-diffusion lengths of the gate and channel instead of the photolithographic resolution of the processing masks that define polysilicon strips, shrinking the JFET to MOSFET levels, following Moore's law, is not possible. As a result, considering the demand for increasingly dense ICs, MOSFETs enjoy the popularity JFETs cannot hope to draw. However, because the channel is well beneath the gate, away from surface irregularities and imperfections, the JFET generates less $1/f$ noise, which means JFETs find a niche market in low-noise applications. Additionally, since JFET resistances can be very high, designers sometimes use them as *pinched resistors*.

3.7 Absolute and Relative Accuracies

In a silicon chip, all microelectronic devices share a common substrate. The fabrication process therefore starts with a silicon wafer. At elevated temperatures, gaseous *donor* and *acceptor* dopant atoms (that ultimately donate electrons and holes with which to accept electrons) first diffuse into the silicon to define P- and N-type regions. The process then *ion-implants* dopants into the wafer to define highly doped regions on or below the surface. Thin and thick oxide regions *oxidize* and grow above the surface after that. *Deposition* of polysilicon and metals follow, where *photoresist* coatings and *etching* acids define the geometries that IC designers prescribe. For reference, the depths of a 0.8-μm CMOS process can be 0.5 μm for the first layer of metal, 1 μm for thick oxide, 14 nm for thin oxide, 200 nm for highly doped diffusions, and 1.2 μm for wells.

Unfortunately, diffusion and deposition rates are not consistent across wafers and lots. The process of defining geometries also includes imperfections. Not surprisingly, diffused resistances can vary by ±20–±50 percent, and polysilicon resistances and oxide capacitances can vary by ±20 percent, as Table 3.1 summarizes. Trimmed thin-film resistors are more accurate, but also more costly because adjusting them with a laser beam requires a post-processing step in the manufacturing process.

Since variations across a single die are considerably lower than across wafers and lots, relative accuracy can be substantially better than absolute accuracy. Polysilicon edges can match so well that resistors and capacitors that use them can match within 0.5 or 1 percent. Because out-diffusion is not as well controlled, diffused resistors do not match as well. Plus, diffused resistors include PN junctions that pinch them, so they suffer from higher voltage coefficients.

After resistors and capacitors, BJTs are next in the matching hierarchy. They suffer from process gradients across the die and mismatched geometries. Normally, base–emitter voltages can match within 3–5 mV. MOSFETs similarly suffer from threshold-voltage and transconductance-parameter variations across the die and mismatched geometries, but to a greater extent. As a result, gate–source voltages normally match within no less than 5–15 mV.

3.8 Summary

The resistance of resistors, including MOSFETs and JFETs, depends on the physical characteristics and dimensions of the current-conducting medium, that is, on resistivity, length, width, and depth. The capacitance of capacitors, on the other hand, depends on the surface area of the facing plates, the distance between them, and the permittivity of the material separating them. A diode, unlike resistors and capacitors, is a diffusion device that conducts current only when forward-biased, or in breakdown, and its relationship to voltage is exponential. BJTs exploit this behavior by exposing diffused carriers to a field that sweeps and collects them, where the base voltage across the forward-biased junction controls the current. A voltage at the gate of MOSFETs and JFETs, on the other hand, adjusts the channel resistance between their terminals, which means gate voltage controls current. MOSFETs and JFETs are therefore voltage-driven devices, unlike their bipolar counterparts, whose base currents drive them.

Components	Range of Values	Absolute Accuracy	Relative Accuracy	Temperature Coefficient	Voltage Coefficient
MOS C	0.3–15 fF/μm²	±20%	±0.5–±1%	+20 ppm/°C	±50 ppm/V
Poly–poly C	0.3–15 fF/μm²	±20%	±0.5–±1%	+25 ppm/°C	±10 ppm/V
Base diffused R	100–200 Ω/sq.	±20%	±2%	+1750 ppm/°C	–
Emitter diffused R	2–10 Ω/sq.	±20%	±2%	+600 ppm/°C	–
S/D diffused R	10–100 Ω/sq.	±20%	±2%	+1500 ppm/°C	200 ppm/V
N-well R	1–10 kΩ/sq.	±40%	±5%	+8000 ppm/°C	10 kppm/V
Pinched base R	2–10 kΩ/sq.	±50%	±10%	+2500 ppm/°C	Poor
Pinched epitaxial R	2–5 kΩ/sq.	±50%	±7%	+3000 ppm/°C	Poor
Implanted R	0.5–2 kΩ/sq.	±20%	±2%	+400 ppm/°C	800 ppm/V
Poly R	30–500 Ω/sq.	±20%	±0.5–±1%	+1500 ppm/°C	100 ppm/V
Thin-film R	0.1–2 kΩ/sq.	±5–±20%	±0.2–±2%	+10–200 ppm/°C	–

Values in the table correspond to typical three-sigma variations.

TABLE 3.1 Typical Absolute and Relative Accuracies of On-Chip Resistors and Capacitors

In the world of analog signal processing, engineers induce small changes in voltages to produce linear translations in current. Accordingly, the small-signal variations that transconductances, input and output resistances, and capacitances model are arguably the most important analog parameters of a device. They describe how a device converts voltages to currents and back to voltages across frequency. In doing so, they also predict the frequency beyond which transistors no longer amplify signals.

Understanding how these devices work is the first step in building robust integrated circuits. However, the effects of extrinsic parasitic factors on a system, when not minimized through careful physical and electrical design, can be fatal. Ultimately, the intrinsic and extrinsic limits of the physical components that comprise a circuit set the operational limits of the system that they serve, especially when exposed to process and temperature extremes. Chapter 4, as a follow-up step in the review process of analog circuit design, discusses how to combine these devices to build the basic building blocks with which IC engineers design analog subsystems like operational amplifiers and voltage regulators.

3.9 Review Questions

1. What parameters can a design engineer usually adjust to change the resistance of a resistor?

2. What parasitic electrical components do resistors usually incorporate?

3. When trying to match resistors, what layout techniques improve matching performance?

4. Which plate determines the area of the overlapping plates in a poly–poly capacitor?

5. What parasitic electrical components do capacitors usually incorporate?

6. What basic mechanism induces charge flow in a diode?

7. Highly doped PN junctions break down at *lower/higher* voltages than their lightly doped counterparts.

8. What does emitter injection efficiency describe in BJTs?

9. What does base-transport factor describe in BJTs?

10. Is the performance of a BJT in light saturation substantially poorer than in forward active?

11. Does a vertical BJT in reverse-active mode perform as well as in forward active?

12. Decreasing the doping concentration of the base *extends/shortens* the effective base width.

13. Decreasing the doping concentration of the base *magnifies/reduces* base-width modulation.

14. What are the transconductance and input and output resistances at room temperature of a BJT whose Early voltage is 50 V and low-frequency current gain is 75 A/A when it conducts 10 μA?

15. In what region does a P-channel MOSFET operate when its gate voltage is high with respect to its source?

16. Drain–source voltage v_{DS} is *positive/negative* in a P-channel MOSFET.

17. What main mechanism drives MOSFET current flow in subthreshold?

18. Raising the bulk terminal voltage of an N-channel MOSFET *raises/lowers* its threshold voltage.

19. What components does the gate–source capacitance of a MOSFET in triode include?

20. What components does the gate–source capacitance of a MOSFET in saturation include?

21. Are the bulk terminals of welled MOSFETs necessarily tied to the substrate?

22. Do dummy polysilicon resistor strips work as well as dummy transistors when trying to improve the matching performance of MOSFETs?

23. Channel resistances of long, narrow P-channel MOSFETs in strong inversion can be on the order of *kilo-ohms/mega-ohms/giga-ohms*.

24. The effective threshold voltage of an N-channel JFET is normally *positive/negative*.

25. What parasitic components do JFETs normally incorporate?

26. How do JFETs outperform MOSFETs?

27. What is the typical tolerance of resistors and capacitors built with polysilicon strips?

28. Between resistors, BJTs, and MOSFETs, which normally match better?

29. Between BJTs and MOSFETs, which normally match better?

CHAPTER 4

Single-Transistor Primitives

Analog and mixed-signal integrated circuits (ICs) use microelectronic devices to sense, condition, filter, and regulate signals of all sorts. For this, transistors buffer, amplify, and at times, attenuate currents and voltages and convert currents into voltages and voltages into currents. This chapter describes how single transistors can process incoming currents and voltages. The underlying objective is to use these primitives to build analog and mixed-signal blocks that implement higher-level functions for amplification, regulation, conversion, and other signal-conditioning purposes.

Reducing circuits to simpler two-port networks is an intrinsic part of this process because insight and innovation are the fruits of simplified and decomposed concepts and circuits. Understanding how shunting and bypass capacitors subtract and add energy and slow and accelerate signals is also important because transistors cannot respond instantly to stimulation. So before exploring the different ways that designers use transistors, this chapter starts with discussions on two-port models, frequency response, and signal flow in transistors.

4.1 Two-Port Models

Two-port models use one or two dependent sources and two resistors to describe the input and output characteristics of any circuit. In the case of the Norton-equivalent network of Fig. 4.1, for example, a dependent current and a resistor in parallel describe how current and voltage into and across the two terminals relate. Similarly, the Thevenin equivalent in the figure relies on a dependent voltage source and a resistor in series to predict current flow and voltage response. Since both networks emulate the behavior of a circuit equally well, engineers typically derive whichever is easier and more convenient to use.

For the network to mimic the operation of a circuit, the resistor should not model what the source emulates, and vice versa. So to

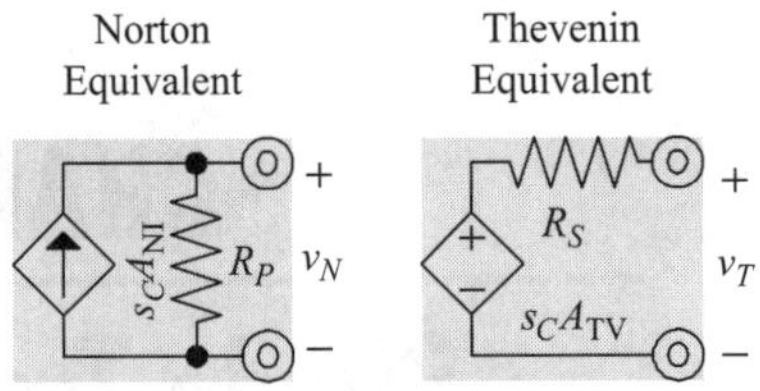

FIGURE 4.1 Norton and Thevenin two-port models.

avoid redundancies of this sort, engineers nullify the effects of one before extracting the other. In the Norton circuit, for instance, current gain A_{NI} refers to the gain when no current flows into parallel resistance R_P, which only happens when Norton-equivalent output voltage v_N is zero. In other words, when deriving gain A_{NI}, v_N must be zero. Similarly, no current can flow through the current source when deriving R_P, so control signal s_C must be zero when extracting R_P from a circuit.

The Thevenin-equivalent circuit is no different in this respect. Thevenin voltage gain A_{TV} is the gain when series resistance R_S drops zero volts, which only happens when no current flows through R_S. Accordingly, when deriving A_{TV}, the terminals of the Thevenin circuit must float in open-circuit fashion. Similarly, A_{TV} should drop zero volts when deriving R_S, so control signal s_C must be zero when extracting R_S. For reference, Table 4.1 summarizes all these test conditions.

Example 4.1 State the test conditions to apply to a circuit when deriving the Norton-equivalent input and Thevenin-equivalent output illustrated in Fig. 4.2.

Solution

1. When deriving R_{IN}, $i_O A_G$ must be zero, so v_O must float in open-circuit fashion: $i_O \equiv 0$.

2. When deriving A_G, R_{IN}'s current must be zero, so v_{IN} must short to ground: $v_{IN} \equiv 0$.

3. When deriving A_V, R_O should drop zero volts, so v_O must float in open-circuit fashion: $i_O \equiv 0$.

4. When deriving R_O, $v_{IN} A_V$ should drop zero volts, so v_{IN} must short to ground: $v_{IN} \equiv 0$.

	Norton Equivalent		Thevenin Equivalent	
Parameter	A_{NI}	R_P	A_{TV}	R_S
Test Condition	Short circuit: $v_N \equiv 0$	$s_C \equiv 0$	Open circuit: $i_R \equiv 0$	$s_C \equiv 0$

TABLE 4.1 Test Conditions for Extracting Two-Port Parameters

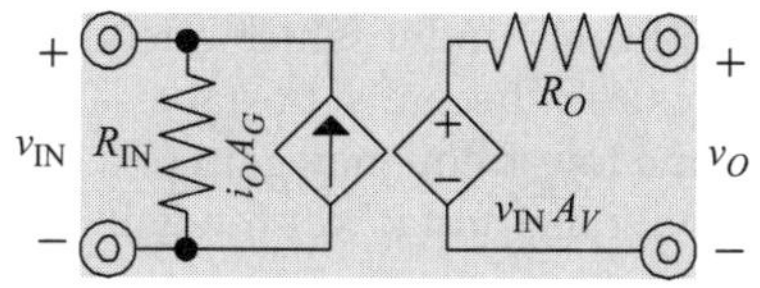

FIGURE 4.2 Sample two-port model of a circuit.

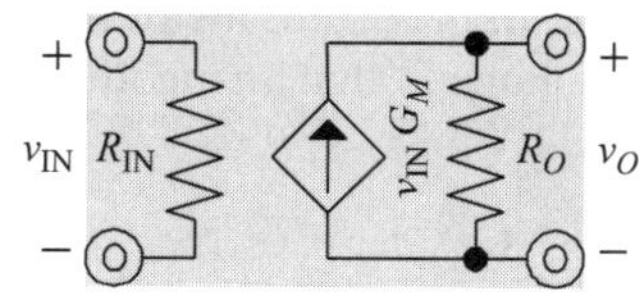

FIGURE 4.3 Popular two-port model.

Because paralleling electrical components is often easier to manage than cascading them, many engineers opt for Norton-equivalent circuits whenever possible. Also, since output signals normally do not propagate backwards to the inputs, inputs often exclude the dependent source that two-port primitives include. In other words, as Fig. 4.3 shows, most designers model the input of a circuit with an input resistance R_{IN} and the output with a Norton circuit that the input voltage v_{IN} controls. Because no other elements model the input, deriving R_{IN} requires no test conditions. However, to derive short-circuit transconductance gain G_M, the current through output resistance R_O must be zero, so v_O must short to ground. Similarly, v_{IN} must short to ground to nullify the effects of $v_{IN}G_M$ when extracting short-circuit output resistance R_O. Note that this basic circuit is the one that models the small-signal response of bipolar-junction transistors (BJTs), metal–oxide–semiconductor field-effect transistors (MOSFETs), and junction FETs (JFETs) in Chap. 3.

4.2 Frequency Response

Unfortunately, real-life devices cannot respond instantly to propagating signals. For one, current into a capacitor raises its voltage slowly, so capacitors slow incoming signals. Similarly, a voltage across an inductor increases its current slowly, so inductors, too, delay signals. In analog ICs, however, the effects of capacitances and resistances normally overwhelm those of inductances, except in radio-frequency (RF) applications where operating frequencies are high in the gigahertz range.

One way of describing how a system behaves across time is in the frequency domain. Here, low and high frequency corresponds to slow- and fast-changing signals, and frequency response describes how a circuit responds to both. Illustrating signal dynamics this way is useful in not only describing the performance of a circuit but also determining the conditions under which a feedback loop, on which many analog systems rely, becomes unstable. In essence, frequency response describes the low-frequency gain of a circuit and identifies the *zeros* and *poles* that raise and lower the gain across frequency.

4.2.1 Poles

To begin the discussion on poles, first consider that a parallel capacitor like C_{EQ} in Fig. 4.4 shunts current away from a resistor R_{EQ}. This means C_{EQ} lowers the voltage v_O that an incoming current i_{IN} produces across R_{EQ}:

$$v_O = i_{IN}\left(R_{EQ} \| \frac{1}{sC_{EQ}}\right) = \frac{i_{IN}R_{EQ}}{1+sR_{EQ}C_{EQ}} = \frac{i_{IN}R_{EQ}}{1+\dfrac{s}{2\pi p_C}} \tag{4.1}$$

where $1/sC_{EQ}$ is C_{EQ}'s impedance and s corresponds to frequency in the s domain. This reduction in voltage, however, only happens when C_{EQ}'s impedance nears or falls below R_{EQ}, because C_{EQ}'s current is negligibly small otherwise. In other words, as frequency rises, the combined parallel impedance falls with $1/sC_{EQ}$ past the frequency where $1/sC_{EQ}$ equals R_{EQ}:

$$\left.\frac{1}{sC_{EQ}}\right|_{f=\frac{1}{2\pi R_{EQ}C_{EQ}}\equiv p_C} = R_{EQ} \tag{4.2}$$

so v_O also falls past this point. In fact, a tenfold rise in frequency corresponds to a tenfold drop in $1/sC_{EQ}$, and as a result, in v_O, but only past pole p_C. This is why the effect of C_{EQ} is to lower the gain past p_C by 20 dB per decade, as Fig. 4.4 also shows. In general, a pole appears when a capacitor shunts the total resistance across its terminals, which is an insightful way of saying at the frequency where their impedances match.

Because C_{EQ}'s current is negligible well below p_C, C_{EQ} does not delay or affect the circuit at low frequency. Near p_C, however, C_{EQ}'s

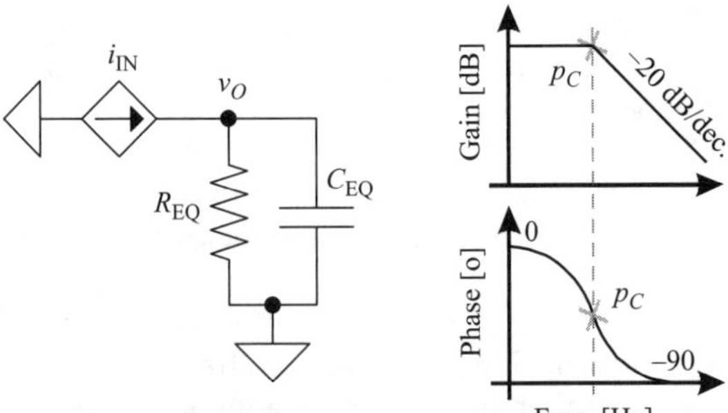

FIGURE 4.4 A shunting capacitor and its corresponding frequency response.

current is appreciable, and so is its impact on v_O, which means C_{EQ} slows v_O. A decade past p_C, when $1/s_{CEQ}$ is 10 times lower than R_{EQ}, R_{EQ} no longer impacts delay, so C_{EQ} alone sets the delay. Another name for this delay is *phase shift* between sinusoids because two out-of-phase sine waves at the same frequency is the same as saying one sine wave is delayed with respect to the other. In the case of a pole, phase shift is nearly zero a decade below p_C, –45° at p_C, and nearly –90° a decade past, and to describe that more precisely, phase in the *Bode plot* of Fig. 4.4 is

$$\text{Phase}_{\text{Pole}} = -\tan^{-1}\left(\frac{f}{p_C}\right) \tag{4.3}$$

where f refers to frequency.

4.2.2 Zeros

Feed-Forward Zeros

In addition to shunting energy, bypass capacitors also feed energy forward. Consider feed-forward capacitors C_{FF} in Fig. 4.5, for example, which bypass noninverting and inverting circuits $+A_V$ and $-A_V$. For starters, C_{FF}, like input capacitor C_{IN}, steers energy away from input resistance R_{IN}, so C_{FF} and C_{IN} together introduce a pole at the input v_{IN}. Unlike output capacitor C_O, however, C_{FF} delivers the energy it steals from R_{IN} to output resistance R_O, so output voltage v_O rises, which is the opposite effect from that of a pole. This only happens, however, when C_{FF}'s feed-forward current i_{FF} nears and surpasses G_M's current i_{GM}. And to be more precise, the short-circuit

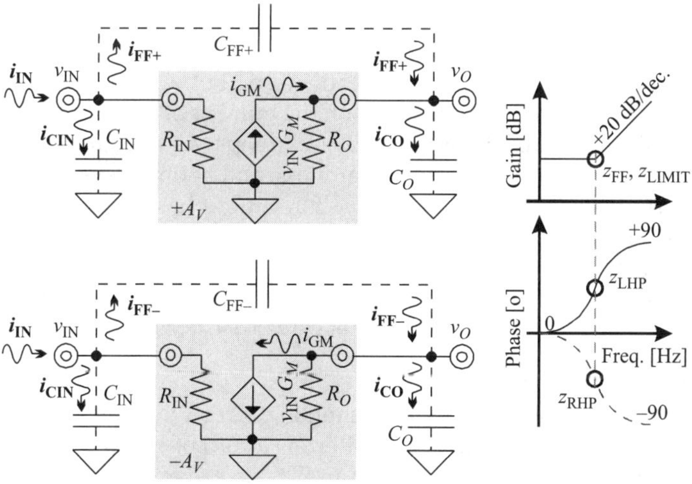

FIGURE 4.5 Bypass feed-forward capacitors and their corresponding responses across frequency.

transconductance gain incorporates a zero z_{FF} that appears at the frequency where i_{FF} equals i_{GM}, when v_O is zero:

$$i_{FF} = \frac{v_{IN} - v_O}{Z_C} = v_{IN} s C_{FF} \Big|_{z_{FF} = \frac{G_M}{2\pi C_{FF}}} = i_{GM} = v_{IN} G_M \quad (4.4)$$

where Z_C is C_{FF}'s impedance $1/sC_{FF}$. This means that, while C_{FF} has no impact on the overall gain of the circuit below z_{FF}, as Fig. 4.5 illustrates, C_{FF} raises the gain past z_{FF}. So as before, because C_{FF}'s impedance falls linearly with frequency, i_{FF} rises 10 times with a tenfold increase in frequency, which is why gain rises 20 dB per decade past a zero.

In the case of an in-phase feed-forward capacitor like C_{FF+}, the zero cancels the effects of a pole. In other words, because feed-forward current i_{FF+} and output current i_O are in phase, C_{FF} recovers the energy and phase that C_O loses. This means the phase shift that an in-phase zero z_{LHP} produces is equal and opposite to that of a pole: close to 0° a decade below z_{LHP}, +45° at z_{LHP}, and nearly +90° a decade past z_{LHP}:

$$\text{Phase}_{\text{LHP Zero}} = \tan^{-1}\left(\frac{f}{z_{\text{LHP}}}\right) \quad (4.5)$$

where LHP in control theory stands for *left-half plane*. Interestingly, an out-of-phase feed-forward capacitor like C_{FF-} sources current when the circuit sinks current, and vice versa. So in addition to adding energy to v_O, C_{FF-} inverts the polarity of v_O, which is what control theoreticians call a *right-half-plane* (RHP) *zero*. In other words, while out-of-phase zeros raise gain, they also reduce phase:

$$\text{Phase}_{\text{RHP Zero}} = -\tan^{-1}\left(\frac{f}{z_{\text{RHP}}}\right) \quad (4.6)$$

To understand the implications of in- and out-of-phase zeros, first consider that designers often rely on the number of inversions across a circuit to stabilize inverting feedback loops. As a result, in-phase feed-forward capacitors can be harmless, and at times, even helpful. Unfortunately, the same is not true for out-of-phase capacitors. The problem with these latter capacitors is that propagating signals undergo one more inversion past z_{RHP}. All else being the same, this means an inverting feedback loop becomes noninverting past z_{RHP}, which is highly undesirable in most control systems.

Current-Limiting Zeros

Ultimately, a pole results because shunting current through a capacitor rises with frequency. This is why inserting a current-limiting resistor in series with the capacitor, as R_{LIMIT} in Fig. 4.6 exemplifies, eventually offsets and cancels the effects of the pole that the capacitor established in the first place. In this example, C_{SHUNT} and R_{LIMIT} shunt energy away

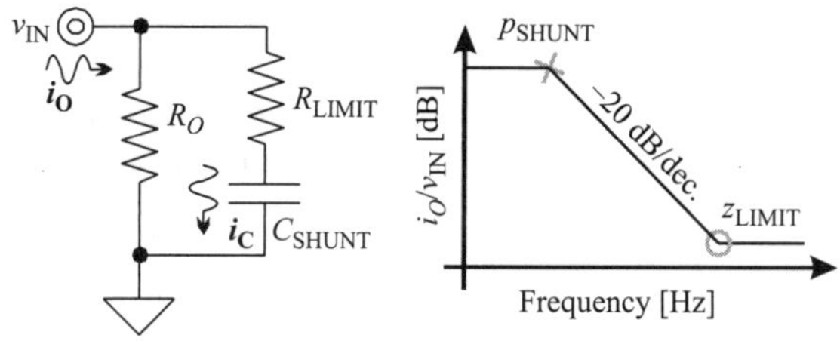

FIGURE 4.6 A current-limiting resistor and its corresponding frequency response.

from R_O in the form of capacitor current i_C to produce the effects of a pole p_{SHUNT} when $1/sC_{SHUNT}$ nears or falls below the resistance across its terminals, which in this case is $R_O + R_{LIMIT}$. Past p_{SHUNT}, the parallel combination reduces to C_{SHUNT} in series with R_{LIMIT}. As a result, i_C climbs with frequency as $1/sC_{SHUNT}$ drops until $1/sC_{SHUNT}$ nears or falls below R_{LIMIT}:

$$\left.\frac{1}{sC_{SHUNT}}\right|_{f=\frac{1}{2\pi R_{LIMIT}C_{SHUNT}}\equiv z_{LIMIT}} = R_{LIMIT} \tag{4.7}$$

beyond which frequency C_{SHUNT} shorts and i_C stops rising at v_{IN}/R_{LIMIT}. Canceling the effects of a pole this way constitutes a zero, and in this case, a current-limiting zero z_{LIMIT}. Because no signals invert in a series configuration of this sort, this zero is in phase and in the left-half plane.

4.2.3 Miller Split

Besides introducing a right-half-plane zero, an out-of-phase feed-forward capacitor across an inverting voltage amplifier, as C_M illustrates in Fig. 4.7, incorporates other idiosyncrasies. As with any capacitor, a current into C_M raises its voltage v_C. However, because v_O falls more quickly than v_{IN} climbs across an inverting amplifier, v_{IN} hardly increases when v_C rises. This is equivalent to saying that C_M's current changes v_{IN} as though v_{IN} had higher capacitance to ground, which is what engineers call the *Miller effect*.

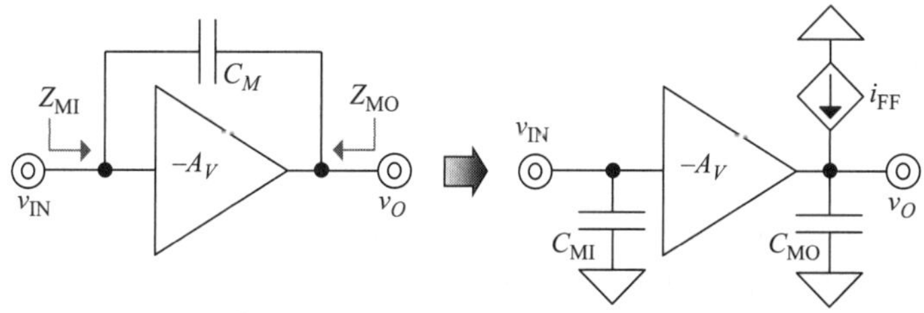

FIGURE 4.7 Miller-multiplying out-of-phase feed-forward capacitor.

To model this effect, consider C_M's effective input impedance Z_{MI}, which is the ratio of v_{IN} to C_M's current i_C. Because the voltage across C_M is $v_{IN} - v_O$, i_C is this voltage into C_M's impedance $1/sC_M$ and v_O is v_{IN}'s amplified counterpart $-v_{IN}A_V$, so Z_{MI} simplifies to

$$Z_{MI} \equiv \frac{v_{IN}}{i_C} = \frac{v_{IN}}{(v_{IN} - v_O)sC_M} = \frac{v_{IN}}{(v_{IN} + v_{IN}A_V)sC_M} = \frac{1}{s(1+A_v)C_M} \equiv \frac{1}{sC_{MI}} \quad (4.8)$$

This $1/s$ relationship means Z_{MI} is the impedance of a capacitor C_{MI} whose equivalent capacitance is $1 + A_V$ times higher than C_M:

$$C_{MI} = (1+A_V)C_M \approx A_V C_M \quad (4.9)$$

which with considerable gain approximates to $A_V C_M$. In other words, A_V Miller-multiplies the effects across C_M on v_{IN} to the extent that a higher capacitance to ground can model the dynamics of C_M.

Interestingly, because a variation in v_O corresponds to a small fluctuation in v_{IN}, v_O does not experience the same multiplication. Here, C_M's output impedance Z_{MO} is v_O/i_C, C_M's voltage is $v_O - v_{IN}$, i_C is that voltage into $1/sC_M$, and v_{IN} is $-v_O/A_V$, so Z_{MO} simplifies to

$$Z_{MO} \equiv \frac{v_O}{i_C} = \frac{v_O}{(v_O - v_{IN})sC_M} = \frac{v_O}{\left(v_O + \dfrac{v_O}{A_V}\right)sC_M} = \frac{1}{s\left(1+\dfrac{1}{A_V}\right)C_M} \equiv \frac{1}{sC_{MO}} \quad (4.10)$$

This $1/s$ relationship implies Z_{MO} is the impedance of a capacitor C_{MO} whose capacitance is

$$C_{MO} = \left(1+\frac{1}{A_V}\right)C_M \approx C_M \quad (4.11)$$

which with considerable gain reduces to C_M. In other words, A_V hardly alters the effects across C_M on v_O, which means v_O's equivalent capacitance to ground is nearly equivalent to C_M.

The purpose of splitting C_M into its equivalent input and output capacitances C_{MI} and C_{MO} is to simplify the analysis and more easily ascertain the impact of C_M on a propagating signal. This way, the designer can readily determine to what extent C_{MI} and C_{MO} help establish poles at v_{IN} and v_O. Notice, however, C_M still feeds an out-of-phase current i_{FF} to v_O, which is why the model of Fig. 4.7 includes i_{FF}. This current, of course, can be high enough to invert v_O and introduce a right-half-plane zero.

4.2.4 Capacitors-Shunting-Resistors Method

Understanding the impact of several capacitors on a circuit can be challenging. Deriving equations help, but not always because drawing physical insight from third- and higher-order polynomials can be difficult. The purpose of this subsection is therefore to develop an insightful and intuitive method that engineers can use to determine the effects of individual capacitors on the frequency response of a circuit. Equations, of course, can later help corroborate and validate conclusions.

To begin, recall that a capacitor is an open circuit at low frequency, its impedance drops as frequency rises, and a pole occurs when it shunts the resistance across its terminals at $1/2\pi R_{EQ}C_{EQ}$. This means, the low-frequency gain A_0 of a circuit is independent of capacitors, so designers can exclude all capacitors in the circuit when determining A_0. This also says that a capacitor is, for all practical purposes, absent at frequencies well below the pole it establishes, and the resistance it shunts is absent at frequencies well above that pole. In addition, the capacitor with the highest time constant $R_{EQ}C_{EQ}$ establishes the lowest frequency pole.

So, finding the lowest frequency pole p_1 in a system amounts to finding the capacitor with the highest time constant. Since this first capacitor C_1 shunts its parallel resistance at lower frequency, the capacitor with the highest time constant when C_1's parallel resistance disappears is therefore responsible for the next pole p_2. Similarly, since this last capacitor C_2 shunts after that, the capacitor with the highest time constant when the parallel resistances of both C_1 and C_2 disappear establishes the next pole p_3. Modifying and simplifying the circuit to accommodate for each successive resistance that disappears and deriving the next highest time constant allows the engineer to extract all other poles, until there are no capacitors left to consider.

Note, however, that Norton and Thevenin test conditions can short or disconnect resistances that capacitors would otherwise shunt. In other words, two-port resistances do not always model all the resistances that a capacitor actually shunts. Also notice that, of possible capacitors, nonparasitic capacitors added intentionally to the circuit, base–emitter and gate–source capacitors C_π and C_{GS}, Miller-multiplied capacitors C_{MI}, and loading capacitors C_{LOAD} are usually the highest in a circuit, so they typically shunt first, but only when their parallel resistances are moderate to high, that is, on par with r_o and r_{ds} or higher, and maybe even with r_π. Normally, these capacitors do not shunt low source resistances R_S and g_m resistances $1/g_m$ until frequencies are fairly high.

Ultimately, this method assumes poles are not sufficiently close to one another for their corresponding impedances to affect each other. This assumption is not always a problem because analog designers typically seek to establish one dominant low-frequency pole. So finding that no single time constant dominates simply tells the engineer to adjust the circuit until one does dominate. Finding that the next two time constants are almost the same indicates that those two capacitors establish two poles near the frequency their averaged time constants predict:

$$f_P \approx \frac{1}{2\pi\,[\mathrm{Avg}(R_{EQ}C_{EQ})]} \tag{4.12}$$

where N is the number of capacitors whose time constants are similar and f_P is the approximate frequency where N poles appear. Note that when summing all time constants, the highest time constant dominates, and f_P becomes p_1, and becomes p_2 after C_1 shunts its resistance if C_2 dominates after C_1.

4.3 Signal Flow

4.3.1 Inputs and Outputs

Not all terminals in a transistor are practicable inputs, nor are they all useful outputs. Consider that, like water through a faucet, the fundamental energy-carrying signal in a transistor is current. Because base–emitter and gate–source voltages in BJTs, MOSFETs, and JFETs control collector and drain currents, bases, gates, emitters, and sources are all good input terminals. In contrast, collectors and drains are not good inputs because their effects on current are minimal. In the same vein, collectors, drains, emitters, and sources are viable outputs because they all carry the currents with which transistors convey energy and information. Bases and gates, on the other hand, experience little to no change in response to voltage and current variations in the other terminals, so bases and gates are poor outputs.

4.3.2 Polarity

As an input, a rise in base or gate voltage v_B or v_G raises collector or drain current i_C or i_D, as Fig. 4.8 illustrates, so the transistor pulls current from the collector or drain and feeds it into the emitter or source. As a result, emitter and source voltages v_E and v_S rise, whereas collector and drain voltages v_C and v_D fall. In other words, emitter and source voltages change in phase with base and gate voltages, and collector and drain voltages respond in the opposite way: out of phase by 180°. Or in more figurative terms, base-to-collector and gate-to-drain signals invert and base-to-emitter and gate-to-source signals do not.

As a complementary input, a fall in emitter or source voltage v_E or v_S raises collector or drain current i_C or i_D. This means the transistor pulls more current from the collector or drain, and collector or drain voltage v_C or v_D falls as a result. Collector and drain voltages are therefore in phase with emitter and source voltages. In other words, only base-to-collector and gate-to-drain signals invert across a transistor.

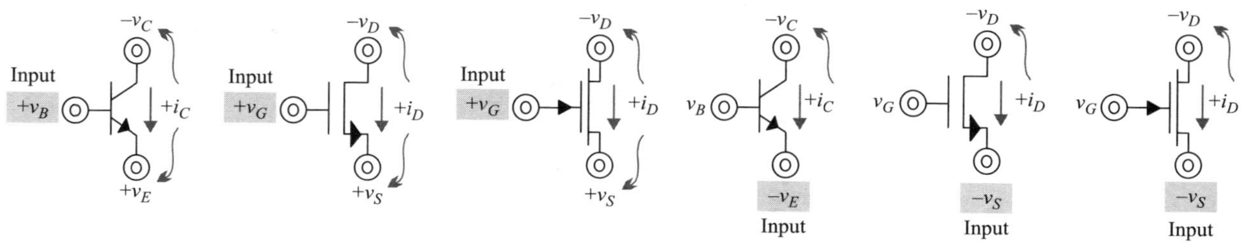

FIGURE 4.8 Signal-flow paths in transistors.

4.3.3 Single-Transistor Primitives

Extending these concepts into circuit formation reveals all the possible ways in which an engineer can use a transistor. Bases and gates, for example, can drive and invert collectors and drains. Textbooks refer to this first single-transistor configuration as the *common-emitter* or *common-source* circuit because the emitter or source in this case is neither an input nor an output. Similarly, emitters and sources can drive collectors and drains in phase as *common-base* and *common-gate* stages. Bases and gates can also drive emitters and sources in phase as *common-collector* and *common-drain* transistors. Other than adding series resistors to the common terminals, no other configuration is possible. This means all circuits ultimately decompose into a combination of these three basic single-transistor primitives, which is why understanding each configuration is the key to deciphering and designing analog circuits with more transistors.

4.4 Common-Emitter/Source Transconductor

Because collector and drain currents are strong functions of base and gate voltages and weak functions of collector and drain voltages, bases and gates in common-emitter (CE) and common-source (CS) transistors Q_{CE} and M_{CS} in Fig. 4.9 are input terminals and collectors and drains are outputs. The load at the output is typically one or a combination of resistors, capacitors, other CE and CS transistors, and common-base and common-gate stages. Irrespective of its specific nature, the load ultimately presents an equivalent impedance and a steady-state current that a Norton-equivalent circuit can model.

4.4.1 Large-Signal Operation

When base or gate input voltage v_{IN} is low, the transistor is off, so biasing load current I_{Bias} in Fig. 4.9 raises output v_O high toward the

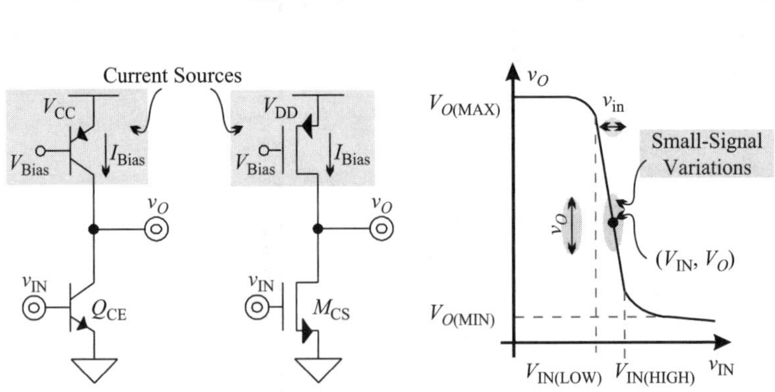

Figure 4.9 CE and CS transconductors and their corresponding large-signal response.

positive power-supply voltage V_{CC} or V_{DD}. As v_{IN} climbs, transistor current rises to increase the voltage dropped across the load, but not until the current nears and surpasses I_{Bias}. In other words, the input voltage $V_{IN(LOW)}$ above which the output starts to fall corresponds to the steady-state input voltage necessary to sustain I_{Bias}. After that, excess current flows through the load to pull v_O toward ground. To what extent v_O rises and falls ultimately depends on the load. In the case of the CE and CS current-source loads shown, for example, v_O climbs to the positive supply when Q_{CE} and M_{CS} are off and drops within $V_{CE(MIN)}$ and $V_{DS(SAT)}$ of ground when driven to sink more current than the current sources can supply. Note that v_O is high when v_{IN} is low and vice versa, which is equivalent to saying base-to-collector and gate-to-drain signals invert.

In the case of BJTs, raising v_{IN} eventually forward-biases the base–collector diode to such an extent that it steals considerable current from the base–emitter junction, thereby clamping v_O to $v_{BE} - v_{BC}$, which corresponds to 0.7 – 0.3 V, or 0.4 V, and a $V_{CE(MIN)}$ of 0.2–0.3 V. Field-effect transistors (FETs), on the other hand, transition from saturation into triode when v_O is within $V_{DS(SAT)}$ of ground. So as v_{IN} continues to rise, M_{CS}'s channel resistance R_{TRIODE} drops, and as a result, so does $I_{Bias}R_{TRIODE}$ and v_O, but only incrementally with respect to the power-supply voltages because R_{TRIODE} is already substantially low at this point.

4.4.2 Small-Signal Model

In most analog applications, designers bias the amplifying transistor at a point in its large-signal range that produces the highest gain. In the case of CE and CS transistors, the highest variation in v_O in response to a small fluctuation in v_{IN} happens when v_O transitions from $V_{O(MAX)}$ to $V_{O(MIN)}$ in Fig. 4.9. Analog designers therefore bias Q_{CE} and M_{CS} somewhere in the transition, when Q_{CE} is in forward-active or slightly saturated and M_{CS} is in saturation. Here, small input signals v_{in} about steady-state input bias voltage V_{IN} produce considerable output variations v_o about output bias voltage V_O.

To determine the gain in this region, consider the small-signal low-frequency models of the transistors and how they map into the popular two-port circuit described in Sec. 4.1 and shown again in Fig. 4.10, where emitter and source voltages v_e and v_s are zero. Here, input resistance R_{IN} is r_π for the BJT, which can be on the order of kilo-ohms, and infinity for the MOSFET, transconductance gain G_M is g_m, and output resistance R_O is r_o for the BJT and r_{ds} for the MOSFET. When combined, $v_{in}G_M$ pulls current from R_O and loading resistance R_{LOAD} to establish an inverting voltage v_o whose ratio with v_{in} is the low-frequency small-signal voltage gain of the circuit:

$$A_{V0} \equiv \frac{v_o}{v_{in}} = \frac{-v_{in}G_M(R_O \parallel R_{LOAD})}{v_{in}} = -g_m(r_o \parallel R_{LOAD}) \qquad (4.13)$$

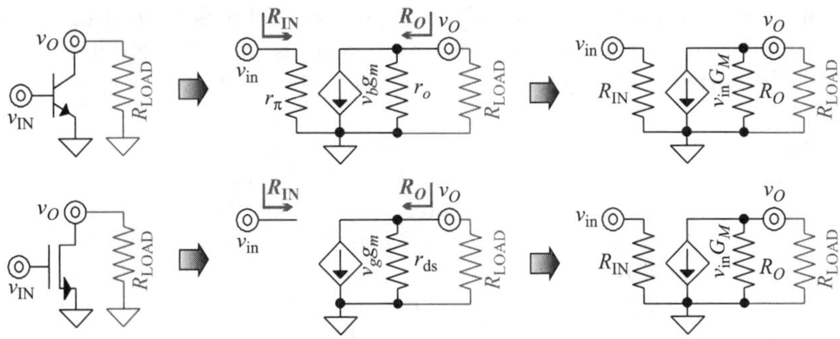

FIGURE 4.10 Small-signal and two-port low-frequency model translations of CE and CS transistors.

where r_o is r_{ds} in the case of FETs. Since R_{LOAD} can only reduce the effective output resistance of the circuit, A_{V0} cannot exceed its unloaded counterpart $-g_m r_o$, which engineers call *intrinsic* or *maximum voltage gain* A_{V0}'. Since g_m can be on the order of hundreds of microamps per volt and r_o and r_{ds} mega-ohms, A_{V0}' can be 100 V/V, or equivalently, 40 dB. This high voltage gain is the reason why engineers often think of CE and CS transistors as voltage amplifiers. More fundamentally, however, CE and CS stages convert voltages to currents that resistors transform into voltages, so CE and CS BJTs and FETs are more transconductors than voltage amplifiers.

SIGNAL PROPAGATION AND GAIN ANALYSIS *The first step in determining the gain across a circuit in an insightful manner is tracing the path that a propagating signal traverses from input to output. Decomposing the circuit in the path into the single-transistor primitives that comprise it is next. Ultimately, each of these transistors translates one signal, be it a voltage or a current, into another. Resistors and capacitors similarly convert currents into voltages and voltages into currents. So tracking and understanding how these voltage–voltage, current–current, voltage–current, and current–voltage conversions occur is the way many engineers derive the gain across multitransistor circuits. For instance, small variations in an unloaded common-emitter base voltage* Δv_B *or* v_b *produce fluctuations in collector current* Δi_C *or* i_c *that* r_o *then converts into changes in output voltage* Δv_O *or* v_o*. Here,* g_m *translates* v_b *into* i_c *and* r_o *converts* i_c *to* v_o *for a combined gain of* $-g_m r_o$*, where the inversion results because* i_c *pulls current from* r_o*. For expanded examples of this approach, see Examples 4.2, 4.3, 4.4, 4.5, and 4.6.*

Example 4.2 Determine the low-frequency small-signal voltage gain v_{o2}/v_{in} across the cascaded N-channel CS and PNP CE stages in Fig. 4.11.

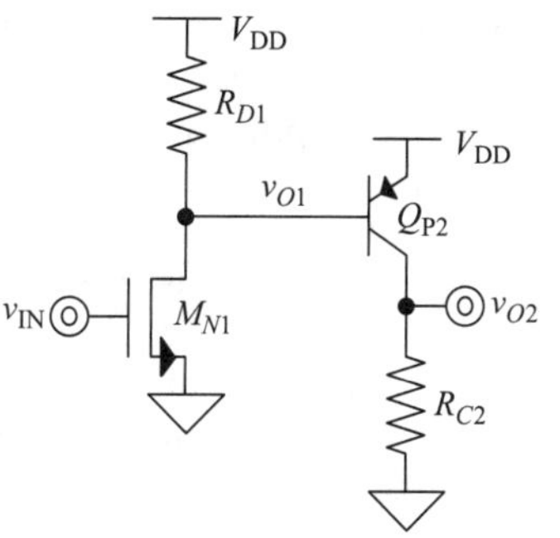

FIGURE 4.11 Cascaded CS and CE example.

Solution

Small input signals Δv_{IN} produce fluctuations in drain current Δi_{D1} that R_{D1} together with M_{N1}'s r_{ds1} then convert into output voltage changes Δv_{O1}. Δv_{O1} then causes variations in collector current Δi_{C2} that R_{C2} with Q_{P2}'s r_{o2} translate into output voltage fluctuations Δv_{O2}. The input signal therefore undergoes the following small-signal voltage–current conversions:

$$\Delta v_{IN} \rightarrow \Delta i_{D1} \rightarrow \Delta v_{O1} \rightarrow \Delta i_{C1} \rightarrow \Delta v_{O2} = v_{in} \rightarrow i_{d1} \rightarrow v_{o1} \rightarrow i_{c1} \rightarrow v_{o2}$$

Since M_{N1}'s $G_{M(CS)}$ or g_{m1} translates v_{in} to i_{d1}; M_{N1}'s $R_{O(CS)}$ or r_{ds1}, Q_{P2}'s $R_{IN\,(CE)}$ or $r_{\pi 2}$, and R_{D1} convert i_{d1} to v_{o1}; Q_{P2}'s $G_{M(CE)}$ or g_{m2} translates v_{o1} to i_{c2}; Q_{P2}'s R_O or r_{o2} and R_{C2} convert i_{c2} to v_{o2}; and g_{m1} and g_{m2} pull currents from their respective loads to invert their outputs, the small-signal voltage gain across the circuit is

$$\begin{aligned} A_{VO} &\equiv \frac{v_{o2}}{v_{in}} = \left(\frac{i_{d1}}{v_{in}}\right)\left(\frac{v_{o1}}{i_{d1}}\right)\left(\frac{i_{c2}}{v_{o1}}\right)\left(\frac{v_{o2}}{i_{c2}}\right) \\ &= (-G_{M(CS)})(R_{O(CS)} \,\|\, R_{D1} \,\|\, R_{IN(CE)})(-G_{M(CE)})(R_{O(CE)} \,\|\, R_{C2}) \\ &= (-g_{m1})(r_{ds1} \,\|\, R_{D1} \,\|\, r_{\pi 2})(-g_{m2})(r_{o2} \,\|\, R_{C2}) \end{aligned}$$

4.4.3 Frequency Response

Parasitic capacitors in transistors and loads shape the frequency response of a circuit. In the case of CE and CS transistors, the only difference between BJTs and FETs is that BJTs incorporate base–emitter resistance r_π, as Fig. 4.12 illustrates. Bulk-effect current $v_{bs} g_{mb}$ is absent in FETs because both bulk and source terminals connect to

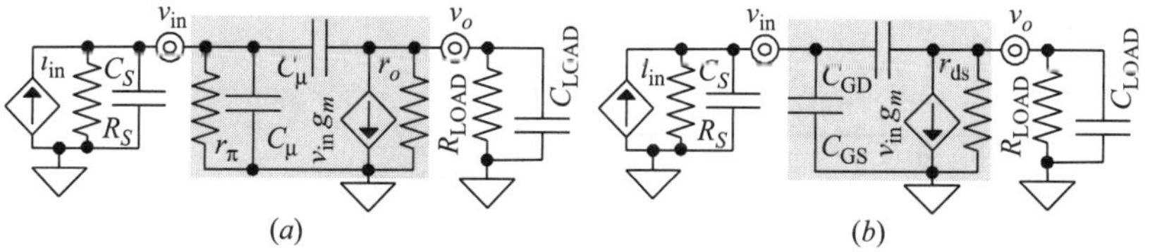

FIGURE 4.12 Wide-spectrum small-signal models for (a) CE and (b) CS transistors.

low-impedance points, which means v_{bs} is zero. Other than that, the effects of C_π, C_μ, and r_o in BJTs are equivalent to those of C_{GS}, C_{GD}, and r_{ds} in FETs. Plus, since neither input is impervious to the influence of capacitors, both equivalent circuits must account for source resistance and capacitance R_S and C_S.

Interestingly, since CE and CS transistors invert signals, C_μ and C_{GD} are out-of-phase feed-forward capacitors. As such, they feed out-of-phase currents to the output that introduce a right-half-plane zero z_{RHP} when their currents equal and exceed the transistors' g_m currents at and past $g_m/2\pi C_\mu$:

$$i_{C\mu} = (v_{be} - v_{ce})sC_\mu = (v_{be} - 0)sC_\mu\Big|_{z_{RHP}=\frac{g_m}{2\pi C_\mu}} \equiv i_{gm} = v_{be}g_m \tag{4.14}$$

where the voltage across r_o and R_{LOAD} is zero when $i_{C\mu}$ equals $v_{in}g_m$ current i_{gm}. C_μ also steers current away from v_{in} and v_o, so C_μ helps establish poles at those nodes. Miller-splitting C_μ into C_{MI} at v_{in} and C_{MO} at v_o is more revealing of this fact:

$$C_{MI} = (1 + A_V)C_\mu = \left[1 + G_M(R_O \parallel R_{LOAD})\right]C_\mu \tag{4.15}$$

and

$$C_{MO} = \left(1 + \frac{1}{A_V}\right)C_\mu = \left[1 + \frac{1}{G_M(R_O \parallel R_{LOAD})}\right]C_\mu \tag{4.16}$$

But since the low-frequency voltage gain across the CE and CS transistors is high, C_{MI} approximates to $G_M(R_O \parallel R_{LOAD})C_\mu$ or $g_m(r_o \parallel R_{LOAD})$ and C_{MO} to C_μ.

Once split, the number of equivalent capacitors in the circuit reduces to C_{MI} in parallel with C_S and C_π and C_{MO} with C_{LOAD}. So if $(R_S \parallel R_{IN})(C_S + C_\pi + C_{MI})$ is higher than $(R_O \parallel R_{LOAD})(C_{MO} + C_{LOAD})$, C_S, C_π, and C_{MI} shunt first to establish Miller input pole p_{IN}:

$$\frac{1}{s(C_S + C_\pi + C_{MI})}\Bigg|_{p_{IN}=\frac{1}{2\pi(C_S+C_\pi+C_{MI})(R_S \parallel R_{IN})}} \equiv R_S \parallel R_{IN} \tag{4.17}$$

After C_S, C_π, and C_μ shunt, their parallel resistances R_S and R_{IN} and related time constants disappear. C_{LOAD} in parallel with the series combination of C_μ and $(C_S + C_\pi)$, which reduces to C_{LOAD} and C_μ because C_π is considerably higher than C_μ, shunt G_M's equivalent resistance R_{GM}, R_O, and R_{LOAD} at p_O:

$$\frac{1}{s\{C_{LOAD} + [C_\mu \oplus (C_S + C_\pi)]\}} \approx \frac{1}{s(C_{LOAD} + C_\mu)}\Bigg|_{p_O \approx \frac{\left(\frac{G_M C_\mu}{C_S + C_\pi + C_\mu}\right)}{2\pi(C_{LOAD} + C_\mu)}}$$

$$\equiv R_{GM} \parallel R_O \parallel R_{LOAD} \approx \frac{C_S + C_\pi + C_\mu}{G_M C_\mu} \tag{4.18}$$

Here, however, because C_μ and $(C_S + C_\pi)$ voltage-divide v_o to v_{in}, the G_M current is a linear function of the voltage across it, like current through a resistor is, so R_{GM} reduces to

$$R_{GM} \equiv \frac{v_o}{i_{GM}} = \frac{v_o}{v_o\left(\dfrac{C_\mu}{C_S + C_\pi + C_\mu}\right)G_M} = \frac{C_S + C_\pi + C_\mu}{G_M C_\mu} \tag{4.19}$$

Since G_M is g_m in CE and CS transistors and $1/g_m$ is typically tens of kilo-ohms, $R_{GM} \,||\, R_O \,||\, R_{LOAD}$ often reduces to R_{GM} and p_O to $G_M C_\mu / 2\pi C_{LOAD}(C_\mu + C_S + C_\pi)$.

Recall that these relations assume R_S is not low. In the case where $(R_S \,||\, R_{IN})(C_S + C_\pi + C_{MI})$ is lower than $(R_O \,||\, R_{LOAD})(C_{MO} + C_{LOAD})$, C_{MO} and C_{LOAD} shunt $R_O \,||\, R_{LOAD}$ first at $p_O{}'$:

$$\left.\frac{1}{s(C_{MO} + C_{LOAD})}\right|_{p_O{}' \approx \frac{1}{2\pi(R_O||R_{LOAD})(C_{MO}+C_{LOAD})}} \equiv R_O \,||\, R_{LOAD} \tag{4.20}$$

R_O and R_{LOAD} therefore disappear at higher frequencies, and because C_μ is considerably lower than C_{LOAD} and C_π, C_S and C_π in parallel with the series combination of C_μ and C_{LOAD} reduce to C_S and C_π. As a result, C_S and C_π now shunt R_S and R_{IN} at a higher frequency:

$$\frac{1}{s[C_S + C_\pi + (C_m \oplus C_{LOAD})]} \approx \left.\frac{1}{s(C_S + C_\pi + C_\mu)}\right|_{p_{IN}{}' \approx \frac{1}{2\pi(C_S+C_\pi+C_\mu)(R_S||R_{IN})}} \equiv R_S \,||\, R_{IN} \tag{4.21}$$

Validation When combining C_S and C_π into $C_{IN}{}'$, R_S and R_{IN} into $R_{IN}{}'$, and R_O and R_{LOAD} into $R_O{}'$ and using conductances $G_{IN}{}'$ for $1/R_{IN}{}'$ and $G_O{}'$ for $1/R_O$ and C_O for C_{LOAD} in Fig. 4.12*a*, input current i_{in} splits into $G_{IN}{}'$, $C_{IN}{}'$, and C_μ:

$$i_{in} = v_{in}G_{IN}{}' + v_{in}sC_{IN}{}' + (v_{in} - v_o)sC_\mu \tag{4.22}$$

and C_μ's current $(v_{in} - v_o)sC_\mu$ splits into G_M, $G_O{}'$, and C_O:

$$(v_{in} - v_o)sC_\mu = v_{in}G_M + v_oG_O{}' + v_osC_O \tag{4.23}$$

Solving for v_{in} in the latter equation and substituting it into the former reveals that i_{in} is

$$\begin{aligned} i_{in} &= v_{in}(G_{IN}{}' + sC_{IN}{}' + sC_\mu) - v_osC_\mu \\ &= \left[\frac{v_o(G_O{}' + sC_O + sC_\mu)}{-(G_M - sC_\mu)}\right](G_{IN}{}' + sC_{IN}{}' + sC_\mu) - v_osC_\mu \end{aligned} \tag{4.24}$$

which solving for transconductance gain v_o/i_{in} shows

$$\frac{v_o}{i_{in}} = \frac{-(G_M - sC_\mu)}{s^2(C_O C_{IN}' + C_O C_\mu + C_\mu C_{IN}') + s(G_M C_\mu + G_O' C_{IN}' + G_O' C_\mu + G_{IN}' C_O + G_{IN}' C_\mu) + G_O' G_{IN}'}$$

$$= \frac{-R_{IN}' R_O' G_M \left(1 - \frac{sC_\mu}{G_M}\right)}{s^2 R_{IN}' R_O'(C_O C_{IN}' + C_O C_\mu + C_\mu C_{IN}') + s\{R_{IN}'[C_\mu(G_M R_O' + 1) + C_{IN}'] + R_O'(C_O + C_\mu)\} + 1}$$

$$\approx \frac{R_{IN}'(-G_M R_O')\left(1 - \frac{sC_\mu}{G_M}\right)}{s^2 R_{IN}' R_O' C_O(C_{IN}' + C_\mu) + s\{R_{IN}'[C_\mu(G_M R_O' + 1) + C_{IN}']\} + 1} \tag{4.25}$$

Although not apparent at first, this relationship reveals all conclusions drawn earlier.

First, low-frequency gain is v_o/i_{in} when s, which represents frequency, is negligibly low, or $-R_{IN}'G_M R_O'$, or equivalently, $-(R_S \parallel R_{IN}) G_M(R_O \parallel R_{LOAD})$, where $R_S \parallel R_{IN}$ converts i_{in} into v_{in} and $-G_M(R_O \parallel R_{LOAD})$ or A_{V0} amplifies v_{in}. Second, v_o/v_{in} rises and inverts polarity when sC_μ/G_M nears and exceeds 1, past right-half-plane zero $G_M/2\pi C_\mu$. At low frequencies, the s term in the denominator overwhelms the quadratic counterpart, which means the latter disappears at low frequencies. So if R_S is not small and $R_{IN}'[C_\mu(G_M R_O' + 1) + C_{IN}']$ or $(R_{IN} \parallel R_S)(C_{MI} + C_S + C_\pi)$ overwhelms $R_O'(C_O + C_\mu)$ or $(R_O \parallel R_{LOAD})(C_{LOAD} + C_\mu)$, C_{MI}, C_S, and C_π shunt first at p_{IN}, as predicted earlier. At higher frequencies, near the next pole, both s and s^2 terms overwhelm 1, so 1 disappears and s factors to reveal the next pole. Here, assuming C_O or C_{LOAD} is normally much higher than C_μ, $G_M R_O'$ in C_{MI} is similarly greater than 1 and C_{MI} likewise is greater than C_S and C_π. R_{IN}' and R_O' therefore cancel when solving for s to establish a pole at p_O, at $G_M C_\mu/2\pi C_O(C_\mu + C_{IN}')$ or $G_M C_\mu/2\pi C_{LOAD}(C_\mu + C_S + C_\pi)$.

On the other hand, if R_S is small, $R_O'(C_O + C_\mu)$ or $(R_O \parallel R_{LOAD})(C_{LOAD} + C_\mu)$ overwhelms $R_{IN}'[C_\mu(G_M R_O' + 1) + C_{IN}']$ or $(R_{IN} \parallel R_S)(C_{MI} + C_S + C_\pi)$, so C_{LOAD} and C_μ shunt first at p_O' or $1/2\pi(R_O \parallel R_{LOAD})(C_\mu + C_{LOAD})$ and v_o/i_{in} simplifies to

$$\frac{v_o}{i_{in}} \approx \frac{R_{IN}'(-G_M R_O')\left(1 - \frac{sC_\mu}{G_M}\right)}{s^2 R_{IN}' R_O' C_O C_{IN}' + sR_O' C_O + 1} \tag{4.26}$$

At higher frequencies, near the next pole, both s and s^2 terms overwhelm 1, so 1 disappears and s factors to reveal the next pole. Here, assuming C_O or C_{LOAD} and C_{IN}' or C_S and C_π are considerably higher than C_μ, R_O' and C_O cancel when solving for s to establish a pole at p_{IN}', at $1/2\pi(R_S \parallel R_{IN})(C_S + C_\pi)$.

4.4.4 Emitter/Source Degeneration

In the case where the common emitter and source terminals connect to ground through a series resistance like R_{DEG} in Fig. 4.13, base–emitter and gate–source voltages v_{be} and v_{gs} only drop a fraction of the input voltage v_{in}. As a result, g_m currents fall, and so do the transconductance and voltage gains they produce. This reduction in gain is the reason why engineers call these series resistors *emitter-* and *source-degenerating resistors*.

Small-Signal Model

Degenerating resistance R_{DEG} is also in series with r_π, so two-port input equivalent resistance R_{IN} is higher than its nondegenerated counterpart. In fact, because base current i_b in a BJT is only a fraction of emitter current i_e that flows through R_E in Fig. 4.13, the BJT magnifies the effects of R_{DEG} on R_{IN} to

$$R_{IN} = \frac{v_{in}}{i_{in}} = \frac{v_{be} + v_e}{i_b} \approx \frac{i_b r_\pi + i_e R_{DEG}}{i_b} \approx \frac{i_b r_\pi + [i_b + (i_b r_\pi) g_m] R_{DEG}}{i_b}$$
$$= r_\pi + (1 + r_\pi g_m) R_{DEG} \tag{4.27}$$

where v_{be} is the voltage i_b produces across r_π and i_e is roughly i_b plus g_m current $v_{be} g_m$. Since $g_m r_\pi$ is β_0, which can be 50–100 A/A, R_{IN} can be 50 to 100 times higher than R_{DEG}, and if R_{DEG} is r_o, 50 to 100 times higher

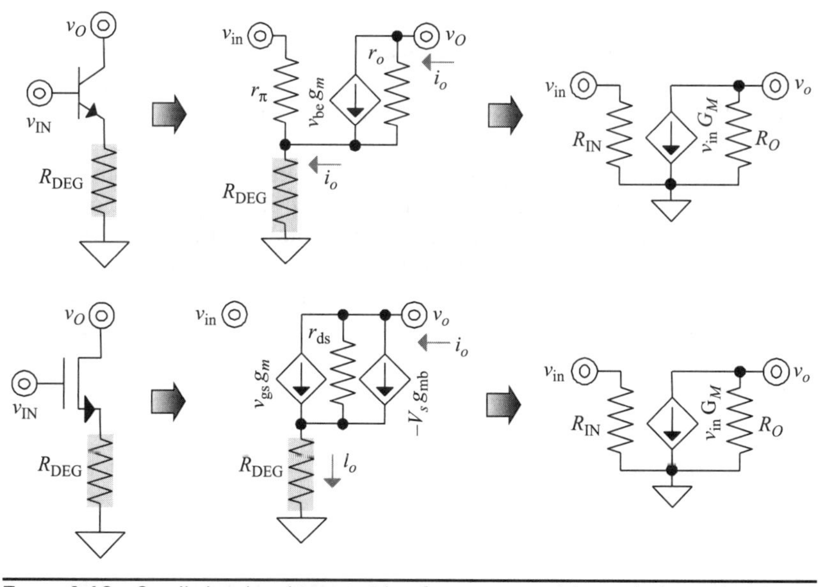

FIGURE 4.13 Small-signal and two-port low-frequency model translations of emitter/source-degenerated CE and CS transistors.

than r_o. Of course, because R_{IN} in the nondegenerated CS transistor already approaches infinity, R_{DEG} has no noticeable effects on the input resistance of FETs.

Since R_{DEG} degenerates v_{be} to a fraction of v_{in}, $v_{be}g_m$ current is lower in the degenerated CE transistor than in the nondegenerated counterpart. To be more precise, first consider that the test condition for two-port transconductance G_M is to shunt v_o to ground. As a result, g_m and r_π currents flow through R_{DEG} and r_o to establish the voltage v_e that degenerates v_{in} in G_M:

$$v_e = v_{be}\left(g_m + \frac{1}{r_\pi}\right)(R_{DEG} \| r_o) \approx v_{be}g_m(R_{DEG} \| r_o) \tag{4.28}$$

where v_e/r_π is r_π's current. Because output current i_o is the combined current that g_m and r_o conduct into v_e, v_{in} drops across v_{be} and v_e, and r_o is normally much higher than $1/g_m$, G_M reduces to

$$G_M\Big|_{v_o=0} \equiv \frac{i_o}{v_{in}} = \frac{v_{be}g_m - \dfrac{v_e}{r_o}}{v_{be} + v_e} \approx \frac{v_{be}g_m - \dfrac{v_{be}g_m(R_{DEG} \| r_o)}{r_o}}{v_{be} + v_{be}g_m(R_{DEG} \| r_o)}$$

$$= \frac{g_m\left(\dfrac{r_o}{r_o + R_{DEG}}\right)}{1 + g_m(R_{DEG} \| r_o)} = \frac{g_m}{1 + \left(\dfrac{1}{r_o} + g_m\right)R_{DEG}} \approx \frac{g_m}{1 + g_m R_{DEG}} \tag{4.29}$$

which as expected, is lower than the nondegenerated g_m.

In the case of FETs, r_π disappears and bulk-effect current $v_{bs}g_{mb}$ appears. To gain insight into the impact of g_{mb} on G_M, consider that gate voltage v_g in $(v_g - v_s)g_m$ "generates" and source voltage v_s "degenerates" drain current i_o. In contrast, while no term in $v_{bs}g_{mb}$ "generates" i_o because the bulk is at ground, v_s "degenerates" i_o. In other words, in addition to the degenerating effect of v_s in $v_{gs}g_m$, v_s in $v_{bs}g_{mb}$ or $-v_sg_{mb}$ also degenerates G_M. To see the quantitative effect of g_{mb} on G_M, first recall that v_o is at ground for this derivation. Then notice that g_{mb} and r_{ds} currents both flow to the grounded output in parallel and both are linear functions of v_s at v_sg_{mb} and v_s/r_{ds}. This means g_{mb} presents a resistance $1/g_{mb}$ that is in parallel with r_{ds}, so G_M in MOSFETs reduces to

$$G_M\Big|_{v_o=0} \equiv \frac{i_o}{v_{in}} = \frac{g_m}{1 + \left[\left(\dfrac{1}{r_o} + g_{mb}\right) + g_m\right]R_{DEG}} \approx \frac{g_m}{1 + (g_{mb} + g_m)R_{DEG}} \tag{4.30}$$

Note that v_{bs} is zero when the bulk and source terminals connect to the same point, in which case the effects of g_{mb} on G_M disappear.

When considering output resistance, R_{DEG} is in series with the transistor, so equivalent two-port output resistance R_O is higher for the degenerated CE transistor than the nondegenerated counterpart. In fact, because $v_{be}g_m$ steals current away from r_o, the BJT amplifies the effects of r_o on R_O when R_{DEG} is present. To be more explicit, since the test condition for deriving R_O is grounding v_{in}, v_{be} reduces to $-v_e$ and r_π and R_{DEG} are in parallel. What is more, as Fig. 4.13 illustrates, i_o into g_m and r_o must also flow through r_π and R_{DEG} to establish v_e, which in series with the voltage that i_o minus $-v_e g_m$ generates across r_o sets R_O to

$$R_O\Big|_{v_{in}=0} \equiv \frac{v_o}{i_o} = \frac{[i_o-(-v_e)g_m]r_o + v_e}{i_o} = \frac{[i_o + i_o(R_{DEG}\,||\,r_\pi)g_m]r_o + i_o(R_{DEG}\,||\,r_\pi)}{i_o}$$
$$= r_o + g_m r_o(R_{DEG}\,||\,r_\pi) + (R_{DEG}\,||\,r_\pi) \tag{4.31}$$

In the case of an FET, r_π is absent, so $R_{DEG}\,||\,r_\pi$ reduces to R_{DEG}. Plus, since v_{in} is zero, both v_{gs} and v_{bs} reduce to $-v_s$, which means g_{mb} and g_m combine and R_O becomes

$$R_O = r_{ds} + (g_m + g_{mb})r_{ds}R_{DEG} + R_{DEG} \tag{4.32}$$

Since intrinsic voltage gain $g_m r_{ds}$ can be on the order of 100 V/V, R_O can be 100 times higher than R_{DEG}, and if R_{DEG} is another r_{ds}, two orders of magnitude higher than r_{ds}. In the case of BJTs, r_π limits the effect of R_{DEG} on R_O to $g_m r_o r_\pi$, and since $g_m r_\pi$ is β_0, R_O can be β_0 times higher than r_o.

With the two-port model derived, the small-signal low-frequency voltage gain across the emitter- and source-degenerated CE and CS transistors is similar in form to the nondegenerated counterpart, but usually lower at

$$A_{V0} \equiv \frac{v_o}{v_{in}} = \frac{-v_{in}G_M(R_O\,||\,R_{LOAD})}{v_{in}} \approx \left(\frac{-g_m}{1+g_m R_{DEG}}\right)(R_{LOAD}) \tag{4.33}$$

Here, g_{mb} adds to the denominator's g_m to further degenerate G_M, but normally not to a great extent. R_O can also be so high that R_{LOAD} can limit the impact of R_O on gain. The main point here, though, is that $1 + g_m R_{DEG}$ degenerates the gain, and again, that results because v_{be} and v_{gs} only drop a fraction of v_{in}.

Example 4.3 Determine the low-frequency small-signal voltage gain v_o/v_{in} across the degenerated CE NPN BJT in Fig. 4.14.

Solution

Small input signals Δv_{IN} produce fluctuations in base voltage Δv_B, which cause variations in collector current Δi_C that M_{PL}'s output resistance R_L or $r_{sd(PL)}$ with Q_N's

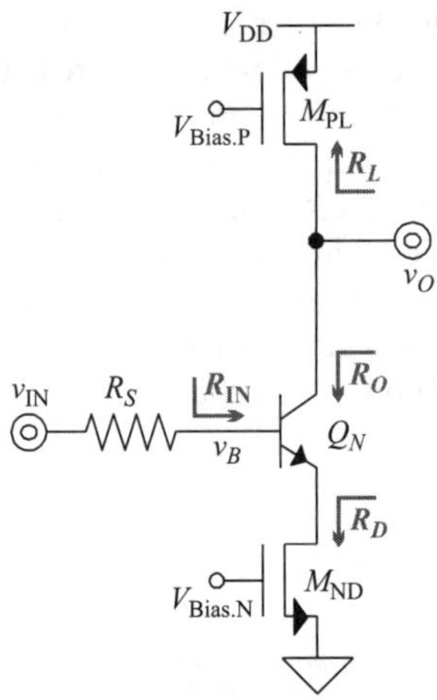

FIGURE 4.14 Degenerated CE example.

counterpart R_O convert into output voltage fluctuations Δv_O. The input signal therefore undergoes the following small-signal voltage–current conversions:

$$\Delta v_{IN} \to \Delta v_B \to \Delta i_C \to \Delta v_O = v_{in} \to v_b \to i_c \to v_o$$

which translates to small-signal low-frequency voltage gain

$$A_{VO} \equiv \frac{v_o}{v_{in}} = \left(\frac{v_b}{v_{in}}\right)\left(\frac{i_c}{v_b}\right)\left(\frac{v_o}{i_c}\right)$$

Here, R_S and Q_N's degenerated input resistance $R_{IN(DCE)}$ voltage-divide v_{in} into v_b, where M_{ND}'s output resistance R_D or $r_{ds(ND)}$ raises $R_{IN(DCE)}$ so much that $R_{IN(DCE)}$ overwhelms an R_S that is low to moderate, so the voltage-divider ratio reduces to nearly 1:

$$\frac{v_b}{v_{in}} = \frac{R_{IN(DCE)}}{R_S + R_{IN(DCE)}} = \frac{r_\pi + (1 + g_m r_\pi)R_D}{R_S + r_\pi + (1 + g_m r_\pi)R_D}$$

$$= \frac{r_\pi + (1 + g_m r_\pi) r_{ds(ND)}}{R_S + r_\pi + (1 + g_m r_\pi) r_{ds(ND)}} \approx \frac{\beta_0 r_{ds(ND)}}{R_S + \beta_0 r_{ds(ND)}} \approx 1$$

Q_N's transconductance $G_{M(DCE)}$ then translates v_b to i_c, and M_{ND}'s R_D or $r_{ds(ND)}$ degenerates short-circuit $G_{M(DCE)}$ to the extent $G_{M(DCE)}$ reduces to

$$\frac{i_c}{v_b} \equiv G_{M(DCE)} \approx \frac{g_m}{1 + \left(g_m + \dfrac{1}{r_o}\right)R_D} = \frac{g_m}{1 + \left(g_m + \dfrac{1}{r_o}\right) r_{ds(ND)}} \approx \frac{1}{r_{ds(ND)}}$$

After that, $G_{M(DCE)}$ pulls current from M_{PL}'s R_L or $r_{sd(PL)}$ and Q_N's $R_{O(DCE)}$ to invert v_o, except M_{ND}'s R_D or $r_{ds(ND)}$ raises $R_{O(DCE)}$ to such a degree that M_{PL}'s nondegenerated R_L or $r_{sd(PL)}$ dominates and $R_{O(DCE)}$ disappears:

$$\frac{v_o}{i_c} \equiv -(R_L \parallel R_{O(DCE)}) = -\{R_L \parallel [r_o + g_m r_o (r_\pi \parallel R_D) + (r_\pi \parallel R_D)]\} \approx -R_L = -r_{sd(PL)}$$

In other words, A_{V0} reduces to

$$A_{V0} \equiv \frac{v_o}{v_{in}} = \left(\frac{v_b}{v_{in}}\right)\left(\frac{i_c}{v_b}\right)\left(\frac{v_o}{i_c}\right) \approx (1)\left(\frac{1}{r_{ds(ND)}}\right)(-r_{sd(PL)}) \approx -\left(\frac{r_{sd(PL)}}{r_{ds(ND)}}\right)$$

Frequency Response

Like in the nondegenerated case, C_μ in the degenerated CE stage feeds forward out-of-phase signals to v_O, so C_μ introduces a right-half-plane zero z_{RHP} at $G_M/2\pi C_\mu$. Since C_μ is still across an inverting gain stage, C_μ Miller-splits into C_{MI} and C_{MO} to help source and load capacitors C_S and C_{LOAD} establish input and output poles p_{IN} and p_O. The fundamental difference here is the impact of degeneration on input and output resistances R_{IN} and R_O and transconductance G_M, and their effects on z_{RHP}, p_{IN}, and p_O. Because degeneration magnifies the effects of R_{IN} and R_O, R_{IN} and R_O are normally so much higher than source and load resistances R_S and R_{LOAD} that the effects of R_{IN} and R_O on p_{IN} and p_O are minimal. A degenerated G_M, however, pulls z_{RHP} to lower frequencies and reduces C_μ's Miller multiplication to push p_{IN} to higher frequencies. Of course, C_{GD} in source-degenerated CS stages produces similar effects.

4.5 Common-Base/Gate Current Buffer

Because collectors and drains are poor input terminals, the emitters and sources of common-base (CB) and common-gate (CG) transistors Q_{CB} and M_{CG} in Fig. 4.15 are input terminals and collectors and drains are outputs. The load at the output is typically one or a combination of resistors, capacitors, CE and CS transconductors, and common-collector and common-drain transistors. Irrespective of its specific nature, the load ultimately presents an equivalent impedance and a steady-state current that a Norton-equivalent circuit can model.

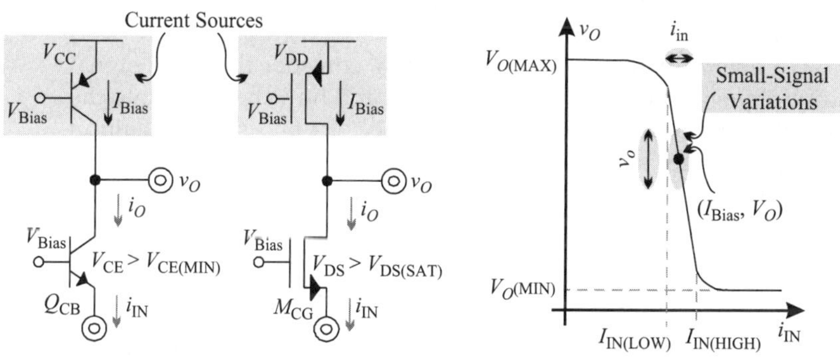

FIGURE 4.15 CB and CG current buffers and their corresponding large-signal response.

Interestingly, output current i_O in this configuration is, for the most part, input current i_{IN}, which is why designers normally use these transistors as *current buffers*.

4.5.1 Large-Signal Operation

When emitter or source input current i_{IN} is zero, the transistor is off, so biasing load current I_{Bias} in Fig. 4.15 raises output v_O high toward the positive power-supply voltage V_{CC} or V_{DD}. As i_{IN} rises, the voltage dropped across the load increases, but not until i_{IN} nears and surpasses I_{Bias}. In other words, the input current $I_{IN(LOW)}$ above which the output starts to fall corresponds to bias current I_{Bias}. After that, excess current flows through the load to pull v_O toward ground, but not all the way to ground. Here, Q_{CB} and M_{CG} enter deep saturation and triode, respectively, when v_O drops about 0.3–0.4 V and V_T below base- and gate-bias voltage V_{Bias}. To what extent v_O rises, ultimately depends on the load. In the case of the CE and CS current-source loads shown, for example, v_O reaches the positive supply V_{CC} or V_{DD} when Q_{CB} and M_{CG} are off, and within $V_{CE(MIN)}$ and $V_{DS(SAT)}$ of input voltage v_{IN} when driven to sink more current than the current sources can supply. Note that v_O is low when i_{IN} pulls considerable current from the input and high otherwise, which is equivalent to saying low emitter and source voltages pull v_O low and high otherwise. In other words, emitter-to-collector and source-to-drain signals are in phase and do not invert.

4.5.2 Small-Signal Model

In most analog applications, designers bias the transistor at a point in its large-signal range that produces the highest gain. In the case of CB and CG transistors, the highest output voltage variation that results from small changes in i_{IN} happens when v_O is between $V_{O(MIN)}$ and $V_{O(MAX)}$ in Fig. 4.15. Analog designers therefore bias Q_{CB} and M_{CG} somewhere in this range. This way, small input signals i_{in} about bias current I_{Bias} produce nearly linear output variations v_o about output bias voltage V_O.

To determine the gain in this region, consider the small-signal low-frequency models of the transistors and how they map into the popular two-port circuit described in Sec. 4.1 and shown again in Fig. 4.16. Here, since base and gate voltages v_b and v_g are zero, base–emitter and gate–source voltages reduce to $-v_e$ and $-v_s$. Plus, since bulk voltage v_b is also zero in FETs, v_{bs} similarly reduces to $-v_s$, and g_m and g_{mb} currents therefore combine into one effective $g_{m(eff)}$ that is equivalent to $g_m + g_{mb}$. Also, since negative currents flowing out of R_{LOAD} are the same as positive currents flowing into R_{LOAD}, G_M currents in the equivalent two-port models shown flow out into R_O and R_{LOAD}, which is another way of saying CB and CG transistors do not invert signals.

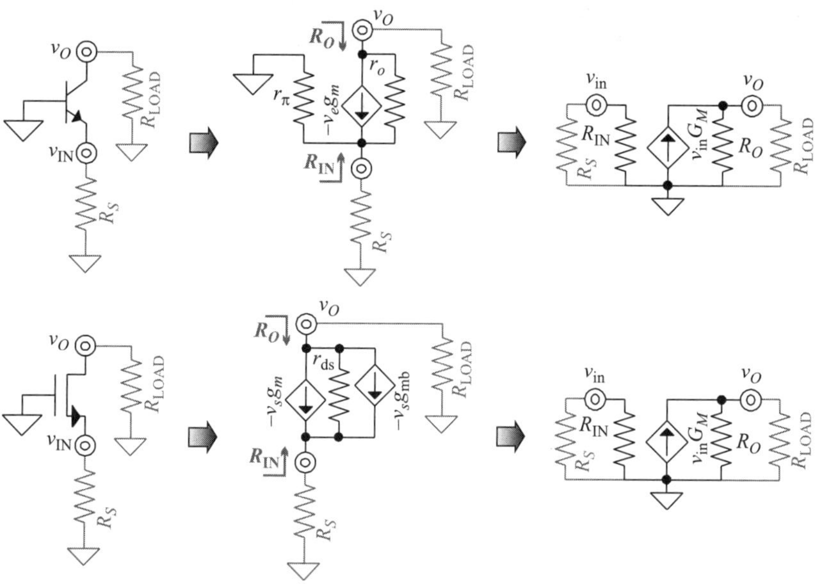

FIGURE 4.16 Small-signal and two-port low-frequency model translations of CB and CG current buffers.

With reference to equivalent input resistance R_{IN}, notice that R_{LOAD} is in series with the transistor, which means R_{LOAD} loads R_{IN}. To determine the extent of this effect, first observe that R_{IN} is the parallel combination of r_π and the resistance R_I that g_m and r_o present. Therefore, current i_i into R_I splits between g_m and r_o and combines back to i_i before flowing into R_{LOAD}. As a result, input voltage v_{in} is the voltage that i_i establishes across r_{ds} and R_{LOAD}:

$$v_{in} = (i_i - v_{in} g_m) r_o + i_i R_{LOAD} \tag{4.34}$$

and solving for R_I, which is v_{in}/i_i, yields

$$R_I \equiv \frac{v_{in}}{i_i} = \frac{r_o + R_{LOAD}}{1 + g_m r_o} > \frac{1}{g_m} \tag{4.35}$$

So, while R_I is approximately $1/g_m$ and $2/g_m$ when R_{LOAD} is well below and near r_o:

$$R_I = \left.\frac{r_o + R_{LOAD}}{1 + g_m r_o}\right|_{R_{LOAD}=r_o} \approx \frac{2}{g_m} \tag{4.36}$$

R_I rises to $r_\pi \parallel r_o$ or r_π when R_{LOAD} is the output resistance of a CE transistor that is degenerated by a transistor's r_o:

$$R_I = \left.\frac{r_o + R_{LOAD}}{1 + g_m r_o}\right|_{R_{LOAD} \approx g_m r_o (r_\pi \parallel r_o)} \approx \frac{r_o + g_m r_o (r_\pi \parallel r_o)}{1 + g_m r_o} \approx r_\pi \parallel r_o \approx r_\pi \tag{4.37}$$

In other words, R_I and R_{IN} are, for the most part, low near $1/g_m$ and considerably greater than that only when loaded with resistances far greater than r_o and r_{ds}:

$$R_{IN} = r_\pi \parallel R_I = r_\pi \parallel \left(\frac{r_o + R_{LOAD}}{1 + g_m r_o}\right) > \frac{1}{g_m} \tag{4.38}$$

where r_π disappears and $g_m + g_{mb}$ replaces g_m in the case of FETs:

$$R_{IN} = R_I = \frac{r_{ds} + R_{LOAD}}{1 + (g_m + g_{mb})r_{ds}} > \frac{1}{g_m + g_{mb}} \tag{4.39}$$

Since the test condition for extracting transconductance G_M is shorting v_o to ground, g_m current $v_{in}g_m$ and r_o current v_{in}/r_o flow to v_o to comprise output current i_o and establish G_M:

$$G_M\Big|_{v_o=0} \equiv \frac{i_o}{v_{in}} = \frac{v_{in}g_m + \left(\frac{v_{in}}{r_o}\right)}{v_{in}} = g_m + \frac{1}{r_o} \approx g_m \tag{4.40}$$

where g_m current is typically orders of magnitude greater than r_o's counterpart, and again, $g_m + g_{mb}$ replaces g_m in the case of FETs. Output resistance R_O is another short-circuit parameter, except v_{in} is the node that shorts to ground. In this latter case, g_m currents are zero and R_O reduces to r_o or r_{ds}. So when combined, $v_{in}G_M$ pushes current into R_O and R_{LOAD} to establish a noninverting voltage v_o whose ratio to v_{in} is low-frequency small-signal voltage gain A_{V0}:

$$A_{V0} \equiv \frac{v_o}{v_{in}} = \frac{v_{in}G_M(R_O \parallel R_{LOAD})}{v_{in}} = \left(g_m + \frac{1}{r_o}\right)(r_o \parallel R_{LOAD}) \approx g_m(r_o \parallel R_{LOAD}) \tag{4.41}$$

What is perhaps more insightful to note is current and transresistance gains A_{I0} and A_{R0}:

$$\begin{aligned} A_{I0} &\equiv \frac{i_o}{i_{in}} = \left(\frac{v_{in}}{i_{in}}\right)\left(\frac{i_o}{v_{in}}\right) = (R_S \parallel R_{IN})(G_M) \\ &= \left[R_S \parallel r_\pi \parallel \left(\frac{r_o + R_{LOAD}}{1 + g_m r_o}\right)\right]\left(g_m + \frac{1}{r_o}\right) \approx \left(\frac{1}{g_m}\right)(g_m) = 1 \end{aligned} \tag{4.42}$$

and

$$A_{R0} \equiv \frac{v_o}{i_{in}} = \left(\frac{i_o}{i_{in}}\right)\left(\frac{v_o}{i_o}\right) = \left[(R_S \parallel R_{IN})G_M\right](R_O \parallel R_{LOAD}) \approx r_o \parallel R_{LOAD} \tag{4.43}$$

because in the presence of a moderate to high R_S and a nondegenerated R_{LOAD}, the CB and CG transistor basically buffers i_{in} into R_{LOAD}. Admittedly, R_S steers current away from R_{IN}, and R_{LOAD} loads R_{IN}, but not by much when R_{IN} is nearly $1/g_m$ and R_{LOAD} is in the vicinity of r_o.

Example 4.4 Determine the low-frequency small-signal transresistance gain v_o/i_{in} across the cascaded N-channel CS and NPN CB stages in Fig. 4.17.

Solution

Small input signals Δi_{IN} or i_{in} produce fluctuations in gate voltage Δv_G or v_g that CS transistor M_{CS} translates into changes in drain current Δi_D or i_d. M_{CS}'s $r_{ds(CS)}$ and Q_{CB}'s $R_{IN(CB)}$ then convert Δi_D into variations in emitter voltage Δv_E or v_e that Q_{CB} translates into emitter and collector current Δi_C or i_c, which M_{PL}'s R_L and Q_{CB}'s $R_{O(CB)}$ ultimately convert into output voltage fluctuations Δv_O or v_o. The input signal therefore undergoes the following small-signal current–voltage conversions:

$$\Delta i_{IN} \to \Delta v_G \to \Delta i_D \to \Delta v_E \to \Delta i_C \to \Delta v_O = i_{in} \to v_g \to i_d \to v_e \to i_c \to v_o$$

which translates to small-signal low-frequency transresistance gain A_{R0} or

$$A_{R0} \equiv \frac{v_o}{i_{in}} = \left(\frac{v_g}{i_{in}}\right)\left(\frac{i_d}{v_g}\right)\left(\frac{v_e}{i_d}\right)\left(\frac{i_c}{v_e}\right)\left(\frac{v_o}{i_c}\right)$$

Here, R_S converts i_{in} to v_g and M_{CS}'s $g_{m(CS)}$ translates v_g into i_d, so v_g/i_{in} is R_S and i_d/v_g is $g_{m(CS)}$. i_d then flows through M_{CS}'s $r_{ds(CS)}$ and Q_{CB}'s $R_{IN(CB)}$ to establish and invert v_e, where M_{PL}'s R_L loading effect on $R_{IN(CB)}$ reduces $R_{IN(CB)}$ to roughly $2/g_{m(CB)}$ to shunt and eliminate the effects of $r_{ds(CS)}$ and $r_{\pi(CB)}$ on v_e/i_d:

$$\frac{v_e}{i_d} = \frac{i_d(r_{ds(CS)} \parallel R_{IN(CB)})}{i_d} = r_{ds(CS)} \parallel r_{\pi(CB)} \parallel \left(\frac{r_{o(CB)} + R_L}{1 + g_{m(CB)} r_{o(CB)}}\right)$$

$$= r_{ds(CS)} \parallel r_{\pi(CB)} \parallel \left(\frac{r_{o(CB)} + r_{sd(PL)}}{1 + g_{m(CB)} r_{o(CB)}}\right) \approx \frac{2}{g_{m(CB)}}$$

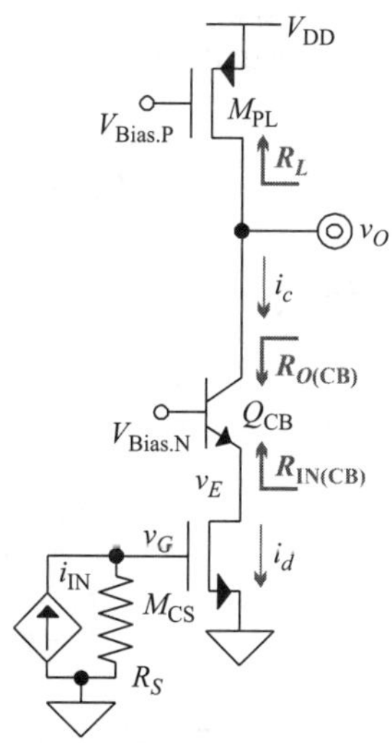

FIGURE 4.17 Cascaded CS and CB example.

Q_{CB}'s transconductance $G_{M(CB)}$ or $g_{m(CB)} + 1/r_{o(CB)}$ then translates v_e to i_c and M_{PL}'s R_L or $r_{sd(PL)}$ and Q_{CB}'s $r_{o(CB)}$ then convert i_c to v_o, reducing A_{R0} to

$$A_{R0} \approx (R_S)(-g_{m(CS)})\left(\frac{2}{g_{m(CB)}}\right)\left(g_{m(CB)} + \frac{1}{r_{o(CB)}}\right)(R_L \parallel r_{o(CB)})$$
$$= -2R_S g_{m(CS)}(r_{sd(PL)} \parallel r_{o(CB)})$$

Direct Translation Excluding the two-port model when analyzing the current buffer reveals insight into the internal mechanics of the circuit. In this case, the current that flows into the buffer's loaded input resistance is the current that flows out of the collector or drain. From this perspective, input current i_{in} establishes a voltage across input-source resistance R_S and the resistance R_I that g_m and r_o introduce, and the fraction of current that flows through R_I reaches the load R_{LOAD} to drop output voltage v_o. In other words, A_{R0} is how much current R_I absorbs to drop v_o across R_{LOAD}:

$$A_{R0} = \left(\frac{v_{in}}{i_{in}}\right)\left(\frac{i_o}{v_{in}}\right)\left(\frac{v_o}{i_o}\right) = (R_S \parallel r_\pi \parallel R_I)\left(\frac{1}{R_I}\right)R_{LOAD} \tag{4.44}$$

where R_{LOAD} loads $1/g_m$ in R_I. With respect to the two-port model, $r_\pi \parallel R_I$ describes R_{IN} and $1/R_I$ consolidates the effects of G_M and R_O, which combined produce the same result.

4.5.3 Frequency Response

In this configuration, as Fig. 4.18 shows, the circuit incorporates two sets of capacitors: C_S and C_π at the input and C_{LOAD} and C_μ at the output. Although both of these sets of capacitors can be equally high, their parallel resistances are usually not. In fact, the transistor's input resistance R_{IN} is typically low, on the order of $1/g_m$, and R_S and r_π

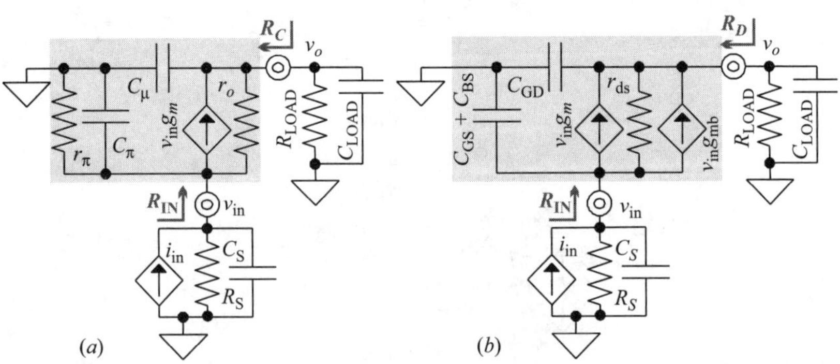

FIGURE 4.18 Wide-spectrum small-signal models for (*a*) CB and (*b*) CG current-buffer transistors.

degenerate the base–emitter voltage of the transistor to raise its collector resistance R_C to considerably higher values than r_o. As a result, C_{LOAD} and C_μ in this configuration normally shunt R_C and R_{LOAD} first at output pole p_O:

$$\left.\frac{1}{s(C_\mu + C_{LOAD})}\right|_{p_O=\frac{1}{2\pi(R_C\|R_{LOAD})(C_\mu+C_{LOAD})}} \equiv R_C \| R_{LOAD}$$

$$= [r_o + g_m r_o(r_\pi \| R_S) + (r_\pi \| R_S)] \| R_{LOAD} \tag{4.45}$$

Here, R_C is not the transistor's equivalent two-port output resistance R_O because R_O is a short-circuit parameter that shorts the degenerating effects of R_S and r_π. And as before, r_π disappears and $g_m + g_{mb}$ replaces g_m in the case of FETs.

Because R_{IN} in current buffers is usually low, C_S and C_π shunt R_S and R_{IN} next at p_{IN}:

$$\left.\frac{1}{s(C_S + C_\pi)}\right|_{p_{IN}\approx\frac{1}{2\pi\left(R_S\|\frac{1}{g_m}\right)(C_S+C_\pi)}} \equiv R_S \| R_{IN} = R_S \| r_\pi \| \left(\frac{r_o + Z_{LOAD}}{1 + g_m r_o}\right) \approx R_S \| \frac{1}{g_m} \tag{4.46}$$

where R_{LOAD} disappears because p_O's capacitances shunt R_{LOAD} past p_O and Z_{LOAD} is therefore less than R_{LOAD} at p_{IN}, so R_{IN} reduces to $1/g_m$. In addition, since engineers normally ensure R_S in this configuration is moderately high, the effects of R_S disappear. As a result, p_{IN} normally asserts its influence on the circuit's response at high frequency, well past p_O. Note that $g_m + g_{mb}$ again replaces g_m and C_{BS} adds to C_S and C_{GS} in the case of FETs.

4.5.4 Base Degeneration

Although not normally on purpose, the voltage source V_{Bias} that biases common-base transistor Q_{CB} in Fig. 4.15 introduces a base resistance R_B in series with the base of NPN BJTs in Figs. 4.16, 4.17, and 4.18. As a result, r_π and R_B voltage-divide v_{in} to impress a fraction across base–emitter voltage v_{be}. In other words, R_B base-degenerates v_{be}, and with respect to v_{in}, a lower effective transconductance g_m' replaces g_m in all previous CB derivations in this section:

$$g_m' = \frac{g_m r_\pi}{r_\pi + R_B} < g_m \tag{4.47}$$

This means g_m' is lower than g_m, and in consequence, R_I and R_{IN} are higher and R_C is lower, and p_{IN} is correspondingly lower and p_O is higher. R_B also current-limits C_π to introduce a left-half-plane zero z_B. Still, R_B is parasitic and normally about an order of magnitude lower than r_π, so R_B's degenerating effects are often minimal, and z_B is often negligibly high in practice.

4.6 Common-Collector/Drain Voltage Follower

Since bases and gates are poor outputs, the emitters and sources of common-collector (CC) and common-drain (CD) transistors Q_{CC} and M_{CD} in Fig. 4.19 are output terminals and bases and gates are inputs, because the influence of base and gate voltages on drain and collector currents is significant. The load at the output is typically one or a combination of resistors, capacitors, other CC and CD circuits, and CE and CS transconductors. Irrespective of its specific nature, the load ultimately presents an equivalent impedance and a steady-state current that a Norton-equivalent circuit can model.

4.6.1 Large-Signal Operation

When base or gate input voltage v_{IN} is low, the transistor is off, so biasing load current I_{Bias} pulls output v_O low toward ground. As v_{IN} climbs, transistor current rises to increase the voltage dropped across the load, but not until the current nears and surpasses I_{Bias}. In other words, the input voltage $V_{IN(LOW)}$ above which the output starts to rise corresponds to the minimum steady-state input voltage necessary to sustain I_{Bias}. After that, excess current flows through the load to pull v_O toward the positive power-supply V_{CC} or V_{DD} voltage. Base–emitter and gate–source voltages v_{BE} and v_{GS}, however, which normally fall anywhere from 0.4 to 1 V, limit how high v_O can rise. To what extent v_O falls, on the other hand, ultimately depends on the load. In the case of the CE and CS current-source loads shown, v_O is zero when Q_{CC} and M_{CD} are off and within a v_{BE} or v_{GS} of the positive supply V_{CC} or V_{DD} when v_{IN} is near the positive supply. Note that v_O is low when v_{IN} is low and high otherwise, which is equivalent to saying base-to-emitter and gate-to-source signals are in phase and do not invert.

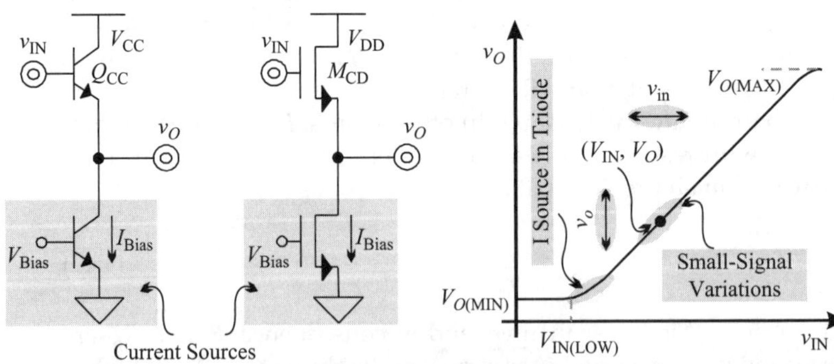

FIGURE 4.19 CC and CD voltage followers and their corresponding large-signal response.

Since variations in collector–emitter and drain–source voltages cause minimal changes in collector and drain currents, the base–emitter and gate–source voltages necessary to sustain I_{Bias} are nearly constant across input voltage v_{IN}'s range. This means the voltage difference between v_{IN} and v_O when Q_{CC} and M_{CD} conduct is also constant, so v_O climbs and falls in unison with v_{IN}. This explains why engineers call these transistors *emitter* and *source followers*, *voltage followers*, and *buffers*, because v_O follows variations in v_{IN} with a gain of nearly 1.

4.6.2 Small-Signal Model

In most analog applications, designers bias the transistor at a point in its large-signal range that produces the highest gain. In the case of CC and CD transistors, the highest gain is almost 1, and that happens when v_O is between $V_{O(MIN)}$ and $V_{O(MAX)}$. Analog designers therefore bias Q_{CC} and M_{CD} somewhere in this range. Since v_O cannot rise above a base–emitter or gate–source voltage below their bases and gates, collector–emitter and drain–source voltages in single-supply systems are never low enough to push Q_{CE} into saturation and M_{CS} into triode. This means small input signals v_{in} about steady-state input bias voltage V_{IN} produce nearly linear output variations v_o about output bias voltage V_O.

To determine the gain in this region, consider the small-signal low-frequency models of the transistors and how they map into the popular two-port circuit described in Sec. 4.1 and shown again in Fig. 4.20. Interestingly, the input of the voltage follower is similar to

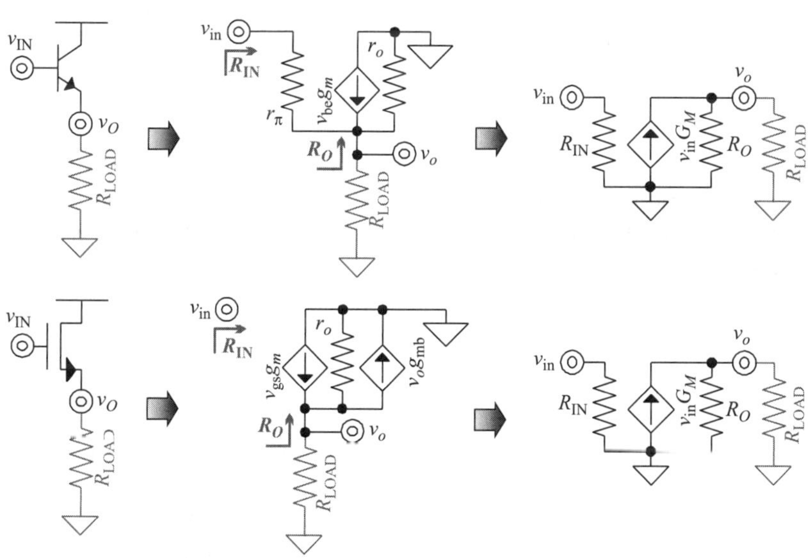

FIGURE 4.20 Small-signal and two-port low-frequency model translations of CC and CD voltage followers.

that of the emitter-degenerated CE transistor, except load resistance R_{LOAD} and the transistor's output resistance r_o play the role of degenerating resistance R_{DEG}. In other words, R_{IN} for the CC case is

$$R_{IN} = r_\pi + (1 + g_m r_\pi)(R_{LOAD} \| r_o) \tag{4.48}$$

Note that R_{LOAD} loads R_{IN}, which is to say R_{LOAD} affects R_{IN} in the two-port equivalent circuit of Fig. 4.20. As before, the resistance into the gate of a FET approaches infinity, so R_{IN} for CD transistors also tends to infinity, irrespective of R_{LOAD} and r_o.

The test condition for extracting transconductance G_M is shorting v_o to ground. The effects of R_{LOAD} and r_o therefore disappear and only r_π and g_m currents contribute to G_M. So since base and g_m currents flow toward v_O, G_M current in the two-port model of Fig. 4.20 also flows to v_O and G_M becomes

$$G_M\Big|_{v_o=0} = \frac{i_o}{v_{in}} = \frac{\frac{v_{be}}{r_\pi} + v_{be} g_m}{v_{be}} = \frac{1}{r_\pi} + g_m \approx g_m \tag{4.49}$$

but because $1/r_\pi$ is β_0 times smaller than g_m, G_M reduces to g_m. In the case of FETs, r_π is absent and bulk-effect current $v_{bs}g_{mb}$ is present. And since bulk voltage v_b is zero and source voltage is v_o, and v_o is zero when extracting G_M, $v_{bs}g_{mb}$ is also zero. As a result, G_M in FETs reduces to g_m.

Output resistance R_O is another short-circuit parameter, except v_{in} is now the one that shorts to ground. This means R_O is the parallel combination of r_o, $v_{be}g_m$'s R_{GM}, and r_π:

$$R_O\Big|_{v_{in}=0} = r_o \| R_{GM} \| r_\pi = r_o \| \frac{1}{g_m} \| r_\pi \approx \frac{1}{g_m} \tag{4.50}$$

Because shorting v_{in} to ground means v_{be} drops $-v_o$, g_m current not only flows into the emitter in response to positive v_o voltages but is also a linear function of its own two terminals, just like a resistor is, so R_{GM} is $1/g_m$:

$$R_{GM} = \frac{v_o}{i_{gm}} = \frac{v_e}{-v_{be}g_m} = \frac{v_e}{v_e g_m} = \frac{1}{g_m} \tag{4.51}$$

which is β_0 times lower than r_π and roughly one more order of magnitude lower than r_o, so R_O reduces to $1/g_m$. Again, in the case of MOSFETs, r_π is absent and $v_{bs}g_{mb}$ is present. And since v_b is zero and the source is at v_o, g_{mb} current flows into the source with positive v_o voltages, just like g_m's current, so $g_m + g_{mb}$ replaces g_m in R_{GM} and R_O.

When combined, $v_{in}G_M$ pushes current into R_O and R_{LOAD} to establish a noninverting voltage v_o whose ratio with v_{in} is low-frequency small-signal voltage gain A_{VO}:

$$A_{V0} \equiv \frac{v_o}{v_{\text{in}}} = \frac{v_{\text{in}} G_M (R_O \parallel R_{\text{LOAD}})}{v_{\text{in}}}$$

$$= \left(g_m + \frac{1}{r_\pi}\right)\left(r_o \parallel \frac{1}{g_m} \parallel r_\pi \parallel R_{\text{LOAD}}\right) \approx g_m \left(R_{\text{LOAD}} \parallel \frac{1}{g_m}\right) \quad (4.52)$$

If R_{LOAD} is considerably higher than $1/g_m$, which is not uncommon, A_{V0} is nearly 1, which is why CC and CD transistors deserve the appellation of voltage followers. Note, however, R_{LOAD} can only reduce the equivalent parallel resistance at v_o, so A_{V0} can never exceed 1.

Example 4.5 Determine the low-frequency small-signal transresistance gain v_o/i_{in} across the CC NPN in Fig. 4.21.

Solution

Small input signals Δi_{IN} produce fluctuations in base voltage Δv_B, which cause variations in emitter current Δi_E that M_{NL}'s output resistance R_L or $r_{\text{ds(NL)}}$ with Q_N's CC counterpart $R_{\text{O(CC)}}$ convert into output voltage fluctuations Δv_O. The input signal therefore undergoes the following small-signal current–voltage conversions:

$$\Delta i_{\text{IN}} \rightarrow \Delta v_B \rightarrow \Delta i_E \rightarrow \Delta v_O = i_{\text{in}} \rightarrow v_b \rightarrow i_e \rightarrow v_o$$

which translates to a transresistance gain of

$$A_{R0} \equiv \frac{v_o}{i_{\text{in}}} = \left(\frac{v_b}{i_{\text{in}}}\right)\left(\frac{i_e}{v_b}\right)\left(\frac{v_o}{i_e}\right)$$

Here, R_S and Q_N's CC input resistance $R_{\text{IN(CC)}}$ convert i_{in} into v_b, where M_{NL}'s output resistance R_L or $r_{\text{ds(NL)}}$ and Q_N's r_o emitter-degenerate Q_N to raise $R_{\text{IN(CC)}}$ to such an extent that a low-to-moderate R_S dominates:

$$\frac{v_b}{i_{\text{in}}} = R_S \parallel R_{\text{IN(CC)}} = R_S \parallel [r_\pi + (1+\beta_0)(R_L \parallel r_o)]$$

$$= R_S \parallel [r_\pi + (1+\beta_0)(r_{\text{ds(NL)}} \parallel r_o)] \approx R_S$$

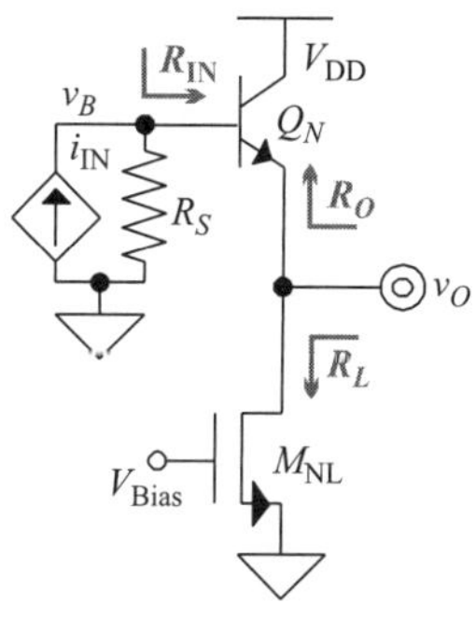

FIGURE 4.21 CC emitter-follower example.

Q_N's CC transconductance $G_{M(CC)}$ or g_m then translates v_b to i_e. After that, Q_N's $R_{O(CC)}$ and M_{NL}'s R_L or $r_{ds(NL)}$ receive and convert i_e into v_o, except $R_{O(CC)}$'s $1/g_m$ is so much lower than r_π, r_o, and R_L or $r_{ds(NL)}$ that $1/g_m$ dominates:

$$\frac{v_o}{i_e} = R_{O(CC)} \,||\, R_L = \left(r_\pi \,||\, \frac{1}{g_m} \,||\, r_o\right) || \, r_{ds(NL)} \approx \frac{1}{g_m}$$

In other words, A_{R0} reduces to roughly R_S:

$$A_{R0} \equiv \frac{v_o}{i_{in}} = \left(\frac{v_b}{i_{in}}\right)\left(\frac{i_e}{v_b}\right)\left(\frac{v_o}{i_e}\right) \approx (R_S)(g_m)\left(\frac{1}{g_m}\right) \approx R_S$$

where the base–emitter voltage gain v_o/v_b across Q_N is roughly 1.

Direct Translation Excluding the two-port model when analyzing the voltage follower reveals insight into the internal mechanics of the circuit. In this case, the equivalent load emitter- or source-degenerates the transistor to establish a current that flows directly into the load. From this perspective, the transconductance $G_{M(DEG)}$ that load resistance R_{LOAD} and the transistor's r_o degenerate produces a current that drops output voltage v_o across R_{LOAD} and r_o. In other words, A_{V0} is how much voltage $G_{M(DEG)}$'s current drops across R_{LOAD} and r_o:

$$A_{V0} \equiv \frac{v_o}{v_{in}} = G_{M(DEG)}(r_o \,||\, R_{LOAD}) = \left[\frac{g_m + \dfrac{1}{r_\pi}}{1 + \left(g_m + \dfrac{1}{r_\pi}\right)(r_o \,||\, R_{LOAD})}\right](r_o \,||\, R_{LOAD}) \tag{4.53}$$

where v_{be}/r_π's contribution to g_m's current is negligible and i_c is nearly i_e. In this interpretation, $G_{M(DEG)}$ consolidates the effects of g_m in G_M and R_O to produce the same result as the two-port model.

4.6.3 Frequency Response

Because every ac node has parasitic capacitance to ground, incoming signals lose energy to what amounts to input and output poles at the base and emitter in the broad-spectrum models of Fig. 4.22. What is perhaps not as apparent at first glance is that C_π feeds forward an in-phase signal from v_{in} to v_o, introducing a left-half plane zero to the mix. As before, because C_π's current is in phase with r_π and g_m currents and helps r_π and g_m drive v_o, the zero is relatively benign and on the left-half plane. The effects of the zero are nonetheless felt when the current through C_π supersedes the combined contribution of g_m and r_π, when C_π's $i_{C\pi}$ just equals the sum of g_m's i_{gm} and r_π's i_b:

$$i_{C\pi} = v_{be} s C_\pi \Big|_{z_{LHP} \approx \frac{g_m}{2\pi C_\pi}} \equiv i_{gm} + i_b = v_{be}\left(g_m + \frac{1}{r_\pi}\right) \approx v_{be} g_m \tag{4.54}$$

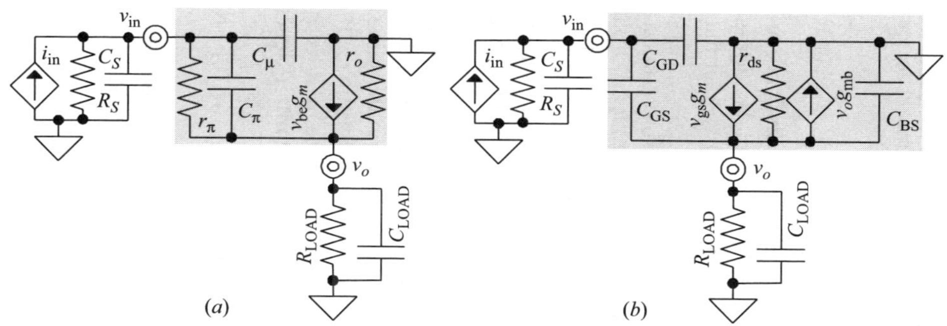

FIGURE 4.22 Wide-spectrum small-signal models for (*a*) CC emitter and (*b*) CD source followers.

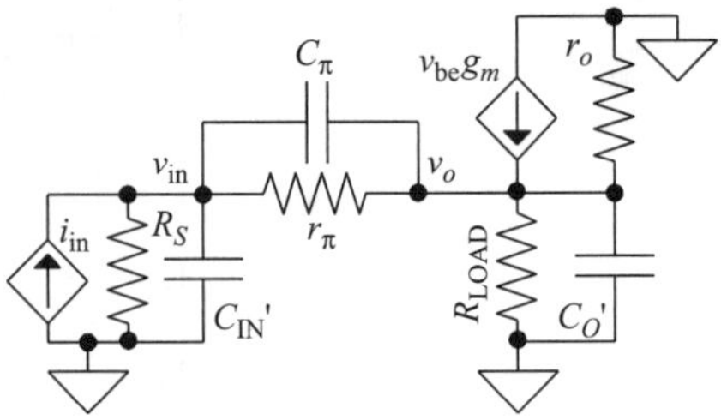

FIGURE 4.23 Simplified wide-spectrum small-signal model for emitter and source followers.

where left-half-plane zero z_{LHP} reduces to $g_m/2\pi C_\pi$ because base current i_b is β_0 times smaller than $v_{be}g_m$. In the metal–oxide–semiconductor (MOS) case, the effects of r_π disappear, so z_{LHP} similarly reduces to $g_m/2\pi C_{GS}$.

Before proceeding with the analysis, first consider that C_μ and C_{GD} add to source capacitance C_S, and C_{BS} to load capacitance C_{LOAD}, so input and output capacitances C_{IN}' and C_O' in Fig. 4.23 represent their combined effects. Also, because bulk–source voltage v_{bs} is $-v_o$, g_{mb} current flows into the emitter or source when v_o rises and is linearly dependent on v_o, which means g_{mb} loads v_o with a resistance R_{mb} that is equivalent to $1/g_{mb}$. In other words, r_π disappears and $r_{ds} \,||\, (1/g_{mb})$ replaces r_o in CD FETs. With these combinations, the analysis is largely the same for both CC and CD circuits.

Since all three sets of capacitors can be on the same order of magnitude, their parallel resistances often determine which set shunts first. The resistance R_{CIN} across C_{IN}', for one, is R_S in parallel with a base-input resistance R_B that R_{LOAD} and r_o emitter-degenerate:

$$R_{CIN} = R_S \,||\, R_B = R_S \,||\, [r_\pi + (1+\beta_0)(R_{LOAD} \,||\, r_o)] \approx R_S \tag{4.55}$$

and if R_S is near r_o or lower, which is not uncommon, R_{IN}' reduces to R_S. The resistance R_{CO} across C_O', on the other hand, is the parallel combination of R_{LOAD}, r_π in series with R_S, and the resistance that g_m and r_o present in the form of R_I:

$$R_{CO} = R_{LOAD} \,||\, (r_\pi + R_S) \,||\, R_I \approx R_I = \frac{r_o}{1+g_m' r_o} \approx \frac{1}{g_m'} = \frac{r_\pi + R_S}{g_m r_\pi} = \frac{r_\pi + R_S}{\beta_0} \tag{4.56}$$

Note that R_I is the $1/g_m$ emitter resistance of the CB transistor whose collector no resistance loads and whose base R_S degenerates. In other words, R_I is typically sufficiently low to shunt R_{LOAD} and r_π in series with R_S, which means R_{CO} can be very low. The last resistance to consider is that of C_π. This resistance is the parallel combination of r_π with the series combination of the resistances connected from v_o to ground and ground to v_{in}, which from the previous two resistances reduce to R_I and R_S:

$$R_{C\pi} = r_\pi \parallel [R_S + (R_I \parallel R_{LOAD})] \approx r_\pi \parallel \left[R_S + \left(\frac{r_\pi + R_S}{g_m r_\pi}\right)\right] \tag{4.57}$$

Normally, R_S is high near r_o, C_S is low near C_μ, and C_{LOAD} high with several C_π's. This means R_{CIN} is high near r_o, R_{CO} is β_0 times smaller than r_π and R_S, and $R_{C\pi}$ is nearly r_π, about an order of magnitude lower than r_o. With such low resistances across C_π and C_{LOAD}, C_{IN}' can shunt first to establish input pole p_{IN}:

$$\left.\frac{1}{s(C_S + C_\mu)}\right|_{p_{IN} \approx \frac{1}{2\pi R_S (C_S + C_\mu)}} \equiv R_S \parallel [r_\pi + (1+\beta_0)(R_{LOAD} \parallel r_o)] \approx R_S \tag{4.58}$$

After C_{IN}' shunts and removes R_S, C_π and the several C_π's that comprise C_{LOAD} can shunt next at p_O:

$$\left.\frac{1}{s\{C_{LOAD} + [C_\pi \oplus (C_S + C_\mu)]\}}\right|_{p_O \approx \frac{g_m}{2\pi(C_{LOAD} + C_\pi)}} \equiv R_{LOAD} \parallel R_I \parallel Z_\pi \approx \frac{Z_\pi + Z_{CIN}}{g_m Z_\pi} \parallel Z_\pi \approx \frac{1}{g_m} \tag{4.59}$$

where r_π and C_π in Z_π and C_μ and C_S in Z_{IN}' base-degenerate g_m, but in light of a low C_S, not by much. Note that in-phase feed-forward zero z_{LHP}, which is $g_m/2\pi C_\pi$, is higher than p_O's $g_m/2\pi(C_\pi + C_{LOAD})$ because $C_\pi + C_{LOAD}$ is greater than C_π, especially when C_{LOAD} includes several C_π's.

If R_S is low near $1/g_m$, on the other hand, R_{CIN}, R_{CO}, and $R_{C\pi}$ are similarly low and the poles they establish correspondingly high. In other words, irrespective of R_S, p_O is normally at high frequencies, but below z_{LHP} if C_{LOAD} includes more than one C_π. Incidentally, since the effect of C_π and C_{GS} is to short base–emitter and gate–source voltages v_{be} and v_{gs}, g_m current falls at higher frequencies. This means emitter and source resistance $1/g_m$ rises with frequency, as an inductor would. This explains why engineers say emitters and sources become *inductive* at high frequencies.

4.7 Small-Signal Generalizations and Approximations

4.7.1 Functionality

Analog IC design hinges on insight and approximations. Understanding basic concepts is key, because at five to six low- and high-frequency equations per transistor, designing operational amplifiers, voltage references, linear regulators, and other analog blocks that incorporate 10 to 100 transistors is virtually impossible. The only way to design these multitransistor circuits effectively is to understand how transistors process voltages and currents and why. CE and CS transistors, for example, are fundamentally transconductors because they convert voltages into currents. CB and CG devices, on the other hand, buffer currents because their input resistances absorb and direct almost all incoming currents to their loads. CC and CD transistors similarly buffer voltages because, with a quasi-constant current, their output voltages follow those of their inputs with a constant voltage between them. Not surprisingly, the input voltages that their loads degenerate establish currents that flow across the same loads to recreate their input voltages.

4.7.2 Resistances

These conclusions result largely because $1/g_m$ resistances are considerably lower than r_π and $1/g_{mb}$, which are in turn substantially lower than r_o and r_{ds}, and those similarly are lower than their degenerated counterparts $g_m r_\pi r_o$ or $\beta_0 r_o$, $g_m r_{ds} r_o$, $g_m r_{ds}^2$, and $g_m r_o^2$. But for these resistances to relate this way, engineers must grasp the effects of current, channel width and length, and terminal voltages on how transistors behave. Higher bias currents, for example, raise g_m and lower r_π and r_o, and higher channel width–length ratios increase g_m and C_{GS}. Also, r_o and r_{ds} are substantial only when their corresponding collector–emitter and drain–source voltages are greater than their respective minimum voltages $V_{CE(MIN)}$ and $V_{DS(SAT)}$. With these concepts in hand, designers can manipulate transistors to operate as described in this chapter and in tandem with other transistors to implement the functional blocks that comprise a microelectronic system.

With these assumptions in play, the input emitter and source resistances of CB and CG current buffers are sufficiently low to absorb almost all incoming currents, and their output resistances are high enough to steer the currents directly into the load. Emitter and source resistances in CC and CD voltage followers are similarly low to ensure emitters and sources can drive and impress the voltages they receive at their base and gate inputs across whatever impedances load them. CE and CS transconductors similarly depend on their high output collector and drain resistances for establishing practicable voltage gains.

Example 4.6 Assuming $1/g_m$ is substantially lower than r_π, r_π is similarly lower than r_o and r_{ds}, and r_o and r_{ds} are likewise lower than $g_m r_\pi r_o$ or $\beta_0 r_o$, use "direct translations" to determine the combined transresistance gain v_o/i_{in} across CS FET M_{CS}, CB NPN Q_{CB}, and CC NPN Q_{CC} from Figs. 4.17 and 4.21, which Fig. 4.24 combines.

Solution

Transconductor M_{CS}'s $g_{m(CS)}$ first converts the voltage that i_{in} drops across R_S into a current $i_{d(CS)}$. The input resistance that $g_{m(CB)}$ and $r_{o(CB)}$ establish as $R_{I(CB)}$ in current-buffer Q_{CB} then absorbs and directs $i_{d(CS)}$ almost entirely to the loading resistances that bias transistor M_{PL} and voltage buffer Q_{CC} present with $r_{sd(PL)}$ and emitter-degenerated input resistance $R_{IN(CC)}$:

$$\frac{v_{o1}}{i_{in}} = (R_S)(-g_{m(CS)})\left(\frac{r_{ds(CS)} \parallel r_{\pi(CB)} \parallel R_{I(CB)}}{R_{I(CB)}}\right)(r_{sd(PL)} \parallel R_{IN(CC)}) \approx -R_S g_{m(CS)} r_{sd(PL)}$$

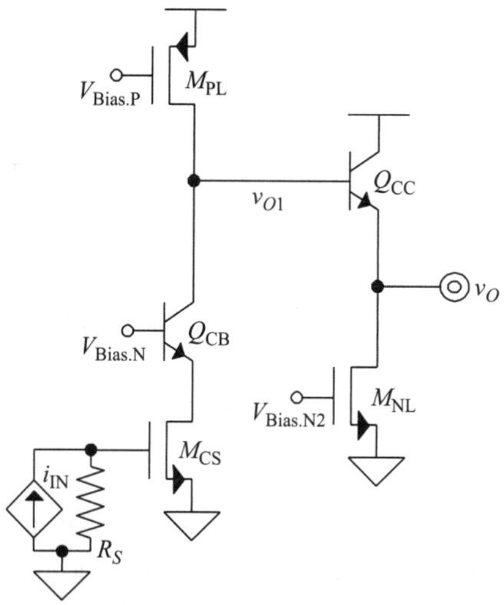

FIGURE 4.24 Cascaded CS transconductor, CB current-buffer, and CC voltage-follower stages.

Here, Q_{CB}'s $R_{I(CB)}$ reduces to $2/g_{m(CB)}$ to absorb most of $g_{m(CS)}$'s current, and $R_{IN(CC)}$ is more than $\beta_{0(CC)}$ times higher than the resistances that M_{NL} and Q_{CC}'s $r_{o(CC)}$ present as $r_{ds(NL)} \parallel r_{o(CC)}$, so $r_{sd(PL)}$ shunts $R_{IN(CC)}$:

$$R_{I(CB)} = \frac{r_{o(CB)} + r_{sd(PL)}}{1 + g_{m(CB)} r_{o(CB)}} \approx \frac{2}{g_{m(CB)}}$$

and

$$\begin{aligned} R_{IN(CC)} &= r_{\pi(CC)} + (1 + g_{m(CC)} r_{\pi(CC)})(r_{ds(NL)} \parallel r_{o(CC)}) \\ &= r_{\pi(CC)} + (1 + \beta_{0(CC)})(r_{ds(NL)} \parallel r_{o(CC)}) \end{aligned}$$

The input voltage v_{o1} and the transcondutance that Q_{CC}'s load resistance $r_{ds(NL)} \parallel r_{o(CC)}$ degenerates establish a current that flows across the same load to recreate v_{o1} in v_o:

$$\frac{v_o}{v_{o1}} = \left[\frac{g_{m(CC)}}{1 + g_{m(CC)}(r_{ds(NL)} \parallel r_{o(CC)})}\right](r_{ds(NL)} \parallel r_{o(CC)}) \approx 1$$

In other words, the overall transresistance gain of the circuit, which corresponds to the product of the two gains just derived, reduces to roughly $-R_S g_{m(CS)} r_{sd(PL)}$.

Although these approximations are important when first conceptualizing a circuit, they often lack the accuracy necessary to predict nonidealities and to fine-tune performance. Thankfully, because transistors are fundamentally three-terminal devices that input and output resistances and transconductances can model, the effects of degeneration and loads on the resistances of BJTs, MOSFETs, and JFETs are generally similar. As a result, engineers only need a handful of equations to forecast and optimize performance.

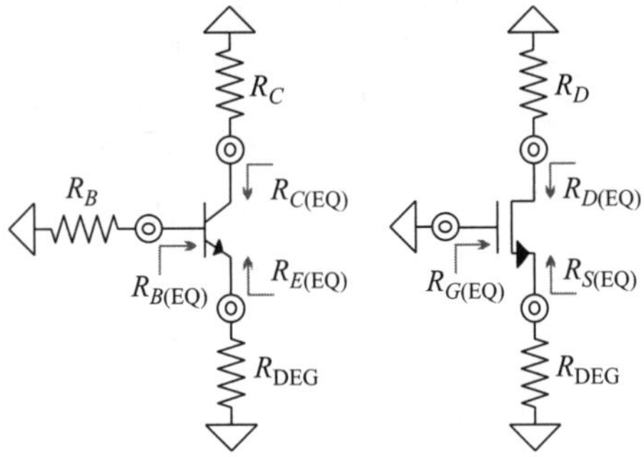

FIGURE 4.25 Relevant resistances in transistor circuits.

How series resistors at the base of BJTs like R_B in Fig. 4.25 degenerate emitter transconductance $g_m{}'$ is one basic relation, as is how resistors $R_{C/D}$ at the collector and drain terminals load equivalent emitter and source resistances $R_{E/S(EQ)}$:

$$g_m{}' = \frac{g_m r_\pi}{r_\pi + R_B} \tag{4.60}$$

and

$$R_{E/S(EQ)} = \left[\frac{r_o + R_{C/D}}{1+(g_m{}' + g_{mb})r_o}\right] \| (r_\pi + R_B) \tag{4.61}$$

where r_π is absent and g_{mb} is present in FETs. The effects of emitter- and source-degenerating resistors R_{DEG} on equivalent base and collector/drain resistances $R_{B(EQ)}$ and $R_{C/D(EQ)}$ and base/gate and emitter/source transconductances $G_{M(EQ)}$ are three other common relationships:

$$R_{B(EQ)} = r_\pi + (1 + g_m r_\pi)R_{DEG} = r_\pi + (1+\beta_0)R_{DEG} \tag{4.62}$$

$$R_{C/D(EQ)} = r_o + (g_m{}' + g_{mb})r_o[R_{DEG} \| (r_\pi + R_B)] + [R_{DEG} \| (r_\pi + R_B)] \tag{4.63}$$

and

$$i_{c/d/s} = v_{be/gs} G_{M(EQ)} \approx \frac{v_{b/g} g_m}{1+(g_m + g_{mb})R_{DEG}} = v_{e/s}(g_m{}' + g_{mb}) \approx i_e \tag{4.64}$$

where the equivalent resistance $R_{G(EQ)}$ into the gate of FETs approaches infinity because little to no current flows into gates. Remembering these equations, however, does not exempt engineers from retaining circuit insight and intuition. The effects of bulk-effect transconductance g_{mb}, for instance, disappear when bulk and source terminals connect, because the small-signal bulk–source voltage v_{bs} that results is zero. Drain resistance, to cite another example, is $R_{D(EQ)}$ for the two-port equivalent circuit of the CS transconductor when degenerated with R_{DEG} and r_{ds} for that of the CG counterpart, whose short-circuit test-circuit requirement eliminates the effects of resistances in series with its source.

4.7.3 Frequency Response

Perhaps the most challenging aspect of determining the frequency response of a circuit is the number of variations that result from the same capacitors, but with different values. As a result, other than Miller relationships, remembering equations here adds little value. What is important to realize is that capacitors open at low frequencies and shunt their parallel resistances at high frequencies, and the transitions, which correspond to poles, happen when the impedances across the capacitors equal their respective resistances. Feed-forward capacitors also introduce zeros when their respective currents equal and surpass their bypassed counterparts. And if these currents are in phase, the zeros they produce are often benign and sometimes even helpful (i.e., on the left-half plane) and harmful otherwise when they are out of phase (i.e., on the right-half plane).

An insightful means of applying these concepts is to think of how circuits change across frequency. For example, removing all capacitors simplifies a circuit and allows the engineer to more easily determine low-frequency gain. After that, the lowest-frequency pole corresponds to the first capacitor that shunts its parallel resistance. After removing the shunted resistance, the next capacitor to shunt establishes the succeeding pole, and so on. In this regard, Miller-splitting capacitors, when applicable, helps simplify circuit alterations.

How to determine which capacitor shunts first amounts to finding the capacitor with the highest capacitance–resistance product, which is nothing but its time constant. In this respect, since base–emitter and gate–source capacitors C_π and C_{GS}, intentionally added nonparasitic capacitors, and the loads they establish as C_{LOAD} are normally highest in a circuit, they typically shunt first, but only if their resistances are correspondingly high at r_o or r_{ds} or higher, or r_π if no other time constant is higher. Engineers therefore scan circuits for nodes with resistances that approach or exceed r_o and r_{ds} across capacitances close to or greater than C_π and C_{GS}. In the CS transconductor, for example, the resistance across C_{GS} and C_{GD}'s Miller equivalent C_{MI} can be high, so the gate pole that results can land at very low frequency. In contrast, the emitter resistance of a CB current buffer is often low, so the pole it establishes is normally at high frequency.

4.8 Summary

Analog systems typically rely on the quasi-linear small-signal response of transistors to realize prescribed functions. Engineers therefore model and reduce these circuits to equivalent two-port linear networks. Excluding capacitors further simplifies the networks and the analysis required to extract and, from a designer's perspective, establish a low-frequency gain. What is important here is how transconductors convert voltages to currents and resistors transform currents into voltages. From this point, the capacitor that first shunts its parallel resistance establishes the lowest-frequency pole. Then, eliminating the resistance it and successive capacitors shunt describes how the circuit behaves at higher frequencies, allowing designers to more intuitively ascertain the high-frequency response of a circuit.

Analog circuits ultimately decompose into a series of single-transistor primitives. Because transistors are basically two-terminal resistors with a controlling third terminal, the manner in which a single transistor can viably connect conforms to one of three possible configurations. Of these, CE and CS transistors are fundamentally transconductors because they convert voltages into currents. CB and CG circuits, on the other hand, channel most of the currents they receive to their outputs, so they behave more like current buffers. And because constant bias currents establish largely invariant base–emitter and gate–source voltages, the emitter/source outputs of CC and CD primitives track their base/gate counterparts, like voltage followers would.

Interestingly, resistors in series with bases, emitters, and sources degenerate how much of the input voltage drops across base–emitter and gate–source terminals, so not only do they raise equivalent base, collector, and drain resistances and lower their corresponding poles but they also lower transconductances. Still, emitter and source resistances are normally low near $1/g_m$ and collector and drain resistances are high at or above r_o and r_{ds}. As a result, emitter and source poles oftentimes lie at higher frequencies and collector and drain poles at lower frequencies. Table 4.2 summarizes the impact of these tendencies on all single-transistor primitives.

Unique combinations of these one-transistor primitives constitute the bedrock on which more complex analog systems that perform a wide range of higher-level functions rest, the most pertinent of which for this textbook is regulation. To that end, the next step in the development process is to construct the basic two- to ten-transistor analog building blocks that most analog circuits incorporate. Chapter 5 therefore reviews and describes how current mirrors, differential pairs, differential amplifiers, and their variants and combinations operate and respond to incoming signals.

	CE/CS Transconductor	CB/CG Current Buffer	CC/CD Voltage Follower
v_o Swing	High	Low	Moderate
Relevant Gains	$A_{VO} = -G_M(R_{OUT} \| \| R_{LOAD})$	$A_{IO} \approx (R_S \| \| R_{IN})(g_m + 1/r_o)$ $A_{RO} = A_{IO}(R_{OUT} \| \| R_{LOAD})$	$A_{VO} \approx g_m(1/g_m \| \| r_o \| \| R_{LOAD})$
R_o	High	High	Low
R_{IN}	r_π or ∞	Low	High or ∞
Input Pole p_{IN}	Low if R_S is high High if R_S is low	High	Low if R_S is high High if R_S is low
Output Pole p_o	High if R_S is high Low if R_S is low	Low	High
Degeneration	Emitter/Source	Base	Base
	Raises R_{IN} and R_O Lowers p_{IN}, p_O, and G_M	Raises R_{IN} Lowers p_{IN} and G_M	Lowers p_O
Peculiarities	Right-half-plane zero	Emitter/source is inductive	Left-half-plane zero Emitter/source is inductive

TABLE 4.2 Qualitative Comparison of Single-Transistor Primitives

4.9 Review Questions

1. State the test condition used to extract the output transconductance of two-port models.

2. What is the impact of capacitors that connect to ground on incoming signals?

3. What are the effects of out-of-phase feed-forward capacitors?

4. What are the effects of in-phase feed-forward capacitors?

5. What is the effect of a resistor that is in series with a capacitor that connects to ground on incoming signals?

6. Are the effects of right-half-plane zeros normally benign? Why?

7. As capacitors shunt across frequency, what components disappear from the circuit?

8. Which terminals in a JFET are practicable inputs?

9. Which transistor configuration(s) invert incoming signals?

10. What are the effects of a source-degenerating resistor in a MOSFET?

11. In which way are CB BJTs most useful? Why?

12. Typically, are input poles in CG MOSFETs at low or high frequencies? Why?

13. What are the effects of base degeneration in CB BJTs?

14. Do CC BJTs incorporate zeros? Why?

15. Normally, which resistances are lowest in a circuit?

16. Normally, which resistances are moderate in a circuit?

17. Normally, which resistances are highest in a circuit?

18. Which capacitors typically establish dominant low-frequency poles?

CHAPTER 5

Analog Building Blocks

Analog integrated-circuit (IC) design is the art of erecting practical analog systems from layers of doped semiconductor, polysilicon material, metal, and silicon dioxide. Understanding the objectives of the targeted system is the first step in this process. Realizing what resistors, capacitors, diodes, and transistors are available, and how and to what extent they conduct currents and drop voltages is next. With these goals and concepts in mind, the purpose of this chapter is to assemble and combine the single-transistor primitives discussed in Chap. 4 into circuit blocks that perform practical, general-purpose tasks. The idea is to "conquer" a design by "dividing" it into simpler functional units that 1- to 10-transistor circuits can realize.

The most basic operations that analog circuits perform are to amplify and buffer signals, which is what the common-emitter/source transconductor, common-base/gate current buffer, and common-collector/drain voltage follower achieve. Mirroring currents is next, as is processing differential signals. The reasons for the latter are noise and regulation. Noise because processing only the difference between two signals rejects the noise that is present in both. Feedback control also relies on the difference between two signals to reduce their error and thereby establish an output that is largely independent of variable operating conditions. Not surprisingly, the current mirror, differential pair, and the differential stage they implement are practically ubiquitous in analog circuits.

An important concept to bear in mind in this and other chapters is the duality of N- and P-type transistors. The only functional difference between the two is the direction of current, which flows out of emitters and sources in the former and into emitters and sources in the latter. Therefore, P-type devices can always implement N-type circuits, and vice versa. In fact, P-type circuits are vertical mirror images of their N-type counterparts. So in light of the aims of this chapter, the ensuing presentation focuses on the structural foundation of analog building blocks, consistently developing and discussing N-type blocks for the sake of clarity and progress. The chapter nonetheless comments on and illustrates P-type realizations that incorporate the same concepts developed with the N-type counterparts.

Another aspect to consider is transistor technology. Fortunately, even when bipolar-junction-transistor (BJT) currents change exponentially with respect to base–emitter voltages and field-effect-transistor (FET) currents vary quadratically with respect to gate–source voltages, their response to small variations is similarly linear. This is why their small-signal models are practically the same. The only difference between their equivalent models is the appearance of the base–emitter's resistance r_π in the BJT and the bulk-effect transconductance g_{mb} in the metal–oxide–semiconductor (MOS) FET, but the latter only applies when bulk and source terminals connect to different points, and even then, g_{mb} is typically 5 to 10 times lower than g_m. This chapter therefore focuses on the architectural bedrock of circuits and comments only on differences that possible BJT and FET variants produce.

5.1 Current Mirror

5.1.1 Theory of Operation

Because collector current depends strongly on base–emitter voltage v_{BE} and weakly on collector–emitter voltage v_{CE} in forward active and light saturation, currents through BJTs that share base–emitter connections, as Fig. 5.1 shows, are largely the same:

$$i_C = I_S\left[\exp\left(\frac{v_{BE}}{V_t}\right)-1\right]\left(1+\frac{v_{CE}}{V_A}\right) \approx I_S\left[\exp\left(\frac{v_{BE}}{V_t}\right)-1\right] \approx i_{IN} \approx i_O \tag{5.1}$$

Similarly, even when drain–source voltages v_{DS} differ, FET saturation currents are, for the most part, equivalent when their gate–source voltages v_{GS} match:

$$i_D = \left(\frac{W}{L}\right)K'(v_{GS}-v_T)^2(1+\lambda v_{DS}) \approx \left(\frac{W}{L}\right)K'(v_{GS}-v_T)^2 \approx i_{IN} \approx i_O \tag{5.2}$$

However, these statements only hold when v_{CE} and v_{DS} are sufficiently high to ensure BJTs are in forward active or light saturation and FETs

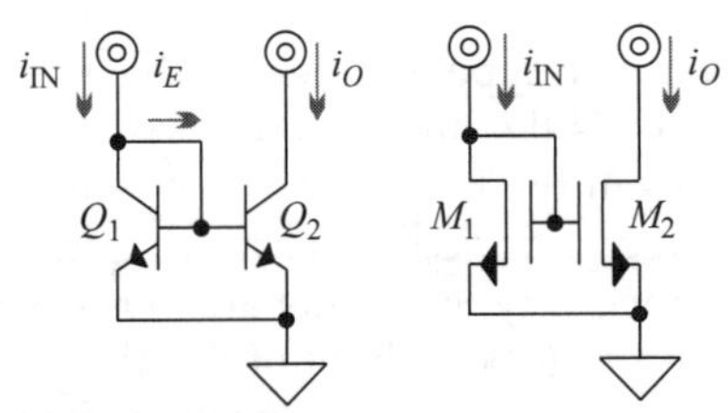

Figure 5.1 BJT and MOSFET current mirrors.

are in saturation. In the case of the basic mirrors shown, for example, output voltage v_O must surpass Q_2's $V_{CE(MIN)}$ and M_2's $V_{DS(SAT)}$ for their output currents i_O to equal Q_1's and M_1's counterparts i_{IN}. Note that input transistors Q_1 and M_1 are automatically in forward active and saturation, respectively, because their v_{CE} and v_{DS} are also v_{BE} and v_{GS}, which are, by default, higher than $V_{CE(MIN)}$ and $V_{DS(SAT)}$.

Gain

Since reverse-saturation current I_S in BJTs is proportional to emitter area A_E, mirroring transistors with larger or smaller but matched A_E's conduct proportionally more or less collector current than their counterparts. If Q_2's A_E in Fig. 5.1, for example, were twice that of Q_1, i_O would be two times higher than i_{IN}. Drain currents in FETs are similarly proportional to the width–length ratios W/L of their channels, so larger ratios produce correspondingly higher currents. While a larger A_E for Q_2 and a higher W/L for M_2 effectively amplify i_{IN}, a smaller A_E and a lower W/L attenuate i_{IN}.

Example 5.1 What is the current-multiplication factor across the BJT mirror in Fig. 5.2?

Solution

Since all transistors conduct the same amount of collector current i_C, input current i_{IN} splits equally into the three input transistors. This means each collector carries a third of i_{IN}, and because two output transistors sink output current i_O, i_O is twice i_C and two-thirds i_{IN}:

$$i_O = 2i_C \approx 2\left(\frac{i_{IN}}{3}\right)$$

Example 5.2 What is the current-multiplication factor across the MOSFET mirror in Fig. 5.3?

Solution

Since all transistors conduct the same amount of drain current i_D and the input transistor carries input current i_{IN}, each drain carries i_{IN} and, because two output transistors sink output current i_O, i_O is twice i_D and therefore twice i_{IN}:

$$i_O = 2i_D \approx 2i_{IN}$$

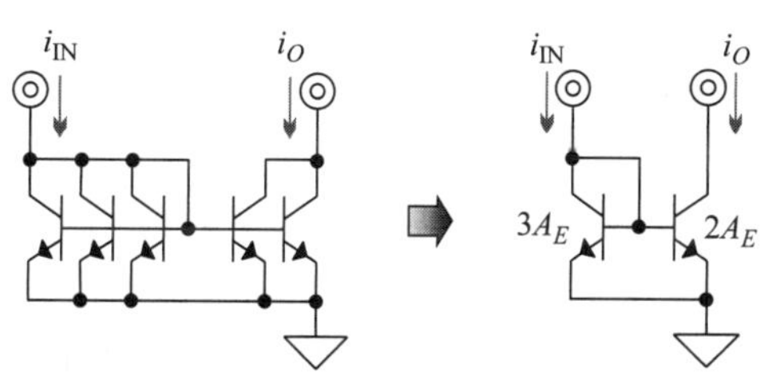

FIGURE 5.2 BJT current-mirror example.

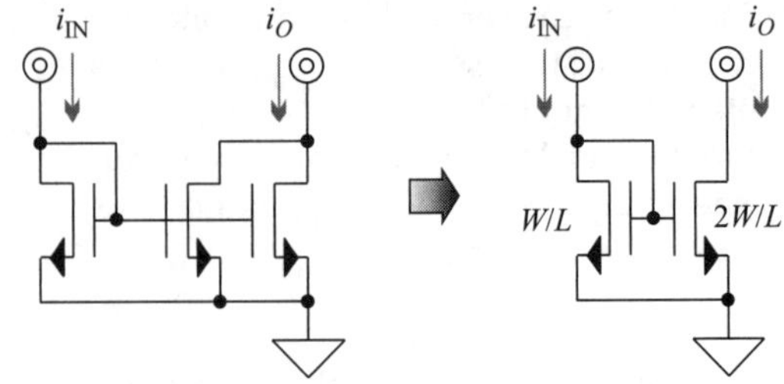

FIGURE 5.3 MOSFET current-mirror example.

Base-Current Error

Although base currents are β_0 times lower, BJT bases steer currents away from input current i_{IN}. This means the collector current i_{C1} of the input transistor Q_1 in Fig. 5.1 is lower than i_{IN} by a base-current error i_E. Since the mirror reproduces i_{C1} in i_O according to the mirror's current gain A_I, i_O includes a gain translation of this error:

$$i_O = A_I i_{C1} = A_I(i_{IN} - i_E) \approx A_I\left[i_{IN} - \left(\frac{i_{IN}}{\beta_0} + \frac{i_O}{\beta_0}\right)\right] \tag{5.3}$$

Error i_E increases with higher gain A_I or more outputs because the number of base currents rises with more output transistors. Luckily, FET gates conduct little to no current, so FETs are exempt from this type of error.

Voltage Error

Although weakly dependent, mismatched collector–emitter and drain–source voltages v_{CE} and v_{DS} between mirroring BJTs and FETs produce an error in output current i_O. Base-width modulation, which Early voltage V_A models, causes this offset in BJTs:

$$A_I = \frac{i_{C2}}{i_{C1}} = \frac{I_{S2}\left[\exp\left(\frac{v_{BE}}{V_t}\right) - 1\right]\left(1 + \frac{v_O}{V_A}\right)}{I_{S1}\left[\exp\left(\frac{v_{BE}}{V_t}\right) - 1\right]\left(1 + \frac{v_{IN}}{V_A}\right)} = \left(\frac{A_{E2}}{A_{E1}}\right)\left(\frac{1 + \frac{v_O}{V_A}}{1 + \frac{v_{BE}}{V_A}}\right) \tag{5.4}$$

where I_{S2}/I_{S1} is equivalent to A_{E2}/A_{E1} and input voltage v_{IN} is the base–emitter voltage of the mirroring transistors. Channel-length modulation, which λ models, imposes similar effects in FETs:

$$A_I = \frac{i_{D2}}{i_{D1}} = \frac{i_O}{i_{IN}} = \frac{\left(\frac{W}{L}\right)_2 K'(v_{GS} - v_T)^2(1 + \lambda v_O)}{\left(\frac{W}{L}\right)_1 K'(v_{GS} - v_T)^2(1 + \lambda v_{IN})} = \left(\frac{S_2}{S_1}\right)\left(\frac{1 + \lambda v_O}{1 + \lambda v_{GS}}\right) \tag{5.5}$$

where S_1 and S_2 are $(W/L)_2$ and $(W/L)_1$. Here, however, transistors with longer channels suffer less from these effects because modulating the length of longer channels constitutes a smaller overall fractional variation. In other words, design engineers can reduce this error by lengthening the channels.

5.1.2 Small-Signal Model

IC designers call BJTs whose collectors and bases short, like Q_1 does in the mirror of Fig. 5.1, *diode-connected transistors* because their terminal voltages drop the diode voltage of the base–emitter junction. Since FETs parallel BJTs in analog circuits, FETs with drains and gates connected share the same appellation. These types of transistors exhibit a couple of characteristics worth mentioning. First, pushing current into a diode-connected transistor raises its terminal-to-terminal voltage until the voltage across the device is sufficiently high to sustain the current. This is an example of negative feedback, as Chap. 6 explains in greater detail, because the transistor adjusts v_{BE} automatically until i_C matches incoming current i_{IN}. Another peculiarity here is the transistor's small-signal resistance R_D, which overall is the parallel combination of r_π, g_m's R_{gm}, and r_o:

$$R_D = r_\pi \,||\, R_{gm} \,||\, r_o = r_\pi \,||\, \frac{1}{g_m} \,||\, r_o \approx \frac{1}{g_m} \tag{5.6}$$

where current $v_{be}g_m$ is a linear translation of its own two terminal voltages v_{be}, like that of a $1/g_m$ resistor, and r_π is absent in FETs. In other words, the resistance of diode-connected devices is normally low at $1/g_m$.

Since the input of a mirror includes a diode-connected transistor, the mirror's input resistance R_{IN} typically reduces to $1/g_m$, and more specifically, to $1/g_{m1}$ in the circuits shown. The circuit's equivalent short-circuit transconductance G_M and output resistance R_O are nothing but Q_2's g_{m2} and r_{o2}, respectively. Ultimately, this means that a source resistance R_S at the input steers current away from R_{IN}, and R_O steals current from the loading resistance R_{LOAD} that the mirror drives. In other words, R_S and R_O reduce the effective low-frequency current gain A_{I0} of the mirror to

$$\begin{aligned} A_{I0} &\equiv \frac{i_{load}}{i_{in}} = \left(\frac{v_{in}}{i_{in}}\right)\left(\frac{i_{gm2}}{v_{in}}\right)\left(\frac{v_o}{i_{gm2}}\right)\left(\frac{i_{load}}{v_o}\right) \\ &= (R_S \,||\, R_{IN})(-G_M)(R_O \,||\, R_{LOAD})\left(\frac{1}{R_{LOAD}}\right) \\ &\approx \left(R_S \,||\, \frac{1}{g_{m1}}\right)(-g_{m2})\left(\frac{r_{o2} \,||\, R_{LOAD}}{R_{LOAD}}\right) \end{aligned} \tag{5.7}$$

Ideal conditions therefore call for high R_S and low R_{LOAD} values, or said differently, good mirrors exhibit low input resistances R_{IN} and high output resistances R_O.

5.1.3 Base-Current Correction

Compensation

The ultimate effect of base currents in a BJT mirror is to reduce the output current i_O. When viewed in this light, compensating the error amounts to incrementing i_O by the mirror-gain translation of the base currents that reduced i_{IN} in the first place. One way to do this is to raise Q_2's base–emitter voltage v_{BE2}. Luckily, Q_1's base current i_{B1} flows into its base, so placing a resistor R_β in series with Q_1's base, as Fig. 5.4 illustrates, drops a voltage that raises v_{BE2} by v_R or $i_{B1}R_\beta$ with respect to Q_1's v_{BE1}, which carries $i_{IN} - i_E$. This way, since base currents in BJTs are so much smaller than collector currents and they therefore amount to small-signal variations, Q_2's g_{m2} translates v_R into a rise in i_O that is equivalent to Q_1's i_{B1} plus Q_2's i_{B2} when R_β is $(1 + A_I)/g_{m2}$:

$$\Delta i_O \approx v_R g_{m2} = i_{B1} R_\beta g_{m2}\Big|_{R_\beta \equiv \frac{1+A_I}{g_{m2}}} \equiv i_E = i_{B1} + i_{B2} = i_{B1}(1 + A_I) \tag{5.8}$$

where A_I is the mirror's gain and R_β is the design variable with which engineers adjust i_O.

A few issues render this solution imperfect, although certainly not useless. For one, v_R and v_{BE2} do not necessarily track across temperature. Secondly, g_{m2} changes with Q_2's bias, so v_R does not track g_{m2} across wide output-current variations. Third, R_β can vary ±20 percent across process, so without trimming, i_O's resulting correction does not track i_E across process. Still, even after all these defects, R_β reduces the error between i_{IN} and i_O, the benefits of which are more profound in mirrors with high gain, whose initial error is correspondingly higher than in low-gain mirrors.

Reduction and Elimination

A more effective means of mitigating the effects of base currents on i_O is to reduce the current drawn from i_{IN} in the first place. Voltage follower Q_B in Fig. 5.5, for example, can reduce the impact of base currents on i_{IN} by a factor of $1 + \beta_0$. Unfortunately, biasing Q_B with the trickle of current that two base currents produce can reduce Q_B's g_m and therefore raise its emitter resistance $1/g_m$ to such an extent that

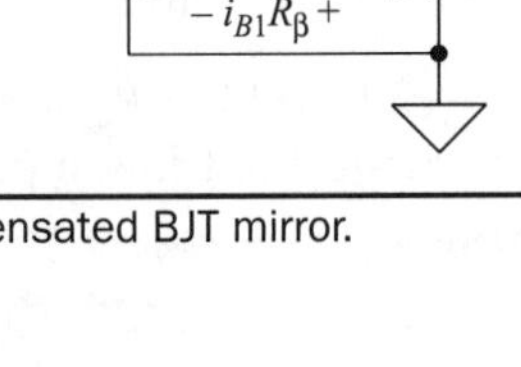

FIGURE 5.4 Base-compensated BJT mirror.

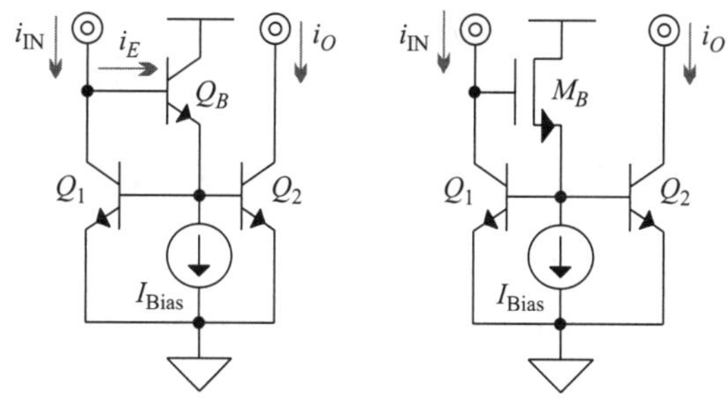

FIGURE 5.5 β-helping transistors in BJT mirrors.

Q_B slows down. This means the mirror cannot respond quickly to either changes in i_{IN} or noise injected into the circuit. This is why designers usually include a bias current I_{Bias} in the mix, to raise g_m and lower $1/g_m$. This way, error current i_E becomes

$$i_E \approx \frac{\frac{i_{IN}}{\beta_0} + \frac{i_O}{\beta_0} + I_{Bias}}{1+\beta_0} \approx \frac{i_{IN}(1+A_I)}{\beta_0^{\ 2}} + \frac{I_{Bias}}{\beta_0} \tag{5.9}$$

Even more than reducing i_E, a FET like the one in Fig. 5.5 altogether eliminates it. This, of course, assumes FETs are available in the process technology used, in which case replacing the entire BJT mirror with its FET counterpart is also an option.

No benefits in the world of design result without drawbacks. In this case, not only does the circuit require more quiescent power but input voltage v_{IN} is now also two diode voltages, or one diode and one gate–source voltage v_{GS}. This means v_{IN} can be anywhere from 0.8 to 2.5 V depending on the operating temperature range and the FET. Still, these input voltages are not always problematic. Similarly, base error in the basic BJT mirror may not be a problem in the application at hand. In other words, like with everything else, fixing what is not broken is a futile exercise that, far from helping, can add risk.

5.1.4 Voltage-Correcting Cascodes

Dissimilar collector–emitter and drain–source voltages v_{CE} and v_{DS} produce an error that, although often acceptable, nonetheless limits mirroring accuracy. Eliminating this error amounts to matching v_{CE}'s and v_{DS}'s in Fig. 5.1, which is the aim of Q_3 and Q_4 and M_3 and M_4 in Fig. 5.6. Since these *cascode transistors* carry equivalent current densities, their base–emitter and gate–source voltages v_{BE} and v_{GS} match. As a result, they impress the voltages across Q_1 and M_1 onto Q_2 and M_2. This way, both v_{CE}'s are at v_{BE1} and both v_{DS}'s are at v_{GS1}.

As is often the case with fixes, adding cascodes is not an ideal solution. For one, the input terminal now drops two diode or gate–source voltages, instead of just one. Plus, the output voltage must now remain

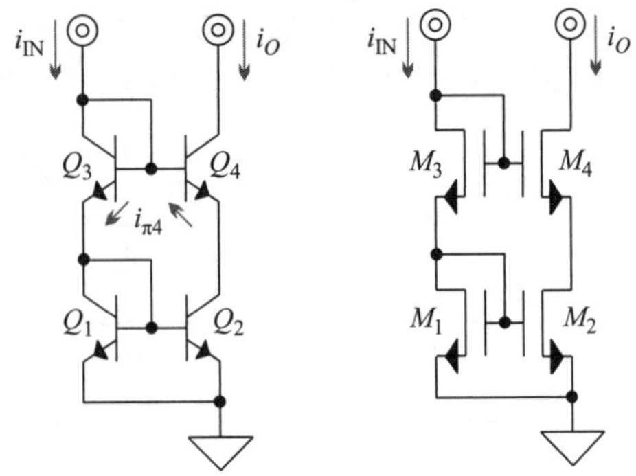

FIGURE 5.6 Voltage-correcting cascodes in BJT and MOSFET mirrors.

above Q_2's and M_2's superimposed v_{BE1} and v_{GS1} by Q_4's and M_4's $V_{CE4(MIN)}$ and $V_{DS4(SAT)}$ to keep Q_4 in forward active or light saturation and M_4 in saturation. In other words, the circuit's minimum output voltage is now higher by a diode or a gate–source voltage. Unfortunately, raising these two voltages limits the headroom under which sources and loads process signals with respect to their supplies, which can be as low as 0.9–2.7 V in battery-operated environments.

In addition, input resistance R_{IN} is now higher. To be more specific, R_{IN} is Q_4's equivalent base resistance R_{B4} in parallel with the series combination of Q_1 and Q_3 whose equivalent resistance R_{C3} represents. Interestingly, Q_4's base current i_{b4} is a β_0 translation of Q_2's collector current, which in turn is a mirrored reproduction of Q_1 and Q_3's current i_{e3}, which means R_{B4} is $\beta_0 + 1$ times higher than R_{C3}:

$$R_{B4} \equiv \frac{v_{in}}{i_{b4}} = v_{in}\left(\frac{1+\beta_0}{i_{e3}}\right) \approx R_{C3}(1+\beta_0) \tag{5.10}$$

As a result, R_{C3} dominates R_{B4} in R_{IN} to reduce R_{IN} to the series resistance of two diode-connected transistors:

$$R_{IN} = R_{C3} \,||\, R_{B4} = R_{C3} \,||\, [R_{C3}(1+\beta_0)] \approx R_{C3} = \frac{1}{g_{m3}} + \frac{1}{g_{m1}} \tag{5.11}$$

Granted, R_{IN} is now higher than without the cascodes, but not by much.

Still more, the voltage-correcting fix is also imperfect. The fact is that the voltages across the cascodes do not necessarily match, so although output current i_O largely matches input current i_{IN}, output voltage variations nonetheless alter i_O, even if just by a small degree. This is another way of saying that the mirror's effective output resistance R_O is finite. To be more precise, R_O is the output resistance of an emitter-degenerated transistor Q_4. In this case, though, Q_2's g_{m2} mirrors

Q_4's $r_{\pi4}$ current $i_{\pi4}$ to establish a degenerating resistance $R_{DEG4}{}'$ that is equivalent to $r_{\pi4} \| r_{\pi4} \| r_{o2}$:

$$R_{DEG4}{}' = \frac{v_{e4}}{i_{r\pi4} + i_{gm2} + i_{ro2}} = \frac{v_{e4}}{2i_{r\pi4} + i_{ro2}}$$

$$= \left[\left(\frac{1}{2}\right)\left(r_{\pi4} + \frac{1}{g_{m3}} + \frac{1}{g_{m1}}\right)\right] \| r_{o2} \approx r_{\pi4} \| r_{\pi4} \| r_{o2} \qquad (5.12)$$

where $r_{\pi4}$ overwhelms the effects of Q_3's and Q_1's diode-connected resistances $1/g_{m3}$ and $1/g_{m1}$. As a result, degenerated R_O reduces to $0.5\beta_0 r_{o4}$ in BJTs and $g_{m4} r_{ds4} r_{ds2}$ in FETs:

$$R_O = r_{o4} + R_{DEG4}{}' + g_{m4} r_{o4} R_{DEG4}{}' \approx g_{m4} r_{o4} R_{DEG4}{}' \approx g_{m4} r_{o4} (r_{o2} \| r_{\pi4} \| r_{\pi4}) \quad (5.13)$$

where $g_{m4} r_{\pi4}$ is β_0 in the former and $r_{\pi4}$ is absent in the latter. Either way, R_O can be 50 to 100 times higher than r_{o2} and r_{ds2}, which means the mirroring imperfection that results is oftentimes a negligible defect in the mirror.

5.1.5 Low-Voltage Cascodes

Separate Bias

A way of reducing the overhead voltage that cascodes in current mirrors require for their output currents to remain accurate is to bias them separately, as Fig. 5.7 shows. This architecture still mirrors current because Q_3 is a common-base transistor that buffers Q_1's current to feed it to Q_1's base. In other words, Q_3 implements the feedback connection that shorting the collector–base terminals otherwise achieves in diode-connected devices. This means v_{IN} is not only one diode or one gate–source voltage but R_{IN} is also $1/g_{m1}$.

But not to digress, the ultimate aim of the low-voltage cascodes is to reduce the voltages across both mirroring transistors Q_1 and Q_2 and M_1 and M_2. In the BJT case, since the base–emitter voltages of the

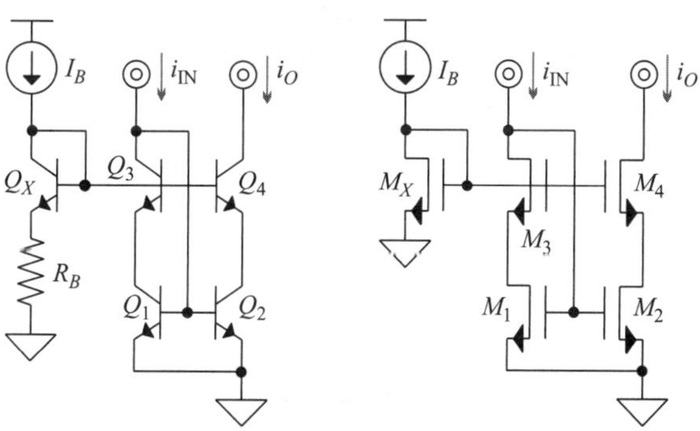

FIGURE 5.7 Separately biased low-voltage cascodes in BJT and MOSFET mirrors.

cascodes and the biasing transistor Q_X for the most part match, Q_X, Q_3, and Q_4 impress R_B's voltage I_BR_B across Q_1's and Q_2's collector–emitter voltages v_{CE1} and v_{CE2}. So for Q_1 and Q_2 to remain in light saturation, I_BR_B must be greater than $V_{CE(MIN)}$. Unfortunately, however, I_BR_B does not track $V_{CE(MIN)}$ across current, temperature, or process, so engineers usually include sufficient margin to ensure that I_BR_B remains above $V_{CE(MIN)}$ across both temperature and process. This ultimately means v_O can approach, but not drop to two $V_{CE(MIN)}$'s.

Luckily, MOSFETs offer an appealing alternative. In the example shown, the voltages across mirroring transistors M_1 and M_2 are the difference between the gate–source voltages of the cascodes and the biasing transistor M_X. In other words, assuming bulk terminals connect to their respective sources, M_X's threshold voltage v_{TN} in v_{GSX} cancels those of the cascodes M_3 and M_4 to impress the difference of their saturation voltages $V_{DSX(SAT)}$ and $V_{DS4(SAT)}$ across M_1 and M_2:

$$v_{DS1} = v_{DS2} = v_{GSX} - v_{GS4} \approx V_{DSX(SAT)} - V_{DS4(SAT)} > V_{DS1(SAT)} = V_{DS2(SAT)} \quad (5.14)$$

So defining a large enough $V_{DSX(SAT)}$ ensures that v_{DS1} and v_{DS2} remain above $V_{DS1(SAT)}$, or equivalently, because they match, $V_{DS2(SAT)}$. What is nice about this is that threshold voltages cancel and transconductance parameter K_N' affects both $V_{DSX(SAT)}$ and $V_{DS4(SAT)}$ in similar ways, which ultimately means the difference can track $V_{DS1(SAT)}$ better. The end result is that less margin may be necessary, and v_O can therefore be closer to two saturation voltages than BJTs can to two $V_{CE(MIN)}$'s.

Self-Bias

One disadvantage with these two last low-voltage circuits is that Q_X and M_X require a bias current I_B that not only dissipates power but also mismatches i_{IN}. A way to circumvent this separate bias requirement is to incorporate the bias generator into the mirror. For example, since Q_1's and M_1's diode connections generate base–emitter and gate–source voltages v_{BE1} and v_{GS1} that track those of their cascodes, inserting a resistor in series with the diode connections, as Fig. 5.8 shows, generates a voltage above v_{BE1} and v_{GS1} that can ensure Q_1 and Q_2 and M_1 and M_2 remain in light saturation and saturation, respectively. Like before, though, R_B's voltages $i_{IN}R_B$ do not track $V_{CE(MIN)}$ or $V_{DS(SAT)}$ across current, temperature, or process, so they must, again, include margin. On the positive side, they rise with i_{IN}, like $V_{DS(SAT)}$, and to a lesser degree, like $V_{CE(MIN)}$. Irrespective of these nuances, this technique reduces v_O's minimum overhead with little to no additional sacrifice in cost, power, or complexity.

5.2 Differential Pair

BJTs and FETs process small signals so similarly that analog circuit topologies that work for one also work for the other. Perhaps the most significant difference between the two technologies is the presence of

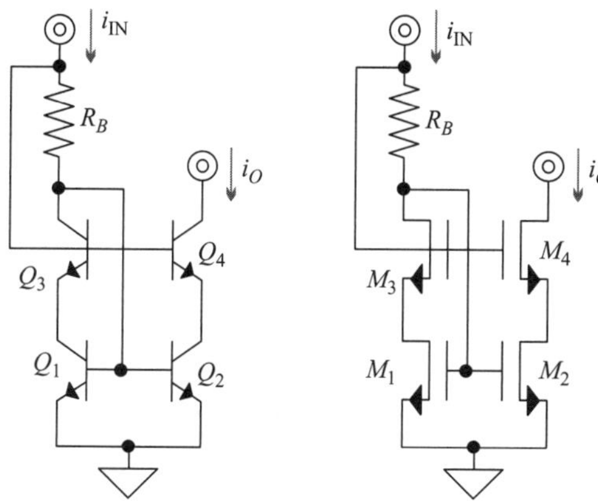

FIGURE 5.8 Self-biased low-voltage cascodes in BJT and MOSFET mirrors.

r_π in BJTs and g_{mb} in FETs, where the latter disappears when bulk and sources connect. Fortunately, removing r_π and adding the effects of g_{mb} to all relations like Chap. 4 shows transforms BJT derivations into their FET counterparts without much trouble. And to generalize further, because the only difference between N- and P-type transistors is the type of charge carrier they process (whose net effect is to produce currents that flow in opposite directions), P-type circuits are the vertical mirror images of N-type circuits. So for simplicity, clarity, and consistency in light of all this, the following discussions use NPN BJTs to illustrate how differential pairs operate. To avoid leaving matters unexplained, this section nonetheless comments on FET realizations, and the next section reverse the strategy, focusing on FET circuits and commenting on their BJT counterparts.

As mentioned earlier in the chapter, differential pairs are important because they reject common-mode noise and can generate the error signal with which feedback loops control and regulate voltages. At its core, the differential pair is an emitter- or source-coupled pair of matched transistors whose combined current a bias current into the common emitter/source node fixes to a constant value. This way, bias tail current I_{Tail} in Fig. 5.9 splits equally between Q_1 and Q_2 when their base voltages match. What is appealing about this is that raising or lowering both base voltages does not change how I_{Tail} splits. Raising one voltage above the other, on the other hand, increases the base–emitter voltage of one transistor as much as it reduces that of the other to establish symmetrical imbalance in currents. In other words, while input *common-mode voltages* v_{IC} and noise produce little to no effect in the circuit, *differential input voltages* v_{ID} produce linear and symmetrical variations in currents.

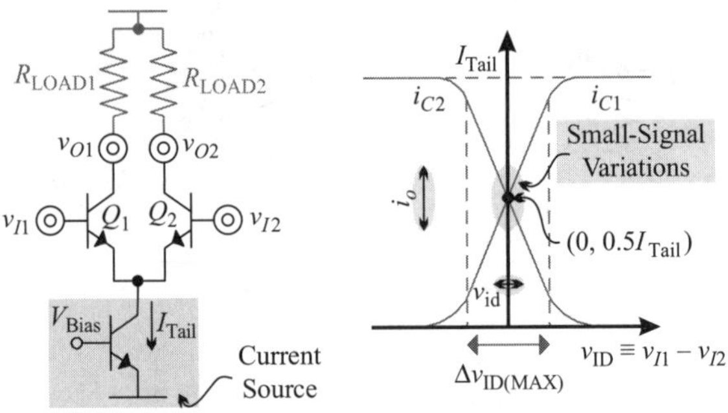

Figure 5.9 Differential NPN BJT pair and its corresponding large-signal response.

5.2.1 Large-Signal Operation

When balanced, as already mentioned, I_{Tail} splits equally, so half of I_{Tail} biases each transistor. Because base–emitter terminals drop a diode voltage, however, their common-mode voltage v_{IC} cannot fall indefinitely without "crushing" the current source that establishes I_{Tail}. This means that v_{IC} should be high enough above the negative supply v_{EE} to drop one base–emitter voltage and the minimum voltage that I_{Tail} can withstand before succumbing to low-voltage effects:

$$v_{EE} + V_{Tail(MIN)} + v_{BE1} < v_{IC} < v_{CC} - v_{LOAD} - V_{CE1(MIN)} + v_{BE1} \tag{5.15}$$

Similarly, v_{IC} should not rise to the extent that one or both transistors in the differential pair "crush" against their loads, at and beyond which point they would enter deep saturation. In other words, v_{IC} should remain low enough to drop the load voltage and one minimum collector–emitter voltage between the positive supply voltage v_{CC} and a base–emitter voltage below v_{IC}. This range, across which transistors remain in forward active or light saturation in BJTs and saturation in FETs, is what engineers call *input common-mode range* (ICMR).

In the case of FETs, v_{IC} must similarly rise above negative supply v_{SS} by the minimum voltage that the tail bias current can sustain without succumbing to low-voltage effects and one gate–source voltage:

$$v_{IC} > v_{SS} + V_{Tail(MIN)} + v_{GS1} = v_{SS} + V_{Tail(MIN)} + v_{TN1} + V_{DS1(SAT)} \tag{5.16}$$

The difference here is that v_{GS1} is the sum of threshold voltage v_{TN1} and $V_{DS1(SAT)}$. On the other end, v_{IC} should remain low enough to drop the load voltage and one saturation voltage between the positive supply voltage v_{DD} and one gate–source voltage below v_{IC}:

$$v_{IC} < v_{DD} - v_{LOAD} - V_{DS1(SAT)} + v_{GS1} = v_{DD} - v_{LOAD} + v_{TN1} \tag{5.17}$$

Here, however, v_{GS1}'s $V_{DS1(SAT)}$ cancels the other $V_{DS1(SAT)}$ to exempt v_{IC}'s upper limit from the effects of saturation voltages in the differential pair.

Irrespective of transistor type, a differential voltage v_{ID} across the input terminals causes one transistor to conduct more current than the other, and as long as v_{IC} remains within the circuit's ICMR, the sum of the currents remains constant at I_{Tail}. So when v_{ID} or $v_{I1} - v_{I2}$ in Fig. 5.9 is negative and substantial in magnitude, Q_1 shuts and Q_2's current i_{C2} carries all of I_{Tail}. Raising v_{ID} does not change the situation until Q_1 begins to engage. At this point, since both currents conduct I_{Tail}, i_{C2} loses what Q_1 conducts. As a result, Q_1's current i_{C1} rises and i_{C2} falls as v_{ID} climbs, until v_{I2} is so far below v_{I1} that Q_2 shuts and i_{C1} carries all of I_{Tail}. In other words, i_{C1} and i_{C2} traverse 0 to I_{Tail} across what amounts to v_{ID}'s maximum *linear range* $\Delta v_{ID(MAX)}$.

Decomposing input voltages v_{I1} and v_{I2} into their common- and differential-mode components is helpful in determining to what extent differential pairs reject v_{IC} and process v_{ID}. With this in mind, since the "common" voltage to both inputs is basically their average $0.5(v_{I1} + v_{I2})$ and subtracting a negative half differential $-0.5v_{ID}$ from a positive counterpart $+0.5v_{ID}$ produces a total difference v_{ID} that is independent of v_{IC}, v_{I1} is equivalent to $v_{IC} + 0.5v_{ID}$ and v_{I2} to $v_{IC} - 0.5v_{ID}$. This means that each input carries v_{IC} and an opposing half of v_{ID}.

5.2.2 Differential Signals

When biased properly, v_{IC}'s and v_{ID}'s steady-state values V_{IC} and V_{ID} are within the circuit's ICMR and linear range $\Delta v_{ID(MAX)}$, respectively. Plus, when only considering small differential signals, v_{IC}'s small-signal components are nil, and v_{ID} only includes small signals v_{id}. In other words, v_{I1} reduces to $0.5v_{id}$ and v_{I2} to $-0.5v_{id}$ like Fig. 5.10 shows.

Since constant currents do not change, the current source in the Norton-equivalent circuit that models the tail current I_{Tail} disappears from the small-signal equivalent circuit. This means Q_1's and Q_2's small-signal currents ultimately flow through the tail current's output

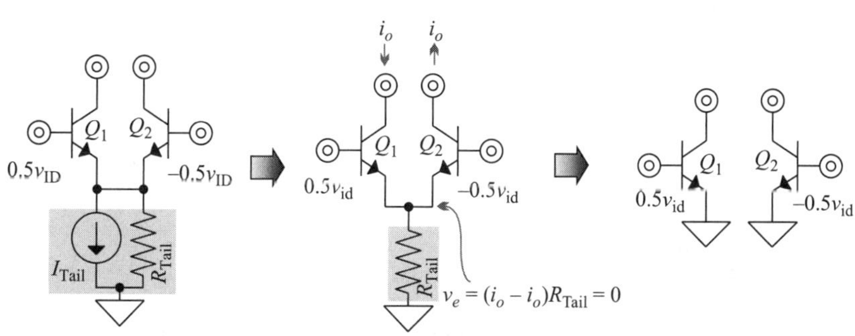

FIGURE 5.10 Differential small-signal half-circuit transformation of the differential pair.

resistance R_{Tail}. But since Q_1's input is as high as Q_2's input is low, or vice versa, their currents are equal and opposite, and as a result, R_{Tail} conducts zero current to drop zero volts. So, because Q_1 and Q_2's common-emitter node is zero, the transistors reduce to two common-emitter transconductors whose base voltages carry $0.5v_{\text{id}}$ and $-0.5v_{\text{id}}$. This transformation is valid only when the transistors carry equal but opposite halves of the differential voltage, that is, when they exclude common-mode components.

The differential input resistance R_{ID} of the two-port circuit that models the differential pair describes how much current i_{id} a small-signal voltage v_{id} produces across the inputs. So since each BJT's r_π receives half of v_{id}, R_{ID} is twice r_π:

$$R_{\text{ID}} \equiv \frac{v_{\text{id}}}{i_{\text{id}}} = \frac{v_{\text{id}}}{\left(\dfrac{0.5v_{\text{id}}}{r_\pi}\right)} = 2r_\pi \tag{5.18}$$

But because gate currents are zero in FETs, R_{ID} for FET pairs approaches infinity. Differential output resistance R_{OD} similarly refers to the current i_{od} that a small-signal voltage v_{od} across the outputs generates. So again, because each output carries half of v_{od}, R_{OD} is twice r_o:

$$R_{\text{OD}} \equiv \frac{v_{\text{od}}}{i_{\text{od}}} = \frac{v_{\text{od}}}{\left(\dfrac{0.5v_{\text{od}}}{r_o}\right)} = 2r_o \tag{5.19}$$

Plus, since Q_1 and Q_2 receive equal but opposite halves of v_{id}, the difference between their currents is two times what $0.5v_{\text{id}}$ produces in each transistor, so differential transconductance G_D is twice $0.5g_m$:

$$G_D \equiv \frac{i_{o1} - i_{o2}}{v_{\text{id}}} = \frac{i_{o1}}{v_{\text{id}}} - \frac{i_{o2}}{v_{\text{id}}} = \frac{(-0.5v_{\text{id}}g_m)}{v_{\text{id}}} - \frac{0.5v_{\text{id}}g_m}{v_{\text{id}}} = -g_m \tag{5.20}$$

When just considering Q_1's output, however, since Q_1 receives half of the input, the differential low-frequency gain A_{D10} is

$$A_{\text{D10}} \equiv \frac{v_{o1}}{v_{id}} = -(0.5g_m)(r_o \,\|\, R_{\text{LOAD}}) \tag{5.21}$$

And because Q_2 receives the opposite half of v_{id}, Q_2's differential low-frequency gain A_{D20} is similar, but in the opposite direction:

$$A_{\text{D20}} \equiv \frac{v_{o1}}{v_{\text{id}}} = -(-0.5g_m)(r_o \,\|\, R_{\text{LOAD}}) = 0.5g_m(r_o \,\|\, R_{\text{LOAD}}) \tag{5.22}$$

This means the differential-to-differential low-frequency voltage gain A_{DD0} across both outputs is twice the gain of each transistor:

$$A_{\text{DD0}} \equiv \frac{v_{\text{od}}}{v_{\text{id}}} \equiv \frac{v_{o1} - v_{o2}}{v_{\text{id}}} = -0.5g_m(r_o \parallel R_{\text{LOAD}}) - 0.5g_m(r_o \parallel R_{\text{LOAD}})$$
$$= -g_m(r_o \parallel R_{\text{LOAD}}) \tag{5.23}$$

As already mentioned, the maximum linear range of the circuit is the differential voltage $\Delta v_{\text{ID(MAX)}}$ that produces a full-range 0-to-I_{Tail} variation in currents. Interestingly, G_D reduces $\Delta v_{\text{ID(MAX)}}$ to $2V_t$ in BJTs and $V_{\text{DS(SAT)}}$ in FETs:

$$\Delta v_{\text{ID(MAX)}}\Big|_{\text{BJT}} \equiv \frac{\Delta i_{\text{C(MAX)}}}{G_D} \approx \frac{I_{\text{Tail}}}{g_m} = \frac{I_{\text{Tail}} V_t}{0.5 I_{\text{Tail}}} = 2V_t \tag{5.24}$$

and

$$\Delta v_{\text{ID(MAX)}}\Big|_{\text{FET}} \equiv \frac{\Delta i_{\text{D(MAX)}}}{G_{\text{D}}} \approx \frac{I_{\text{Tail}}}{g_m} = \frac{I_{\text{Tail}}}{\sqrt{2(0.5I_{\text{Tail}})K'\left(\frac{W}{L}\right)}}$$
$$= \sqrt{\frac{2(0.5I_{\text{Tail}})}{K'\left(\frac{W}{L}\right)}} = V_{\text{DS(SAT)}} \tag{5.25}$$

This means that, for the differential pair to remain linear, input variations must not exceed $2V_t$, or 52 mV, at room temperature for BJTs and $V_{\text{DS(SAT)}}$ for FETs, where I_{Tail} and channel width–length ratio W/L set $\Delta v_{\text{ID(MAX)}}$ in the latter to usually 100–500 mV. The fundamental reason why FETs feature wider linear ranges is lower transconductance, because currents rise quadratically with gate–source voltages and exponentially with base–emitter voltages.

5.2.3 Common-Mode Signals

When biased properly and considering only small common-mode signals, v_{IC}'s steady-state value V_{IC} is within the circuit's ICMR and v_{ID} is zero, so v_{I1} and v_{I2} both reduce to v_{ic}, as Fig. 5.11 illustrates. Like in the differential case, I_{Tail} disappears because I_{Tail} is constant and Q_1's and Q_2's small-signal currents flow through the tail current's output resistance R_{Tail}. Here, however, v_{ic} at both bases establishes equivalent base–emitter voltages that produce equal and in-phase currents. As a result, currents combine and flow through R_{Tail} to establish a voltage $(2i_o)R_{\text{Tail}}$ that one current alone would also produce across twice R_{Tail}. In other words, when only considering small common-mode voltages, the differential pair is equivalent to two separate common-emitter transconductors each degenerated with $2R_{\text{Tail}}$.

Since the common-mode input resistance R_{IC} of the circuit is the parallel combination of two emitter-degenerated common-emitter

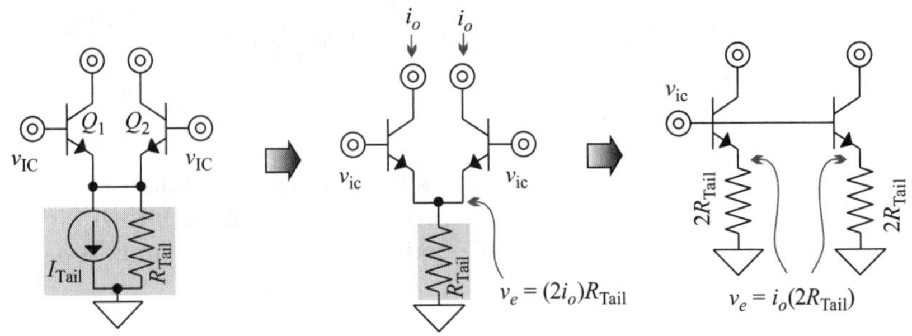

FIGURE 5.11 Common-mode small-signal half-circuit transformation of the differential pair.

transconductors, R_{IC} is half the resistance of each degenerated base and therefore nearly $\beta_0 R_{Tail}$:

$$R_{IC} \equiv R_{B1} \,||\, R_{B2} = 0.5R_B = 0.5[r_\pi + (1+\beta_0)(2R_{Tail})] \approx \beta_0 R_{Tail} \tag{5.26}$$

where R_{Tail} is normally high at or greater than r_o or r_{ds} and R_{IC} for FETs approaches infinity because gate currents are zero. $2R_{Tail}$ similarly degenerates and raises common-mode output resistances R_{OC1} and R_{OC2} to roughly $\beta_0 r_o$:

$$R_{OC1} = R_{OC2} = r_o + g_m r_o (2R_{Tail} \,||\, r_\pi) + (2R_{Tail} \,||\, r_\pi) \approx g_m r_o (2R_{Tail} \,||\, r_\pi) \approx \beta_0 r_o \tag{5.27}$$

where the degenerating effects of r_π disappear in FETs to raise R_{OC1} and R_{OC2} to $2g_m r_{ds} R_{Tail}$. Not surprisingly, $2R_{Tail}$ also degenerates Q_1's and Q_2's transconductances to

$$G_{C1} = G_{C2} \equiv \frac{i_{o1}}{v_{ic}} = \frac{i_{o2}}{v_{ic}} \approx \frac{g_m}{1 + 2R_{Tail} g_m} \tag{5.28}$$

However, since both output currents are equal and in phase, the difference, which is what differential outputs process, is zero. In other words, common-mode signals do not affect the differential output v_{od} of the circuit, which is to say common-to-differential transconductance G_{CD} is zero:

$$G_{CD} \equiv \frac{i_{od}}{v_{ic}} \equiv \frac{i_{o1} - i_{o2}}{v_{ic}} = \frac{i_{o1}}{v_{ic}} - \frac{i_{o2}}{v_{ic}} = G_{C1} - G_{C2} = 0 \tag{5.29}$$

Since both half circuits are mirror images of one another, the common-mode low-frequency voltage gains A_{C10} and A_{C20} to Q_1's and Q_2's outputs are equally low at

$$A_{C10} \equiv \frac{v_{o1}}{v_{ic}} = \frac{v_{o2}}{v_{ic}} \equiv A_{C20} \approx -\left(\frac{g_m}{1 + 2R_{Tail} g_m}\right)(R_{OC1} \,||\, R_{LOAD}) \tag{5.30}$$

This gain is merely the degenerated counterpart of the differential-to-differential gain of the circuit A_{DD0}. Differential outputs, however, process the voltage difference across transistor outputs, so equal common-mode output voltages in common-to-differential low-frequency gain A_{CD0} cancel to reduce A_{CD0} to zero:

$$A_{CD0} \equiv \frac{v_{od}}{v_{ic}} \equiv \frac{v_{o1} - v_{o2}}{v_{ic}} = \frac{v_{o1}}{v_{ic}} - \frac{v_{o2}}{v_{ic}} = 0 \tag{5.31}$$

which is another way of saying differential pairs reject common-mode signals.

This cancellation, of course, assumes transistors match perfectly, which is impossible in practice. Engineers gauge this imperfection with a parameter that describes how much a circuit favors differential signals over their common-mode counterparts, which they define as the ratio of differential gain A_{DD} to common-mode gain A_{CD} and call *common-mode rejection ratio* (CMRR). Although low-frequency CMRR approaches infinity under ideal conditions,

$$\text{CMRR}_0 \equiv \frac{A_{DD0}}{A_{CD0}} = \frac{A_{DD0}}{0} \to \infty \tag{5.32}$$

mismatches between Q_1 and Q_2 and their loads normally raise A_{CD0} enough to lower CMRR_0 to about 60–80 dB.

5.2.4 Emitter/Source Degeneration

Designers oftentimes extend the linear range of BJT differential pairs whose $\Delta v_{ID(MAX)}$ is normally low at 52 mV at room temperature by degenerating their transconductance. They can afford to reduce the gain this way because the differential voltage gain across the pair can be higher than needed at 50–60 dB. For this purpose, Fig. 5.12 shows how matched resistors R_{DEG1} and R_{DEG2} can degenerate Q_1 and Q_2. In the first case, I_{Tail} splits between R_{DEG1} and R_{DEG2} to establish steady-state and small-signal voltages across R_{DEG1} and R_{DEG2}. In the second example, matched bias currents $0.5I_{Tail}$ pull currents directly from Q_1's

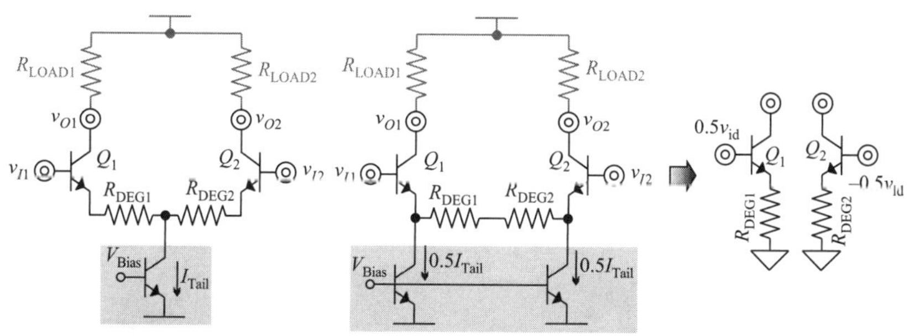

FIGURE 5.12 Degenerated pairs and their differential small-signal equivalent circuit.

and Q_2's respective emitters. Here, since the steady-state voltage across both R_{DEG1} and R_{DEG2} is zero, they do not conduct steady-state currents, and as a result, do not drop steady-state voltages.

Considering that constant currents do not affect small signals, the topological difference between the two examples shown disappears in their small-signal equivalent circuit. That is to say, both pairs reduce to the degenerated common-emitter transconductor shown in Fig. 5.12. As such, R_{DEG1}, which matches R_{DEG2} and is equivalent to R_{DEG}, degenerates the combined differential transconductance G_D' of the pair to:

$$G_D' \approx \frac{g_m}{1+g_m R_{DEG}} \tag{5.33}$$

and extends the linear range $\Delta v_{ID(MAX)}'$ that G_D' establishes by a factor of $1 + g_m R_{DEG}$ to

$$\Delta v_{ID(MAX)}' \equiv \frac{\Delta i_{O(MAX)}}{G_D'} \approx \frac{I_{Tail}(1+g_m R_{DEG})}{g_m} = \Delta v_{ID(MAX)}(1+g_m R_{DEG}) \tag{5.34}$$

Although both circuits degenerate to the same transconductance, pulling two currents from the example that includes two $0.5I_{Tail}$ sources means both bias currents must match for the circuit to balance. In other words, this example incorporates more circuit components whose mismatches establish a parasitic unbalance in the circuit. The advantage of this second example, however, is higher ICMR, because the voltages across R_{DEG1} and R_{DEG2} contribute to ICMR in the first example and not in the second.

5.2.5 CMOS Pairs

As mentioned earlier in this section, the differential pair is a structure that most transistors can realize. Figure 5.13, for example, showcases N-channel MOSFET translations of the nondegenerated and degenerated differential pairs already discussed. The fundamental difference between them and their BJT counterparts is the extent to which g_m in

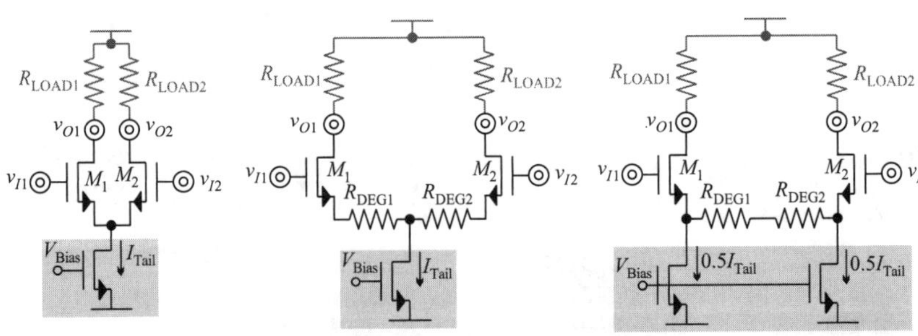

FIGURE 5.13 Nondegenerated and degenerated N-channel MOSFET differential pairs.

BJTs and FETs and the absence of r_π in FETs and g_{mb} in BJTs affect performance, not functionality. For example, voltage gain is lower and linear range is wider in FETs because g_m is lower than in BJTs of similar size and currents, and that results because FETs are square-law devices and BJTs are exponential. Input resistances in FET circuits approach infinity because gate currents are zero. Although gate–source and saturation voltages in FETs relate more clearly than base–emitter and minimum collector–emitter voltages in BJTs, transistors in the structure affect ICMR in similar ways. In all, the differences between BJT and FET circuits are merely numerical, and practically inconsequential with respect to functionality.

5.3 Base/Gate-Coupled Pair

Emitter- and source-coupled pairs process differential signals because, with common emitters and sources, base and gate voltages dictate how much more current one transistor conducts than the other. Similarly, the emitter and source voltages of base- and gate-coupled pairs, an example of which Fig. 5.14 illustrates with transistors Q_1 and Q_2, establish a differential output current. This means base/gate-coupled pairs can also implement differential input stages.

5.3.1 Large-Signal Operation

The input that Q_1 in the base-coupled pair of Fig. 5.14 receives essentially sets the base voltage about which Q_2 operates. More specifically, Q_1 drops the emitter–base voltage necessary to conduct bias current I_{Bias}. If Q_1 and Q_2 match and Q_2's emitter input voltage v_{I2} equals Q_1's v_{I1}, Q_2 also conducts I_{Bias}, and output current i_O is zero. When v_{I2} rises above this point, or in other words, when differential input voltage $v_{I2} - v_{I1}$ or v_{ID} climbs above zero, Q_2's current i_{C2} rises to raise i_O. If on the other hand, v_{ID} falls below zero, i_{C2} drops below I_{Bias} and i_O therefore reverses direction. When v_{I2} decreases well below v_{I1}, Q_2 shuts and i_O saturates to $-I_{Bias}$. The same, however, is not true when v_{ID} rises well above zero.

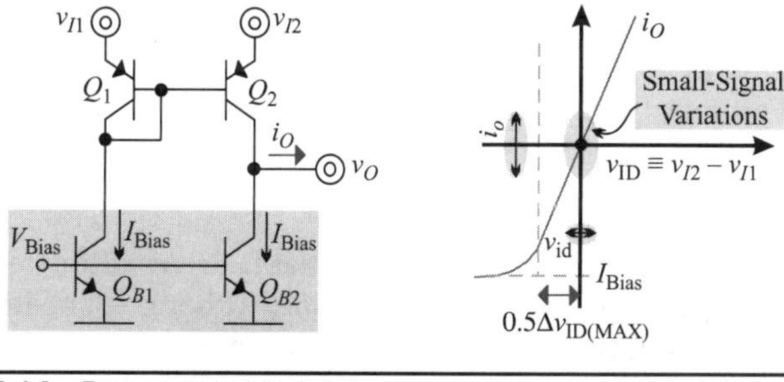

FIGURE 5.14 Base-coupled PNP BJT pair and its corresponding large-signal response.

Regardless, small variations in v_{ID} produce nearly linear variations in both i_{C2} and i_O. From this perspective, saturating i_O to I_{Bias} in the negative direction limits the circuit's linear range Δv_{ID} to $\pm v_{ID(MIN)}$ or

$$\Delta v_{ID(MAX)} = 2v_{ID(MIN)} = 2\left(\frac{i_{O(MIN)}}{G_D}\right) = 2\left(\frac{I_{Bias}}{G_D}\right) \tag{5.35}$$

where G_D is the circuit's equivalent two-port transconductance.

Input Common-Mode Range

Input common-mode voltage v_{IC} in Fig. 5.14 must be high enough to accommodate Q_1's emitter–base voltage v_{EB1} and the minimum voltage $V_{Bias(MIN)}$, or in this case, $V_{CE(MIN)}$ that bias transistor Q_{B1} can tolerate above negative supply voltage v_{EE} before succumbing to low-voltage effects:

$$v_{IC} > v_{EE} + V_{Bias(MIN)} + v_{EB1} = v_{EE} + V_{CE(MIN)} + v_{EB1} \tag{5.36}$$

Luckily, v_{IC} has no limitation in the other direction, except for junction-breakdown voltages. The fact that the ICMR can rise above the circuit's positive supply is one of the most appealing features of this circuit. MOSFETs, of course, enjoy similar benefits:

$$v_{IC} > v_{SS} + V_{Bias(MIN)} + v_{SG1} = v_{SS} + V_{DS(SAT)} + \left|v_{TP}\right| + V_{SD(SAT)} \tag{5.37}$$

where source–gate voltage v_{SG} decomposes into threshold voltage v_{TP} and saturation source–drain voltage $V_{SD(SAT)}$. So as long as v_{IC} stays above its lower limit, i_O remains a function of differential input voltage v_{ID}.

Output Swing

For bias transistor Q_{B2} to remain lightly saturated or forward active, output v_O must not fall within a minimum collector–emitter $V_{CE(MIN)}$ of the negative supply:

$$v_{EE} + V_{CE(MIN)} < v_O < v_{IC} - V_{CE(MIN)} \tag{5.38}$$

v_O must likewise remain below input common-mode voltage v_{IC} by one minimum collector–emitter $V_{CE(MIN)}$ to keep Q_2 in the high-gain region. Fortunately, v_{IC} in the applications that call for this circuit is usually high, so v_O's rising limit is typically not a major source of concern.

Slew Rate

Output current i_O and the capacitance that i_O drives ultimately limit how fast output voltage v_O can rise and fall. Although parasitic base–collector and collector–substrate capacitances at v_O under the guise of equivalent capacitance C_{EQ} may not be disabling in this regard, base–emitter, loading, and intentionally added capacitances can be.

Since load capacitance C_{LOAD} usually includes one or more base–emitter capacitances and maybe even additional capacitance, C_{LOAD} can be substantial enough to limit the circuit's response because i_O can discharge or charge C_{EQ} and C_{LOAD} to the extent that v_O slews down or up across time at a rate of:

$$\text{SR} \equiv \frac{\Delta v_O}{\Delta t} = \frac{dv_O}{dt} = \frac{i_O}{C_{LOAD} + C_{EQ}} = \frac{i_2 - I_{Bias}}{C_{LOAD} + C_{EQ}} \tag{5.39}$$

where *slew rate* (SR) is in volts per second.

Generally speaking, SR describes how fast the output of a circuit can rise or fall in response to large-signal events, like when transistors fully engage or completely shut. Of several possible capacitors in a circuit, the one with the highest capacitance will normally slew before others have a chance to follow suit. In this circuit, lowering v_{I2} can shut Q_2, so i_O can sink up to I_{Bias}, but no more. Luckily, no bias current limits how much Q_2 can source when v_{I2} climbs, so i_O can rise well above I_{Bias} to source considerably more current into the load than I_{Bias} can sink. This means the falling slew rate SR^- is typically more limiting at roughly I_{Bias}/C_{LOAD} than the rising counterpart SR^+ at $(i_{2(MAX)} - I_{Bias})/C_{LOAD}$.

More to the point, Q_2 is the only transistor in the circuit that can conduct considerably more current than I_{Bias}, so Q_2's current is the strength behind the faster slew-rate limit. In fact, mirroring Q_2's current can shift which slew rate the circuit favors. In the case of Fig. 5.15, for example, because Q_{B1} and Q_1 share emitter and base terminals, Q_1 mirrors and sources Q_{B1}'s I_{Bias} and Q_4–Q_3 mirrors and sinks Q_2's i_2. Since Q_1 and Q_2 still share a base connection, currents balance when v_{I1} equals v_{I2} to produce no output current, and differential voltages produce an imbalance that i_O manifests. This way, Q_2 via Q_4–Q_3 sets SR^- to $(i_{2(MAX)} - I_{Bias})/C_{LOAD}$ and Q_1 limits SR^+ to I_{Bias}/C_{LOAD}, so SR^- can be considerably greater than SR^+. The purpose of presenting this circuit is simply to say that base- and gate-coupled pairs can favor either slew-rate limit.

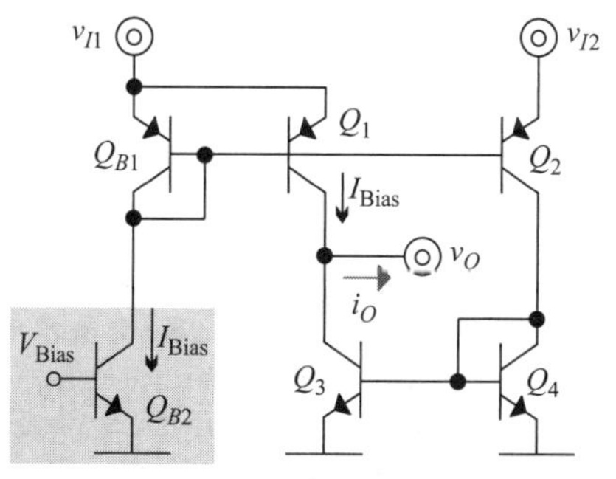

FIGURE 5.15 Folded base-coupled PNP BJT pair.

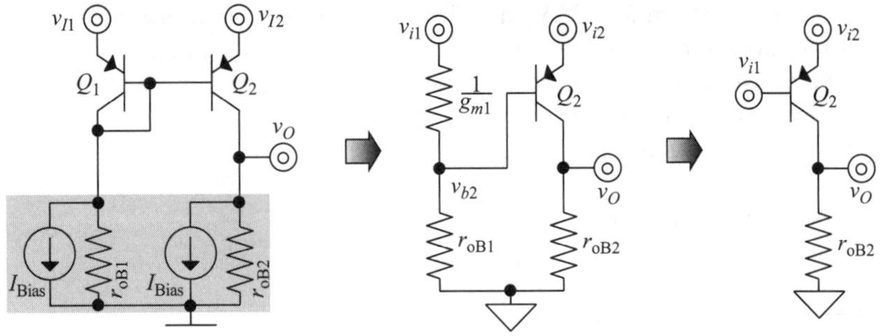

Figure 5.16 Small-signal transformation of the base-coupled pair.

5.3.2 Small-Signal Response

Small-Signal Model

Because bias currents do not carry small signals and power-supply voltages do not change with variations in current, current sources in Q_{B1} and Q_{B2} from Fig. 5.14 disappear when considering small signals and the negative supply voltage becomes ground. Plus, since the resistance across a diode-connected transistor is roughly $1/g_m$, a $1/g_{m1}$ resistance emulates the effects of Q_1 on small signals in the equivalent circuit of Fig. 5.16. In fact, $1/g_{m1}$ is so small relative to Q_{B1}'s r_{oB1} and Q_2's base resistance R_{B2} that v_{b2} practically follows v_{i1}:

$$v_{b2} = \frac{v_{i1}(r_{oB1} \,||\, R_{B2})}{R_{I1}} = \frac{v_{i1}(r_{oB1} \,||\, R_{B2})}{\dfrac{1}{g_{m1}} + (r_{oB1} \,||\, R_{B2})} \approx \frac{v_{i1}(r_{oB1} \,||\, R_{B2})}{r_{oB1} \,||\, R_{B2}} = v_{i1} \tag{5.40}$$

where R_{I1} is Q_1's input resistance. This means Q_1's input voltage v_{i1} essentially drives the base of Q_2. This result is more conclusive in FETs because the absence of $r_{\pi 2}$ removes the effects of R_{B2} in v_{b2} entirely.

Q_1's input resistance R_{I1} is the series combination of Q_1's $1/g_{m1}$ resistance and the parallel combination of Q_{B1}'s r_{oB1} and Q_2's base resistance R_{B2}:

$$R_{I1} = \frac{1}{g_{m1}} + (r_{oB1} \,||\, R_{B2}) \approx r_{oB1} \,||\, R_{B2} = r_{oB1} \,||\, [r_{\pi 2} + (1 + g_{m2} r_{\pi 2}) R_{SI2}] \tag{5.41}$$

where again, $1/g_{m1}$ is so small with respect to Q_{B1}'s r_{oB1} and R_{B2} that $1/g_{m1}$ effectively disappears from R_{I1}. Note that v_{I2}'s source resistance R_{SI2} appears in R_{I1} because R_{SI2} emitter-degenerates Q_2. In other words, r_{oB1} and $r_{\pi 2}$ together with R_{SI2} raise R_{I1} to moderate and maybe even high levels, anywhere from $r_{\pi 2}$ to r_{oB1} depending on R_{SI2}. Note that R_{I1} in Fig. 5.15's folded rendition of the circuit is also high because fixing Q_1's base–emitter voltage to Q_{B1}'s keeps Q_1's current from changing in response to variations in v_{I1}.

Q_2's input resistance R_{I2} is entirely different. Here, Q_2's emitter resistance R_{E2} shunts the effects of Q_2's $r_{\pi 2}$, Q_1's $1/g_{m1}$, v_{I1}'s source resistance R_{SI1}, and Q_{B1}'s r_{oB1} to reduce Q_2's input resistance R_{I2} to R_{E2}, to roughly $2/g_{m2}'$:

$$R_{I2} = \left\{ r_{\pi 2} + \left[r_{oB1} \,||\, \left(\frac{1}{g_{m1}} + R_{SI1} \right) \right] \right\} || R_{E2} \approx R_{E2} = \frac{r_{o2} + r_{oB2}}{1 + g_{m2}' r_{o2}} \approx \frac{2}{g_{m2}'} \quad (5.42)$$

where $1/g_{m1}$ and R_{SI1} base-degenerate g_{m2} to g_{m2}'. In the case of FETs, $r_{\pi 2}$'s R_{B2} disappears from R_{I1} to reduce R_{I1} to r_{oB1}, and $r_{\pi 2}$ opens in R_{I2} to keep R_{I2} near $2/g_{m2}$. Irrespective of the technology, though, R_{I1} is high and R_{I2} is low.

Output resistance R_O is the parallel combination of Q_{B2}'s r_{oB2} and Q_2's emitter-degenerated collector resistance R_{C2}, except r_{oB2} normally shunts R_{C2} to reduce R_O to r_{oB2}:

$$R_O = r_{oB2} \,||\, R_{C2} \approx r_{oB2} \quad (5.43)$$

So, because Q_2's emitter–base voltage in Fig. 5.16 is ultimately $v_{i2} - v_{i1}$ or differential voltage v_{id}, differential transconductance G_D is simply Q_2's g_{m2}:

$$G_D \equiv \frac{i_o}{v_{id}} = \frac{i_o}{v_{i2} - v_{i1}} = g_{m2} \quad (5.44)$$

As a result, G_D's current flows through R_O and load resistance R_{LOAD} to establish a differential low-frequency voltage gain A_{D0} that is roughly equivalent to that of the differential pair at

$$A_{D0} \equiv \frac{v_o}{v_{id}} = G_D(R_O \,||\, R_{LOAD}) \quad (5.45)$$

Frequency Response

In their most basic form, base- and gate-coupled pairs incorporate four nodes: inputs v_{I1} and v_{I2}, common base V_{B2}, and output v_O. Of these, v_O is the one with the highest capacitance and resistance, so parasitic capacitances C_{EQ} and load capacitance C_{LOAD} at v_O shunt R_O and load resistance R_{LOAD} first to produce output pole p_O:

$$\left. \frac{1}{s(C_{LOAD} + C_{EQ})} \right|_{p_o \approx \frac{1}{2\pi(R_O||R_{LOAD})(C_{LOAD}+C_{EQ})}} \equiv R_O \,||\, R_{LOAD} \quad (5.46)$$

Q_1's parasitic capacitance shunts R_{I1} next to produce pole p_{I1}, but if v_{I1}'s R_{SI1} is low, which is normally the case, p_{I1} appears at high frequency. Similarly, Q_2's parasitic capacitance shunts R_{I2} at high frequency because Q_2's emitter reduces R_{I2} to low levels. Q_1's $C_{\pi 1}$ feeds forward an in-phase zero to v_{B2} that parasitic capacitance at v_{B2} offsets with a corresponding pole. All this, though, at high frequencies because Q_1's diode-connected resistance $1/g_{m1}$ is low.

5.3.3 Input-Referred Offset and Noise

Offset

When perfectly balanced, zero volts across the inputs of the base/gate-coupled pair produces a current in Q_2 that matches Q_{B2}'s bias I_{Bias} exactly. Unfortunately, transistors do not match perfectly, so identical base–emitter voltages in Q_1 and Q_2 and Q_{B1} and Q_{B2} establish currents that are not exactly equal. Plus, a collector–emitter voltage difference between Q_1 and Q_2 and Q_{B1} and Q_{B2}, which Δv_C depicts in Fig. 5.17, also produces a current imbalance. The net result is that a finite, nonzero *offset voltage* V_{OS} across the inputs balances the circuit like zero volts would under perfectly matched conditions.

Three-sigma base- and gate-referred mismatches between BJTs and FETs can be 1–3 mV and 5–15 mV, respectively, which are small enough to produce linear effects in the circuit like small signals would. For the circuit to respond linearly to differential input voltages v_{ID}, v_O must also reside at a level that ensures both Q_2 and Q_{B2} are in light saturation or forward active, so what amounts to a small-signal voltage across the bases of Q_1 and Q_2 can produce Δv_C. In other words, the equivalent small-signal model of the base/gate-coupled pair can predict the effects of mismatches between Q_1 and Q_2 in base-referred offset Δv_{12}, Q_{B1} and Q_{B2} in base-referred offset Δv_{B12}, and collector voltages in Δv_C.

While base- and gate-referred mismatches are random, collector and drain voltage differences are not. This means V_{OS} incorporates both *random* and *systemic* components. For statistically meaningful results, systemic variations add to the square-rooted sum of the squares of the random components. So input-referred offset V_{OS} includes the systemic collector-induced voltage $V_{OS(S)}$ that, when amplified by differential voltage gain A_D or $g_{m2}(r_{oB2} \,||\, R_{LOAD})$, cancels Δv_C:

$$V_{OS} = V_{OS(S)} \pm V_{OS}^{*} \approx \frac{\Delta v_C}{g_{m2}(r_{oB2} \,||\, R_{LOAD})} \pm \sqrt{\Delta v_{12}^{\,2} + \left(\frac{\Delta v_{B12} g_{mB2}}{g_{m2}}\right)^2} \tag{5.47}$$

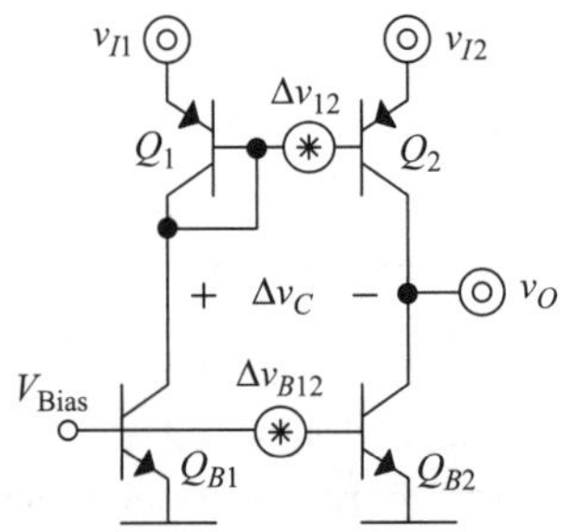

FIGURE 5.17 Offset-producing mismatches in base-coupled pairs.

Similarly, random component V_{OS}^* includes the voltage that, when multiplied by the base-coupled pair's transconductance g_{m2} and divided by Q_{B1} and Q_{B2}'s counterpart g_{mB2}, offsets Q_{B1} and Q_{B2}'s original mismatch Δv_{B12}. However, since Q_1 and Q_2's mismatch Δv_{12} is already across the inputs of the base-coupled pair, no gain translation is necessary for Δv_{12}. All this means that high gains A_V and high pair-to-bias transconductance ratios g_{m2}/g_{mB2} reduce the effects of mismatched voltages Δv_C and Δv_{B12} to yield low input-referred offsets in V_{OS}. Of course, matching Q_1 to Q_2 and Q_{B1} to Q_{B2} and collector voltages is more fundamental and therefore an engineer's first line of attack.

Noise

Surface irregularities across the channel of MOSFETs trap a fraction of moving electrons to cause currents to vary slightly. Fortunately, fewer electrons suffer from this effect when the operating frequency of the current rises, so the noise they produce drops correspondingly. Interestingly, P-type MOSFETs exhibit less of this *one-over-frequency* (1/*f*) *noise* than their N-type counterparts. And because BJT and JFET currents flow well below the silicon surface, BJTs and JFETs do not suffer from 1/*f* noise. Still, charges collide when they cross PN junctions to generate *shot noise*. Similarly, temperature energizes and propels charges in resistive materials and channels, which both BJTs and FETs incorporate, to collide against one another and in the process produce *thermal noise*. Both shot and thermal noise amount to *white noise*, in the sense that they appear uniformly across frequency.

When designing circuits, engineers typically refer these noise effects to the gates and bases of transistors. Because these input-referred signals are on the order of nanovolts per root hertz, they amount to small-signal variations whose propagating effects small-signal equivalent models aptly describe. So to analyze their impact on circuits, engineers first account for all noise generators in the form of gate- and base-referred noise voltages.

In the case of the base-coupled pair, the bases of Q_1, Q_2, Q_{B1}, and Q_{B2} generate noise. Q_{B2}'s transconductance g_{mB2} converts Q_{B1} and Q_{B2}'s combined noise v_{B12}^* to current, and Q_1 and Q_2's transconductance g_{m2} refers that noise current back to an equivalent input noise voltage $v_{B12}^* g_{mB2}/g_{m2}$. Since Q_1 and Q_2's combined contribution v_{12}^* is already at the input, combined *input-referred noise* v_N^* across base/gate-coupled pairs reduces to the square root of the squares of v_{12}^* and $v_{B12}^* g_{mB2}/g_{m2}$:

$$v_N^* = \sqrt{(v_{12}^*)^2 + \left(\frac{v_{B12}^* g_{mB2}}{g_{m2}}\right)^2} \tag{5.48}$$

Note that, because small base- and gate-referred signals model noise in transistors and mismatches between devices, the input-referred effects of noise and offsets in a circuit manifest in similar ways.

5.4 Differential Stage

As before, FETs and BJTs can implement most analog circuits with the same topology because they process small signals in similar ways. P-type circuits are also vertical mirror images of their N-type counterparts because the only physical difference between them is that currents flow in the opposite direction. So like in the previous section, the ensuing discussions illustrate the large- and small-signal behavior of the differential stage with a MOSFET realization. The text nevertheless highlights differences that BJT implementations exhibit. Plus, a sample problem at the end of the section exemplifies how an N-type input stage maps into its P-type complement.

The differential stage discussed here is the outgrowth of the current mirror and differential pair discussed in previous sections. These basic building blocks are largely incomplete on their own because they process currents and voltages without context. The most important use of the former is to mirror currents and the latter is to convert differential voltages into differential currents. But what to do with the differential currents generated? Feed them into a current mirror, like Fig. 5.18 shows, that folds one current into the other and loads the combined result with enough resistance to generate one output voltage that is substantially higher than the differential voltage v_{ID} that fed

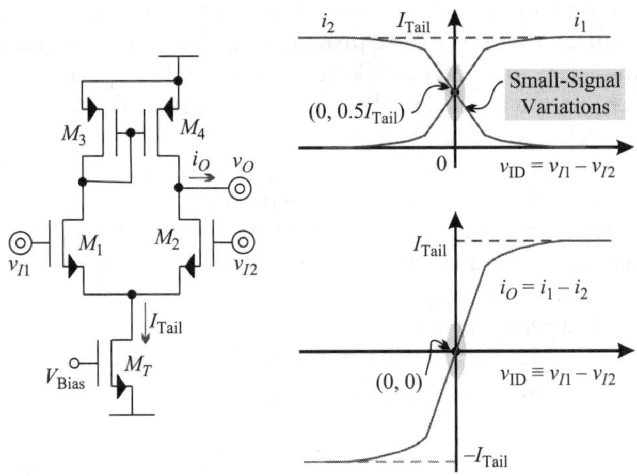

FIGURE 5.18 N-channel metal–oxide–semiconductor-input differential stage and its corresponding large-signal response.

the circuit in the first place. In other words, loading a differential pair with a current mirror amplifies and converts *double-ended differential input signals* into a *single-ended output*.

For the individual blocks to work as prescribed, their constituent transistors must match: M_1 to M_2 in the differential pair and M_3 to M_4 in the mirror. Note that the mirror here serves several functions. First, it replicates and folds M_1's current into that of M_2 for the overall circuit to output the difference. This is important because one current i_O now represents the differential signal that two transistor currents i_1 and i_2 otherwise would. The mirror also derives the currents that bias the differential pair from the pair itself. This way, the mirror supplies just enough current to satisfy the tail current source I_{Tail}. The third and last function of the mirror is to load the differential pair, that is, to convert its combined currents in i_O into a voltage v_O. Perhaps the most notable purpose of the mirror is *double- to single-ended conversion* of differential signals.

5.4.1 Large-Signal Operation

As in the differential pair, because both transistors combined carry I_{Tail}, a voltage difference v_{ID} across the inputs induces one transistor to carry more current by the same amount that the other transistor conducts less. Therefore, a large negative v_{ID} shuts M_1 and prompts M_2 to carry I_{Tail}. Raising v_{ID} eventually engages M_1 to carry current that M_2 loses until M_1's current i_1 peaks at I_{Tail} and M_2 shuts. The fundamental difference between the differential stage and the differential pair is that i_O, which carries i_1 and i_2's difference, climbs twice as much as either i_1 or i_2, from $-I_{Tail}$ to $+I_{Tail}$.

Input Common-Mode Range

Tail current I_{Tail} splits between M_1 and M_2 only if the circuit that generates I_{Tail} has sufficient headroom. So like before, the input's common-mode voltage v_{IC} must be high enough to accommodate, in this case, M_1's gate–source voltage v_{GS1} and the saturation voltage $V_{DST(SAT)}$ of the tail current source M_T above negative supply voltage v_{SS}:

$$v_{IC} > v_{GS1} + V_{Tail(MIN)} + v_{SS} = v_{TN1} + V_{DS1(SAT)} + V_{DST(SAT)} + v_{SS} \tag{5.49}$$

v_{IC} must also be sufficiently low to fit the mirror's source–gate voltage v_{SG3} and the differential pair's $V_{DS1(SAT)}$ between the positive power supply v_{DD} and a gate–source voltage v_{SG1} below v_{IC}:

$$v_{IC} < v_{DD} - v_{SG3} - V_{DS1(SAT)} + v_{GS1} = v_{DD} - \left|v_{TP3}\right| - V_{SD3(SAT)} + v_{TN1} \tag{5.50}$$

except v_{GS1}'s $V_{DS1(SAT)}$ cancels the other one in the relationship. So as long as v_{IC} stays within this ICMR, I_{Tail} remains constant and splits between M_1 and M_2 as just described.

Output Swing

For all transistors in the circuit to remain in their high-gain regions, output v_O must not rise within a saturation voltage $V_{SD4(SAT)}$ of the positive supply v_{DD}:

$$v_{IC} - v_{GS2} + V_{DS2(SAT)} = v_{IC} - v_{TN2} < v_O < v_{DD} - V_{SD4(SAT)} \tag{5.51}$$

v_O must likewise remain sufficiently high to keep M_2 saturated, except M_2's source voltage is one gate–source voltage v_{GS2} below v_{IC}, so M_2's $V_{DS2(SAT)}$ is not a constraint in this regard. v_{IC}, on the other hand, is a constraint for v_O, which is not desirable when considering that a wide and largely independent ICMR is desirable.

Slew Rate

No matter how quickly transistors steer more or less current, tail current I_{Tail} limits how fast the output voltage v_O can rise and fall in the presence of capacitance. Although parasitic gate–drain and drain–bulk capacitances may not be disabling in this regard, gate–source, loading, and intentionally added capacitances can be. In a differential stage, load capacitance C_{LOAD}, which usually includes one or more gate–source capacitances and maybe even additional capacitance, can be substantial enough to limit the circuit's response. Since M_2 can sink and M_4 mirror and source no more than I_{Tail} when driven to extremes, I_{Tail} discharges or charges C_{LOAD} to the extent that v_O slews down or up across time at a rate of

$$SR \equiv \frac{dv_O}{dt} = \frac{I_{Tail}}{C_{LOAD} + C_{EQ}} \tag{5.52}$$

where slew rate SR is in volts per second and C_{EQ} is the equivalent parasitic capacitance at v_O.

5.4.2 Differential Signals

Small-Signal Model

When biased properly and considering only small differential signals, v_{IC}'s and v_{ID}'s steady-state values V_{IC} and V_{ID} are within the circuit's ICMR and linear range $\Delta v_{ID(MAX)}$, respectively, v_{IC}'s small-signal components are nil, and v_{ID} only includes small signals v_{id}. In other words, like before, the differential stage reduces to the pair of common-source transconductors in Fig. 5.19, whose inputs carry $0.5v_{id}$ and $-0.5v_{id}$. Since supply voltages hardly change with variations in current, which is another way of saying they do not generate small-signal voltages, they constitute ground connections to small signals. This means the mirror effectively draws current from ground in the small-signal equivalent of Fig. 5.19.

The differential input resistance R_{ID} of the two-port equivalent circuit corresponds to that of the differential pair, which is nearly infinite

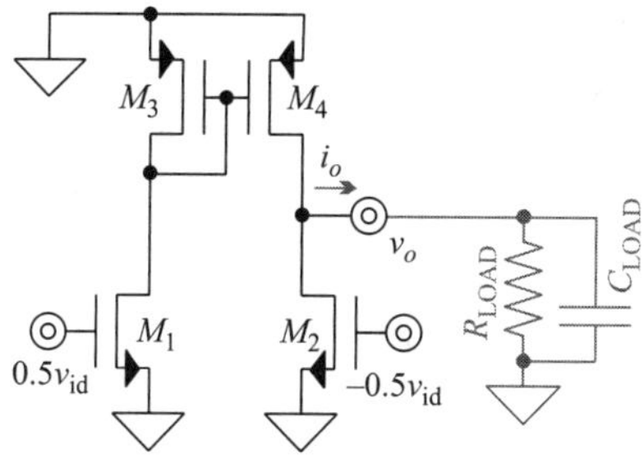

FIGURE 5.19 Differential small-signal equivalent circuit of the differential stage.

for FETs and $2r_\pi$ for BJTs. Output resistance R_O here is the combined parallel resistances that M_2 and M_4 present:

$$R_O = r_{ds2} \,||\, r_{sd4} \tag{5.53}$$

Short-circuit transconductance G_M is the ratio of output current i_o, which is the difference between M_1's and M_2's i_1 and i_2, to v_{id} when v_o is zero. Because M_1 carries v_{id}'s positive half and M_2 the opposite half, both halves combine to g_{m1}:

$$G_M \equiv \frac{i_o}{v_{id}} = \frac{i_1 - i_2}{v_{id}} = \frac{i_1}{v_{id}} - \frac{i_2}{v_{id}} = \frac{0.5v_{id}g_{m1}}{v_{id}} - \frac{(-0.5v_{id})g_{m2}}{v_{id}} = g_{m1} \tag{5.54}$$

where as before, M_1 and M_2's currents balance and match, so g_{m1} equals g_{m2}. As a result, the low-frequency voltage gain A_{V0} across the circuit is

$$A_{V0} \equiv \frac{v_o}{v_{id}} = G_M(R_O \,||\, R_{LOAD}) = g_{m1}(r_{ds2} \,||\, r_{sd4} \,||\, R_{LOAD}) \tag{5.55}$$

which can be 20–40 dB when R_{LOAD} is r_{ds} or higher.

The reason why A_{V0} is noninverting is because v_{ID} here is $v_{I1} - v_{I2}$. To be more specific, M_2 is a common-source amplifier that inverts M_2's input v_{I2} to output v_O. Similarly, M_1 also inverts its input v_{I1}, but so does M_4 in the mirror. In other words, v_{I1} signals invert once across M_1 and again across M_4, so v_{I1} is ultimately the noninverting input of the amplifier and v_{I2} the inverting counterpart.

Example 5.3 Assuming threshold voltages are 0.5 V, N- and P-type transconductance parameters are 100 and 40 μA/V², channel-length modulation $1/\lambda$ is 50 V, and all width–length aspect ratios are 1, what is the low-frequency differential gain of a PMOSFET differential pair loaded with a basic negative-supply NMOSFET current mirror when 10 μA biases the pair?

Solution

$$A_{V0} \equiv \frac{v_o}{v_{id}} = G_M(r_{ds2} \,||\, r_{sd4}) = g_{m1}(0.5r_{ds2}) = \sqrt{2(0.5I_{Tail})K_P'\left(\frac{W}{L}\right)}\left(\frac{0.5}{0.5I_{Tail}\lambda}\right)$$

$$= \sqrt{2(5\,\mu A)(40\,\mu A/V^2)(1)}\left[\frac{0.5(50\,V)}{5\,\mu A}\right] = 100\text{ V/V} = 40\text{ dB}$$

Frequency Response

Since loading capacitances C_{LOAD} often include one or more gates, r_{ds} or higher resistances across C_{LOAD} and gate–source capacitances C_{GS} usually shunt first across frequency. So if R_{LOAD} is r_{ds} or higher, which is a typical case in integrated circuits, C_{LOAD} and M_2's and M_4's parasitic capacitances at v_o shunt $R_O \,||\, R_{LOAD}$ first to establish the dominant bandwidth-limiting pole at output pole p_O:

$$\left.\frac{1}{s(C_{LOAD}+C_{GD2}+C_{DB2}+C_{GD4}+C_{DB4})}\right|_{p_o \approx \frac{1}{2\pi(R_O||R_{LOAD})C_{EQO}}} \equiv R_O \,||\, R_{LOAD}$$

$$= r_{ds2} \,||\, r_{sd4} \,||\, R_{LOAD} \tag{5.56}$$

where C_{EQO} is the equivalent capacitance at v_o. The only other set of capacitances to shunt is at the gate of the mirror, which happens at mirror pole p_M when M_3's, M_4's, and M_1's parasitic capacitances shunt M_3's diode-connected $1/g_{m3}$ resistance and other parallel resistances:

$$\frac{1}{s(C_{GS3}+C_{GS4}+C_{GD4}+C_{GD1}+C_{DB3}+C_{DB1})} \approx \left.\frac{1}{s(2C_{GS3})}\right|_{p_M \approx \frac{g_{m3}}{2\pi(2C_{GS3})}}$$

$$\equiv \frac{1}{g_{m3}} \,||\, r_{sd3} \,||\, r_{ds1} \approx \frac{1}{g_{m3}} \tag{5.57}$$

where C_{GS} overwhelms gate–drain and drain–bulk capacitances C_{GD} and C_{DB}, C_{GS3} equals C_{GS4} because M_3 matches M_4, and $1/g_{m3}$ is substantially lower than r_{sd3} and r_{ds1}. So although $2C_{GS3}$ is significant, $1/g_{m3}$ is so low that p_M typically resides at high frequency.

To be more specific, the effect of p_M is to reduce M_1's contribution to i_o in G_M. This means that i_o and therefore G_M drop past p_M, as Fig. 5.20

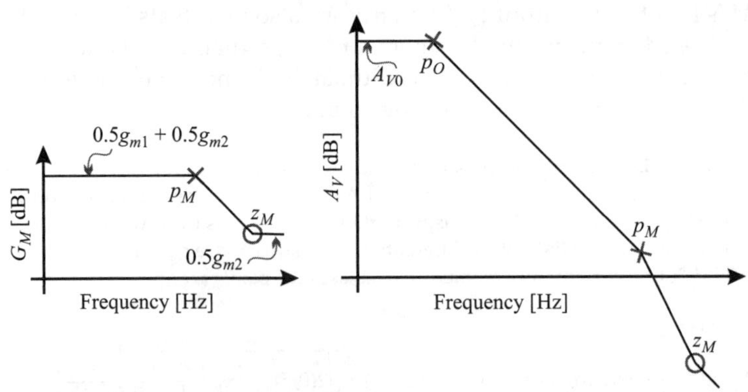

Figure 5.20 Frequency response of the circuit's transconductance and voltage gain.

illustrates, because i_1 in i_o falls. However, when i_1 in i_o is so low that i_o reduces to $-i_2$, i_o stops falling, which amounts to the onset of an in-phase zero. This left-half-plane mirror zero z_M therefore occurs when G_M drops to M_2's $0.5g_{m2}$ at twice p_M:

$$G_M = \frac{0.5g_{m1}}{\left(1+\dfrac{s}{2\pi p_M}\right)} + 0.5g_{m2} = \frac{g_{m1}\left[1+\dfrac{s}{2\pi(2p_M)}\right]}{\left(1+\dfrac{s}{2\pi p_M}\right)} \equiv \frac{g_{m1}\left(1+\dfrac{s}{2\pi z_M}\right)}{\left(1+\dfrac{s}{2\pi p_M}\right)} \tag{5.58}$$

where like before, g_{m1} matches g_{m2}. All this means that voltage gain A_V first drops –20 dB per decade from A_{V0} past bandwidth pole p_O, and –40 dB per decade past p_M, before z_M recovers phase and reduces the drop back to –20 dB per decade. Fortunately, p_M is considerably higher than p_O, and z_M recovers what p_M loses soon after p_M.

5.4.3 Common-Mode Signals

Small-Signal Model

When biased properly and considering only small common-mode signals, v_{IC}'s and v_{ID}'s steady-state values V_{IC} and V_{ID} are within the circuit's ICMR and linear range $\Delta v_{ID(MAX)}$, respectively, v_{ID}'s small-signal components are nil, and v_{IC} only includes small signals v_{ic}. In other words, the differential stage reduces to the pair of degenerated common-source transconductors in Fig. 5.21 with v_{ic} at their inputs. Here, the degenerating impedance is twice that of the tail current, which is equivalent to $2R_{Tail}$, like before, and half the parasitic capacitance $0.5C_{Tail}$ is similarly present at each source-coupled node. The circuit includes this capacitance because M_1's and M_2's bulk–substrate capacitances

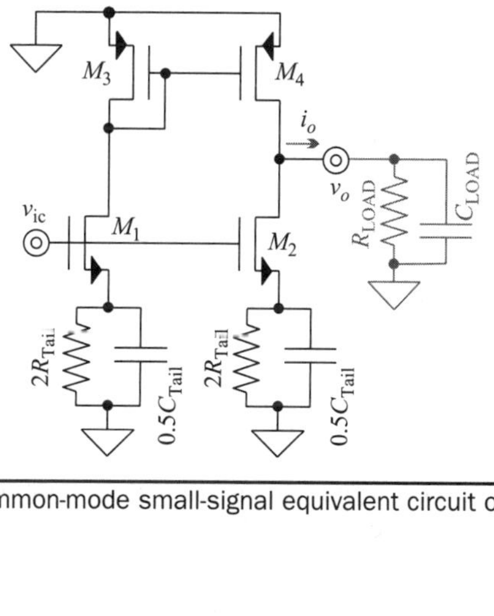

Figure 5.21 Common-mode small-signal equivalent circuit of the differential stage.

can be significant when isolated bulk terminals and sources connect. And the mirror, like before, effectively draws current from ground because supply voltages hardly change in response to small variations in current.

The common-mode input resistance R_{IC} of the two-port equivalent circuit corresponds to that of the differential pair, which is nearly infinite for FETs and roughly $\beta_0 R_{Tail}$ for BJTs. Common-mode output resistance R_{OC} here is the combined parallel resistances that M_2 and M_4 present:

$$R_{OC} = R_{D2} \,||\, r_{sd4} = (r_{ds1} + 2R_{Tail} g_{m1} r_{ds1} + 2R_{Tail}) \,||\, r_{sd4} \approx r_{sd4} \tag{5.59}$$

where $2R_{Tail}$ in this case is $2r_{dsT}$, or higher when using a cascoded or degenerated current source, and M_2's degenerated drain resistance R_{D2} is so much higher than M_4's r_{sd4} that r_{sd4} dominates.

Short-circuit common-mode transconductance G_C is the ratio of i_o, which is the difference between M_1's and M_2's i_1 and i_2, to v_{ic} when v_o is zero, but because v_{ic} causes M_1 and M_2 to conduct equal currents, their difference in G_C reduces to zero. This only happens, however, when i_1 and i_2 match perfectly, which is not always the case. Although the gate–source voltages of M_1 and M_2 in the differential pair and M_3 and M_4 in the mirror match, for example, their drain–source voltages v_{DS} may not. G_C therefore represents the error that v_{DS} mismatches between M_1 and M_2 and between M_3 and M_4 produce in i_1 and then in i_4 with respect to i_2:

$$G_C = \frac{\Delta i_O}{v_{ic}} = \frac{i_4 - i_2}{v_{ic}} = \left(\frac{1}{v_{ic}}\right)\left[i_2\left(\frac{i_1}{i_2}\right)\left(\frac{i_4}{i_1}\right) - i_2\right] = \frac{i_2}{v_{ic}}\left[\left(\frac{1+\lambda v_{DS1}}{1+\lambda v_{DS2}}\right)\left(\frac{1+\lambda v_{SD4}}{1+\lambda v_{SD3}}\right) - 1\right]$$
$$= G_{C2} E_\lambda = \left(\frac{g_{m2}}{1+2R_{Tail} g_{m2}}\right) E_\lambda \tag{5.60}$$

where G_{C2} is M_2's degenerated transconductance and E_λ describes the error that channel-length modulation produces with respect to i_2. Because of this, low-frequency common-mode gain A_{C0} is also not zero and is similarly dependent on E_λ:

$$A_{C0} \equiv \frac{v_o}{v_{ic}} = G_C(R_{OC} \,||\, R_{LOAD}) = G_{C2} E_\lambda (R_{OC} \,||\, R_{LOAD})$$
$$\approx \left(\frac{g_{m2}}{1+2R_{Tail} g_{m2}}\right) E_\lambda (r_{sd4} \,||\, R_{LOAD}) \tag{5.61}$$

Note, however, that not only does $2R_{Tail}$ at $2r_{dsT}$ degenerate g_{m2} considerably but channel-length modulation parameter λ can be so low at 20–50 mV^{-1} that E_λ is also markedly low. In other words, G_C and A_{C0} may not be zero, but they can nonetheless be extremely low.

Frequency Response

As before, because C_{LOAD} usually has at least one gate–source capacitance, C_{LOAD} and M_2's and M_4's parasitic capacitances at v_o shunt $R_{OC} \| R_{LOAD}$ first at p_O':

$$\left.\frac{1}{s(C_{LOAD}+C_{GD2}+C_{DB2}+C_{GD4}+C_{DB4})}\right|_{p_o' \approx \frac{1}{2\pi(R_{OC}\|R_{LOAD})C_{EQO}}} \equiv R_{OC} \| R_{LOAD} \approx r_{sd4} \| R_{LOAD} \quad (5.62)$$

which is near output pole p_O. Like in the differential case, M_3's, M_4's, and M_1's parasitic capacitances then shunt M_3's diode-connected $1/g_{m3}$ resistance and other parallel resistances at p_M, except lowering i_4's contribution in i_o raises the error from $i_4 - i_2$ to $-i_2$. In fact, the error is 3 dB from reaching i_2 at p_M, as Fig. 5.22 indicates, which means G_C starts climbing from $G_{C2}E_\lambda$ at mirror zero z_M' toward G_{C2} below p_M. In other words, z_M' is the frequency from which $G_{C2}E_\lambda$ rises to reach G_{C2} at p_M:

$$G_C = G_{C2}E_\lambda\left.\left(1+\frac{s}{2\pi z_M'}\right)\right|_{f>>z_M'} \approx \left.\frac{sG_{C2}E_\lambda}{2\pi z_M'}\right|_{f=p_M} \equiv G_{C2} \quad (5.63)$$

so z_M' is E_λ times lower than p_M at $E_\lambda p_M$. Another way of deriving this is by realizing that zeros raise gains linearly across frequency because increasing $G_{C2}E_\lambda$ by a factor of E_λ means z_M' asserted its influence at an equivalent factor of p_M, at $E_\lambda p_M$. Note that z_M' can be at low to moderate frequencies.

The only other capacitance present here is C_{Tail}, and past the frequency z_D where $0.5C_{Tail}$ shunts $2R_{Tail}$, the degenerating effects of R_{Tail} begin to disappear from G_{C2}:

$$\left.\frac{1}{s(0.5C_{Tail})}\right|_{z_D \approx \frac{1}{2\pi R_{Tail}C_{Tail}}} \equiv 2R_{Tail} \quad (5.64)$$

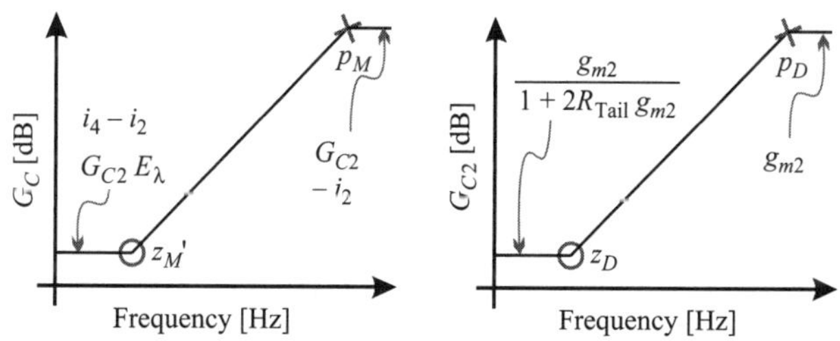

FIGURE 5.22 Effects of the mirror and tail current on common-mode transconductance.

which can happen at moderate frequencies because R_{Tail} is normally high. Past z_D, G_{C2} rises and reaches its nondegenerated counterpart g_{m2} at degenerating pole p_D when G_{C2} equals g_{m2}:

$$G_{C2} = \left(\frac{g_{m2}}{1+2R_{Tail}g_{m2}}\right)\left(1+\frac{s}{2\pi z_D}\right)\Bigg|_{f>>z_D}$$
$$\approx \left(\frac{g_{m2}}{1+2R_{Tail}g_{m2}}\right)\left(\frac{s}{2\pi z_D}\right)\Bigg|_{p_D=(1+2R_{Tail}g_{m2})z_D} \equiv g_{m2} \tag{5.65}$$

where p_D is so much higher than z_D that the s term overwhelms 1. Again, because zeros raise gains linearly across frequency, raising G_{C2} by a factor of $1 + 2R_{Tail}g_{m2}$ to p_D requires p_D to be $1 + 2R_{Tail}g_{m2}$ times higher than z_D.

When combining all these effects, common-mode gain A_C falls from A_{C0} past p_O', as Fig. 5.23 shows. Degenerating and mirror zeros z_D and z_M' not only arrest but also reverse A_C's fall at moderate frequencies. Past mirror pole p_M, A_C stops rising and flattens. Degenerating pole p_D can be so much higher than z_D that p_D can assert its influence past the transitional frequency f_T of one or several transistors in the circuit. This means p_D is oftentimes above practical frequencies of interest.

Interestingly, $g_{m2}(r_{sd4} \,||\, R_{LOAD})$ from A_{C0} in common-mode gain A_C is so close to $g_{m2}(r_{ds2} \,||\, r_{sd4} \,||\, R_{LOAD})$ from A_{V0} in differential voltage gain A_V that they effectively cancel in *common-mode rejection ratio* CMRR, in A_V/A_C:

$$\text{CMRR} \equiv \frac{A_V}{A_C} = \frac{A_{V0}\left(1+\frac{s}{2\pi z_M}\right)\left(1+\frac{s}{2\pi p_O'}\right)\left(1+\frac{s}{2\pi p_M}\right)\left(1+\frac{s}{2\pi p_D}\right)}{A_{C0}\left(1+\frac{s}{2\pi z_D}\right)\left(1+\frac{s}{2\pi z_M'}\right)\left(1+\frac{s}{2\pi p_O}\right)\left(1+\frac{s}{2\pi p_M}\right)}$$
$$\approx \frac{g_{m2}(r_{ds2} \,||\, r_{sd4} \,||\, R_{LOAD})\left(1+\frac{s}{2\pi z_M}\right)\left(1+\frac{s}{2\pi p_D}\right)}{\left[\frac{E_\lambda g_{m2}(r_{sd4} \,||\, R_{LOAD})}{1+2R_{Tail}g_{m2}}\right]\left(1+\frac{s}{2\pi z_D}\right)\left(1+\frac{s}{2\pi z_M'}\right)}$$

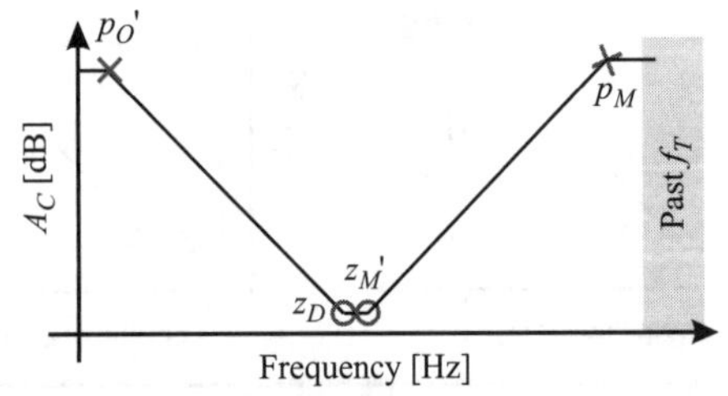

FIGURE 5.23 Common-mode gain across frequency.

$$\approx \frac{2R_{\text{Tail}}g_{m2}\left(1+\frac{s}{2\pi z_M}\right)\left(1+\frac{s}{2\pi p_D}\right)}{E_\lambda\left(1+\frac{s}{2\pi z_D}\right)\left(1+\frac{s}{2\pi z_M{}'}\right)}$$

$$= \frac{\text{CMRR}_0\left(1+\frac{s}{2\pi z_M}\right)\left(1+\frac{s}{2\pi p_D}\right)}{\left(1+\frac{s}{2\pi z_D}\right)\left(1+\frac{s}{2\pi z_M{}'}\right)} \qquad (5.66)$$

As a result, low-frequency ratio CMRR_0 reduces to $2R_{\text{Tail}}g_{m2}/E_\lambda$, which can be high at 50–90 dB, as Fig. 5.24 shows. Since mirror pole p_M appears in both A_V and A_C, p_M cancels in CMRR. Also, since r_{sd4} in $p_O{}'$ is so close to $r_{\text{ds2}} \| r_{\text{sd4}}$ in output pole p_O, $p_O{}'$ and p_O similarly cancel. But because degenerating and mirror zeros z_D and $z_M{}'$ in A_C are usually below mirror zero z_M in A_V and degenerating pole p_D in A_C, CMRR drops at moderate frequencies past z_D and $z_M{}'$. At higher frequencies, near and oftentimes above frequencies of interest, z_M in A_V eventually slows CMRR's drop.

5.4.4 Input-Referred Offset and Noise

Offset

When perfectly balanced, zero volts across the inputs of the differential stage splits constant tail current I_{Tail} equally between M_1 and M_2. Unfortunately, transistors do not match perfectly, so identical gate–source voltages in M_1 and M_2 and M_3 and M_4 establish currents that are not exactly equal. Plus, as already mentioned, a drain–source voltage difference between M_1 and M_2 and M_3 and M_4, as Δv_D depicts

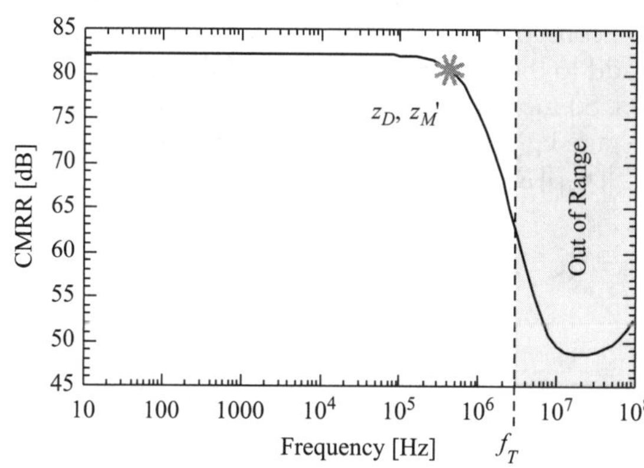

Figure 5.24 Sample CMRR performance.

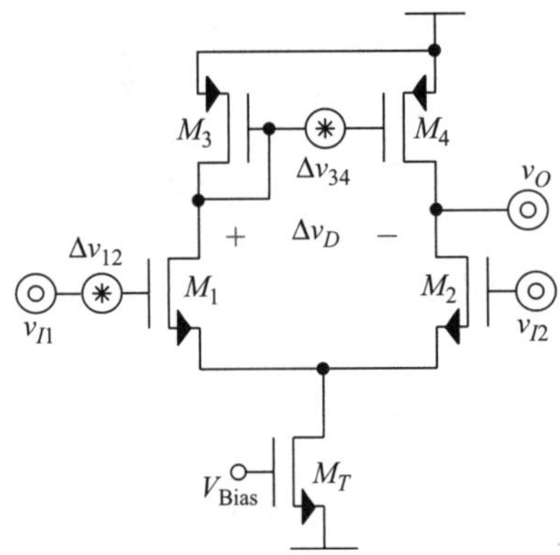

FIGURE 5.25 Offset-producing mismatches in a differential stage.

in Fig. 5.25, also produces a current imbalance. The net result is that a finite, nonzero offset voltage V_{OS} across the inputs balances the circuit like zero volts would under perfectly matched conditions.

Three-sigma base- and gate-referred mismatches between BJTs and FETs can be 1–3 mV and 5–15 mV, respectively, which are small enough to produce linear effects in the circuit like small signals would. For the circuit to respond linearly to differential input voltages v_{ID}, v_O must also reside at a level that ensures both M_2 and M_4 are in saturation, so what amounts to a small-signal voltage across the gates of M_1 and M_2 can produce Δv_D. In other words, the equivalent small-signal model of the differential pair can predict the effects of mismatches between M_1 and M_2 in gate-referred offset Δv_{12}, M_3 and M_4 in gate-referred offset Δv_{34}, and drain voltages in Δv_D.

While base- and gate-referred mismatches are random, differences in drain voltages are not. This means V_{OS} incorporates both random and systemic components. For statistically meaningful results, systemic variations add to the square-rooted sum of the squares of the random components. So input-referred offset V_{OS} includes the systemic drain-induced voltage $V_{OS(S)}$ that, when amplified by differential voltage gain A_V or $g_{m1}(r_{ds2} \| r_{sd4} \| R_{LOAD})$, cancels Δv_D:

$$V_{OS} = V_{OS(S)} \pm V_{OS}^{*}$$

$$= \frac{\Delta v_D}{g_{m1}(r_{ds2} \| r_{sd4} \| R_{LOAD})} \pm \sqrt{\Delta v_{12}^{\,2} + \left(\frac{\Delta v_{34}\, g_{m4}}{g_{m1}}\right)^2} \tag{5.67}$$

Similarly, random component V_{OS}^{*} includes the voltage that, when multiplied by the differential pair's transconductance g_{m1} and divided by M_3 and M_4's counterpart g_{m4}, offsets M_3 and M_4's original mismatch Δv_{34}. However, since M_1 and M_2's mismatch Δv_{12} is already across the inputs of the differential stage, no gain translation is necessary for Δv_{12}. All this means that high gains A_V and high pair-to-mirror transconductance ratios g_{m1}/g_{m4} reduce the effects of mismatched voltages and mirror transistors Δv_D and Δv_{34} on input-referred offset V_{OS}. Of course, matching M_1 to M_2 and M_3 to M_4 and drain voltages is more fundamental and therefore an engineer's first line of attack.

Example 5.4 Assuming threshold voltages are 0.5 V, N- and P-type transconductance parameters are 100 and 40 μA/V², channel-length modulation 1/λ is 100 V, all aspect ratios are one, three-sigma gate-referred offsets are 10 mV, and drain voltages mismatch by 1 V, what is the input-referred offset of a PMOSFET differential pair loaded with a basic negative-supply NMOSFET current mirror when 10 μA biases the pair?

Solution

$$V_{OS(S)} = \frac{\Delta v_D}{g_{m1}(r_{ds2} \parallel r_{sd4})} = \frac{1\text{ V}}{\sqrt{2(5\,\mu\text{A})(40\,\mu\text{A/V}^2)(1)}\left[\dfrac{0.5(100\text{ V})}{5\,\mu A}\right]}$$

$$= \frac{1\text{ V}}{200\text{ V/V}} = 5\text{ mV}$$

$$V_{OS}^* = \sqrt{\Delta v_{12}^{\;2} + \left(\frac{\Delta v_{34}\,g_{m4}}{g_{m1}}\right)^2}$$

$$= \sqrt{(10\text{ mV})^2 + \left[(10\text{ mV})\left(\frac{\sqrt{100\,\mu\text{A/V}^2}}{\sqrt{40\,\mu\text{A/V}^2}}\right)\right]^2} \approx 19\text{ mV}$$

$$V_{OS} = V_{OS(S)} \pm V_{OS}^* \approx 5\text{ mV} \pm 19\text{ mV}$$

Noise

In the case of the differential stage, the gates of M_1, M_2, M_3, M_4, and M_T generate noise, except M_T's noise current splits equally between M_1 and M_2 to cause little to no impact on differential signals. M_4's transconductance g_{m4} converts M_3 and M_4's combined noise v_{34}^* to current, and M_1 and M_2's transconductance g_{m1} refers that noise current back to an equivalent input noise voltage $v_{34}^* g_{m4}/g_{m1}$. Since M_1 and M_2's combined contribution v_{12}^* is already at the input, combined input-referred noise v_N^* across the differential stage reduces to the square root of the squares of v_{12}^* and $v_{34}^* g_{m4}/g_{m1}$:

$$v_N^* = \sqrt{(v_{12}^*)^2 + \left(\frac{v_{34}^*\,g_{m4}}{g_{m1}}\right)^2} \tag{5.68}$$

5.4.5 Power-Supply Rejection

Microelectronic systems today incorporate diverse functions that amplify, regulate, convert, transmit, process, and drive signals to on- and off-chip components like other chips, filters, and antennae. Many circuits not only respond on demand but also switch periodically at kilohertz, megahertz, and even gigahertz. To perform as prescribed, these systems can draw hundreds of milliwatts or even watts, but not always, because not all subsystems operate continuously. In fact, these same systems can also draw nanowatts when idling and hundreds of microwatts when engaged to perform light functions. Such a diverse and, if not erratic, unpredictable load pulls a vast range of periodic and asynchronous currents from imperfect power supplies, producing in the process ripples on the order of kilohertz to megahertz in the supply voltages. This is why it is critical to understand how analog circuits react to noisy supplies.

In this light, *power-supply rejection* (PSR) refers to the ability of a circuit to reject fluctuations in the supply, or said differently, to the circuit's inability to amplify supply ripples, which is why PSR is the inverse of supply gain A_{SUPPLY}, of s_O/v_{SUPPLY}:

$$\text{PSR} \equiv \frac{1}{A_{SUPPLY}} \equiv \frac{v_{SUPPLY}}{s_O} \tag{5.69}$$

where output signal s_O can be a voltage, current, or time and v_{SUPPLY} any positive or negative supply in the circuit like v_{DD}, v_{SS}, v_{CC}, or v_{EE}. Similarly, though not the same, *power-supply rejection ratio* (PSRR) refers to how much a circuit favors processed signals over supply ripples, which is the ratio of forward gain A_{FWD} to supply gain A_{SUPPLY}:

$$\text{PSRR} \equiv \frac{A_{FWD}}{A_{SUPPLY}} = \left(\frac{s_O}{s_I}\right)\left(\frac{v_{SUPPLY}}{s_O}\right) \tag{5.70}$$

where again, input signal s_I can be a voltage, current, or time. Interestingly, power-supply designers sometimes use *power-supply ripple rejection* to refer to PSR, which is unfortunate because the resulting acronym is indistinguishable from that of power-supply rejection ratio. This book therefore uses PSR for supply rejection and PSRR exclusively for rejection ratio.

For insight, evaluating supply gain A_{SUPPLY} is oftentimes more straightforward and intuitive than PSR or PSRR. And in this light, the output stage ultimately determines what fraction of the supply ripple reaches s_O. In the case of the differential stage, the mirror plays an important role because it replicates supply ripples in its input path. Consider, for example, positive supply ripples v_{dd} in the MOSFET embodiment of Fig. 5.26. Since supply gain A_{VDD} describes how v_{dd} affects output v_o in the absence of incoming signals v_{I1} and v_{I2}, v_{I1} and v_{I2} reduce to zero. With only a zero-volt common-mode component to consider, the differential stage further decomposes into two source-degenerated MOSFETs.

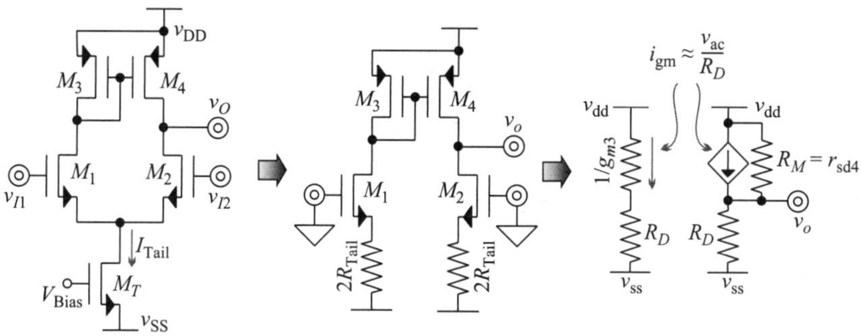

FIGURE 5.26 Small-signal decomposition of the differential pair loaded with a positive-supply P-type mirror when considering supply noise.

Since the small-signal resistance of a diode-connected transistor is roughly $1/g_m$ and a mirroring transistor duplicates its input current, M_3 reduces to a $1/g_{m3}$ resistor and M_4 to a transconductor current i_{gm} that is equivalent to M_3's current in parallel with M_4's r_{sd4}. For the sake of simplicity and generality, R_M here represents M_4's mirror resistance r_{sd4} and R_D the resistance across the matched degenerated MOSFETs. This way, when applying superposition, the effect of v_{dd} on v_o because of R_M in the absence of i_{gm} and v_{ss} is voltage-divided fraction $v_{dd}R_D/(R_D + R_M)$:

$$A_{VDD} \equiv \frac{v_o}{v_{dd}} = \frac{\left(\frac{v_{dd}R_D}{R_D + R_M}\right) + [i_{gm}(R_D \,||\, R_M)]}{v_{dd}}$$

$$\approx \frac{\left(\frac{v_{dd}R_D}{R_D + R_M}\right) + \left[\left(\frac{v_{dd}}{R_D}\right)(R_D \,||\, R_M)\right]}{v_{dd}} = 1 \tag{5.71}$$

where a supply gain of 1 means v_{dd} appears entirely in v_o. Since $1/g_{m3}$ is considerably smaller than R_D, M_3's current is roughly v_{dd}/R_D, so M_4 pushes v_{dd}/R_D to v_o's resistances R_D and R_M, where R_M connects to small-signal ground as a result of superposition. Therefore, R_M and i_{gm} both add supply noise components to v_o that, combined, replicate v_{dd} in v_o.

Interestingly, with respect to negative supply ripple v_{ss} in the absence of v_{dd}, i_{gm}, or v_{ss}/R_D flows in the opposite direction to counter ripples injected through R_D. As a result, i_{gm} eradicates v_{ss} from v_o:

$$A_{VSS} \equiv \frac{v_o}{v_{ss}} = \frac{\left(\frac{v_{ss}R_M}{R_D + R_M}\right) - [i_{gm}(R_D \,||\, R_M)]}{v_{ss}}$$

$$\approx \frac{\left(\frac{v_{ss}R_M}{R_D + R_M}\right) - \left[\left(\frac{v_{ss}}{R_D}\right)(R_D \,||\, R_M)\right]}{v_{ss}} = 0 \tag{5.72}$$

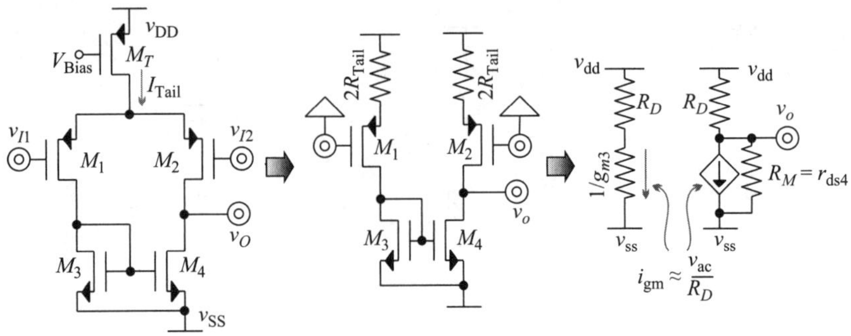

FIGURE 5.27 Small-signal decomposition of the differential pair loaded with a negative-supply N-type mirror when considering supply noise.

In other words, a positive-supply P-type mirror reproduces positive supply ripples and eliminates negative supply ripples. Similarly, a negative-supply N-type mirror replicates negative supply ripples and cancels positive supply ripples. In the vertical mirror image of Fig. 5.26, for example, which the P-type pair and N-type mirror in Fig. 5.27 implement, the mirror pulls v_{dd}/R_D from v_o to annihilate v_{dd} from v_o and feeds v_{ss}/R_D to v_o to impress together with M_4's R_M v_{ss} on v_o.

In summary, current mirrors generally favor ripples from their supplies and reject ripples from opposing supplies. This means engineers can design circuits to reject ripples from either supply. In the case of battery-supplied systems, for example, where ground planes are typically less noisy than battery voltages, designers usually construct their mirrors to reject more positive supply noise than ground noise.

5.4.6 Folding Cascodes

Large-Signal Operation

An N-type differential pair need not always drive a positive-supply P-type mirror. In fact, since the outputs of the pair are currents, P-type common-gate current buffers M_5 and M_6 in Fig. 5.28 can steer and fold pair currents i_1 and i_2 into a negative-supply N-type mirror. For this, however, additional current sources M_9 and M_{10} must supply sufficient current to drive both the differential pair and the mirror via M_5 and M_6. To ensure that the current buffers and the mirror transistors never starve, M_9 and M_{10} usually source 1.5 times tail current I_{Tail}. This way, even when i_1 or i_2 drops to zero, neither buffer nor mirroring transistor shuts completely. This is important because reengaging transistors from their off state requires additional time. Since the mirror cancels the common currents that M_9 and M_{10} source, and each

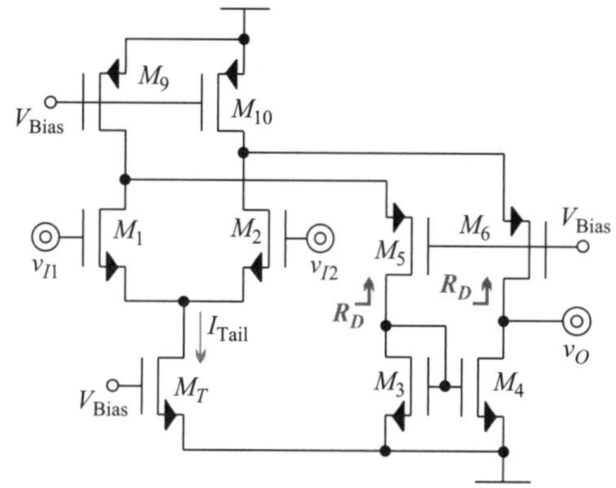

FIGURE 5.28 Folding differential currents with common-gate current buffers M_5 and M_6.

transistor in the differential pair can carry all of I_{Tail}, the structure can fold and source current into the load C_{LOAD} as much current as it can sink to slew, like before, at $I_{\text{Tail}}/(C_{\text{LOAD}} + C_{\text{EQ}})$ in both directions.

Because the output resistance of the folding current buffer M_6 at v_O is that of a source-degenerated cascode transistor, IC engineers often refer to this and derivatives of this differential stage as *folded-cascode amplifiers*. This circuit topology incorporates a few peculiarities worth mentioning. First, current sources M_9 and M_{10} require less voltage across them than the input of a positive-supply P-type mirror would. So if M_5 and M_6's gate bias voltage is just high enough to ensure M_9 and M_{10} drop a little more than their saturation voltages, input common-mode voltage v_{IC} can rise higher toward the positive supply v_{DD} than with a positive-supply P-type mirror, whose overhead is a source–gate voltage. To be more specific, v_{IC} can rise higher toward v_{DD} before "crushing" the differential pair against M_9 and M_{10}:

$$v_{\text{IC}} < v_{\text{DD}} - v_{\text{SD9}} - V_{\text{DS1(SAT)}} + v_{\text{GS1}} = v_{\text{DD}} - v_{\text{SD9}} + v_{\text{TN1}} \quad (5.73)$$

where current buffers M_5 and M_6 can push M_9's and M_{10}'s v_{SD9} and v_{SD10} close to their saturation point. Notice, by the way, v_{IC} no longer constrains how far v_O swings. In this case, v_O can drop within M_4's saturation voltage of v_{SS}, and if optimally biased, up to within two saturation voltages of v_{DD}, those of M_6 and M_{10}:

$$v_{\text{SS}} + V_{\text{DS4(SAT)}} < v_O < v_{\text{DD}} - v_{\text{SD10}} - V_{\text{SD6(SAT)}} \quad (5.74)$$

Small-Signal Model

The small-signal output resistance R_O of this folding structure is nearly twice that of its nonfolding predecessor because the drain resistance R_{D6} of the degenerated current buffer M_6 is so much higher than the mirroring transistor's r_{ds4} that the latter dominates:

$$R_O = R_{D6} \parallel r_{ds4} = [r_{sd6} + g_{m6} r_{sd6}(r_{sd10} \parallel r_{sd2}) + (r_{sd10} \parallel r_{sd2})] \parallel r_{ds4} \approx r_{ds4} \quad (5.75)$$

This means that the low-frequency gain A_{V0} of the circuit is correspondingly higher by up to two times, or 6 dB, because differential currents i_1 and i_2 flow through M_5 and M_6 into M_3 and M_4 to establish, like before, a gain of $[0.5g_{m1} - (-0.5g_{m2})](R_O \parallel R_{LOAD})$, or $g_{m1}(R_O \parallel R_{LOAD})$. Luckily, inserting cascode transistors M_7 and M_8 into the mirror, like Fig. 5.29 shows, raises the output resistance of the mirror from r_{ds4} to roughly $g_{m8} r_{ds8} r_{ds4}$. This way, both R_O and A_{V0} can be substantially higher at maybe 10–100 MΩ and 60–80 dB.

Power-Supply Rejection

Given the symmetry of the circuit, equivalent resistances R_D feed the negative-supply N-type mirror that M_3 and M_4 implement in Figs. 5.28 and 5.29. This means M_3 and M_4, like in Fig. 5.27, cancel v_{dd} ripples and reinforce v_{ss} ripples to reproduce v_{ss} in v_o. In other words, positive PSR and PSRR approach infinity and their negative counterparts reduce to 1 and A_V, respectively.

Frequency Response

Shunting capacitances at the output v_O, at M_3's mirroring gate v_{G3}, and at M_5's and M_6's source nodes v_{S5} and v_{S6} introduce poles. But since v_{S5} and v_{S6} each carry half of the differential signal, their combined effect is to introduce one pole to the overall gain. So, since v_O's resistance R_O

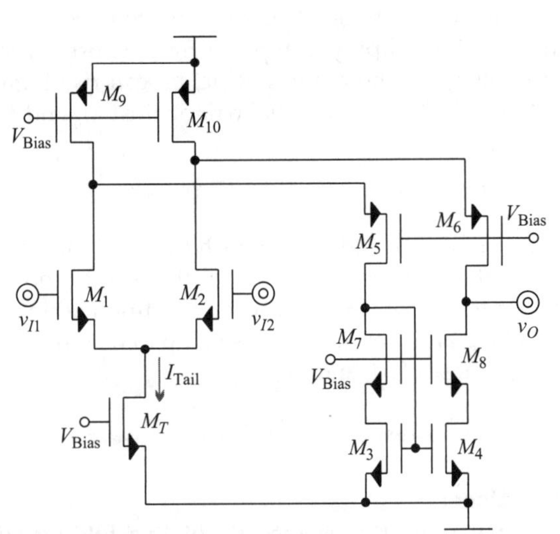

FIGURE 5.29 Folded-cascode differential amplifier.

and load capacitance C_{LOAD} are both high, and equivalent or lower capacitances shunt the $1/g_m$ resistances that M_3's diode connection and M_5's and M_6's source terminals establish, the output pole p_O is usually low and dominant at about $1/2\pi R_O(C_{LOAD} + C_{EQ})$, where C_{EQ} is the parasitic capacitance at v_O. With two gate–source capacitances, the mirror pole p_M is usually next at $g_{m3}/2\pi(2C_{GS})$. Like before, though, a phase-saving mirror zero z_M appears at $2p_M$ when M_4's mirrored impression of M_5 and M_1's current is so low that the circuit can only output M_6's impression of M_2's current, which corresponds to half of the differential current. And because C_{LOAD} shunts v_O's loading effect on M_6's $1/g_{m6}$ at higher frequency, one gate–source capacitance shunts M_5 and M_6's $1/g_{m6}$ resistance near z_M at $p_{S5,6}$, at $g_{m6}/2\pi C_{GS6}$.

Input-Referred Offset

One disadvantage here is that this circuit includes more transistors whose mismatched imperfections raise input-referred offset V_{OS}. Thankfully, M_1, M_2, M_9, M_{10}, M_3, and M_4 in Fig. 5.29 degenerate cascoding current buffers M_5, M_6, M_7, and M_8, so the gate-referred mismatches of the latter produce noticeably lower offset currents than those of the former. In other words, mismatches between M_1 and M_2, M_3 and M_4, and M_9 and M_{10} dominate V_{OS}'s random component V_{OS}^*:

$$V_{OS} = V_{OS(S)} \pm V_{OS}^*$$

$$\approx \frac{\Delta v_D}{g_{m1}(R_O \parallel R_{LOAD})} \pm \sqrt{\Delta v_{12}^2 + \left(\frac{\Delta v_{34}\, g_{m4}}{g_{m1}}\right)^2 + \left(\frac{\Delta v_{9/10}\, g_{m9}}{g_{m1}}\right)} \tag{5.76}$$

where Δv_D, like before, is the drain-voltage difference between v_O and the input of the mirror and $\Delta v_{9/10}$ is the gate-referred mismatch between M_9 and M_{10} that g_{m9} converts into a current and g_{m1} refers back to the input. Notice that Δv_D has a lower impact on V_{OS} because the voltage gain is higher here than in the previous circuits. Also note that lowering M_9 and M_{10}'s transconductance g_{m9} reduces the degrading effect of their mismatch $\Delta v_{9/10}$ on V_{OS}^*.

Design Note

Realizing that cascode transistors buffer currents is powerful. Consider folding current buffers M_5 and M_6 in Fig. 5.30, for example. Here, M_5 buffers M_3's current back to its gate to implement a diode connection around M_3. This means that M_4 mirrors whatever current M_3 conducts, including the bias current that M_9 establishes. In addition, M_5 steers and folds M_1's current i_1 into the input of the mirror that M_3 and M_4 implement. As a result, M_4 sources i_1 and M_2 subtracts i_2 from M_6 to steer the difference $0.5v_{id}g_{m1} - (-0.5v_{id})g_{m2}$ or $v_{id}g_{m1}$ into v_o, which is what the differential stages in Figs. 5.18, 5.25, 5.26, 5.27, 5.28, and 5.29

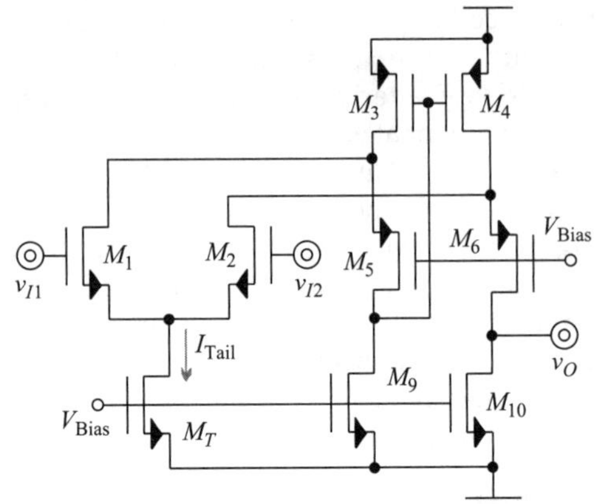

FIGURE 5.30 Folding N-type differential currents with current buffers into a P-type mirror.

ultimately achieve. The advantage of this topology over its nonfolding counterpart in Fig. 5.18 is extended input common-mode range. The reason for this is that M_3 here only drops a source–drain voltage, whereas M_3 in its nonfolding predecessor drops a source–gate voltage.

Example 5.5 Refer to the P-type differential pair in Fig. 5.31, whose currents N-type common-gate transistors buffer and fold into a positive-supply P-type current mirror, to derive expressions for the input common-mode range, output swing limits, and positive and negative supply-ripple gains.

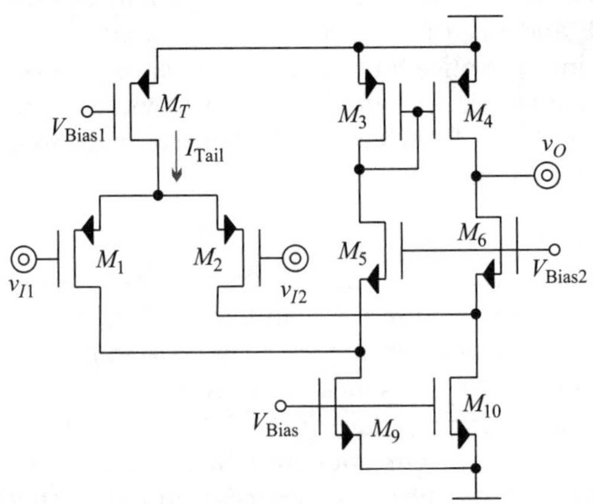

FIGURE 5.31 Folding P-type differential currents with N-type current buffers into a P-type mirror.

Solution

1. ICMR: $v_{SS} + v_{DS9} - |v_{TP}| < v_{IC} < v_{DD} - V_{SDT(SAT)} - |v_{TP}| - V_{SD1(SAT)}$
2. v_O Swing: $v_{SS} + v_{DS10} + V_{DS6(SAT)} < v_O < v_{DD} - V_{SD4(SAT)}$

   where

$$v_{DS9,10} = V_{Bias2} - v_{GS6} - v_{SS} = V_{Bias2} - v_{TN} - V_{DS6(SAT)} - v_{SS} \geq V_{DS9,10(SAT)}$$

3. A_{VDD}:

$$A_{VDD} = \frac{\left(\dfrac{v_{dd} R_{D6}}{R_{D6} + r_{sd4}}\right) + [i_{gm}(R_{D6} \parallel r_{sd4})]}{v_{dd}}$$

$$\approx \frac{\left(\dfrac{v_{dd} R_{D6}}{R_{D6} + r_{sd4}}\right) + \left[\left(\dfrac{v_{dd}}{R_{D5}}\right)(R_{D6} \parallel r_{sd4})\right]}{v_{dd}} = 1$$

where R_{D5} and R_{D6} are the matched resistances between the drains of M_5 and M_6 and v_{SS}.

4. A_{VSS}:

$$A_{VSS} = \frac{\left(\dfrac{v_{ss} r_{sd4}}{R_{D6} + r_{sd4}}\right) - [i_{gm}(R_{D6} \parallel r_{sd4})]}{v_{ss}}$$

$$\approx \frac{\left(\dfrac{v_{ss} r_{sd4}}{R_{D6} + r_{sd4}}\right) - \left[\left(\dfrac{v_{ss}}{R_{D5}}\right)(R_{D6} \parallel r_{sd4})\right]}{v_{ss}} = 0$$

5.5 Summary

Arguably, the most commonly used building blocks in analog IC design are current mirrors and differential pairs. Current mirrors are important because they sense, replicate, amplify, fold, and convey the fundamental medium through which analog circuits process and channel information. Differential pairs are popular because, among other features, they reject substrate and coupled noise from which mixed-signal systems that share a common silicon substrate often suffer. They also generate the error signal with which feedback loops often control and regulate voltages and currents. Together, in fact, they implement differential stages that not only amplify analog signals and reject common noise but also fold differential currents into one easy-to-process output. Not surprisingly, these circuits are almost ubiquitous in analog systems, from operational amplifiers, linear regulators, and bias circuits to data converters, switching dc–dc converters, and energy-harvesting chargers and supplies.

Like with everything else, however, these circuits are not perfect. Current mirrors, for instance, suffer from base-current, voltage-mismatch, and transistor-mismatch errors. Luckily, although not always necessary, base-adjusting resistors and transistors and voltage-correcting cascodes can reduce these nonidealities to negligible levels. Emitter/source- and base/gate-coupled differential pairs and the differential stages they and current mirrors implement similarly suffer from finite ICMR, output-voltage swing, slew rate, differential gain, linear range, bandwidth-limiting poles, CMRR, input-referred offset and noise, and PSR and PSRR. When needed, though, emitter/source-degenerating resistors and folding current buffers can extend their linear range and ICMR, respectively, and cascodes can raise their differential gain. Plus, since current mirrors cancel ripples from their opposing supplies, engineers can tailor their circuits to reject ripples from the noisiest supply.

Ultimately, analog systems decompose into the analog building blocks presented here and the single-transistor primitives discussed in Chap. 4. This is because a design engineer, like a painter, composes his or her art from these basic circuit structures, drawing assistance from the integrated resistors, capacitors, diodes, and complementary BJT, MOSFET, and JFET technologies described in Chap. 3. From here, though, implementing higher-level functions like regulation requires a foundation on negative feedback. Chapter 6 therefore introduces negative-feedback loops and describes how they operate. This way, with microelectronic devices, single-transistor circuits, analog building blocks, and feedback concepts in hand, engineers can design bias circuits, operational amplifiers, linear regulators, data converters, switching dc–dc supplies, and other analog systems of comparable complexity.

5.6 Review Questions

1. What is the current gain of a BJT mirror with five equally sized transistors at the input and four at the output?

2. If β_0 is 50 and input current is 10 μA in a base-corrected BJT current mirror, how much current do the bases of the transistors in the mirror of Ques. 1 draw?

3. Neglecting base-current error and assuming the base–emitter voltage is 0.7 V, Early voltage is 50 V, and the output of the mirror in Ques. 1 connects to a 5-V source, what is the gain error of the current mirror in Ques. 1?

4. How can engineers reduce base-current errors in BJT mirrors?

5. How can engineers reduce voltage-mismatch errors in MOSFET mirrors?

6. What is the linear range of BJT differential pairs?

7. What is the linear range of MOSFET differential pairs?

8. Under which conditions does a BJT differential pair reduce to a pair of nondegenerated common-emitter transistors?

9. How can engineers extend the linear range of a JFET differential pair?

10. Which transistors typically enter deep saturation in a basic nonfolding BJT differential stage when their common-mode input voltage surpasses ICMR limits?

11. What limits how close the output of a nonfolding differential stage with a P-type differential pair loaded with a negative-supply N-type current mirror can rise with respect to the positive power supply before pushing a transistor into triode?

12. What determines the response time of a circuit?

13. What is/are the purpose(s) of a current mirror in a differential stage?

14. Which node typically establishes the bandwidth-limiting pole in a differential stage?

15. How does the mirror affect the frequency response of a differential stage?

16. What is the impact of parasitic capacitance at the source-coupled node of a MOSFET differential pair on the common-mode gain of the differential stage?

17. What is the common-mode transconductance gain of a MOSFET differential stage whose input pair and load-mirror transistors match perfectly?

18. Which poles and zeros affect CMRR?

19. What noise mechanisms are present in MOSFETs?

20. What is the power-supply rejection of a P-type differential pair loaded with a negative-supply N-type mirror?

21. Which transistors in a differential stage that folds current via current buffers into a cascoded mirror do not noticeably degrade the input-referred offset voltage of the circuit?

CHAPTER 6

Negative Feedback

Negative feedback is probably one of the most pervasive and revealing examples of how the physical universe works. Humans, for one, would not avoid injuries and survive without perceiving and reacting to their constantly changing environment. Stopping at the door, for example, requires the brain to "counter" its forward motion when it receives a visual indication that the door is near. Feeding images, sounds, and odors that prompt the brain to "stop" this way constitutes "negative feedback" because feeding back "positive" commands when the body nears the door would otherwise impel the body to accelerate and crash into the door. Similarly, maneuvering hands to write, draw, paint, drive, and perform nearly any task imaginable requires feedback information about how fingers respond to instructions from the brain. Not surprisingly, analog designers similarly employ negative feedback to *regulate* voltages and currents and the local and external parameters they ultimately *control*, from temperature and humidity to motion of mechanical systems.

This chapter combines the single-transistor primitives and analog building blocks discussed in Chaps. 4 and 5 to implement feedback circuits that regulate voltages and currents to remain feedback translations of their inputs. Negative feedback not only controls and desensitizes signals but also alters impedances, bandwidth, and noise. But to respond in time and remain in control, the feedback loop must manage to avoid ringing effects that approach or end in unstable operating conditions. This chapter discusses all this: the characteristics, analysis, and design of negative-feedback circuits, the principles of which dictate how engineers design linear low-dropout regulators and other analog systems.

6.1 Feedback Loop

6.1.1 Loop Composition

In its most basic form, feedback circuits implement a signal-processing *loop*, as Fig. 6.1*a* illustrates. The loop is insignificant on its own. It is only useful when processing an input or outputting a signal that

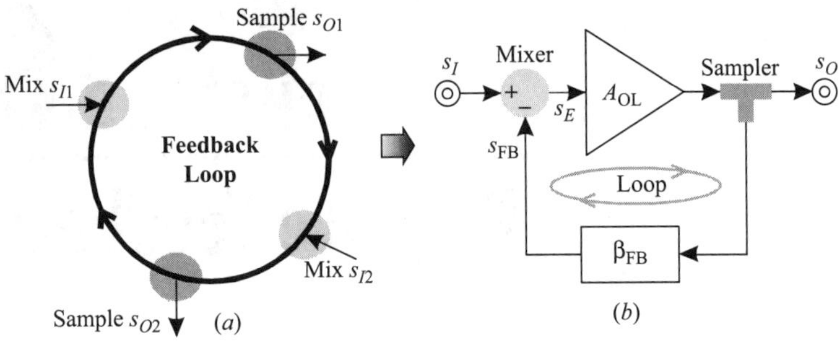

FIGURE 6.1 (*a*) Conceptual and (*b*) symbolic representations of a feedback loop.

feeds another circuit. In this context, injection points "mix" input signals like s_{I1} and s_{I2} in Fig. 6.1*a* into the loop, and extraction points "sense" signals like s_{O1} and s_{O2} so external circuits or devices can process or monitor them.

To be of any use, feedback loops must incorporate one or several *samplers* with which to interface with other circuits. Although oscillators, which rely on positive feedback, do not always receive input signals, negative-feedback circuits almost always do, so they also include *mixers* that mix an incoming signal s_I with an output-derived fed-back signal s_{FB}, as Fig. 6.1*b* shows. The gain from a mixer to a sampler constitutes the *forward open-loop gain* A_{OL} associated with the signals being mixed and sampled. Similarly, β_{FB} refers to the *feedback factor* between the sampler and mixer. Combined, *loop gain* A_{LG} is the gain across the loop, which is nothing but $A_{OL}\beta_{FB}$.

The loop signal s_{FB} with which s_I mixes must be of the same type as s_I, so if s_I is a current, s_{FB} must be also. Because the loop here implements negative feedback, the gain across the loop is inverting. This is why the mixer includes an inversion when both A_{OL} and β_{FB} are positive or negative. The negative sign becomes positive when either A_{OL} or β_{FB} is negative. Although feedback loops need not always be inverting, they nearly always are when used for regulation. Accordingly, because of its vast application to analog circuits and, more specifically, operational amplifiers and regulators, this chapter focuses exclusively on negative feedback.

6.1.2 Regulation

The ability to control actions is the reason negative feedback is so ubiquitous in the universe. And all this because the loop inverts and amplifies. Inversion propels the loop to resist perturbations, and gain determines to what extent. In Fig. 6.1*b*, for example, high loop gain means the system stabilizes when the difference between input and feedback signals s_I and s_{FB}, which error signal s_E represents, is close to nil.

Otherwise, the loop amplifies and inverts s_E, whose net effect is to further amplify and again invert s_E, and so on until the system saturates and loses gain. When stable, the loop therefore opposes the effects of external influences on s_{FB} until s_E is so small that s_{FB} practically mirrors s_I. In other words, the loop gain A_{LG} monitors and amplifies small errors between s_I and s_{FB} to such a degree that the system maintains s_{FB} near s_I.

Quantitatively, since s_E is the difference between s_I and s_{FB}, and s_{FB} is $A_{OL}\beta_{FB}$ or A_{LG} times greater than s_E, s_E is nearly A_{LG} times smaller than s_I:

$$s_E = s_I - s_{FB} = s_I - s_E A_{OL}\beta_{FB} = s_I - s_E A_{LG} = \frac{s_I}{1+A_{LG}} \approx \frac{s_I}{A_{LG}} \to 0 \qquad (6.1)$$

High loop gain A_{LG} on the order of 100 to 1000 or greater will therefore reduce s_E to negligible levels. So when A_{LG} is substantially higher than 1, s_{FB} is practically equal to s_I:

$$s_{FB} = s_E A_{OL}\beta_{FB} = s_E A_{LG} = (s_I - s_{FB})A_{LG} = \frac{s_I A_{LG}}{1+A_{LG}} \approx s_I \qquad (6.2)$$

which means s_{FB} is a *virtual mirror translation* of s_I.

6.1.3 Output Translation

The purpose of feedback factor β_{FB} is to translate s_{FB} to a level s_O that is serviceable to other circuits. So, since s_{FB} is nearly s_I and s_O is only a β_{FB} translation of s_{FB}, s_O is close to s_I/β_{FB}:

$$s_O = (s_I - s_{FB})A_{OL} = (s_I - s_O\beta_{FB})A_{OL} = \frac{s_I A_{OL}}{1+A_{OL}\beta_{FB}} \approx \frac{s_I}{\beta_{FB}} \qquad (6.3)$$

This means that the *closed-loop gain* A_{CL} between s_I and s_O reduces to a $1/\beta_{FB}$ translation when loop gain A_{LG}, or $A_{OL}\beta_{FB}$, is substantially greater than 1:

$$A_{CL} \equiv \frac{s_O}{s_I} = \frac{A_{OL}}{1+A_{OL}\beta_{FB}} = A_{OL} \parallel \frac{1}{\beta_{FB}} \approx \frac{1}{\beta_{FB}} \qquad (6.4)$$

Interestingly, when also considering A_{OL} as a forward translation of s_I to s_O, the effective gain between s_I and s_O is nearly the smaller of the two translations: A_{OL} and $1/\beta_{FB}$, and equivalent to the parallel combination of the two when treating them like impedances.

In all this, notice that inaccuracies and variations in β_{FB} ultimately appear in s_O. The reason for this unfortunate effect is that high loop gain only works to ensure s_{FB} mirrors s_I. This is why feedback-factor blocks are almost always accurate and wide-bandwidth, as in the case of shorts, current-buffer transistors, voltage dividers, and current mirrors.

6.2 Feedback Effects

6.2.1 Sensitivity

Regulating s_{FB} to remain near s_I is another way of saying the loop desensitizes s_{FB} to external influences. In other words, negative feedback renders s_{FB} insensitive to disturbances in the system. To quantify this effect, first consider that the sensitivity of the forward open-loop gain A_{OL} is the amount A_{OL} varies with respect to its nominal value:

$$S_{A_{OL}} \equiv \frac{\Delta A_{OL}}{A_{OL}} = \frac{dA_{OL}}{A_{OL}} \tag{6.5}$$

where ΔA_{OL} and dA_{OL} are A_{OL}'s variation and derivative. The sensitivity of the closed-loop gain A_{CL} is similar, except A_{CL} is the loop-gain reduced counterpart of A_{OL}. So evaluating and dividing A_{CL}'s derivative dA_{CL} by A_{CL} reveals that A_{CL}'s sensitivity is also the loop-gain reduced counterpart of A_{OL}'s:

$$S_{A_{CL}} \equiv \frac{dA_{CL}}{A_{CL}} = \frac{d\left(\dfrac{A_{OL}}{1+A_{OL}\beta_{FB}}\right)}{\left(\dfrac{A_{OL}}{1+A_{OL}\beta_{FB}}\right)} = \frac{\dfrac{(1+A_{OL}\beta_{FB})dA_{OL}-(A_{OL})\beta_{FB}dA_{OL}}{(1+A_{OL}\beta_{FB})^2}}{\dfrac{A_{OL}}{1+A_{OL}\beta_{FB}}}$$

$$= \frac{dA_{OL}}{A_{OL}(1+A_{OL}\beta_{FB})} = \frac{S_{A_{OL}}}{1+A_{OL}\beta_{FB}} = \frac{S_{A_{OL}}}{1+A_{LG}} \tag{6.6}$$

In other words, A_{CL} is roughly A_{LG} times less sensitive to perturbations than A_{OL} is. This insensitivity is the reason engineers often employ negative feedback when designing precision amplifiers for applications that demand predictable, reliable, and well-defined gains.

Example 6.1 Use 20-dB gain stages with 3.3-percent sensitivities to design a 20-dB amplifier with less than 0.1-percent sensitivity.

Solution Employ negative feedback to reduce sensitivity to 0.1 percent. Since negative feedback also reduces forward open-loop gain A_{OL}, cascade sufficient 20-dB gain stages to raise A_{OL} to the extent that A_{OL} is high enough to recover A_{CL} back to 10, or 20 dB:

$$A_{CL} = \frac{A_{OL}}{1+A_{LG}} = \frac{A_{OL}}{1+A_{OL}\beta_{FB}} = \frac{A_V{}^N}{1+A_V{}^N\beta_{FB}} = \frac{10^N}{1+10^N\beta_{FB}} \equiv 10$$

where A_V is the gain of each gain stage and N the number of stages. Since $A_{OL(MAX)}$ is $A_{V(MAX)}{}^N$ and $A_{V(MAX)}$ is $A_V + \Delta A_V$ or $A_V(1 + S_V)$, $A_{OL(MAX)}$ is $A_V{}^N(1 + S_V)^N$ and A_{OL}'s sensitivity S_{OL} is

$$S_{OL} = \frac{\Delta A_{OL}}{A_{OL}} = \frac{A_{OL(MAX)} - A_{OL}}{A_{OL}} = \frac{A_{OL(MAX)}{}^N - A_V{}^N}{A_V{}^N}$$

$$= \frac{A_V{}^N(1+S_V)^N - A_V{}^N}{A_V{}^N} = (1+S_V)^N - 1$$

Therefore, to reduce this S_{OL} to less than 0.1 percent, loop gain A_{LG} or $A_{OL}\beta_{FB}$ must be correspondingly high:

$$S_{CL} = \frac{S_{OL}}{1+A_{OL}\beta_{FB}} = \frac{S_{OL}}{1+A_{LG}} = \frac{(1+S_V)^N - 1}{1+A_{LG}} \leq 0.001$$

With two stages, A_{OL} is 100 and, to ensure that A_{CL} is 10, A_{LG} must be 9, so S_{CL} is 0.67 percent, which is above the target. With three stages, A_{OL} is 1000 and A_{LG} must be 99, so S_{CL} is 0.10 percent, which is good. With four stages, A_{OL} is 10,000 and A_{LG} must be 999, so S_{CL} is 0.01 percent, which is even better.

6.2.2 Impedance

Resistance, or more generally, impedance essentially describes how much current changes in response to variations in voltage, or in reverse, how much voltage varies when current changes. From this perspective, regulating the voltage or current of a signal alters its impedance characteristics. This means negative feedback changes the impedance of every external signal its loop touches, that is to say, of all mixed inputs s_I and sampled outputs s_O.

Mixed Voltage

For voltages to add or subtract, they must combine in *series*. Therefore, to subtract feedback voltage v_{FB} from input voltage v_I in the feedback loop of Fig. 6.1, error voltage v_E must be in series with v_{FB}, as Fig. 6.2 illustrates. Therefore, since negative feedback ensures variations in v_I appear in v_{FB}, the current flowing through the open-loop impedance $Z_{I.OL}$ across the two terminals hardly changes when v_I varies. This means closed-loop input impedance $Z_{I.CL}$, which Ohm's law dictates is the ratio of voltage variations to changes in current, is higher at $Z_{I.CL}$ with *series* feedback than without at $Z_{I.OL}$ because input current variations are lower with feedback.

To quantify how much the loop raises open-loop input impedance $Z_{I.OL}$, consider how loop gain A_{LG} amplifies v_E to establish v_{FB}. Since $Z_{I.CL}$ is the ratio of v_I to input current i_{IN}, $Z_{I.OL}$ the ratio of v_E to i_{IN}, and v_{FB} the loop-gain translation of v_E, $Z_{I.CL}$ is $1 + A_{LG}$ times greater than $Z_{I.OL}$:

$$Z_{I.CL} \equiv \frac{v_I}{i_{IN}} = \frac{v_E + v_{FB}}{i_{IN}} = \frac{v_E + v_E A_{LG}}{i_{IN}} = \left(\frac{v_E}{i_{IN}}\right)(1+A_{LG})$$

$$= Z_{I.OL}(1+A_{LG}) = Z_{I.OL} + Z_{SERIES} \quad (6.7)$$

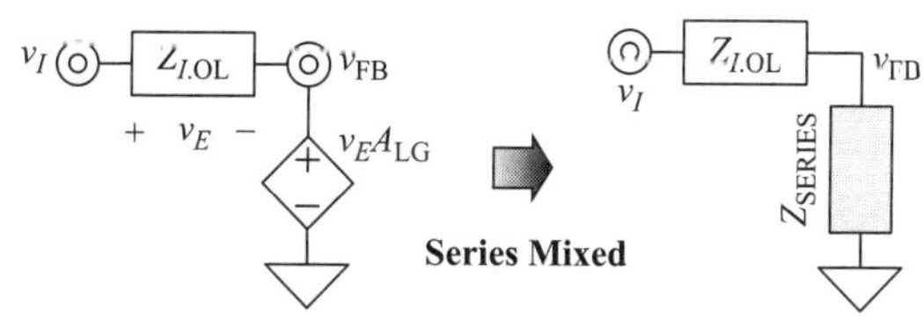

FIGURE 6.2 Voltage-mixed impedance.

In other words, negative feedback introduces a series impedance Z_{SERIES} to the input that is A_{LG} times greater than whatever impedance was there in the first place as $Z_{I.\text{OL}}$.

Mixed Current

For currents to add or subtract, they must combine in parallel. Therefore, to subtract feedback current i_{FB} from input current i_I in the feedback loop of Fig. 6.1, error current i_E must be in parallel with i_{FB}, as Fig. 6.3 illustrates. As such, since negative feedback ensures that variations in i_I appear in i_{FB}, the voltage that a small i_E establishes across the open-loop impedance $Z_{I.\text{OL}}$ of the circuit hardly changes when i_I varies. This means closed-loop input impedance $Z_{I.\text{CL}}$, which Ohm's law dictates is the ratio of voltage variations to changes in current, is lower at $Z_{I.\text{CL}}$ with *shunt* feedback than without at $Z_{I.\text{OL}}$ because input voltage variations are lower with feedback.

To quantify how much the loop reduces open-loop input impedance $Z_{I.\text{OL}}$, consider how loop gain A_{LG} amplifies i_E to establish i_{FB}. Since $Z_{I.\text{CL}}$ is the ratio of input voltage v_{IN} to i_I, $Z_{I.\text{OL}}$ the ratio of v_{IN} to i_E, and i_{FB} the loop-gain translation of i_E, $Z_{I.\text{CL}}$ is $1 + A_{\text{LG}}$ times smaller than $Z_{I.\text{OL}}$:

$$Z_{I.\text{CL}} \equiv \frac{v_{\text{IN}}}{i_I} = \frac{v_{\text{IN}}}{i_E + i_{\text{FB}}} = \frac{v_{\text{IN}}}{i_E + i_E A_{\text{LG}}} = \left(\frac{v_{\text{IN}}}{i_E}\right)\left(\frac{1}{1 + A_{\text{LG}}}\right)$$
$$= \frac{Z_{I.\text{OL}}}{1 + A_{\text{LG}}} = Z_{I.\text{OL}} \parallel Z_{\text{SHUNT}} \tag{6.8}$$

In other words, negative feedback shunts the input with an impedance Z_{SHUNT} that is A_{LG} times lower than whatever impedance was there in the first place as $Z_{I.\text{OL}}$.

Sampled Voltage

Since a sampled output voltage v_O is a β_{FB} translation of feedback signal s_{FB} and β_{FB} is a predictable and well-defined function, negative feedback regulates v_O nearly as well as it controls s_{FB}. The loop will therefore supply or sink whatever current is necessary to keep v_O in Fig. 6.4 near its target, like a voltage source would. In other words, irrespective of how much output current i_{OUT} the circuit receives or supplies,

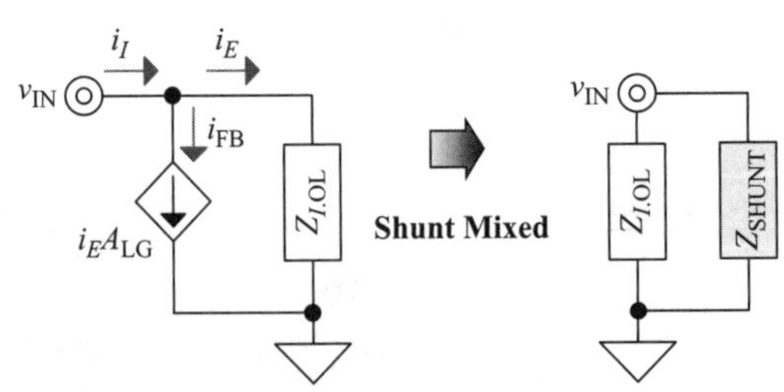

FIGURE 6.3 Current-mixed impedance.

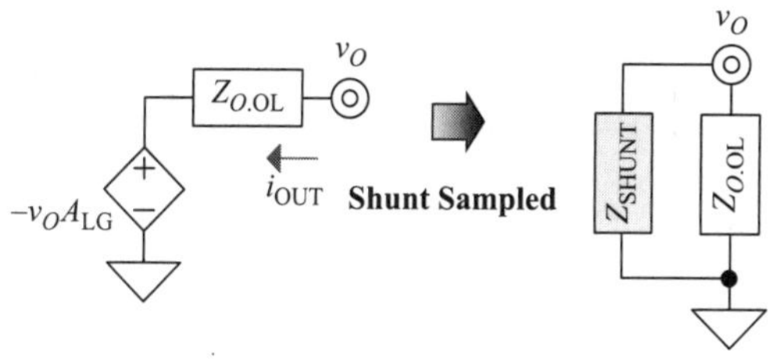

FIGURE 6.4 Voltage-sampled impedance.

v_O hardly changes. This means closed-loop output impedance $Z_{O.CL}$, which Ohm's law dictates is the ratio of voltage variations to changes in current, is lower at $Z_{O.CL}$ with shunt feedback than without at $Z_{O.OL}$ because output voltage variations are lower with feedback.

To quantify how much the loop reduces open-loop output impedance $Z_{O.OL}$, consider how loop gain A_{LG} samples and amplifies v_O to feed a loop-gain translation of v_O to $Z_{O.OL}$. Notice that the polarity of the voltage source in the model of Fig. 6.4 is in phase with v_O because a positive error signal s_E raises v_O. Since v_O is the voltage that i_{OUT} establishes across $Z_{O.OL}$ above the inverting loop-gain translation of v_O, which is $-v_OA_{LG}$, v_O is

$$v_O = i_{OUT}Z_{O.OL} - v_OA_{LG} = i_{OUT}\left(\frac{Z_{O.OL}}{1+A_{LG}}\right) \tag{6.9}$$

and because $Z_{O.CL}$ is the ratio of v_O to i_{OUT}, $Z_{O.CL}$ is $1 + A_{LG}$ times smaller than $Z_{O.OL}$:

$$Z_{O.CL} \equiv \frac{v_O}{i_{OUT}} = \frac{Z_{O.OL}}{1+A_{LG}} = Z_{O.OL} \parallel Z_{SHUNT} \tag{6.10}$$

In other words, negative feedback shunts the output with an impedance Z_{SHUNT} that is A_{LG} times lower than whatever impedance was there in the first place as $Z_{O.OL}$. Note that, like a voltmeter, sampling a voltage requires the β_{FB} block to connect in parallel.

Sampled Current

Since a sampled output current i_O is a β_{FB} translation of feedback signal s_{FB} and β_{FB} is a predictable and well-defined function, negative feedback regulates i_O nearly as well as it controls s_{FB}. The loop will therefore supply or sink whatever current i_{LG} is necessary to keep i_O in Fig. 6.5 near its target, like a current source would. In other words, irrespective of how much output voltage v_{OUT} varies, i_O hardly changes. This means closed-loop output impedance $Z_{O.CL}$, which Ohm's law dictates is the ratio of voltage variations to changes in current, is higher at $Z_{O.CL}$ with series feedback than without at $Z_{O.OL}$ because output current variations are lower with feedback.

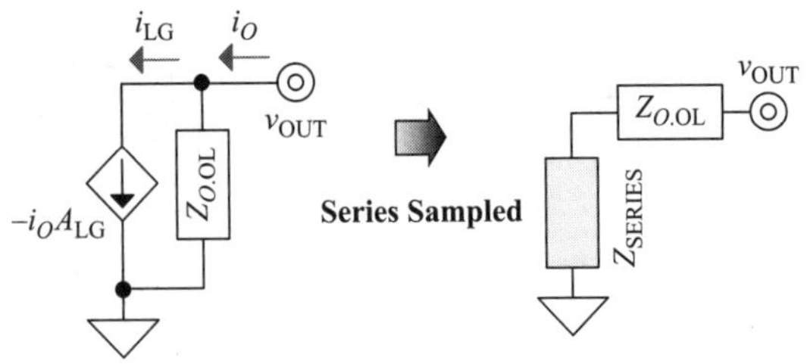

FIGURE 6.5 Current-sampled impedance.

To quantify how much the loop raises open-loop output impedance $Z_{O.OL}$, consider how loop gain A_{LG} samples and amplifies i_O to sink loop-gain current i_{LG}. Notice that A_{OL} in the model of Fig. 6.5 sinks i_{LG} from v_{OUT} because a positive error signal s_E raises i_O and i_O flows into the circuit. Since v_{OUT} is the voltage that i_O and i_{LG} establish across $Z_{O.OL}$ and i_{LG} is an inverting loop-gain translation of i_O, v_{OUT} is

$$v_{OUT} = (i_O - i_{LG})Z_{O.OL} = [i_O - (-i_O A_{LG})]Z_{O.OL} = i_O(1 + A_{LG})Z_{O.OL} \tag{6.11}$$

and because $Z_{O.CL}$ is the ratio of v_{OUT} to i_O, $Z_{O.CL}$ is $1 + A_{LG}$ times greater than $Z_{O.OL}$:

$$Z_{O.CL} \equiv \frac{v_{OUT}}{i_O} = Z_{O.OL}(1 + A_{LG}) = Z_{O.OL} + Z_{SERIES} \tag{6.12}$$

In other words, negative feedback introduces a series impedance Z_{SERIES} to the output that is A_{LG} times higher than whatever impedance was there in the first place as $Z_{O.OL}$. Notice that, like an ammeter, sampling a current requires the β_{FB} block to connect in series.

6.2.3 Frequency Response

Deciphering the frequency response of an inverting feedback circuit is ultimately a sensitivity exercise. From this perspective, closed-loop gain A_{CL} approaches feedback-factor translation $1/\beta_{FB}$ whenever loop gain A_{LG} rises above 1, regardless of what happens to forward open-loop gain A_{OL}. As A_{LG} falls below 1, however, which corresponds to A_{OL} dropping below $1/\beta_{FB}$, feedback signal s_{FB} loses its effect on error signal s_E, so s_E nears s_I and A_{CL} approaches A_{OL}. In the case of Fig. 6.6, for example, A_{CL} nears $1/\beta_{FB}$ between $p_{CL(L)}$ and $p_{CL(H)}$ because A_{OL} is greater than $1/\beta_{FB}$, and A_{CL} approaches A_{OL} below $p_{CL(L)}$ and above $p_{CL(H)}$ because A_{OL} falls below $1/\beta_{FB}$ at those frequencies. In other words, zeros and poles in A_{OL} have little to no effect on A_{CL} when A_{OL} is higher than $1/\beta_{FB}$, except to determine when A_{OL} crosses $1/\beta_{FB}$, where A_{CL} transitions between A_{OL} and $1/\beta_{FB}$.

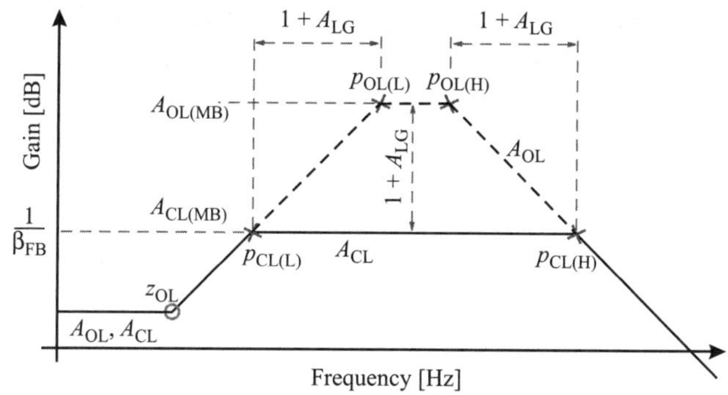

Figure 6.6 Frequency response of a closed-loop system.

Zeros

Since closed-loop gain A_{CL} is nearly feedback-factor translation $1/\beta_{FB}$ when loop gain A_{LG} is well above 1, a zero z_{OL} in the forward open-loop gain A_{OL} has little effect on A_{CL} when A_{LG} is greater than 1:

$$A_{CL} = \frac{A_{OL}}{1+A_{LG}} = \frac{A_{OL}}{1+A_{OL}\beta_{FB}} = \left.\frac{A_{OL0}\left(1+\frac{s}{2\pi z_{OL}}\right)}{1+\left[A_{OL0}\left(1+\frac{s}{2\pi z_{OL}}\right)\right]\beta_{FB}}\right|_{A_{OL}\beta_{FB}>>1} \approx \frac{1}{\beta_{FB}} \tag{6.13}$$

where A_{OL0} is A_{OL}'s low-frequency gain. Whenever A_{LG} falls below 1, however, loop-gain translation 1 + A_{LG} approaches 1, and A_{CL} therefore approaches A_{OL}, so zeros in A_{OL} appear unaffected in A_{LG}. In other words, zeros in A_{OL} remain in A_{CL} when A_{LG} is below 1 like z_{OL} illustrates in Fig. 6.6 and disappear otherwise when A_{LG} rises above 1.

Poles

Reducing A_{OL} by a loop-gain translation across frequency is equivalent to saying negative feedback lowers the impedance that sets A_{OL} at any particular frequency by $1 + A_{LG}$. Since poles occur when capacitors shunt resistors, reducing the impedance by 1 + A_{LG} before the onset of a pole shifts the frequency when the opposing impedance dominates. In other words, negative feedback shifts poles by a factor of $1 + A_{LG}$.

Shunting Capacitors A shunting pole $p_{OL(H)}$, for example, arises when a capacitor C_S shunts the parallel resistance across its terminals R_P. Since R_P dominates before the onset of $p_{OL(H)}$, negative feedback reduces R_P by $1 + A_{LG}$ and $1/sC_S$ shunts the loop-gain reduced resistance at a higher frequency, past a pole $p_{CL(H)}$ that is $1 + A_{LG}$ times higher than the open-loop counterpart $p_{OL(H)}$ from Fig. 6.6:

$$\left.\frac{1}{sC_S}\right|_{p_{CL(H)}=\frac{1}{2\pi\left(\frac{R_P}{1+A_{LG}}\right)C_S}=p_{OL(H)}(1+A_{LG})} = \frac{R_P}{1+A_{LG}} \tag{6.14}$$

To validate this, consider how an open-loop pole ω_{OL} in radians per second in the forward open-loop gain A_{OL} affects the closed-loop gain A_{CL} of the system:

$$A_{CL} = \frac{\frac{A_{OL0}}{1+s/\omega_{OL}}}{1+\left(\frac{A_{OL0}}{1+s/\omega_{OL}}\right)\beta_{FB}} = \frac{A_{OL0}}{1+\frac{s}{\omega_{OL}}+A_{OL0}\beta_{FB}}$$

$$= \frac{\frac{A_{OL0}}{1+A_{OL0}\beta_{FB}}}{1+\frac{s}{\omega_{OL}(1+A_{OL0}\beta_{FB})}} = \frac{A_{CL0}}{1+\frac{s}{\omega_{CL}}} \tag{6.15}$$

where A_{OL0} and A_{CL0} are the low-frequency gains of A_{OL} and A_{CL}. The net result is that negative feedback extends the bandwidth of the open-loop response.

Coupling Capacitors Not surprisingly, the ultimate impact of feedback on the effects of a coupling capacitor C_C is the same, though the mechanics of the process differ slightly. In this case, $1/sC_C$ overwhelms the series resistance R_O that the output presents at low frequency. Negative feedback therefore reduces $1/sC_C$ by $1 + A_{LG}$, which is equivalent to raising C_C by the same factor, and R_O therefore dominates the loop-gain reduced impedance at a lower frequency, at a corner-frequency pole $p_{CL(L)}$ that is $1 + A_{LG}$ times lower than the open-loop counterpart $p_{OL(L)}$ in Fig. 6.6:

$$\left.\frac{1}{sC_C(1+A_{LG})}\right|_{p_{CL(L)}=\frac{1}{2\pi R_O C_C(1+A_{LG})}=\frac{p_{OL(L)}}{1+A_{LG}}} = R_O \tag{6.16}$$

To validate this, first note that the voltage gain across a coupling capacitor is practically nil at low frequencies, rises with frequency, and flattens to $A_{OL(HF)}$ past open-loop corner pole $p_{OL(L)}$, which is equivalent to $\omega_{OL(L)}$ in radians per second in

$$A_{OL} = \frac{A_{OL(HF)}\left(\frac{s}{2\pi p_{OL(L)}}\right)}{\left(1+\frac{s}{2\pi p_{OL(L)}}\right)} = \frac{A_{OL(HF)}\left(\frac{s}{\omega_{OL(L)}}\right)}{\left(1+\frac{s}{\omega_{OL(L)}}\right)} \tag{6.17}$$

The closed-loop gain A_{CL} of the system is therefore

$$A_{CL} = \frac{\left[\dfrac{A_{OL(HF)}(s/\omega_{OL(L)})}{1+s/\omega_{OL(L)}}\right]}{1+\left[\dfrac{A_{OL(HF)}(s/\omega_{OL(L)})}{1+s/\omega_{OL(L)}}\right]\beta_{FB}} = \frac{A_{OL(HF)}\left(\dfrac{s}{\omega_{OL(L)}}\right)}{1+\left(\dfrac{s}{\omega_{OL(L)}}\right)+A_{OL(HF)}\beta_{FB}\left(\dfrac{s}{\omega_{OL(L)}}\right)}$$

$$= \frac{A_{OL(HF)}\left(\dfrac{s}{\omega_{OL(L)}}\right)}{1+\dfrac{s(1+A_{OL(HF)}\beta_{FB})}{\omega_{OL(L)}}} = \frac{\left(\dfrac{A_{OL(HF)}}{1+A_{OL(HF)}}\right)\left[\dfrac{s(1+A_{OL(HF)})}{\omega_{OL(L)}}\right]}{1+\dfrac{s(1+A_{OL(HF)})}{\omega_{OL(L)}}} = \frac{A_{CL(HF)}\left(\dfrac{s}{2\pi p_{CL(L)}}\right)}{1+\dfrac{s}{2\pi p_{CL(L)}}} \tag{6.18}$$

which demonstrates $p_{CL(L)}$ is $1 + A_{LG}$ times lower than $p_{OL(L)}$. So like before, negative feedback extends the bandwidth of the open-loop system.

Band-Pass Response Extending the feedback concept to a band-pass response reveals similar conclusions. In this case, because a low-frequency coupling capacitor C_C dominates at low frequency, negative feedback reduces $1/sC_C$, and the equivalent output resistance R_O starts to dominate the loop-gain reduced impedance of C_C at a lower corner frequency $p_{CL(L)}$ in Fig. 6.6 than the open-loop counterpart $p_{OL(L)}$. Similarly, since R_O dominates past $p_{CL(L)}$, a high-frequency shunting capacitor C_S does not establish a pole $p_{CL(H)}$ until $1/sC_S$ drops past R_O's loop-gain reduced resistance, so $p_{CL(H)}$ is at a higher frequency than the open-loop counterpart $p_{OL(H)}$. So like before, poles shift and gain drops by a factor of $1 + A_{LG}$, and their ultimate effect is to extend bandwidth.

6.2.4 Noise

Since embedded in-band noise in the input is indistinguishable from the actual signal, input signals must exceed this noise floor for a system to be able to process information. Unfortunately, power supplies and components in microelectronic systems inject noise into the signal stream. This means that the output of the amplifier in Fig. 6.7 incorporates not only the amplified translation of the input signal $s_I A_X$ but

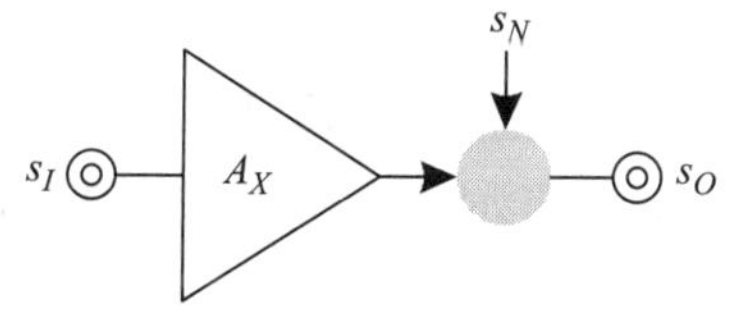

FIGURE 6.7 Output noise in an open-loop system.

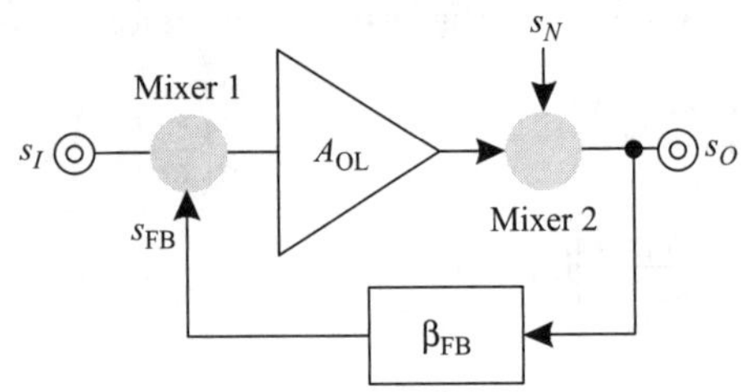

FIGURE 6.8 Output noise in a closed-loop system.

also noise s_N. Since the amplifier multiplies s_I, the *signal-to-noise ratio* (SNR) at the output, which is a measure of how much the system favors input signals s_I over noise s_N, is the forward gain across the amplifier A_X:

$$\text{SNR}_{\text{OL}} \equiv \frac{A_I}{A_N} = \frac{A_X}{1} = A_X \tag{6.19}$$

where A_I and A_N are how much the system amplifies s_I and s_N, respectively, and A_N is 1.

Since negative feedback reduces forward open-loop gain A_{OL}, A_{OL} in Fig. 6.8 should be $1 + A_{LG}$ times greater than A_X to realize a closed-loop gain A_{CL} that is equivalent to A_X:

$$A_{\text{CL}} = \frac{A_{\text{OL}}}{1 + A_{\text{OL}}\beta_{\text{FB}}} \equiv A_X \tag{6.20}$$

With this in mind, injecting noise s_N into s_O amounts to mixing s_N into the loop. This means that the forward open-loop gain from s_N to s_O, or $A_{OL(N)}$ for short, is 1 and the closed-loop gain is therefore $1 + A_{LG}$ times lower than 1:

$$A_{\text{CL(N)}} \equiv \frac{s_O}{s_N} = \frac{A_{\text{OL(N)}}}{1 + A_{\text{LG}}} = \frac{1}{1 + A_{\text{LG}}} \tag{6.21}$$

The closed-loop signal-to-noise ratio SNR_{CL} therefore reduces to A_{OL}, which is $1 + A_{LG}$ times higher than the open-loop counterpart of A_X at SNR_{OL}:

$$\text{SNR}_{\text{CL}} \equiv \frac{A_I}{A_N} = \frac{A_{\text{CL}}}{A_{\text{CL(N)}}} = \frac{\left(\dfrac{A_{\text{OL}}}{1 + A_{\text{LG}}}\right)}{\left(\dfrac{1}{1 + A_{\text{LG}}}\right)}$$

$$= A_{\text{OL}} = A_X(1 + A_{\text{LG}}) = \text{SNR}_{\text{OL}}(1 + A_{\text{LG}}) \tag{6.22}$$

In other words, negative feedback effectively suppresses output noise by a factor of $1 + A_{LG}$.

6.2.5 Linearity

Linearity performance ultimately refers to the effect of external perturbations on the output. As a result, in regulating feedback signal s_{FB}, and through a β_{FB} translation, output signal s_O, negative feedback opposes the effects of *distortion* on s_{FB} and s_O. Consider, for example, fluctuations of gain across signal strength, and more specifically, gain saturation. First recall that circuits exhibit the highest gain when their transistors operate in light saturation in the case of bipolar-junction transistors (BJTs) and saturation in the case of field-effect transistors (FETs). Unfortunately, output voltages, to cite an example, can swing so much that they can push transistors toward triode, where gain is lower. This variation in gain causes distortion because the open-loop amplifier multiplies the input s_I by A_{OL} when output s_O is halfway across its range and by a saturated gain $A_{OL(SAT)}$ at the extremes, at $S_{O(MAX)}$ and $S_{O(MIN)}$ in Fig. 6.9. The net result is that gain varies across signal strength:

$$\frac{\Delta A_{OL}}{A_{OL}} = \frac{A_{OL} - A_{OL(SAT)}}{A_{OL}} \tag{6.23}$$

While reducing the operating range of the system improves linearity, doing so also reduces signal-to-noise ratio at the output, which is not desirable.

Example 6.2 Determine the percent voltage-gain variation across the two cascaded 10× amplifiers in Fig. 6.10, which saturate to half their nominal gain at the extremes of their output range.

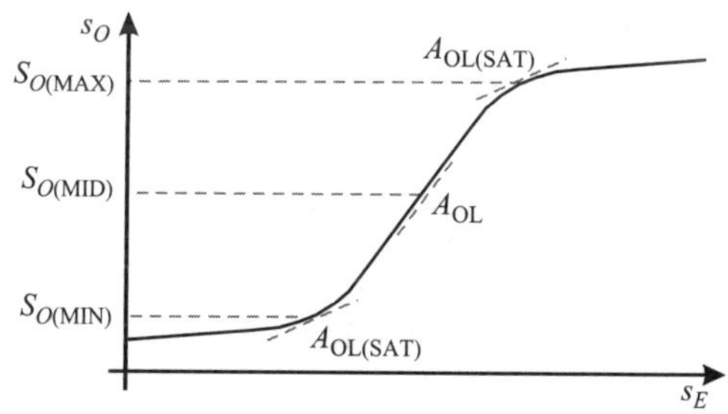

FIGURE 6.9 Gain variation across the output range of an open-loop amplifier.

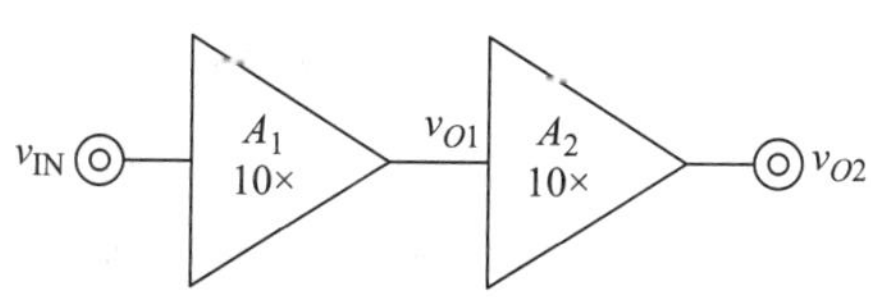

FIGURE 6.10 Cascaded 10× amplifiers that saturate to 5× at the extremes of their output range.

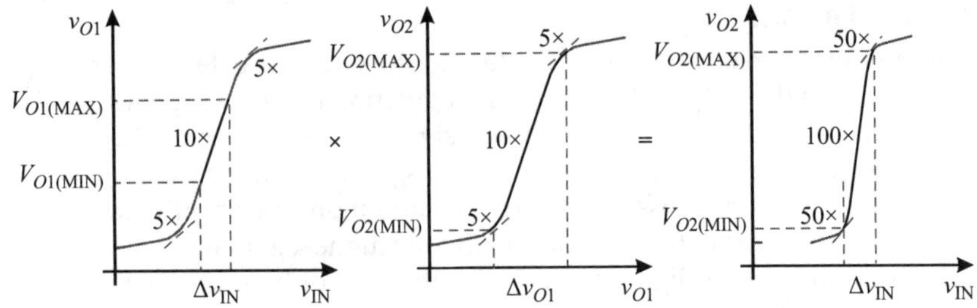

Figure 6.11 Signal swing and gain across the two amplifiers in Fig. 6.10.

Solution Since the output voltage v_{O2} of the second amplifier A_2 is nominally 10 times higher than the output of the first v_{O1}, v_{O2} swings 10 times more than v_{O1}. So when v_{O2} saturates, v_{O1} is 10 times lower and therefore well within the first amplifier's extreme limits, as Fig. 6.11 demonstrates. In other words, the overall gain of the system at v_{O2}'s limits is (10)(5) or 50 and at midrange (10)(10) or 100, which means gain varies by 50 percent:

$$\frac{\Delta A_V}{A_V} = \frac{100-50}{100} = 50\%$$

Negative feedback lowers the forward open-loop gain across the entire range, from $S_{O(MIN)}$ to $S_{O(MAX)}$ in Fig. 6.12. Luckily, loop gain at signal extremes $S_{O(MIN)}$ and $S_{O(MAX)}$, where gain saturates, is lower at $A_{OL(SAT)}\beta_{FB}$, or $A_{LG(SAT)}$, than at midrange at $A_{OL}\beta_{FB}$, or A_{LG}. This means negative feedback reduces midrange gain A_{OL} more than it does saturated gain, $A_{OL(SAT)}$, which is lower than A_{OL} in the first place. The difference between the two closed-loop gains A_{CL} and $A_{CL(SAT)}$ is therefore lower by 1 plus the saturated loop gain $A_{LG(SAT)}$:

$$\frac{\Delta A_{CL}}{A_{CL}} = \frac{\left(\dfrac{A_{OL}}{1+A_{OL}\beta_{FB}}\right)-\left(\dfrac{A_{OL(SAT)}}{1+A_{OL(SAT)}\beta_{FB}}\right)}{\left(\dfrac{A_{OL}}{1+A_{OL}\beta_{FB}}\right)}$$

$$= \frac{\left(\dfrac{A_{OL}-A_{OL(SAT)}}{A_{OL}}\right)}{1+A_{OL(SAT)}\beta_{FB}} = \frac{\left(\dfrac{\Delta A_{OL}}{A_{OL}}\right)}{1+A_{OL(SAT)}\beta_{FB}} \tag{6.24}$$

In other words, the distortion that a saturating gain produces is lower by a factor of $1 + A_{LG(SAT)}$.

Example 6.3 Use negative feedback to realize a 100× voltage-amplifying system that reduces the distortion produced by individual amplifiers in the system when they saturate to half their prescribed gains at the extremes of their output range.

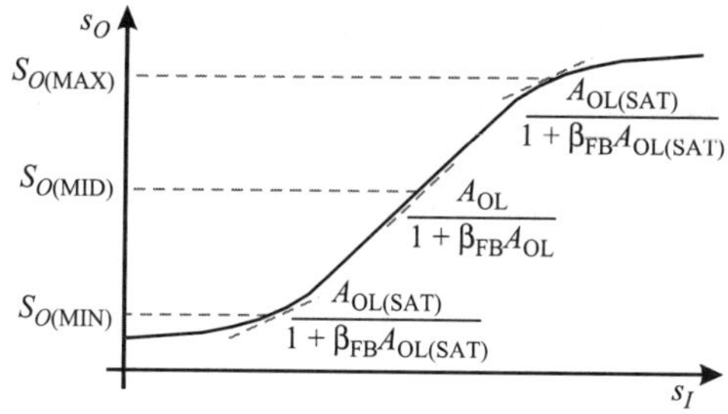

Figure 6.12 Gain variation across the output range of a closed-loop amplifier.

Solution Negative feedback around the output stage reduces the distortion that saturating gains produce in an amplifier system. Since gain climbs to its midrange level as soon as the output falls within 100 to 300 mV of its extremes, using negative feedback to divide the gain of the 10× amplifier A_2 in Fig. 6.13 to 2× keeps the swing of the preceding amplifier A_1 two times below its extremes, well within its saturating limits, as Fig. 6.14 demonstrates. To realize a closed-loop gain of 2× and an overall gain of 100×, the feedback factor β_{FB} should be 0.4:

$$A_{CL2} = \frac{A_2}{1+A_2\beta_{FB}} = \frac{10}{1+10\beta_{FB}} \equiv 2$$

and A_1 should be 50×. This way, the saturated loop gain $A_{LG(SAT)}$ is (5)(0.4) and the nonsaturated counterpart A_{LG} is (10)(0.4), so 5× in A_2 reduces to 1.67× in A_{CL2}

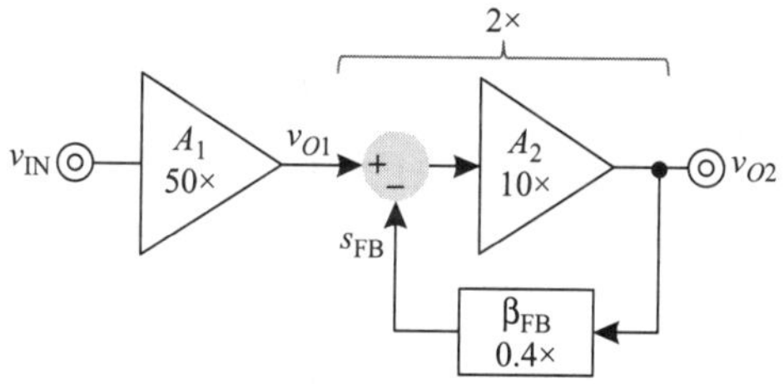

Figure 6.13 100× amplifier system with negative feedback.

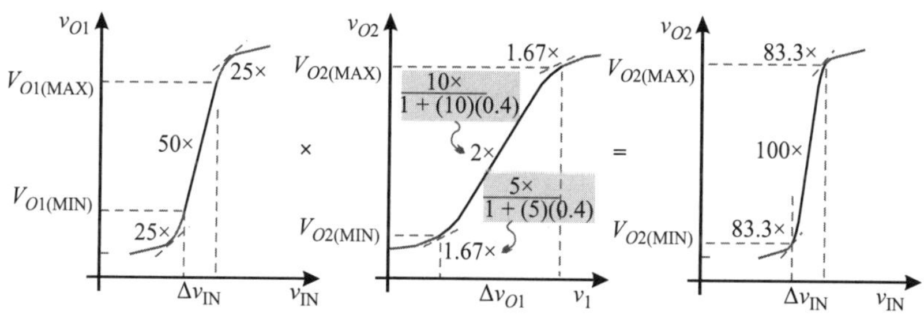

Figure 6.14 Signal swing and gain across the 100× amplifier system in Fig. 6.13.

at the boundaries and 10× to 2× away from the boundaries. In other words, the overall gain at the limits is (50)(1.67) or 83.3 and at midrange (50)(2) or 100, which means gain varies by 16.7 percent, $1 + A_{LG(SAT)}$ times lower than A_2's initial 50-percent variation from Example 6.2:

$$\frac{\Delta A_V}{A_V} = \frac{100 - 83.3}{100} = 16.7\%$$

6.3 Configurations

Negative feedback relates each input–output combination in a loop in a particular way. A feedback loop that mixes a voltage and samples a current, for example, implements the function of a transconductance amplifier. Similarly, voltage–voltage, current–current, and current–voltage feedback networks realize voltage, current, and transimpedance amplifiers. Understanding which signals the loop mixes and samples is important because the network, the driving source, and the load affect them differently. The following subsections therefore model how each of these configurations mixes and samples signals, and with the models, they describe the impact of drivers and loads on gain.

6.3.1 Transconductance Amplifier

Connectivity

A feedback loop that mixes an input voltage v_I and samples an output current i_O relates the latter to the former via a closed-loop *transconductance gain* $A_{G.CL}$. Since the forward open-loop transconductance gain $A_{G.OL}$ in the network amplifies the difference between v_I and the feedback voltage v_{FB} that the feedback factor β_{FB} generates, v_I, $A_{G.OL}$, and β_{FB} must connect in series, like Fig. 6.15 shows. This explains why closed-loop input impedance $Z_{I.CL}$ is higher than its open-loop counterpart. Plus, to monitor $A_{G.OL}$'s i_O, β_{FB}, like an ammeter, must also connect in series with $A_{G.OL}$, so closed-loop output impedance $Z_{O.CL}$ is likewise higher than its open-loop counterpart.

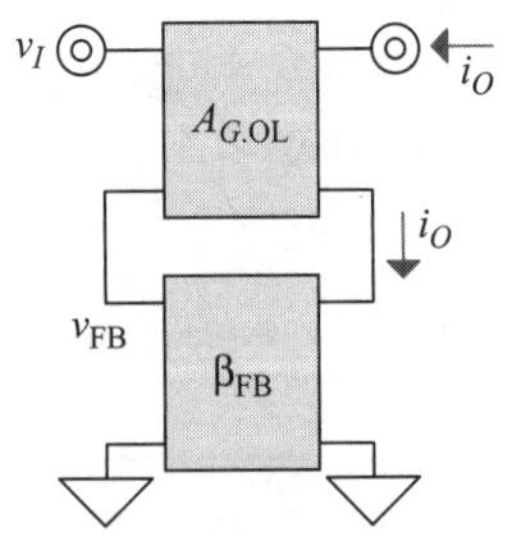

FIGURE 6.15 Voltage-mixing and current-sampling feedback amplifier.

Two-Port Model

Because resistance is normally sufficient to predict the loading effects of most microelectronic circuits on a driving circuit, modeling the input of a feedback network with only a resistance R_I is common practice. The same is not true for the output because the model must also incorporate the gain of the network. In this respect, although both Thevenin- and Norton-equivalent circuits can model electrical networks equally well, the Norton equivalent from Fig. 6.16 is more convenient when emulating the output of a feedback network that controls current. Therefore, to avoid modeling redundancies, R_O's current must be zero when deriving $A_{G.CL}$, and $A_{G.CL}$'s current must be zero otherwise when extracting R_O. In other words, output voltage v_{OUT} must short to ground when deriving $A_{G.CL}$ and input voltage v_I must similarly connect to ground when extracting R_O.

Gain Losses

In practice, the circuit that drives the input of the feedback amplifier incorporates an equivalent source resistance R_S, as Fig. 6.17 shows. Unfortunately, this resistance drops a portion of the incoming voltage v_{IN}, leaving a voltage-divided fraction across the input v_I of the feedback network. Similarly, the load that the feedback loop drives also presents a resistance R_L. In this case, the output resistance of the loop R_O steers current away from R_L, so R_O, like R_S, also reduces the overall transconductance gain of the system A_G to

$$A_G = \frac{i_{OUT}}{v_{IN}} = \left(\frac{R_I}{R_S + R_I}\right) A_{G.CL} \left(\frac{R_O \parallel R_L}{R_L}\right) = \left(\frac{R_I}{R_S + R_I}\right) A_{G.CL} \left(\frac{R_O}{R_O + R_L}\right) \tag{6.25}$$

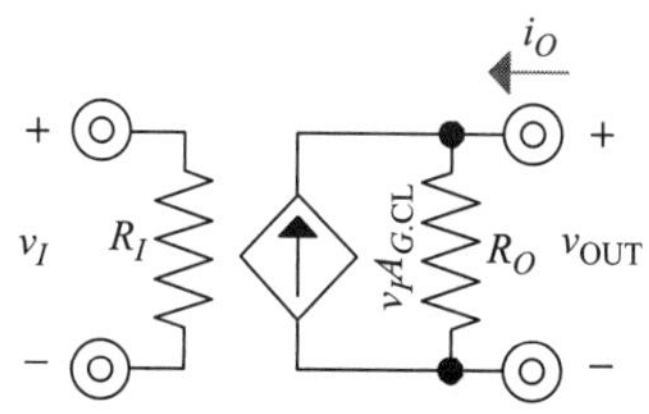

Figure 6.16 Two-port model of a transconductance feedback amplifier.

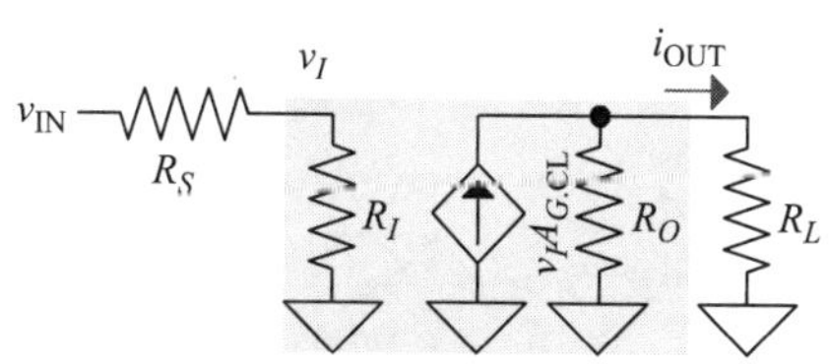

Figure 6.17 Transconductance feedback amplifier under practical operating conditions.

If loop gain is high, however, series mixing raises input resistance and series sampling increases output resistance to such a degree that most of v_{IN} drops across R_I and little current flows through R_O. This establishes a system gain A_G that is nearly equal to the closed-loop gain of the loop at $A_{G.CL}$.

6.3.2 Voltage Amplifier

Connectivity

A feedback loop that mixes an input voltage v_I and samples an output voltage v_O relates the latter to the former via a closed-loop *voltage gain* $A_{V.CL}$. Since the forward open-loop voltage gain $A_{V.OL}$ in the network amplifies the difference between v_I and the feedback voltage v_{FB} that the feedback factor β_{FB} generates, v_I, $A_{V.OL}$, and β_{FB} must connect in series, like Fig. 6.18 shows. This explains why closed-loop input impedance $Z_{I.CL}$ is higher than its open-loop counterpart. Plus, to monitor $A_{V.OL}$'s v_O, β_{FB}, like a voltmeter, must connect in parallel with $A_{V.OL}$, so closed-loop output impedance $Z_{O.CL}$ is likewise lower than its open-loop counterpart.

Two-Port Model

As discussed earlier, because resistance is normally sufficient to predict the loading effects of most microelectronic circuits on a driving circuit, modeling the input of a feedback network with only a resistance R_I is common practice. Since the same is not true for the output, and a Thevenin-equivalent circuit is convenient when emulating how a circuit regulates output voltage, the output of the model in Fig. 6.19

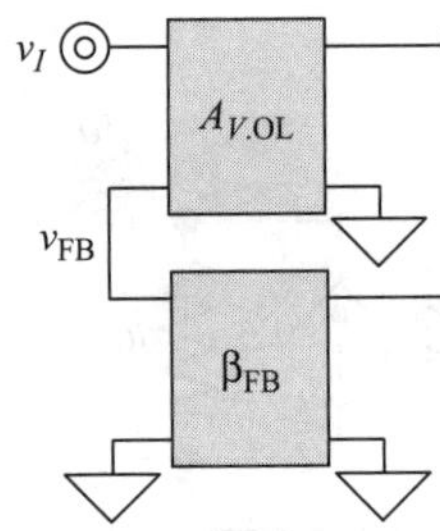

FIGURE 6.18 Voltage-mixing and voltage-sampling feedback amplifier.

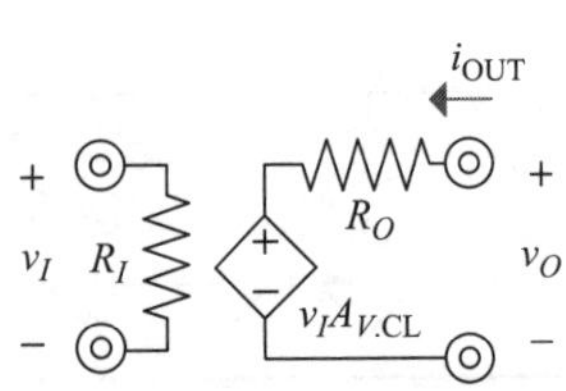

FIGURE 6.19 Two-port model of a voltage feedback amplifier.

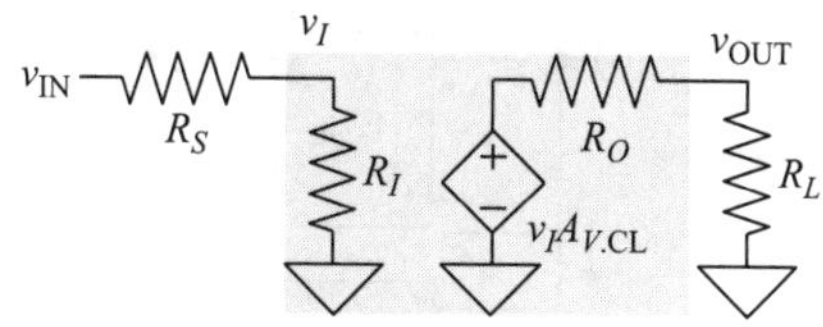

FIGURE 6.20 Voltage feedback amplifier under practical operating conditions.

implements a Thevenin network. So to avoid modeling redundancies, R_O's voltage must be zero when deriving $A_{V.CL}$, and $A_{V.CL}$'s voltage must be zero otherwise when extracting R_O. In other words, output current i_{OUT} must be zero when deriving $A_{V.CL}$, which means no load should appear at the output, and input voltage v_I should short to ground when extracting R_O.

Gain Losses

In practice, the circuit that drives the input of the feedback amplifier incorporates an equivalent source resistance R_S, as Fig. 6.20 shows. Unfortunately, this resistance drops a portion of the incoming voltage v_{IN}, leaving a voltage-divided fraction across the input v_I of the feedback network. Likewise, the load that the feedback loop drives also presents a resistance R_L. Here, the output resistance of the loop R_O drops some of the voltage that $A_{V.CL}$ generates, so R_O, like R_S, also reduces the overall voltage gain of the system A_V to

$$A_V = \frac{v_{OUT}}{v_{IN}} = \left(\frac{R_I}{R_S + R_I}\right) A_{V.CL} \left(\frac{R_L}{R_O + R_L}\right) \tag{6.26}$$

If loop gain is high, however, series mixing raises input resistance and shunt sampling lowers output resistance to such a degree that most of v_{IN} drops across R_I and most of the voltage that $A_{V.CL}$ outputs appears across R_L to establish a system gain A_V that is nearly equal to the closed-loop gain of the loop at $A_{V.CL}$.

6.3.3 Current Amplifier

Connectivity

A feedback loop that mixes an input current i_I and samples an output current i_O relates the latter to the former via a closed-loop *current gain* $A_{I.CL}$. Since the forward open-loop current gain $A_{I.OL}$ in the network amplifies the difference between i_I and the feedback current i_{FB} that the feedback factor β_{FB} generates, $A_{I.OL}$ and β_{FB} must connect in parallel, like Fig. 6.21 shows. This explains why closed-loop input impedance $Z_{I.CL}$ is lower than its open-loop counterpart. Plus, to monitor $A_{I.OL}$'s i_O, β_{FB}, like an ammeter, must connect in series with $A_{I.OL}$, so closed-loop output impedance $Z_{O.CL}$ is likewise higher than its open-loop counterpart.

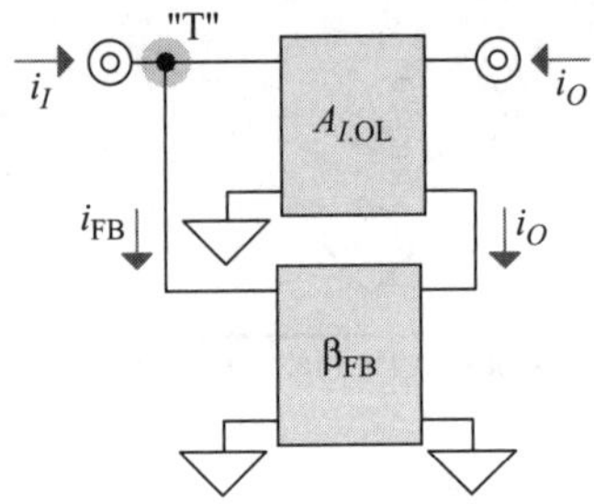

Figure 6.21 Current-mixing and current-sampling feedback amplifier.

Two-Port Model

As before, because resistance is normally sufficient to predict the loading effects of most microelectronic circuits on a driving circuit, modeling the input of a feedback network with only a resistance R_I is common practice. Since the same is not true for the output, and a Norton-equivalent circuit is convenient when emulating how a circuit regulates output current, the output of the model in Fig. 6.22 implements a Norton network. So to avoid modeling redundancies, R_O's current must be zero when deriving $A_{I.CL}$, and $A_{I.CL}$'s current must be zero otherwise when extracting R_O. In other words, output voltage v_{OUT} must short to ground when deriving $A_{V.CL}$, and input current i_I must be zero when extracting R_O.

Gain Losses

In practice, the circuit that drives the input of the feedback amplifier incorporates an equivalent source resistance R_S, as Fig. 6.23 shows. Unfortunately, this resistance steers current away from the input of the feedback network. Likewise, the load that the feedback loop drives also presents a resistance R_L. Here too, the output resistance of the loop R_O steers current away from R_L, so R_O, like R_S, also reduces the overall current gain of the system A_I to

$$A_I = \frac{i_{OUT}}{i_{IN}} = \left(\frac{R_S \parallel R_I}{R_I}\right) A_{I.CL} \left(\frac{R_O \parallel R_L}{R_L}\right) = \left(\frac{R_S}{R_S + R_I}\right) A_{I.CL} \left(\frac{R_O}{R_O + R_L}\right) \quad (6.27)$$

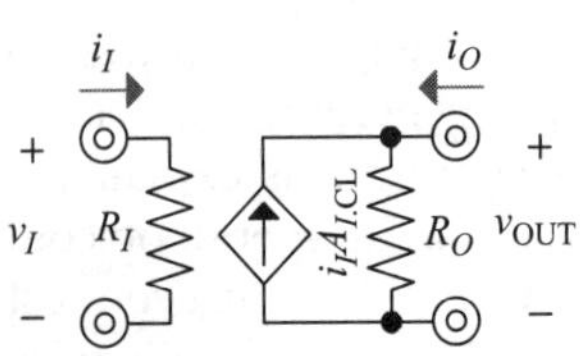

Figure 6.22 Two-port model of a current feedback amplifier.

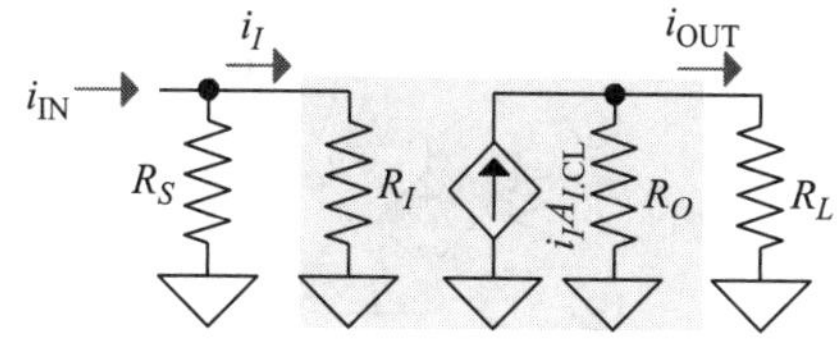

FIGURE 6.23 Current feedback amplifier under practical operating conditions.

If loop gain is high, however, shunt mixing lowers input resistance and series sampling raises output resistance to such a degree that most of i_{IN} reaches R_I and most of the current that $A_{I.CL}$ outputs reaches R_L to establish a system gain A_I that is nearly equal to the closed-loop gain of the loop at $A_{I.CL}$.

6.3.4 Transimpedance Amplifier

Connectivity

A feedback loop that mixes an input current i_I and samples an output voltage v_O relates the latter to the former via a closed-loop *transimpedance gain* $A_{Z.CL}$. Since the forward open-loop transimpedance gain $A_{Z.OL}$ in the network amplifies the difference between i_I and the feedback current i_{FB} that the feedback factor β_{FB} generates, $A_{Z.OL}$ and β_{FB} must connect in parallel, like Fig. 6.24 shows. This explains why closed-loop input impedance $Z_{I.CL}$ is lower than its open-loop counterpart. Plus, to monitor $A_{Z.OL}$'s v_O, β_{FB}, like a voltmeter, must connect in parallel with $A_{Z.OL}$, so closed-loop output impedance $Z_{O.CL}$ is likewise lower than its open-loop counterpart.

Two-Port Model

As before, because resistance is normally sufficient to predict the loading effects of most microelectronic circuits on a driving circuit, modeling the input of a feedback network with only a resistance R_I is common practice. Since the same is not true for the output, and a

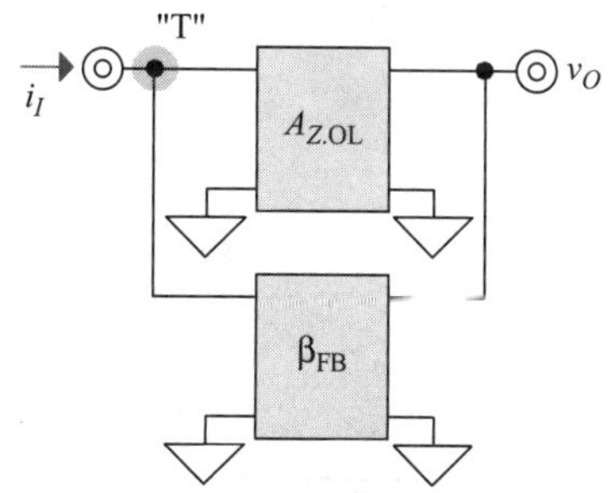

FIGURE 6.24 Current-mixing and voltage-sampling feedback amplifier.

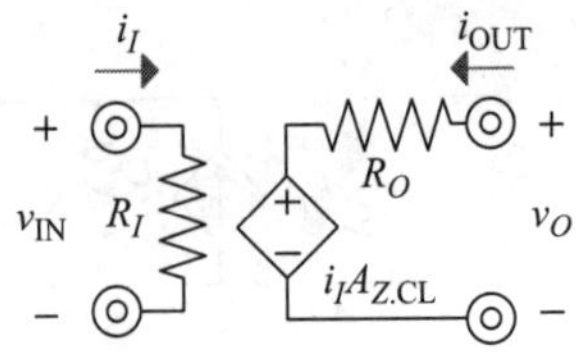

FIGURE 6.25 Two-port model of a transimpedance feedback amplifier.

Thevenin-equivalent circuit is convenient when emulating how a circuit regulates output voltage, the output of the model in Fig. 6.25 implements a Thevenin network. So to avoid modeling redundancies, R_O's voltage must be zero when deriving $A_{I.CL}$, and $A_{Z.CL}$'s voltage must be zero otherwise when extracting R_O. In other words, the load should be absent when deriving $A_{Z.CL}$, and input current i_I should be zero when extracting R_O.

Gain Losses

In practice, the circuit that drives the input of the feedback amplifier incorporates an equivalent source resistance R_S, as Fig. 6.26 shows. Unfortunately, this resistance steers current away from the input of the feedback network. Likewise, the load that the feedback loop drives also presents a resistance R_L. Here, the output resistance of the loop R_O drops some of the voltage $A_{Z.CL}$ produces, so R_O, like R_S, also reduces the overall transimpedance gain of the system A_Z to

$$A_Z = \frac{v_{OUT}}{i_{IN}} = \left(\frac{R_S \parallel R_I}{R_I}\right) A_{Z.CL} \left(\frac{R_L}{R_O + R_L}\right) = \left(\frac{R_S}{R_S + R_I}\right) A_{Z.CL} \left(\frac{R_L}{R_O + R_L}\right) \tag{6.28}$$

If loop gain is high, however, shunt mixing lowers input resistance and shunt sampling reduces output resistance to such a degree that most of i_{IN} reaches R_I, and most of the voltage that $A_{Z.CL}$ outputs drops across R_L to establish a system gain A_Z that is nearly equal to the closed-loop gain of the loop at $A_{Z.CL}$.

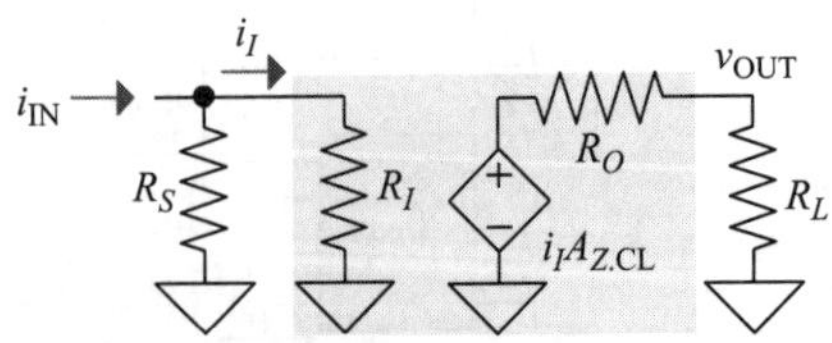

FIGURE 6.26 Transimpedance feedback amplifier under practical operating conditions.

6.4 Analysis

Analyzing circuits that employ negative feedback can take many forms, from strictly algebraic to part algebraic and part insight or fully insightful approaches. This book and this section in particular try to use insight to both draw practical conclusions and derive quantitative expressions that describe the operating and performance limits of feedback loops. The ensuing discussion therefore describes and illustrates the analytical and identification processes with which engineers can analyze transconductance, voltage, current, and transimpedance feedback circuits. Along the way, the text also highlights and explains how some of the single-transistor primitives in Chap. 4 incorporate feedback components.

6.4.1 Analytical Process

Single Loop

The first step in the analytical process is to identify the presence of a feedback loop. For this, engineers trace the small-signal path in the circuit to see if it loops back. No matter where in the path the process starts, negative feedback will always process and feed a signal back to the starting point. For reference, although not always true or obvious, electrical components on the schematics of feedback circuits normally connect and conform to the shape of a visible, though oftentimes contorted loop, like Fig. 6.1 illustrates, which is usually the first indication that feedback is present.

The second step is to identify whether or not the loop is inverting. Retracing the loop to track the number of inversions across the loop reveals this, because negative feedback can only incorporate an odd number of inversions. Bear in mind that most if not all inversions appear across base–collector, gate–drain, and inverting input–output terminals of differential amplifiers.

Identification Next is identifying the mixing and sampling components. Since a loop can mix and output at several points along its path, what is important here is which mixer receives the input and which sampler senses the output. This is why the identification process starts with the input and output terminals of the network. Literally tracing these terminals into the loop discloses which devices ultimately mix the input with a loop signal and sample the output that one or several external circuits use. Ultimately, the types of signals the mixer receives and the sampler outputs indicate which amplifier the feedback loop implements.

Open-Loop Parameters Because closed-loop response already describes how input signal s_I affects the system, s_I should be zero when extracting open-loop parameters. So to begin, forward open-loop gain A_{OL}

is the gain from error signal s_E to output signal s_O, or s_O/s_E. Realizing the input of A_{OL} is the mixed difference of s_I and feedback signal s_{FB} oftentimes circumvents the need for breaking the loop at this point, the mechanics of which can otherwise introduce errors in the form of excluded or approximated impedances. Note that A_{OL} is part of a two-port network, so when deriving A_{OL}, the output must short to ground if A_{OL} drives a current, or the load must be absent otherwise if A_{OL} sources a voltage. Feedback factor β_{FB} follows, which is the gain from s_O to feedback signal s_{FB}, or s_{FB}/s_O.

Open-loop input and output resistances $R_{I.OL}$ and $R_{O.OL}$ are the resistances at s_I and s_O when the loop signals that feed them are absent. In this respect, the most accurate way of excluding loop signals is to zero the current of a Norton source or the voltage of the Thevenin counterpart. This way, $R_{I.OL}$ and $R_{O.OL}$ incorporate the effects of all resistances in the network. Incidentally, for the same reasons, excluding loop signals this way is generally the best way of breaking a feedback loop. Note, by the way, that $R_{O.OL}$ is part of a Norton or Thevenin circuit, so deriving $R_{O.OL}$ requires the Norton current or Thevenin voltage that drives it to be zero.

Closed-Loop Model Next is modeling the feedback network with a two-port equivalent that closed-loop gain A_{CL} and closed-loop input and output resistances $R_{I.CL}$ and $R_{O.CL}$ describe. A_{CL} here, which refers to the gain from s_I to s_O, is $1 + A_{LG}$ times lower than A_{OL}. The resistances of shunt-mixed currents and shunt-sampled voltages are similarly $1 + A_{LG}$ times lower than their open-loop counterparts. Conversely, the resistances of series-mixed voltages and series-sampled currents are $1 + A_{LG}$ times higher than those of the open-loop system.

The most important part of this entire process is the overriding conclusions that result. If loop gain A_{LG}, which is equivalent to $A_{OL}\beta_{FB}$, is substantially greater than 1, for example, s_{FB} mirrors s_I and s_O is a $1/\beta_{FB}$ translation of not only s_{FB} but also s_I. With an accurate and wide-bandwidth feedback factor β_{FB}, this means that the loop regulates s_{FB} and s_O, and it will therefore oppose the effects of external influences like supply variations, noise injection, gain saturation, and other factors on s_{FB} and s_O. But if A_{LG} falls below 1, s_{FB} has little to no effect on s_E, so s_E nears s_I and A_{CL} reduces to A_{OL}.

Example 6.4 Draw closed-loop conclusions from the fact that a feedback network with high loop gain mixes voltages v_A and v_B and samples current i_C, where the latter flows through resistor R_D to establish v_B.

Solution Mixed voltages and a sampled current imply the network series-mixes and series-samples, so the loop implements a transconductance feedback amplifier. Because the loop gain is high, v_B mirrors v_A. Since R_D converts i_C to v_B and v_B is practically v_A, i_C is nearly v_A/R_D, which is another way of saying the closed-loop gain is roughly $1/R_D$. Closed-loop input and output resistances are high because both the input and output of the feedback network connect in series with the forward circuit.

Overall Response Now that a simple two-port circuit models the feedback loop, analyzing the effect of the network on a system is relatively straightforward. The system essentially translates the input it receives into the signal that the feedback loop processes, and similarly converts what it samples from the loop into the signal that it ultimately outputs. From this, note that the input and output terminals of the system s_{IN} and s_{OUT} are not always those of the feedback circuit s_I and s_O. Plus, source and load resistances R_S and R_L around the loop reduce the gain across the loop, but only to the extent that $R_{I.CL}$ and $R_{O.CL}$ allow.

Embedded Feedback Loops

Embedding feedback loops within a larger loop is not uncommon. By default, the loop that receives and samples the signals of the system s_I and s_O is the outer loop, as Fig. 6.27 shows. As a result, internal loops ultimately affect the forward open-loop gain A_{OL} or the feedback factor β_{FB}. From this perspective, analyzing the inner loops first is simpler and more intuitive than extracting A_{OL} and β_{FB} directly. A_{OL} and β_{FB} therefore become the combined closed-loop gains of internal loops and other gain stages. In the case of Fig. 6.27, for example, A_{OL} is $A_X A_{CL.I1}$ and β_{FB} is $A_{CL.I2}\beta_X$. Notice loops that share either the mixer or the sampler with the outer loop, but not both, are still inner loops in the forward path.

Parallel Feedback Loops

In the case of parallel loops, as Fig. 6.28 illustrates, the strongest feedback signal usually overpowers the weakest to render the weaker loop ineffectual. This means that the highest loop gain normally dominates and determines which feedback signal mirrors the input s_I and how the output s_O relates to s_I. But to be more precise, consider that the mixer subtracts all feedback signals from s_I to generate error s_E. Since each feedback signal is a loop-gain translation of s_E, s_E

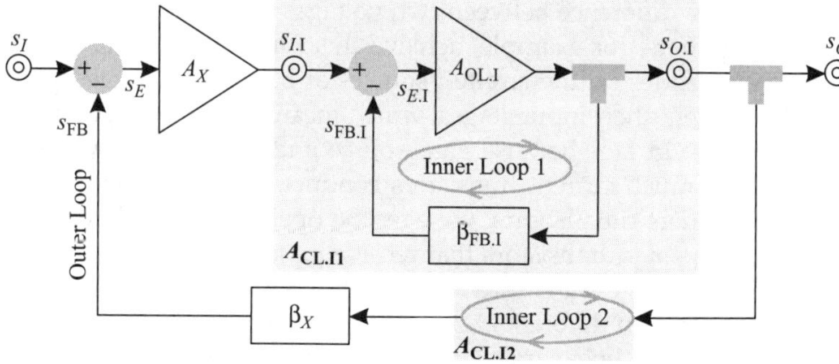

Figure 6.27 Feedback system with embedded feedback loops.

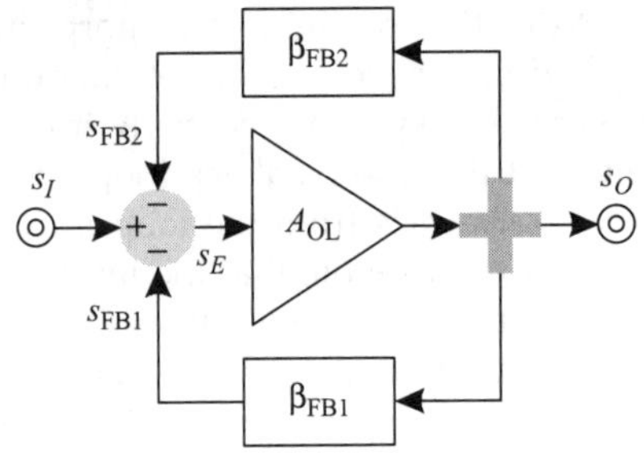

FIGURE 6.28 Feedback system with parallel feedback loops.

is ultimately lower than s_I by a factor that is 1 plus the sum of the parallel loop gains:

$$s_E = s_I - s_{FB1} - s_{FB2} = s_I - s_E A_{OL}\beta_{FB1} - s_E A_{OL}\beta_{FB2}$$
$$= \frac{s_I}{1 + A_{OL}(\beta_{FB1} + \beta_{FB2})} = \frac{s_I}{1 + A_{LG1} + A_{LG2}} \tag{6.29}$$

Therefore, because s_O is a forward-gain translation of s_E, the closed-loop gain A_{CL} that results is similarly lower than the open-loop counterpart A_{OL} by 1 plus the sum of the parallel loop gains:

$$s_O = s_E A_{OL} = \frac{s_I A_{OL}}{1 + A_{OL}(\beta_{FB1} + \beta_{FB2})} = \frac{s_I A_{OL}}{1 + A_{LG1} + A_{LG2}} = s_I A_{CL} \tag{6.30}$$

But if one of the loop gains, or equivalently, one of the feedback signals overwhelms the others, the highest loop gain determines A_{CL} and all other closed-loop parameters in the system.

6.4.2 Mixers

Voltage Mixer

The role of a voltage mixer is to subtract a feedback voltage v_{FB} from an input voltage v_I. From this perspective, any circuit component that processes the difference between two voltages can be a mixer. Operational amplifiers, for example, achieve this functionality. This ability ultimately resides on the emitter/source- or base/gate-coupled pairs that implement their input stages, which means differential pairs also mix voltages. In fact, because gate–source and base–emitter voltages define how much current transistors conduct, transistors like operational amplifiers can also mix voltages. So operational amplifiers, differential pairs, and transistors that receive input and feedback signals across their input terminals, as Fig. 6.29 shows, mix voltages. The only difference between differential pairs and transistors is that the current in transistors is a linear translation of small-signal voltages only, not their large-signal counterparts. This means that operational amplifiers

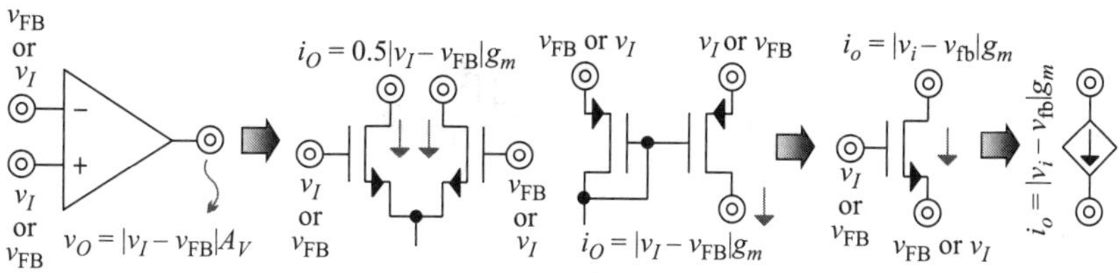

FIGURE 6.29 Voltage mixers.

and differential pairs mix and mirror both small- and large-signal voltages, and transistors mix and mirror small-signal components only.

Since open-loop parameters must account for all impedances in the network, the mixer in the feedback model of Fig. 6.1*b* is an impedance-free component. This means the actual mixer is the Norton- or Thevenin-equivalent source that models the gain of the mixing block. In the case of transistors, for example, the sourcing transconductance g_m is the fundamental mixer that subtracts v_{FB} from v_I.

Current Mixer

Similarly, the role of a current mixer is to subtract a feedback current i_{FB} from an input current i_I. Interestingly, because the sum of all currents into a junction is zero, the third terminal of T connections that receive two currents outputs their sum or difference. In the case of negative feedback, i_{FB} is an inverted translation of error current i_E, and therefore of i_I, so three-terminal junctions that connect i_I and i_{FB} signals output their difference. This means T connections in general can be current mixers, and identifying them amounts to spotting T connections like those of Figs. 6.21 and 6.24. But before concluding that a loop shunt-mixes currents, ruling out the presence of voltage mixers is good practice. Note, by the way, that T connections can mix both small- and large-signal components of i_I.

6.4.3 Samplers

Voltage Sampler

Any component that outputs a signal that is a function of an incoming voltage can connect in parallel with the output to sample its voltage. In fact, voltage mixers can, by default, sample voltages. The difference between a voltage mixer and a voltage sampler is that the mixer processes the difference between two voltages and the sampler does not. So operational amplifiers, emitter/source- and base/gate-coupled differential pairs, transistors, and transconductors that receive no other loop signals as inputs except v_O are voltage samplers, as Fig. 6.30 illustrates. In other words, gates, sources, bases, and emitters can all sample voltages, but only when the other input does not carry a loop signal.

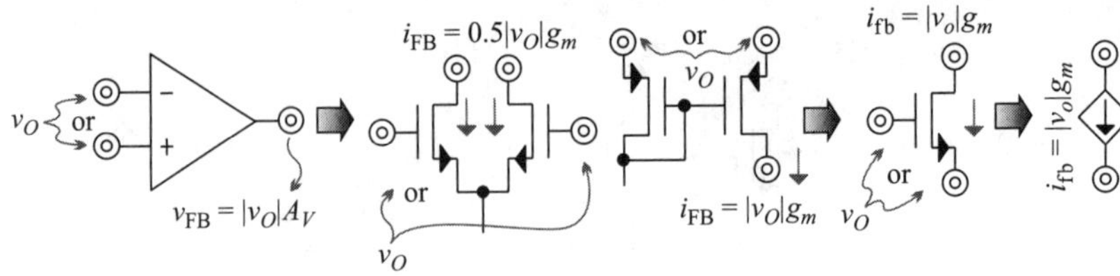

FIGURE 6.30 Voltage samplers.

One way to test this ability in all cases is by applying the *voltage-sampler test*. This test is to ground the output terminal of the network because the output of a feedback factor β_{FB} that samples voltage is always zero when v_O is zero. This is why short circuits and voltage dividers also sample voltages. Note that shorting connections and voltage dividers can sample both small- and large-signal components of the output voltage, and so do operational amplifiers and differential pairs, but only when their opposing input terminals are zero.

Current Sampler

Any component that receives and outputs a current can connect in series with the output to sample output currents. Current mirrors can therefore sample currents. What is more, drains, sources, collectors, and emitters can also sample currents because they can connect in series and conduct the same current that their opposing terminals carry. For this, though, as Fig. 6.31 shows, their output currents must reach feedback factor β_{FB} to establish the feedback signal s_{FB} with which the input signal s_I mixes. Generally, mirrors and transistors can sample both small- and large-signal components of the output current.

Interestingly, sources and emitters can sample either voltages or currents. In these cases, testing whether or not they sample a voltage reveals what they sense. In other words, if grounding the source or

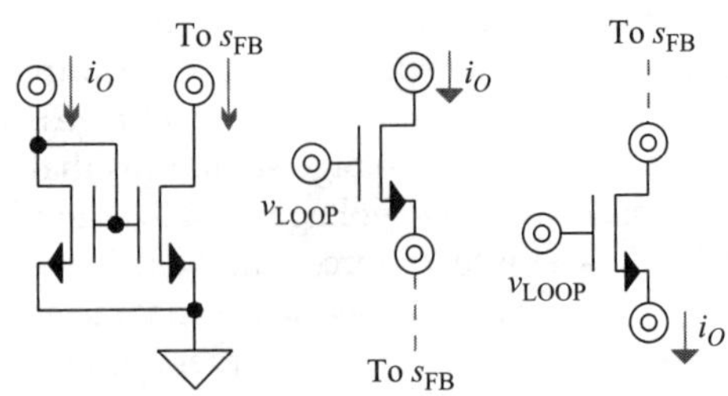

FIGURE 6.31 Current samplers.

emitter terminal does not force the output of the feedback factor β_{FB} to zero, the terminal tested samples current. This is the reason why sources and emitters sample currents only when their gates and bases carry loop signals, as v_{LOOP} shows in Fig. 6.31, and when their drain and collector currents ultimately reach β_{FB} to establish s_{FB}.

6.4.4 Transconductance Amplifiers

Emitter/Source-Degenerated Transistors

Although not immediately apparent, source- and emitter-degenerated transistors from Chap. 4 incorporate negative feedback. To perceive this, consider that a rise in collector current i_O in Fig. 6.32 raises emitter voltage v_{FB}, the effect of which is to lower the transistor's base–emitter voltage, and as a result, reduce collector current i_O. Or to be more general and succinct, the transistor opposes variations in i_O, which means emitter- and source-degenerated transistors are inverting feedback circuits.

Tracing the input into the circuit to find the mixer reveals a receiving base. Since the opposing emitter terminal of the transistor is a loop signal v_{FB}, the transistor's transconductance g_m series-mixes base voltage v_i into the loop. Similarly, following the output into the network to find the sampler discloses a collector. Because a collector terminal can only sense current, and the current here establishes the feedback signal v_{FB}, the collector series-samples output current i_O. In other words, the feedback loop implements a transconductance amplifier.

The forward open-loop transconductance gain $A_{G.OL}$ is the Norton-equivalent gain translation from the series-mixed difference v_e or $v_i - v_{fb}$ to the series-sampled output i_o when the input v_i and the output voltage v_{out} are zero. In the small-signal circuit of Fig. 6.32, g_m's and r_π's currents into the parallel combination of r_o and degenerating

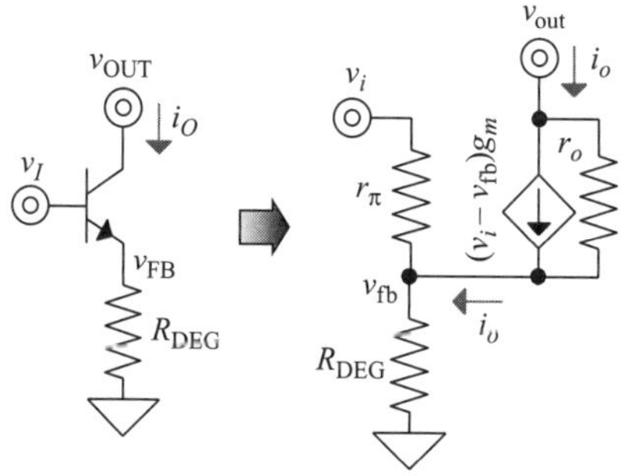

FIGURE 6.32 Emitter-degenerated transistor and its small-signal equivalent circuit.

resistor R_{DEG} at v_{fb} establish v_{fb}, and g_m's current and the current that v_{fb} induces through r_o comprise i_o, so $A_{G.OL}$ is

$$A_{G.OL}\Big|_{v_{out}\equiv 0} \equiv \frac{i_o}{v_e} = \frac{v_e g_m - \dfrac{v_{fb}}{r_o}}{v_e} = \frac{v_e g_m - \dfrac{v_e\left(g_m + \dfrac{1}{r_\pi}\right)(R_{DEG} \parallel r_o)}{r_o}}{v_e}$$
$$\approx \frac{g_m (R_{DEG} \parallel r_o)}{R_{DEG}} \tag{6.31}$$

Feedback factor β_{FB} is the gain from output i_o to feedback v_{fb} when the input v_i is zero. Since i_o into the collector and i_o's base translation i_o/β_0 flow to the emitter, v_{fb} is the voltage these currents establish across R_{DEG}:

$$\beta_{FB}\big|_{v_i\equiv 0} \equiv \frac{v_{fb}}{i_o} = \frac{i_o\left(1+\dfrac{1}{\beta_0}\right)R_{DEG}}{i_o} \approx R_{DEG} \tag{6.32}$$

As a result, closed-loop gain $A_{G.CL}$ is 1 plus loop gain $A_{G.OL}\beta_{FB}$, or equivalently, $1 + g_m(R_{DEG} \parallel r_o)$ times lower than $A_{G.OL}$:

$$A_{G.CL} = \frac{A_{G.OL}}{1 + A_{G.OL}\beta_{FB}} \approx \frac{g_m\left(\dfrac{R_{DEG} \parallel r_o}{R_{DEG}}\right)}{1 + g_m(R_{DEG} \parallel r_o)} = \frac{g_m}{\left(\dfrac{R_{DEG}}{R_{DEG} \parallel r_o}\right) + g_m R_{DEG}}$$
$$= \frac{g_m}{\left(\dfrac{R_{DEG}}{r_o} + 1\right) + g_m R_{DEG}} = \frac{g_m}{1 + \left(\dfrac{1}{r_o} + g_m\right)R_{DEG}} \approx \frac{g_m}{1 + g_m R_{DEG}} \tag{6.33}$$

where r_o is usually substantially greater than r_π. Since removing transconductance g_m eliminates the feedback source that feeds the input and leaves only r_π and R_{DEG}, open-loop input resistance $R_{I.OL}$ is

$$R_{I.OL} \approx r_\pi + R_{DEG} \tag{6.34}$$

With a series-mixed input, closed-loop input resistance $R_{I.CL}$ is $1 + g_m(R_{DEG} \parallel r_o)$ times greater than $R_{I.OL}$ at

$$R_{I.CL} = R_{I.OL}(1 + A_{G.OL}\beta_{FB}) \approx (r_\pi + R_{DEG})[1 + g_m(R_{DEG} \parallel r_o)] \tag{6.35}$$

Similarly, removing g_m when input v_i is zero eliminates the feedback source that feeds the output. This reveals that Norton-equivalent open-loop output resistance $R_{O.OL}$ is r_o in series with the parallel combination of r_π and R_{DEG}:

$$R_{O.OL}\big|_{v_i\equiv 0} = r_o + (r_\pi \parallel R_{DEG}) \tag{6.36}$$

With a series-sampled output, closed-loop output resistance $R_{O.CL}$ is therefore $1 + g_m(R_{DEG} \| r_o)$ times higher than $R_{O.OL}$ at

$$R_{O.CL} = R_{O.OL}(1 + A_{G.OL}\beta_{FB}) \approx [r_o + (r_\pi \| R_{DEG})][1 + g_m(R_{DEG} \| r_o)] \quad (6.37)$$

As expected, all these closed-loop results are similar to those derived in Chap. 4.

Voltage-Controlled Regulated-Cascode Current Source

A popular way of establishing a current is to drop a user-defined input voltage v_I across a resistor R_I. In the case of Fig. 6.33, differential amplifier A_G and transistor M_O implement a feedback loop that impresses v_I across R_I. The purpose of M_O here is to sample R_I's current v_I/R_I, which is also i_O, and as a current source, sink i_O from a load. Because v_I and R_I establish a current i_O that is practically independent of the load, the network realizes a voltage-controlled current source.

The feedback loop that connects gate and feedback voltages v_G and v_{FB} is inverting because, of the elements in the loop, A_G is the only one that inverts. Interestingly, M_O, as a source-degenerated transistor, also implements an inverting feedback circuit. But of the two loops, the loop that includes A_G is the only one attached to the input and output terminals of the network. This means M_O is an embedded loop whose closed-loop response affects the open-loop parameters of the outer loop.

Tracing the input into the circuit to find the mixer reveals a receiving differential amplifier A_G. Since A_G's opposing input terminal is a loop signal v_{FB}, A_G series-mixes small- and large-signal components of input-terminal voltage v_I into the loop. Similarly, following the output into the network to find the sampler discloses M_O's drain. Because

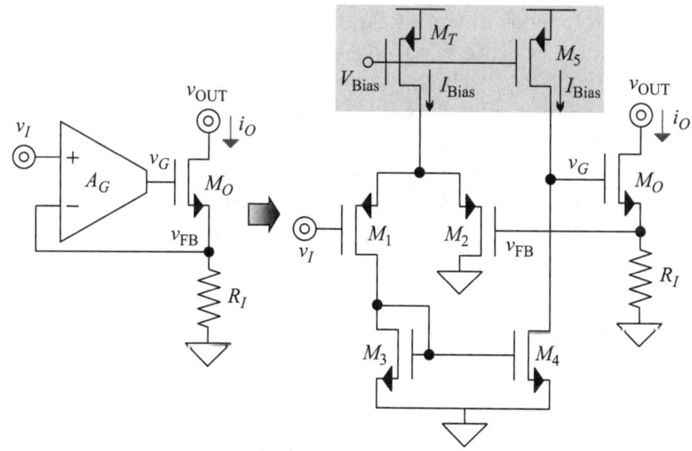

FIGURE 6.33 Voltage-controlled regulated-cascode current source.

a drain terminal can only sense current, and the current here establishes the feedback signal v_{FB}, the drain series-samples small- and large-signal components of output current i_O. In other words, the feedback loop implements a transconductance amplifier.

Before proceeding, recall that open-loop parameters must account for the closed-loop response of the internal source-degenerated transistor. With this in mind, the forward open-loop transconductance gain $A_{G.OL}$ is the Norton-equivalent gain translation from the series-mixed difference v_e or $v_i - v_{fb}$ to the series-sampled output i_o when the input v_i and the output voltage v_{out} are zero. Here, A_G amplifies v_e and outputs a current that flows through A_G's output resistance R_{OA} to establish gate voltage v_g, which M_O's degenerated closed-loop transconductance gain G_{MO} ultimately translates into i_o:

$$A_{G.OL}\Big|_{v_{out}\equiv 0} \equiv \frac{i_o}{v_e} = \left(\frac{v_g}{v_e}\right)\left(\frac{i_o}{v_g}\right) = (A_G R_{OA})G_{MO} \approx \frac{A_G R_{OA} g_{mO}}{1+g_{mO}(R_I \parallel R_{ID})} \tag{6.38}$$

where R_{ID} is A_G's differential input resistance, R_I and R_{ID} source-degenerate M_O, and g_{mO} is M_O's transconductance. Feedback factor β_{FB} is the gain from output i_O to feedback v_{FB} when the input v_i is zero. Since i_O into the drain flows entirely to the source, v_{FB} is the voltage that i_O establishes across R_I and R_{ID}:

$$\beta_{FB}\Big|_{v_I\equiv 0} \equiv \frac{v_{FB}}{i_O} = R_I \parallel R_{ID} \tag{6.39}$$

As a result, closed-loop gain $A_{G.CL}$ is 1 plus loop gain $A_{G.OL}\beta_{FB}$ times lower than $A_{G.OL}$. Since M_O and A_G process both small- and large-signal components of i_O and v_I linearly, v_O is a $1/\beta_{FB}$ translation of v_{FB} and v_I, or $v_I/(R_I \parallel R_{ID})$, when $A_{G.OL}\beta_{FB}$ is considerably greater than 1. Here, removing A_G's Norton current source eliminates the feedback source that ultimately feeds the input. And recalling that M_O is the inner loop reveals R_I and $1/g_{mO}$ are in parallel and in series with R_{ID} in open-loop input resistance $R_{I.OL}$, which means $R_{I.OL}$ is $R_{ID} + R_I \parallel (1/g_{mO})$. With a series-mixed input, closed-loop input resistance $R_{I.CL}$ is $1 + A_{G.OL}\beta_{FB}$, or when $g_{mO}(R_{ID} \parallel R_I)$ is substantially higher than 1, $A_G R_{OA}$ times higher than $R_{I.OL}$. Similarly, removing A_G's Norton current source when input v_i is zero eliminates the feedback source that feeds the output. This reveals that Norton-equivalent open-loop output resistance $R_{O.OL}$ is M_O's degenerated closed-loop drain resistance R_{DO}. With a series-sampled output, closed-loop output resistance $R_{O.CL}$ is therefore $1 + A_G R_{OA}$ times higher than $R_{O.OL}$. Notice, by the way, that $A_G R_{OA}$ multiplies the loop-gain effects already present in the degenerated transistor.

To cite a transistor-level example of A_G, the P-type differential pair M_1–M_2 in Fig. 6.33 series-mixes the input to accommodate input voltages that near ground. Folding one of M_1–M_2's output currents into current source M_5 to establish output voltage v_G allows A_G to drive high voltages without affecting the input pair. As such, A_G reduces to $0.5g_{m1}$, R_{OA} is $r_{ds4} \| r_{sd5}$, and R_{ID} approaches infinity.

Source-Sampling Current Sink

Although not as apparent at first, the circuit in Fig. 6.34 is the source-sampling counterpart of the drain-sampling current source in Fig. 6.33. To compare the circuits, first consider that differential amplifier A_G and transistor M_O in both cases implement a feedback loop that impresses v_I across R_I. Similarly, M_O samples R_I's current i_O to sink it from a load. The difference here is that M_O's source terminal samples i_O.

The feedback loop that connects gate and feedback voltages v_G and v_{FB} is inverting because, of the elements in the loop, M_O is the only one that inverts. Like in the previous circuit, M_O, as a source-degenerated transistor, also implements an inverting feedback circuit. But of the two loops, the loop that includes A_G is the only one attached to the input and output terminals of the network. This means M_O is an embedded loop whose closed-loop response affects the open-loop parameters of the outer loop.

Tracing the input into the circuit to find the mixer reveals a receiving differential amplifier A_G. Since A_G's opposing input terminal is a loop signal v_{FB}, A_G series-mixes small- and large-signal components of input-terminal voltage v_I into the loop. Similarly, following the output into the network to find the sampler discloses M_O's source. Because a source terminal whose gate counterpart carries a loop signal v_G can only sense current, and the current here establishes the feedback signal v_{FB}, the source series-samples small- and large-signal components of output current i_O. In other words, the feedback loop implements a transconductance amplifier.

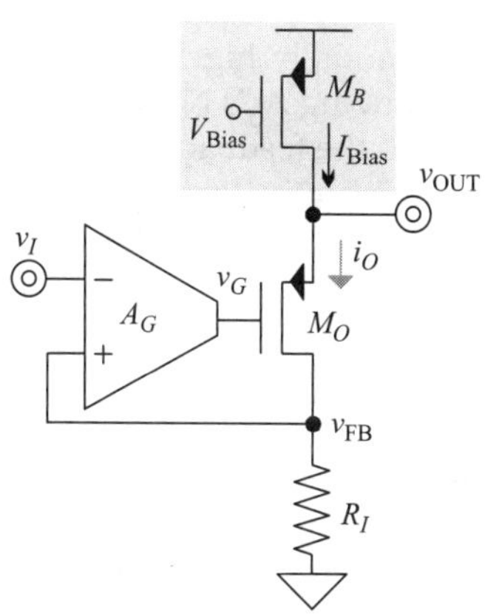

FIGURE 6.34 Source-sampling current sink.

Before anything else, recall that open-loop parameters must account for the closed-loop response of the internal source-degenerated transistor. With this in mind, the forward open-loop transconductance gain $A_{G.OL}$ is the Norton-equivalent gain translation from the series-mixed difference v_e or $v_i - v_{fb}$ to the series-sampled output i_o when the input v_i and the output voltage v_{out} are zero. Here, A_G amplifies v_e and outputs a current that flows through A_G's output resistance R_{OA} to establish gate voltage v_g, which M_O's degenerated closed-loop transconductance gain G_{MO} then translates into i_o:

$$A_{G.OL}\Big|_{v_{out}\equiv 0} \equiv \frac{i_o}{v_e} = \left(\frac{v_g}{v_e}\right)\left(\frac{i_o}{v_g}\right) = (-A_G R_{OA})(-G_{MO}) \approx \frac{A_G R_{OA} g_{mO}}{1 + g_{mO} r_{sdB}} \tag{6.40}$$

where g_{mO} is M_O's transconductance and M_B's r_{sdB} source-degenerates M_O. Feedback factor β_{FB} is the gain from output i_O to feedback v_{FB} when the input v_I is zero. Since i_O into M_O's source flows entirely to the drain, v_{FB} is the voltage that i_O establishes across R_I and A_G's differential input resistance R_{ID}:

$$\beta_{FB}\Big|_{v_I\equiv 0} \equiv \frac{v_{FB}}{i_O} = R_I \parallel R_{ID} \tag{6.41}$$

As a result, closed-loop gain $A_{G.CL}$ is 1 plus loop gain $A_{G.OL}\beta_{FB}$ times lower than $A_{G.OL}$. Since M_O and A_G process both small- and large-signal components of i_O and v_I linearly, i_O is a $1/\beta_{FB}$ translation of v_{FB} and v_I, or $v_I/(R_I \parallel R_{ID})$, when $A_{G.OL}\beta_{FB}$ is considerably greater than 1. Here, removing A_G's Norton current source eliminates the feedback source that ultimately feeds the input. And recalling that M_O is the inner loop reveals R_I and M_O's source-degenerated drain resistance $R_{DO(DEG)}$ are in parallel and in series with R_{ID} in open-loop input resistance $R_{I.OL}$, which means $R_{I.OL}$ is $R_{ID} + R_I \parallel R_{DO(DEG)}$. With a series-mixed input, closed-loop input resistance $R_{I.CL}$ is $1 + A_{G.OL}\beta_{FB}$, or when $g_{mO} r_{sdB}$ is substantially higher than 1, $A_G R_{OA}$ times higher than $R_{I.OL}$. Similarly, removing A_G's Norton current source when the input v_i is zero eliminates the feedback source that feeds the output. This reveals that Norton-equivalent open-loop output resistance $R_{O.OL}$ is M_O's source resistance R_{SO}, which the parallel combination of R_I and R_{ID} loads to

$$R_{O.OL}\big|_{v_i\equiv 0} = R_{SO} = \frac{r_{sdO} + (R_I \parallel R_{ID})}{1 + g_{mO} r_{sdO}} \tag{6.42}$$

With a series-sampled output, closed-loop output resistance $R_{O.CL}$ is therefore about $A_G R_{OA}$ times higher than $R_{O.OL}$. Through all this, note that, while $A_{G.CL}$ is equivalent to that of the previous example when loop gain is high, $R_{O.OL}$ is not, and that is because, here, M_O's source is the current sampler, not M_O's drain.

6.4.5 Voltage Amplifiers

Noninverting Op Amp

The noninverting operational-amplifier configuration in Fig. 6.35 is a classic example of negative feedback. To recognize this, consider that a rise in output voltage v_O raises inverting input voltage v_{FB}, the effect of which is to reduce output voltage v_O. In other words, the amplifier opposes variations in v_O, which means noninverting operational-amplifier configurations implement inverting feedback circuits.

Tracing the input into the circuit to find the mixer reveals a receiving differential amplifier A_V. Since A_V's opposing input terminal is a loop signal v_{FB}, A_V series-mixes small- and large-signal components of input-terminal voltage v_I into the loop. Similarly, following the output into the network to find the sampler discloses A_V's output and R_2. Because A_V's output cannot sense and R_2 voltage-divides v_O to A_V's inverting input, whose opposing input terminal does not carry a loop signal, R_2 buffers and A_V ultimately shunt-samples small- and large-signal components of output voltage v_O. In other words, the feedback loop implements a voltage amplifier.

The forward open-loop voltage gain $A_{V.OL}$ is the Thevenin-equivalent gain translation from the series-mixed difference v_E or $v_I - v_{FB}$ to the shunt-sampled output v_O when the input v_I is zero and the load is absent. In this case, A_V amplifies v_E and drops a portion of the voltage across A_V's output resistance R_{OA} and another fraction across feedback resistors R_2 and R_1 and A_V's differential input resistance R_{ID}, where R_1 and R_{ID} are in parallel. As a result, open-loop voltage gain $A_{V.OL}$ is

$$A_{V.OL}\Big|_{i_{OUT}\equiv 0} \equiv \frac{v_O}{v_E} = \frac{v_O}{v_I - v_{FB}} = \frac{A_V[R_2 + (R_1 \parallel R_{ID})]}{R_{OA} + R_2 + (R_1 \parallel R_{ID})} \approx A_V \tag{6.43}$$

where A_V's R_{OA} is usually, although not always, quite low. Feedback factor β_{FB} is the gain from output v_O to feedback v_{FB} when the input v_I is zero, which is the voltage-divided ratio that R_2, R_1, and A_V's differential input resistance R_{ID} establish:

$$\beta_{FB}\Big|_{v_I\equiv 0} \equiv \frac{v_{FB}}{v_O} = \frac{R_1 \parallel R_{ID}}{(R_1 \parallel R_{ID}) + R_2} \approx \frac{R_1}{R_1 + R_2} \tag{6.44}$$

where R_{ID} is usually much higher than R_1.

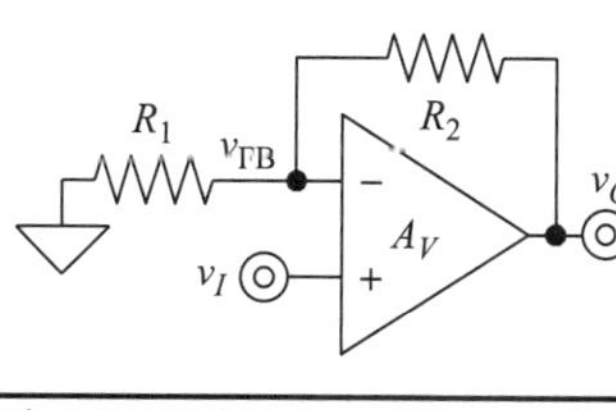

FIGURE 6.35 Noninverting operational-amplifier configuration.

Closed-loop gain $A_{V.CL}$ is therefore 1 plus loop gain $A_{V.OL}\beta_{FB}$ times lower than $A_{V.OL}$. Since the resistor divider and A_G process both small- and large-signal components of v_O and v_I linearly, v_O is a $1/\beta_{FB}$ translation of v_{FB} and v_I, or v_I $(R_1 + R_2)/R_1$, when $A_{V.OL}\beta_{FB}$ is considerably greater than 1, which matches the well-known gain already reported in textbooks. Here, shorting A_V's Thevenin voltage source eliminates the feedback source that ultimately feeds the input. This leaves R_{ID} and the parallel combination of R_1 and $R_2 + R_{OA}$, so open-loop input resistance $R_{I.OL}$ is

$$R_{I.OL} = R_{ID} + [R_1 \parallel (R_2 + R_{OA})] \tag{6.45}$$

With a series-mixed input, closed-loop input resistance $R_{I.CL}$ is $1 + A_{V.OL}\beta_{FB}$ times greater than $R_{I.OL}$. Similarly, shorting A_V's Thevenin voltage source when the input v_I is zero eliminates the feedback source that feeds the output. This reveals that Thevenin-equivalent open-loop output resistance $R_{O.OL}$ is A_V's R_{OA} in parallel with the series combination of R_2 and $R_1 \parallel R_{ID}$:

$$R_{O.OL}\Big|_{v_I \equiv 0} = R_{OA} \parallel [R_2 + (R_1 \parallel R_{ID})] \tag{6.46}$$

With a shunt-sampled output, closed-loop output resistance $R_{O.CL}$ is therefore $1 + A_{V.OL}\beta_{FB}$ times lower than $R_{O.OL}$. In other words, as is also well known, input resistance is high and output resistance is low.

Noninverting Transistor Amp

Since base–emitter terminals amplify the difference between two voltages, the circuit in Fig. 6.36 is a transistor-level, small-signal implementation of the noninverting operational-amplifier configuration in Fig. 6.35. Here, transistors Q_1 and M_2 implement the amplifier,

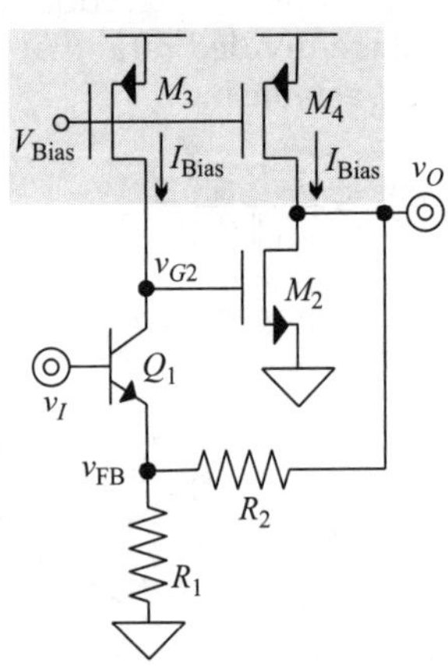

FIGURE 6.36 Noninverting transistor-amplifier configuration.

where Q_1 buffers feedback current to M_2's gate to establish a voltage v_{g2} that M_2 then amplifies. The loop is inverting because, of the three stages in the loop—Q_1, Q_2, and R_2–R_1—M_2 is the only one that inverts. In other words, the loop opposes the influence of external variations, which means the network is an inverting feedback circuit.

Interestingly, since Q_1 is an emitter-degenerated transistor, Q_1 also implements an inverting feedback circuit. Of the two loops present, however, the loop that includes M_2 is the only one attached to the input and output terminals of the network. This means that Q_1 is an embedded loop whose closed-loop response affects the open-loop parameters of the outer loop.

Tracing the input into the circuit to find the mixer reveals in Q_1 a receiving base. Since Q_1's opposing emitter terminal is a loop signal v_{fb}, Q_1 series-mixes small-signal components of input-terminal voltage v_i into the loop. Similarly, following the output into the network to find the sampler discloses M_2's drain and R_2. Because M_2's drain cannot sense voltage and its current only flows to ground, and R_2 voltage-divides v_O to Q_1's emitter, whose opposing input terminal does not carry a loop signal, R_2 buffers and Q_1 ultimately shunt-samples small-signal components of output voltage v_O. In other words, the feedback loop implements a voltage amplifier.

Like before, recall that open-loop parameters must account for the closed-loop response of the internal emitter-degenerated transistor. With this in mind, the forward open-loop voltage gain $A_{V.OL}$ is the Thevenin-equivalent gain translation from the series-mixed difference v_e or $v_i - v_{fb}$ to the shunt-sampled output v_o when the input v_i is zero and the load is absent. Here, Q_1's transconductance g_{m1} amplifies v_e to output a current i_{c1} that drops a voltage v_{g2} across M_3's output resistance r_{sd3} and Q_1's emitter-degenerated output resistance R_{C1}. M_2 then amplifies v_{g2} to generate a current i_{d2} that, together with M_2's and M_4's output resistances r_{ds2} and r_{sd4} and R_2, R_1, and Q_1's emitter resistance R_{E1}, establishes output voltage v_o:

$$A_{V.OL}\Big|_{i_{out}\equiv 0} \equiv \frac{v_o}{v_e} = g_{m1}(r_{sd3} \parallel R_{C1})g_{m2}[r_{ds2} \parallel r_{sd4} \parallel [R_2 + (R_1 \parallel R_{E1})]] \tag{6.47}$$

where $r_{\pi 1}$ loads Q_1's emitter and r_{sd3} loads Q_1's $1/g_{m1}$ resistance, all of which reduces R_{E1} to roughly $2/g_{m1}$ when r_{sd3} and r_{o1} are similarly high:

$$R_{E1} = \left(\frac{r_{sd3} + r_{o1}}{1 + g_{m1} r_{o1}}\right) \parallel r_{\pi 1} \approx \frac{2}{g_{m1}} \tag{6.48}$$

Feedback factor β_{FB} is the gain from output v_o to feedback v_{fb} when the input v_i is zero, which is the voltage-divided ratio that R_2, R_1, and R_{E1} establish:

$$\beta_{FB}\Big|_{v_i \equiv 0} \equiv \frac{v_{fb}}{v_o} = \frac{R_1 \parallel R_{E1}}{(R_1 \parallel R_{E1}) + R_2} \tag{6.49}$$

As a result, closed-loop gain $A_{V.CL}$ is 1 plus loop gain $A_{V.OL}\beta_{FB}$ times lower than $A_{V.OL}$. Since Q_1 only mixes small-signal components of v_I linearly, only v_o is a $1/\beta_{FB}$ translation of v_{fb} and v_i when $A_{V.OL}\beta_{FB}$ is considerably greater than 1. Here, removing M_2's transconductance g_{m2} eliminates the feedback source that ultimately feeds the input. This leaves Q_1's base resistance R_{B1}, which R_1 and the series combination of R_2 and $r_{ds2} \parallel r_{sd4}$ degenerate, so open-loop input resistance $R_{I.OL}$ is

$$R_{I.OL} = R_{B1} = r_{\pi 1} + (1 + g_{m1} r_{\pi 1})\{R_1 \parallel [R_2 + (r_{ds2} \parallel r_{sd4})]\} \tag{6.50}$$

With a series-mixed input, closed-loop input resistance $R_{I.CL}$ is $1 + A_{V.OL}\beta_{FB}$ times greater than $R_{I.OL}$. Similarly, removing M_2's g_{m2} when the input v_i is zero eliminates the feedback source that feeds the output. This reveals that Thevenin-equivalent open-loop output resistance $R_{O.OL}$ is the resistive network that M_2's r_{ds2}, M_4's r_{sd4}, R_2, R_1, and Q_1's R_{E1} realize:

$$R_{O.OL}\Big|_{v_i \equiv 0} = r_{ds2} \parallel r_{sd4} \parallel [R_2 + (R_1 \parallel R_{E1})] \tag{6.51}$$

With a shunt-sampled output, closed-loop output resistance $R_{O.CL}$ is therefore $1 + A_{V.OL}\beta_{FB}$ times lower than $R_{O.OL}$. In other words, with the appropriate substitutions for A_V in Fig. 6.35, $A_{V.CL}$, $R_{I.CL}$, and $R_{O.CL}$ relationships for the circuit of Fig. 6.36 resemble those of the operational-amplifier case in Fig. 6.35.

Gate-Coupled-Pair Amplifier

In the amplifier of Fig. 6.37, M_2 buffers a current from the output v_O that M_3 and A_G amplify to drive M_4 back into v_O, so M_2, M_3, A_G, and M_4 implement a feedback loop. This loop is inverting because, of its three

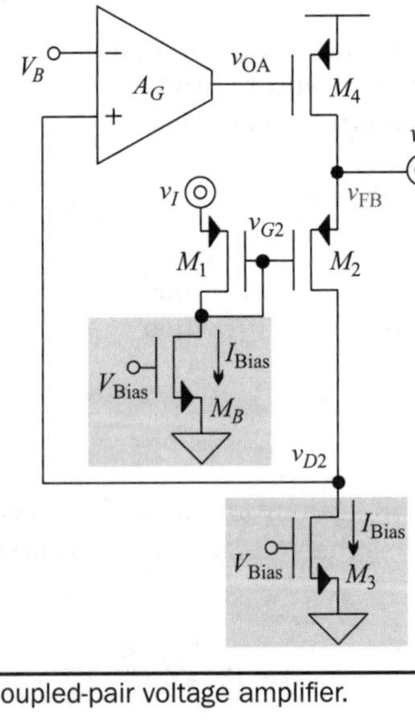

FIGURE 6.37 Gate-coupled-pair voltage amplifier.

constituent stages, M_4 is the only one that inverts. Interestingly, the loop mixes two signals into the loop: v_I and V_B. Of these two signals, though, only v_I carries small-signal information, so the purpose of V_B is to establish the voltage that biases M_3's drain voltage, which V_B does through feedback. With enough loop gain, in fact, A_G series-mixes V_B with M_3's drain voltage and drives an error signal that keeps M_3's drain at V_B.

Tracing v_I into the circuit to find the mixer reveals in gate-coupled pair M_1–M_2 a receiving source. Since M_1–M_2's opposing source terminal is a loop signal v_O, M_1–M_2 series-mixes small- and large-signal components of input-terminal voltage v_I into the loop. In this case, notice that v_O is both the output and the feedback voltage v_{FB}. Following the output into the network to find the sampler discloses M_4's drain and a short to M_2's source. Because M_4's drain cannot sense voltage and its current only flows to the supply, and M_2's opposing gate terminal is not in the loop, the short to M_2 voltage-samples small- and large-signal components of v_O to drive A_G. In other words, the feedback loop implements a voltage amplifier.

Since M_2 is a source-degenerated transistor, open-loop parameters must account for its closed-loop response. With this in mind, the forward open-loop voltage gain $A_{V.OL}$ is the Thevenin-equivalent gain translation from the series-mixed difference v_e or $v_i - v_{fb}$ to the shunt-sampled output v_o when the input v_i is zero and the load is absent. Here, M_1's diode-connected $1/g_{m1}$ and M_B's r_{dsB} voltage-divide v_i into M_2's gate v_{g2}, except $1/g_{m1}$ is so much smaller than r_{dsB} that v_{g2} follows v_i. As a result, M_2's transconductance g_{m2} amplifies $v_{fb} - v_{g2}$, which is equivalent to $v_{fb} - v_i$ and $-v_e$, to output a small-signal current i_{d2} that drops a voltage v_{d2} across M_3's r_{ds3}, A_G's R_{IA}, and M_2's source-degenerated drain resistance R_{D2}. A_G and A_G's output resistance R_{OA} amplify v_{d2} to v_{oa}, and M_4 converts v_{oa} into a current i_{d4} that, together with M_2's source resistance R_{S2} and M_4's r_{sd4}, establish output voltage v_o:

$$A_{V.OL}\Big|_{i_{out}\equiv 0} \equiv \frac{v_o}{v_e} = g_{m2}(r_{ds3} \parallel R_{IA} \parallel R_{D2})A_G R_{OA} g_{m4}(R_{S2} \parallel r_{sd4}) \approx g_{m2} r_{ds3} A_G R_{OA} g_{m4} R_{S2} \tag{6.52}$$

where R_{D2} is typically considerably higher than r_{ds3}, r_{sd4} is similarly greater than R_{S2}, and r_{ds3} and R_{IA} load M_2's $1/g_{m2}$ resistance to set R_{S2} to

$$R_{S2} = \frac{(r_{ds3} \parallel R_{IA}) + r_{sd2}}{1 + g_{m2} r_{sd2}} \approx \frac{2}{g_{m2}} \tag{6.53}$$

Since feedback factor β_{FB} is the gain from output v_O to feedback v_{FB} and v_O in this case is a short, v_O is v_{FB} and β_{FB} is 1.

As a result, closed-loop gain $A_{V.CL}$ is 1 plus loop gain $A_{V.OL}\beta_{FB}$ times lower than $A_{V.OL}$. Since shorts and M_1–M_2 process both small- and large-signal components of v_O and v_I linearly, v_O is a $1/\beta_{FB}$ or unity-gain translation of v_{FB} and v_I when $A_{V.OL}\beta_{FB}$ is considerably greater than 1. Here, removing M_4's transconductance g_{m4} eliminates the feedback source that feeds the input. The differential input resistance that results is M_1's diode-connected resistance in series with M_2's gate–source terminals, which means both open- and closed-loop input resistances $R_{I.OL}$ and $R_{I.CL}$ approach infinity. Similarly, removing g_{m4} when the input v_i is zero eliminates the feedback source that feeds the output. This reveals that Thevenin-equivalent open-loop output resistance $R_{O.OL}$ is the parallel combination of M_4's r_{sd4} and M_2's source resistance R_{S2}, or $r_{sd4} \parallel R_{S2}$. With a shunt-sampled output, closed-loop output resistance $R_{O.CL}$ is therefore $1 + A_{V.OL}\beta_{FB}$ times lower than $R_{O.OL}$.

6.4.6 Current Amplifiers

Regulated-Cascode Current Mirror

The underlying purpose of cascode transistors in current mirrors is to reduce the error that mismatched collector and drain voltages produce. Said differently, their objective is to reduce the sensitivity of the output current to output voltage variations, which is the manifestation of higher output resistance. In fact, these cascodes raise output resistance because the mirrors degenerate them to realize current-sampling feedback loops. But when output resistance is not sufficiently high, designers sometimes enhance their efficacy by amplifying their effective loop gain, which is the purpose of amplifier A_G in Fig. 6.38. The feedback loop that the amplifier implements around the cascoding device is the reason engineers call these transistors *regulated cascodes*. The feedback loop is inverting because, of the two stages in the loop, A_G is the only one that inverts.

Tracing the input into the circuit to find the mixer reveals in Q_2 a receiving base, but since the opposing emitter terminal is not in the loop, Q_2 is not a voltage mixer. Tracing the input further past Q_2

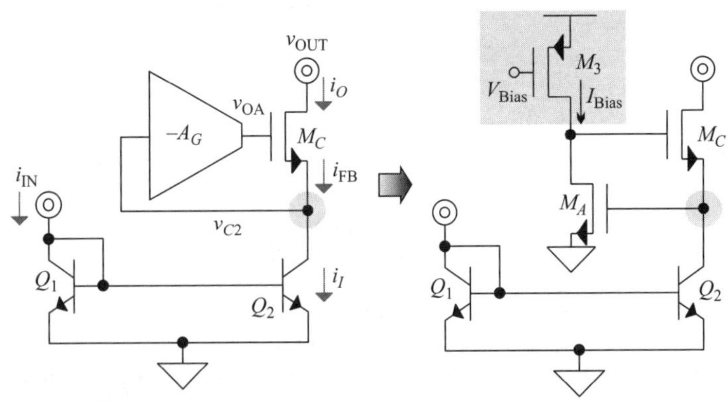

FIGURE 6.38 Regulated-cascode current mirror.

exposes the input of A_G and the source of cascoding device M_C. Since A_G has no other input voltage with which to mix Q_2's collector, A_G is not mixing voltages either. Although M_C's opposing gate terminal carries loop signal v_{OA}, the output of the network loops to Q_2, which is the device that channels signals from the input terminal, before it reaches v_{OA}. This means that v_{OA} is not the feedback signal, so M_C is not a voltage mixer. Exhausting these voltage-mixing possibilities and recognizing the T connection that Q_2, A_G, and M_C implement uncovers the actual mixer, into which Q_2 injects input current i_I. Similarly, following the output into the network to find the sampler discloses in M_C a receiving drain. Because drains can sample currents and output current i_O flows through M_C to establish the feedback current i_{FB} with which i_I mixes, M_C series-samples small- and large-signal components of i_O. In other words, the feedback loop implements a current amplifier.

Since M_C is a source-degenerated transistor, open-loop parameters must account for its closed-loop response. With this in mind, the forward open-loop current gain $A_{I.OL}$ is the Norton-equivalent gain translation from the shunt-mixed difference i_e or $i_i - i_{fb}$ to the series-sampled output i_o when the input i_i that Q_2's transconductance g_{m2} injects is zero and output voltage v_{out} is zero. Here, i_e flows through Q_2's r_{o2} to establish a voltage that A_G amplifies. M_C's effective transconductance G_{MC}, which r_{o2} degenerates, then converts A_G's output voltage v_{oa} into i_o to set $A_{I.OL}$ to roughly $A_G R_{OA}$:

$$A_{I.OL}\Big|_{v_{out}\equiv 0} \equiv \frac{i_o}{i_e} = (-r_{o2})(-A_G)R_{OA}G_{MC} = r_{o2}A_G R_{OA}\left(\frac{g_{mC}}{1+g_{mC}r_{o2}}\right) \approx A_G R_{OA} \tag{6.54}$$

Note that $1/g_{mC}$ does not load i_e because g_{mC} carries part of i_{FB}. Feedback factor β_{FB} is the gain from output i_O to feedback i_{FB} when the input i_I is zero. Since M_C buffers i_O to i_{FB}, β_{FB} is 1:

$$\beta_{FB}\Big|_{i_I\equiv 0} \equiv \frac{i_{FB}}{i_O} = 1 \tag{6.55}$$

As a result, closed-loop gain $A_{I.CL}$ is 1 plus loop gain $A_{I.OL}\beta_{FB}$, or roughly $A_G R_{OA}$ times lower than $A_{I.OL}$. Since M_C and the T connection process both small- and large-signal components of i_O and i_I linearly, i_O is a $1/\beta_{FB}$ or unity-gain translation of i_{FB} and i_I when $A_{I.OL}\beta_{FB}$ is considerably greater than 1. Here, removing A_G's Norton current source eliminates the feedback source that feeds the input. This leaves Q_2's r_{o2} and M_C's loaded source resistance R_{SC}, so open-loop input resistance $R_{I.OL}$ is

$$R_{I.OL} = r_{o2} \parallel R_{SC} = r_{o2} \parallel \left(\frac{R_{LOAD} + r_{dsC}}{1 + g_{mC}r_{dsC}}\right) \tag{6.56}$$

where load resistance R_{LOAD} loads M_C's $1/g_{mC}$. With a shunt-mixed input, closed-loop input resistance $R_{I.CL}$ is $1 + A_{I.OL}\beta_{FB}$, or roughly $A_G R_{OA}$ times lower than $R_{I.OL}$. Similarly, removing A_G's Norton current source when the input i_i is zero eliminates the feedback source that feeds the output. This reveals that Norton-equivalent open-loop output resistance $R_{O.OL}$ is M_C's drain resistance R_{DC}, which Q_2's r_{o2} degenerates to

$$R_{O.OL}\big|_{i_i \equiv 0} = R_{DC} = r_{dsC} + r_{o2} + g_{mC} r_{dsC} r_{o2} \tag{6.57}$$

With a series-sampled output, closed-loop output resistance $R_{O.CL}$ is therefore $1 + A_{I.OL}\beta_{FB}$, or roughly $A_G R_{OA}$ times greater than $R_{O.OL}$, which again, is normally the driving motivation for regulating cascode transistors. Note that the feedback network is only part of the overall circuit, so the overall current gain is the combined gain of the mirror, which Q_1's and Q_2's emitter areas A_{E1} and A_{E2} set, and the closed-loop gain $A_{I.CL}$ of the current-mixed, current-sampled network:

$$\frac{i_O}{i_{IN}} = \left(\frac{i_I}{i_{IN}}\right)\left(\frac{i_O}{i_I}\right) \approx \left(\frac{A_{E2}}{A_{E1}}\right) A_{I.CL} \approx \frac{A_{E2}}{A_{E1}} \tag{6.58}$$

741's Tail-Current Source

The *741 Op Amp* is a classic example of analog design because it incorporates and employs several analog components and concepts, from single-transistor amplifiers, current mirrors, and differential pairs to negative feedback. The input stage alone, which Fig. 6.39 illustrates, includes a differential pair Q_1–Q_2, current-buffer transistors Q_3 and Q_4, an emitter-degenerated current mirror Q_{M1}–Q_{M2}, and the feedback loop that Q_1–Q_2, Q_3–Q_4, and current mirror Q_5–Q_6 realize. Here, the purpose of the loop is to establish Q_1–Q_2's tail current in the form of Q_5's current.

To see the loop more clearly, first consider that connecting the operational amplifier in a feedback configuration normally establishes a virtual short between Q_1's and Q_2's base inputs v_+ and v_-. So for the purposes of the internal loop, v_+ and v_- constitute one terminal with a common-mode bias voltage V_{CM}. Q_1–Q_3 and Q_2–Q_4 are therefore equivalent and in parallel, which is why $Q_{1,2}$ and $Q_{3,4}$ in Fig. 6.39 can model them. In this light, $Q_{1,2}$ buffers and feeds $Q_{3,4}$'s current to mirror Q_5–Q_6, whose output feeds $Q_{3,4}$'s base to complete the loop. The loop is inverting because Q_6 is the only transistor in the loop that inverts. Since the purpose of the loop is to use bias current I_{Bias} to establish Q_1–Q_2's tail current, the input and output of the loop are I_{Bias} and Q_5's current.

Tracing the input into the circuit to find the mixer reveals in $Q_{3,4}$ a receiving base. Since $Q_{3,4}$'s opposing emitter voltage v_{E3} is a loop signal, $Q_{3,4}$ can mix voltages. In this case, however, the output of the

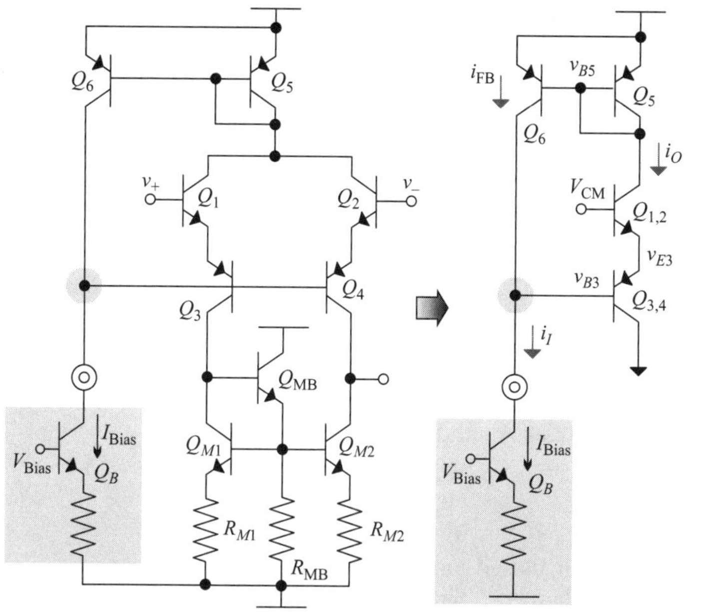

FIGURE 6.39 Input stage of the 741 Op Amp.

network loops to the input where I_{Bias} flows before it reaches $Q_{3,4}$'s emitter, so $Q_{3,4}$'s emitter is not the feedback signal, which means $Q_{3,4}$ is not a voltage mixer. Exhausting these voltage-mixing possibilities and recognizing the T connection that bias generator Q_B, $Q_{3,4}$, and Q_6 create uncovers the actual mixer, into which I_{Bias} injects input current i_I. Similarly, following the output at Q_5 into the network to find the sampler discloses in Q_5–Q_6 a receiving current mirror. Because mirrors can sample current and Q_5's output current i_O flows through Q_5 to establish Q_6's feedback current i_{FB} with which i_I mixes, Q_5 series-samples small- and large-signal components of i_O. In other words, the feedback loop implements a current amplifier.

Since $Q_{1,2}$ emitter-degenerates $Q_{3,4}$, open-loop parameters must account for $Q_{3,4}$'s closed-loop response. With this in mind, the forward open-loop current gain $A_{I.OL}$ is the Norton-equivalent gain translation from the shunt-mixed difference i_e or $i_i - i_{fb}$ to the series-sampled output i_o when the input i_i and the output voltage v_{b5} are zero. As already noted, $Q_{1,2}$'s emitter resistance $1/g_{m1}$ emitter-degenerates $Q_{3,4}$'s transconductance to G_{M3}, which reduces to approximately $0.5g_{m3}$:

$$G_{M3} \approx \frac{g_{m3}}{1 + g_{m3}\left(\dfrac{1}{g_{m1}}\right)} \approx \frac{g_{m3}}{2} \tag{6.59}$$

and base resistance reduces to R_{B3} or roughly $2r_{\pi 3}$:

$$R_{B3} = r_{\pi 3} + (1 + g_{m3} r_{\pi 3})\left(\frac{1}{g_{m1}}\right) \approx 2r_{\pi 3} \tag{6.60}$$

i_e therefore flows through Q_6's r_{o6} and $Q_{3,4}$'s R_{B3} to establish a voltage that $Q_{3,4}$'s G_{M3} converts into the current i_o that Q_5 samples to set $A_{I.OL}$ to roughly $Q_{3,4}$'s β_{03}:

$$A_{I.OL}\big|_{v_{b5}\equiv 0} \equiv \frac{i_o}{i_e} = -(r_{o6} \parallel R_{B3})(-G_{M3}) \approx (r_{o6} \parallel 2r_{\pi 3})\left(\frac{g_{m3}}{2}\right) \approx r_{\pi 3} g_{m3} = \beta_{03} \tag{6.61}$$

Feedback factor β_{FB} is the gain from output i_O to feedback i_{FB} when the input i_I is zero. Since Q_5–Q_6 mirrors i_O to i_{FB}, β_{FB} is the ratio of Q_6's and Q_5's emitter areas A_{E6} and A_{E5}:

$$\beta_{FB}\big|_{i_I\equiv 0} \equiv \frac{i_{FB}}{i_O} \approx \frac{A_{E6}}{A_{E5}} \tag{6.62}$$

As a result, closed-loop gain $A_{I.CL}$ is 1 plus loop gain $A_{I.OL}\beta_{FB}$ or β_{03} times lower than $A_{I.OL}$. Since Q_5–Q_6 and the T connection process both small- and large-signal components of i_O and i_I linearly, i_O is a $1/\beta_{FB}$ or mirror-gain translation of i_{FB} and i_I, or $i_I A_{E6}/A_{E5}$, when $A_{I.OL}\beta_{FB}$ is considerably greater than 1. In other words, i_O is $I_{Bias}A_{E5}/A_{E6}$, and Q_1 and Q_2 split I_{Bias} to bias the input stage of the 741 Op Amp. Removing Q_6's Norton current source eliminates the feedback source that feeds the input. This leaves Q_6's r_{o6} and $Q_{3,4}$'s degenerated base resistance R_{B3}, so open-loop input resistance $R_{I.OL}$ is

$$R_{I.OL} = r_{o6} \parallel R_{B3} \approx r_{o6} \parallel 2r_{\pi 3} \approx 2r_{\pi 3} \tag{6.63}$$

With a shunt-mixed input, closed-loop input resistance $R_{I.CL}$ is $1 + A_{I.OL}\beta_{FB}$ or β_{03} times lower than $R_{I.OL}$. Similarly, removing Q_6's Norton current source when the input i_i is zero eliminates the feedback source that ultimately feeds the output. This reveals that Norton-equivalent open-loop output resistance $R_{O.OL}$ is the parallel combination of Q_5's diode-connected resistance $1/g_{m5}$ and $Q_{1,2}$'s emitter-degenerated collector resistance R_{C1}, except the former shunts the latter to reduce $R_{O.OL}$ to $1/g_{m5}$:

$$R_{O.OL}\big|_{i_i\equiv 0} = \frac{1}{g_{m5}} \parallel R_{C1} \approx \frac{1}{g_{m5}} \tag{6.64}$$

With a series-sampled output, closed-loop output resistance $R_{O.CL}$ is therefore $1 + A_{I.OL}\beta_{FB}$ or β_{03} times greater than $R_{O.OL}$.

Current Amp

Figure 6.40 showcases what is perhaps a more general, and maybe even more practical realization of a current amplifier. In this case, transistors M_1 and M_3 and current mirror M_4–M_5 connect a loop to

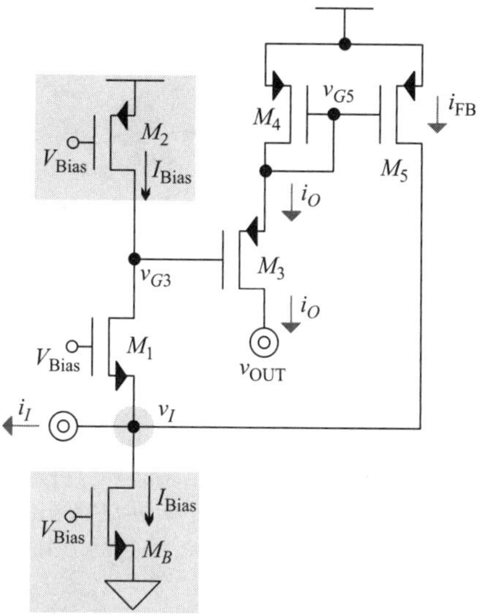

FIGURE 6.40 Current amplifier.

realize an inverting feedback circuit. The loop is inverting because M_5 is the only transistor in the network that inverts.

Tracing the input into the circuit to find the mixer reveals in M_1 a receiving source. Since M_1's opposing gate terminal is not in the loop, M_1 is not a voltage mixer. Exhausting this voltage-mixing possibility, discarding drains because they do not mix voltages, and recognizing the T connection that bias transistor M_B, M_1, and M_5 create uncovers the actual mixer, into which input i_I flows. Similarly, following the output into the network to find the sampler discloses in M_3 a receiving drain. Because drains can only sample current and M_3's output current flows into mirror M_4–M_5 to output the feedback current i_{FB} with which i_I mixes, M_3 series-samples small- and large-signal components of i_O. In other words, the feedback loop implements a current amplifier.

Since M_B and M_5 source-degenerate M_1 and M_4 similarly source-degenerates M_3, open-loop parameters must account for M_1 and M_3's closed-loop response. With this in mind, the forward open-loop current gain $A_{I.OL}$ is the Norton-equivalent gain translation from the shunt-mixed difference i_e or $i_i - i_{fb}$ to the series-sampled output i_o when the input i_i is zero and the output voltage v_{out} is zero. Here, M_4's diode-connected resistance $1/g_{m4}$ source-degenerates M_3's to G_{M3}:

$$G_{M3} \approx \frac{g_{m3}}{1 + g_{m3}\left(\dfrac{1}{g_{m4}}\right)} \tag{6.65}$$

In addition, bias transistor M_2 loads M_1's $1/g_{m1}$ resistance to nearly $2/g_{m1}$:

$$R_{S1} = \frac{r_{ds1} + r_{sd2}}{1 + g_{m1} r_{ds1}} \approx \frac{2}{g_{m1}} \tag{6.66}$$

Therefore, i_e drops a voltage across M_1's source resistance R_{S1}, M_B's r_{dsB}, and M_5's r_{sd5} to establish a current into R_{S1} that M_1 buffers to M_3's gate. This current then flows through M_2's r_{sd2} to drop a voltage that M_3's G_{M3} converts into current i_o to set $A_{I.OL}$ to

$$A_{I.OL}\Big|_{v_{out}\equiv 0} \equiv \frac{i_o}{i_e} = -\left(\frac{r_{dsB} \parallel R_{S1} \parallel r_{sd5}}{R_{S1}}\right) r_{sd2}(-G_{M3}) \approx r_{sd2} G_{M3} \tag{6.67}$$

where R_{S1}'s $2/g_{m1}$ resistance shunts r_{dsB} and r_{sd5}. Feedback factor β_{FB} is the gain from output i_O to feedback i_{FB} when the input i_I is zero. Since M_4–M_5 mirrors i_O to i_{FB}, β_{FB} is the ratio of M_5's and M_4's width–length ratios $(W/L)_5$ and $(W/L)_4$:

$$\beta_{FB}\Big|_{i_I\equiv 0} \equiv \frac{i_{FB}}{i_O} \approx \frac{(W/L)_5}{(W/L)_4} \tag{6.68}$$

As a result, closed-loop gain $A_{I.CL}$ is 1 plus loop gain $A_{I.OL}\beta_{FB}$ or $G_{M3}r_{sd2}$ times lower than $A_{I.OL}$. Since M_4–M_5 and the T connection process both small- and large-signal components of i_O and i_I linearly, i_O is a $1/\beta_{FB}$ or mirror-gain translation of i_{FB} and i_I, or $i_I(W/L)_4/(W/L)_5$, when $A_{I.OL}\beta_{FB}$ is considerably greater than 1. Removing M_5's Norton current source eliminates the feedback source that feeds the input. This leaves M_1's R_{S1}, M_B's r_{dsB}, and M_5's r_{sd5}, where R_{S1} shunts the rest to reduce open-loop input resistance $R_{I.OL}$ to roughly $2/g_{m1}$:

$$R_{I.OL} = R_{S1} \parallel r_{dsB} \parallel r_{sd5} \approx R_{S1} \approx \frac{2}{g_{m1}} \tag{6.69}$$

With a shunt-mixed input, closed-loop input resistance $R_{I.CL}$ is $1 + A_{I.OL}\beta_{FB}$ times lower than $R_{I.OL}$. Similarly, removing M_5's Norton current source when the input i_i is zero eliminates the feedback source that ultimately feeds the output. This reveals that Norton-equivalent open-loop output resistance $R_{O.OL}$ is M_3's drain resistance R_{D3}, which M_4's $1/g_{m4}$ lightly degenerates to

$$R_{O.OL}\Big|_{i_i\equiv 0} = r_{sd3} + \frac{1}{g_{m4}} + \frac{g_{m3} r_{sd3}}{g_{m4}} \approx r_{sd3}\left(1 + \frac{g_{m3}}{g_{m4}}\right) \tag{6.70}$$

With a series-sampled output, closed-loop output resistance $R_{O.CL}$ is therefore $1 + A_{I.OL}\beta_{FB}$ times greater than $R_{O.OL}$. Note that lowering what already is a low input resistance $R_{I.OL}$ at $2/g_{m1}$ and raising what already is a high output resistance $R_{O.OL}$ at nearly r_{sd3} means source and load resistances at the input and output of the feedback network will hardly affect $A_{I.CL}$, which is ultimately why this topology is so practical.

6.4.7 Transimpedance Amplifiers

Diode-Connected Transistor

Interestingly, the diode-connected transistor is also an inverting feedback circuit. To see this more clearly, consider that a rise in the base voltage v_O of the diode-connected transistor in Fig. 6.41 raises the transistor's collector current, the net effect of which is to pull v_O down. In other words, diode-connected transistors oppose the effects of external variations on their base voltages, which is another way of saying they employ negative feedback.

Tracing the input into the circuit to find the mixer reveals in Q_1 a receiving base, but since the opposing emitter terminal is not in the loop, Q_1 does not mix voltages. Exhausting this voltage-mixing possibility and recognizing the T connection that bias transistor M_2, Q_1's transconductance g_{m1}, and resistances in the circuit implement uncovers the actual mixer, into which an external source injects input current i_i. Since v_O is the only node in the circuit, v_O is both the input and output terminal of the circuit. In this light, following the output into the network to find the sampler discloses in Q_1 a receiving base. Because the opposing emitter terminal is not in the loop and Q_1's base sets the feedback current i_{fb} with which i_i mixes, Q_1 shunt-samples small-signal components of v_O. In other words, the feedback loop implements a transimpedance amplifier.

The forward open-loop transimpedance gain $A_{Z.OL}$ is therefore the Thevenin-equivalent gain translation from the shunt-mixed difference i_e or $i_i - i_{fb}$ to the shunt-sampled output v_o when the input i_i is zero and the load is absent. In this case, i_e flows through Q_1's $r_{\pi1}$ and r_{o1} and M_2's r_{sd2} to drop output voltage v_o and set $A_{Z.OL}$ to

$$A_{Z.OL}\Big|_{i_{out}\equiv 0} \equiv \frac{v_o}{i_e} = r_{o1} \parallel r_{\pi1} \parallel r_{sd2} \approx r_{\pi1} \tag{6.71}$$

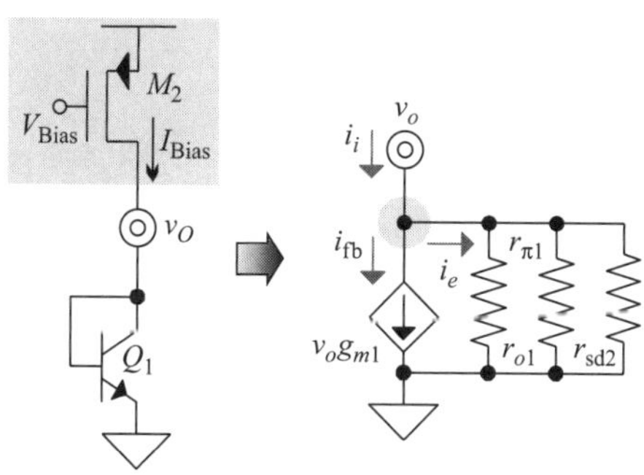

Figure 6.41 Diode-connected transistor.

which reduces to $r_{\pi 1}$. Feedback factor β_{FB} is the gain from output v_o to feedback i_{fb} when the input i_i is zero. Since Q_1's transconductance g_{m1} multiplies v_o to establish i_{fb}, β_{FB} is g_{m1}:

$$\beta_{FB}\Big|_{i_i \equiv 0} \equiv \frac{i_{fb}}{v_o} = g_{m1} \tag{6.72}$$

As a result, closed-loop gain $A_{Z.CL}$ is 1 plus loop gain $A_{Z.OL}\beta_{FB}$, or roughly $g_{m1}r_{\pi 1}$ or β_{01} times lower than $A_{Z.OL}$. Since Q_1's g_{m1} only processes small-signal components of v_o linearly, only v_o is a $1/\beta_{FB}$ or $1/g_{m1}$ translation of i_{fb} and i_i when $A_{Z.OL}\beta_{FB}$ is considerably greater than 1, which is the case here. Here, removing Q_1's Norton current source eliminates the feedback source that feeds both the input and output. This leaves Q_1's $r_{\pi 1}$ and r_{o1} and M_2's r_{sd2}, so open-loop input and output resistances $R_{I.OL}$ and $R_{O.OL}$ are both one and the same at

$$R_{I.OL} = R_{O.OL}\Big|_{i_i \equiv 0} = r_{o1} \parallel r_{\pi 1} \parallel r_{sd2} \approx r_{\pi 1} \tag{6.73}$$

which reduce to $r_{\pi 1}$. With a shunt-mixed input and a shunt-sampled output, closed-loop input and output resistances $R_{I.CL}$ and $R_{O.CL}$ are $1 + A_{Z.OL}\beta_{FB}$, or approximately $g_{m1}r_{\pi 1}$ times lower than $R_{I.OL}$ and $R_{O.OL}$ at roughly $1/g_{m1}$. In other words, diode-connected resistances are $1/g_m$, which is the conclusion that Chap. 5 reaches in the current-mirror section and Chap. 5 and this chapter later use to analyze other circuits.

Inverting Op Amp

Although not obvious at first, the inverting operational-amplifier configuration in Fig. 6.42 is a classic example of a transimpedance negative-feedback amplifier. Here, resistor R_{FB} feeds output signals back to the inverting input of the amplifier to establish negative feedback. The loop is inverting, of course, because the amplifier is the only component in the loop that inverts.

Tracing the input into the circuit to find the mixer reveals in input resistor R_{IN} a buffer that cannot mix voltages and in amplifier A_V a

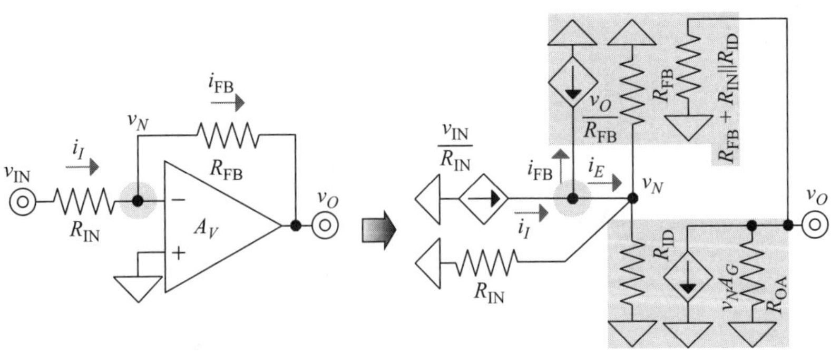

FIGURE 6.42 Inverting operational-amplifier configuration.

possible receiving input. Since A_V's opposing input voltage is not in the loop, A_V is not a voltage mixer. Exhausting these voltage-mixing possibilities and recognizing the T connection that R_{IN}, A_V, and feedback resistor R_{FB} implement uncovers the actual mixer, into which R_{IN} injects input current i_I. Similarly, following the output into the network to find the sampler discloses in R_{FB} a buffer. Because R_{FB}'s current is the feedback current i_{FB} with which i_I mixes, R_{FB} converts output voltage v_O into i_{FB}, which means R_{FB} shunt-samples small- and large-signal components in v_O. In other words, the feedback loop implements a transimpedance amplifier.

Before proceeding, recall that open-loop parameters must account for all impedances in the circuit. For this, all resistances that convert signals into currents or voltages that the loop mixes must decompose into their two-port equivalents. R_{IN}'s Norton-equivalent resistance and current, for example, are the resistance from A_V's inverting input v_N into R_{IN} when the input v_{IN} is zero, which is R_{IN}, and R_{IN}'s current into v_N when v_N is zero, or v_{IN}/R_{IN}, as the small-signal translation in Fig. 6.42 shows. Similarly, R_{FB}'s Norton-equivalent output resistance and transconductance current are the resistance from v_N into R_{FB} when v_O is zero, or R_{FB}, and R_{FB}'s current into v_N when v_N is zero, or v_O/R_{FB}. Lastly, R_{FB}'s input resistance is the resistance from v_O into R_{FB}, or $R_{FB} + (R_{IN} \,||\, R_{ID})$, where A_V's differential input resistance R_{ID} is normally so high that $R_{IN} \,||\, R_{ID}$ reduces to R_{IN}.

With two-port models in place, the forward open-loop transimpedance gain $A_{Z.OL}$ is therefore the Thevenin-equivalent gain translation from the shunt-mixed difference i_E or $i_I - i_{FB}$ to the shunt-sampled output v_O when the input i_I is zero and the load is absent. In this case, i_E flows through R_{IN}, R_{FB}, and A_V's R_{ID} to drop a voltage that A_V amplifies. A_V's inverting Norton-equivalent transconductance A_G then pulls current from A_V's output resistance R_{OA} and R_{FB}'s input resistance $R_{FB} + (R_{IN} \,||\, R_{ID})$ to set $A_{Z.OL}$ to

$$A_{Z.OL}\Big|_{i_{OUT}\equiv 0} \equiv \frac{v_O}{i_E} = (R_{IN} \,||\, R_{FB} \,||\, R_{ID})(-A_G)\{R_{OA} \,||\, (R_{FB} + (R_{IN} \,||\, R_{ID}))\} \quad (6.74)$$

Feedback factor β_{FB} is the gain from output v_O to feedback i_{FB} when the input i_I is zero. Since R_{FB}'s two-port transconductance multiplies v_O to establish i_{FB}, but in the opposite direction in Fig. 6.42, β_{FB} is $-1/R_{FB}$:

$$\beta_{FB}\Big|_{i_I\equiv 0} \equiv \frac{i_{FB}}{v_O} = -\frac{1}{R_{FB}} \quad (6.75)$$

As a result, closed-loop gain $A_{Z.CL}$ is 1 plus loop gain $A_{Z.OL}\beta_{FB}$ times lower than $A_{Z.OL}$. Since the voltage divider and the T connection process both small- and large-signal components of v_O and i_I linearly, v_O is a $1/\beta_{FB}$ translation of i_{FB} and i_I when $A_{Z.OL}\beta_{FB}$ is considerably greater than 1. Removing A_V's Norton current source A_G eliminates the feedback source that ultimately feeds the input. This leaves R_{IN}, R_{FB}, and

A_V's R_{ID}, so open-loop input resistance $R_{I.OL}$ is their combined parallel resistance:

$$R_{I.OL} = R_{IN} \parallel R_{FB} \parallel R_{ID} \tag{6.76}$$

With a shunt-mixed input, closed-loop input resistance $R_{I.CL}$ is 1 + $A_{Z.OL}\beta_{FB}$ times lower than $R_{I.OL}$. Similarly, removing A_V's Norton current source A_G when the input i_I is zero eliminates the feedback source that feeds the output. This reveals that Norton-equivalent open-loop output resistance $R_{O.OL}$ is the parallel combination of A_V's R_{OA} and R_{FB}'s $R_{FB} + R_{IN} \parallel R_{ID}$:

$$R_{O.OL}\Big|_{i_I \equiv 0} = R_{OA} \parallel [R_{FB} + (R_{IN} \parallel R_{ID})] \tag{6.77}$$

With a shunt-sampled output, closed-loop output resistance $R_{O.CL}$ is therefore $1 + A_{Z.OL}\beta_{FB}$ times lower than $R_{O.OL}$. In spite of all this, many applications that employ this configuration are more interested in the voltage gain across the entire network than in the transimpedance counterpart across the feedback circuit, which refers to the Norton conversion of v_{IN} into i_I and the feedback translation of i_I into v_O, or

$$\frac{v_O}{v_{IN}} = \left(\frac{i_I}{v_{IN}}\right)\left(\frac{v_O}{i_I}\right) = \left(\frac{1}{R_{IN}}\right)A_{Z.CL} \approx \left(\frac{1}{R_{IN}}\right)\left(\frac{1}{\beta_{FB}}\right) = -\frac{R_{FB}}{R_{IN}} \tag{6.78}$$

This, of course, is the inverting gain most popularly associated with this circuit.

Inverting Transistor Amp

Since base–emitter terminals amplify the difference between two voltages, the circuit in Fig. 6.43 is a transistor-level, small-signal implementation of the inverting operational-amplifier configuration in Fig. 6.42. Here, transistor Q_1 implements amplifier A_V, whose inverting transconductance A_G is small-signal transconductance g_{m1}, output resistance R_{OA} is the parallel combination of Q_1's r_{o1} and bias

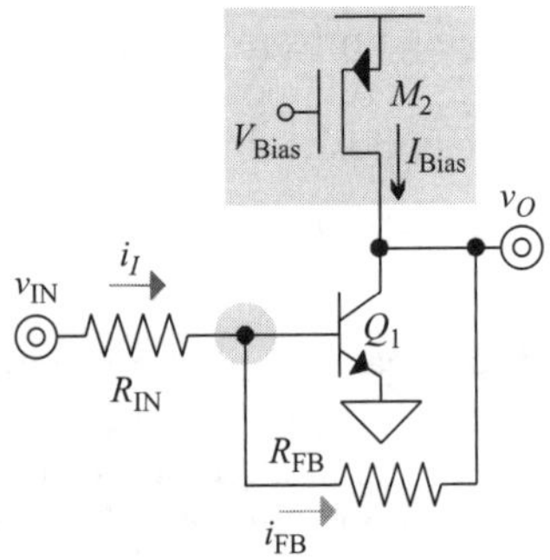

Figure 6.43 Inverting transistor-amplifier configuration.

transistor M_2's r_{sd2}, and differential input resistance R_{ID} is $r_{\pi 1}$. Therefore, with enough loop gain, the small-signal voltage gain across this circuit reduces to roughly $-R_{FB}/R_{IN}$.

Miller Capacitor

Interestingly, although not surprisingly, the Miller effects from Chap. 4 are nothing but the results of negative feedback. To see this, consider that, like in the inverting operational-amplifier configuration, amplifier A_G in Fig. 6.44 inverts and Miller capacitor C_M feeds a noninverting signal back to A_G's input. Together, they comprise a feedback loop that inverts once across A_G.

Tracing the input into the circuit to find the mixer reveals in A_G a receiving input, except A_G has no opposing input voltage with which to mix voltages. Exhausting this voltage-mixing possibility and recognizing the T connection that an incoming current i_I, C_M, and input resistance R_{IN} implement uncovers the actual mixer, into which i_I flows. Similarly, following the output into the network to find the sampler discloses in C_M an impedance Z_M, or $1/sC_M$, that buffers the output v_O. Because Z_M's current is the feedback current i_{FB} with which i_I mixes, C_M converts output voltage v_O into i_{FB}, which means C_M shunt-samples v_O. In other words, the feedback loop implements a transimpedance amplifier.

Like before, all resistances that convert signals into currents or voltages that the loop mixes must decompose into their two-port equivalents. C_M's Norton-equivalent output impedance and transconductance current are therefore the impedance from the input v_{IN} into C_M when v_O is zero, or $1/sC_M$, and C_M's current into v_{IN} when v_{IN} is zero, or $v_O sC_M$. Lastly, C_M's input impedance is the impedance from v_O into C_M, or $1/sC_M + R_{IN}$.

With C_M's two-port model in place, the forward open-loop transimpedance gain $A_{Z.OL}$ is the Thevenin-equivalent gain translation from the shunt-mixed difference i_E or $i_I - i_{FB}$ to the shunt-sampled output v_O when the input i_I is zero and the load is absent. In this case, i_E drops a voltage across R_{IN} and Z_M that A_G converts into a current

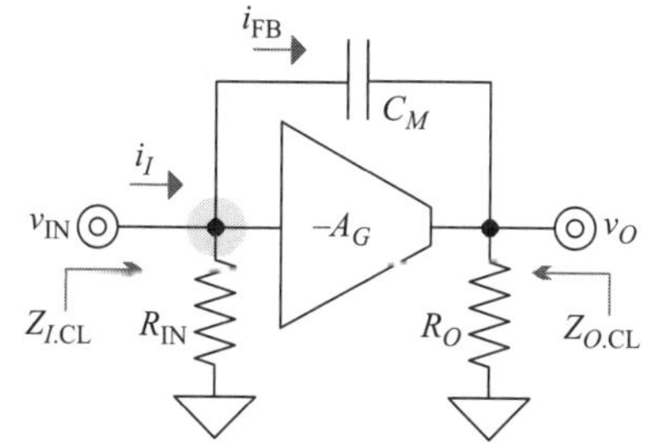

Figure 6.44 Miller capacitor.

and later a voltage across output resistance R_O and C_M's input impedance $1/sC_M + R_{IN}$ to set $A_{Z.OL}$ to

$$A_{Z.OL}\Big|_{i_{OUT}\equiv 0} \equiv \frac{v_O}{i_E} = (R_{IN} \parallel Z_M)(-A_G)[R_O \parallel (Z_M + R_{IN})] \tag{6.79}$$

Feedback factor β_{FB} is the gain from output v_O to feedback i_{FB} when the input i_I is zero. Since C_M's two-port transconductance multiplies v_O to establish i_{FB}, but in the opposite direction in Fig. 6.44, β_{FB} is $-sC_M$:

$$\beta_{FB}\Big|_{i_I\equiv 0} \equiv \frac{i_{FB}}{v_O} = -\frac{1}{Z_M} = -sC_M \tag{6.80}$$

Removing A_G eliminates the feedback source that ultimately feeds the input. This leaves R_{IN} and C_M's Z_M, so open-loop input impedance $Z_{I.OL}$ is $R_{IN} \parallel Z_M$, and with a shunt-mixed input, closed-loop input impedance $Z_{I.CL}$ is $1 + A_{Z.OL}\beta_{FB}$ times lower than $Z_{I.OL}$, or with enough loop gain, $A_{Z.OL}\beta_{FB}$ times lower than $Z_{I.OL}$ at

$$Z_{I.CL} \approx \frac{Z_{I.OL}}{A_{Z.OL}\beta_{FB}} = \frac{R_{IN} \parallel Z_M}{(R_{IN} \parallel Z_M)(-A_G)[R_O \parallel (Z_M + R_{IN})](-sC_M)} = \frac{1}{s(A_V C_M)} \tag{6.81}$$

where A_V is voltage gain v_O/v_{IN}, or $A_G[R_O \parallel (Z_M + R_{IN})]$. In other words, $Z_{I.CL}$ is the impedance of a capacitor that is A_V times greater than C_M. Similarly, removing A_G when the input i_I is zero eliminates the feedback source that feeds the output. This reveals that Norton-equivalent open-loop output impedance $Z_{O.OL}$ is the parallel combination of R_O and C_M's $Z_M + R_{IN}$, and with a shunt-sampled output, closed-loop output impedance $Z_{O.CL}$ is $1 + A_{Z.OL}\beta_{FB}$ times lower than $Z_{I.OL}$, or with enough loop gain, $A_{Z.OL}\beta_{FB}$ times lower than $Z_{O.OL}$ at

$$\begin{aligned} Z_{O.CL} &\approx \frac{Z_{O.OL}\big|_{i_I\equiv 0}}{A_{Z.OL}\beta_{FB}} = \frac{R_O \parallel (Z_M + R_{IN})}{(R_{IN} \parallel Z_M)(-A_G)[R_O \parallel (Z_M + R_{IN})](-sC_M)} \\ &= \frac{1}{(R_{IN} \parallel Z_M)A_G(sC_M)} = \frac{1}{\left(\dfrac{sC_M R_{IN}}{1 + sC_M R_{IN}}\right)A_G} \end{aligned} \tag{6.82}$$

which at high frequency reduces to $1/A_G$. Overall, these feedback assertions say that the input introduces high capacitance to v_{IN} and, after that shunts, the effective output resistance at v_O is low. Stated differently, $Z_{I.CL}$ shunts external input resistances at low frequency and external output capacitances shunt $Z_{O.CL}$ at high frequency, which mirror the Miller-split conclusions reached in Chap. 4.

6.5 Stability

6.5.1 Response

In practice, the gain across a feedback loop is never entirely constant across frequency. In fact, every node in the small-signal path of a circuit incorporates capacitances that shunt signals to ground. This means nodes introduce poles that limit the bandwidth of the loop gain A_{LG} or $A_{OL}\beta_{FB}$. Even the simplest embodiment of A_{LG} will include one node that bandwidth-limits A_{LG} in Fig. 6.45 to pole p_{BW}. Interestingly, dropping at a rate of 20 dB per decade of frequency, or equivalently, at a rate of 10× for every tenfold rise in frequency means A_{LG} falls past p_{BW} by the same amount that frequency rises. The *gain–bandwidth product* (GBW), or $A_{LG}p_{BW}$, is therefore constant past p_{BW} at

$$\text{GBW} = A_{LG}p_{BW} = A_{OL}\beta_{FB}p_{BW} = f_{0dB}\big|_{\text{One Pole}} \tag{6.83}$$

In other words, the frequency where A_{LG} is 1 is GBW over 1. Or stated differently, the *unity-gain frequency* f_{0dB} of a single-pole system is GBW.

Because amplified outputs ultimately reduce to accurate feedback translations of inputs, feedback factor β_{FB} is normally accurate and its bandwidth-limiting pole is high, or at least higher than that of the forward open-loop gain A_{OL}. And since the closed-loop gain that results in A_{CL}, or $1/\beta_{FB}$, can never surpass A_{OL}, $1/\beta_{FB}$ is also lower than A_{OL}, but normally greater than or equal to 1 to be able to amplify or buffer input signals. As a result, β_{FB} is usually less than or equal to 1, which means A_{LG}, or $A_{OL}\beta_{FB}$, is therefore lower than A_{OL} by a factor of $1/\beta_{FB}$. Plus, as Fig. 6.45 illustrates, A_{LG} starts dropping past A_{OL}'s dominant

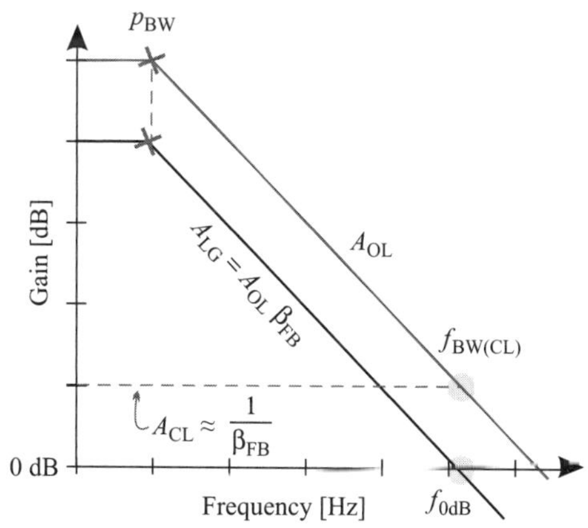

FIGURE 6.45 Loop-gain A_{LG}, forward open-loop A_{OL}, and closed-loop A_{CL} responses.

low-frequency pole p_{BW}. At A_{LG}'s f_{0dB}, in fact, A_{OL} and its closed-loop counterpart A_{CL} are $1/\beta_{FB}$ times higher than A_{LG}. This means that A_{CL}'s closed-loop bandwidth $f_{BW(CL)}$ is A_{LG}'s unity-gain frequency f_{0dB}.

Uncompensated

Since every node in a circuit establishes a pole, and to amplify signals, circuits incorporate more than one node, finding two or more poles in an amplifier is not uncommon. What is more, gain usually results from feeding transconductance currents into high-impedance nodes, so these poles occur when capacitances shunt high resistances. In other words, finding two or more poles in A_{LG} before A_{LG} reaches f_{0dB} like Fig. 6.46 shows is not unreasonable.

Unfortunately, A_{LG}'s phase falls about 90° a decade of frequency after each pole. Therefore, after two poles, A_{LG} reaches f_{0dB} with 180° of phase shift, so closed-loop gain A_{CL} at f_{0dB} approaches infinity:

$$A_{CL}\Big|_{f_{0dB}} = A_{CL}\Big|_{f_{BW(CL)}} = \frac{A_{OL}}{1+A_{OL}\beta_{FB}}\Bigg|_{A_{OL}\beta_{FB}=1\angle 180^\circ} = \frac{A_{OL}}{1-1} \to \infty \qquad (6.84)$$

This means that A_{CL} nears $1/\beta_{FB}$ at low frequency and peaks sharply to infinity at A_{LG}'s f_{0dB}, at A_{CL}'s $f_{BW(CL)}$ like Fig. 6.46 shows. In other words, the closed-loop system amplifies noise at f_{0dB} to the point that the loop loses control of the output, which is highly undesirable.

Compensated

Luckily, as the second pole p_2 climbs toward f_{0dB}, phase shift across the circuit at f_{0dB} falls, and in consequence, $A_{OL}\beta_{FB}$ at f_{0dB} drops to a fraction of –1. In other words, a lower phase shift dampens A_{CL}'s peak at f_{0dB}. Generally, the *stability criterion* is for the loop gain A_{LG} to reach unity gain at f_{0dB} with less than 180° of phase shift. This is why engineers commonly refer to the margin of phase that A_{LG} has before reaching

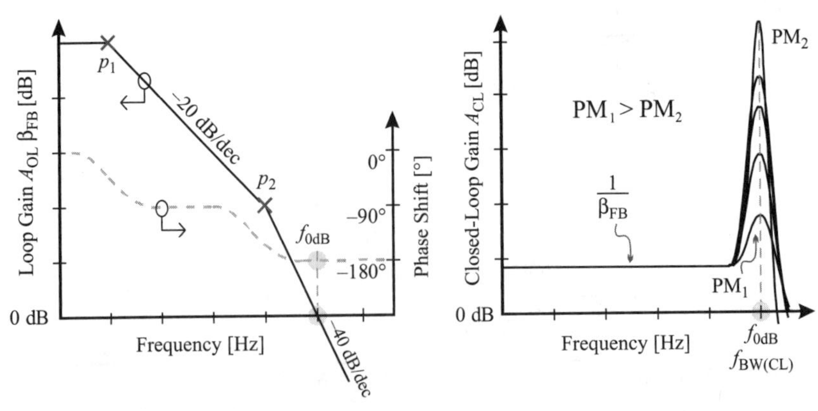

FIGURE 6.46 Uncompensated loop-gain and closed-loop responses.

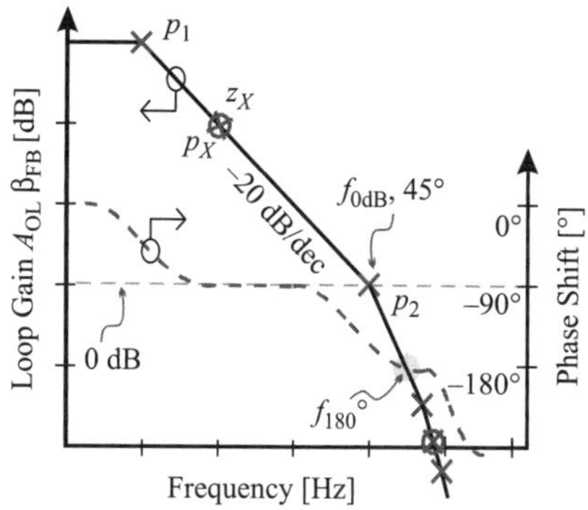

FIGURE 6.47 Compensated loop-gain response.

180° at $f_{180°}$ as *phase margin* PM. Or stated differently, *gain margin* GM is the gain that A_{LG} drops past f_{0dB} to reach a phase shift of 180° at $f_{180°}$.

Most compensated responses for A_{LG} shift the second pole p_2 near or past f_{0dB}, as Fig. 6.47 shows, to keep phase margin at or above 45°. Another way of saying this is A_{LG} reaches f_{0dB} while dropping 20 dB per decade. For this, A_{LG} can accommodate before f_{0dB} as many poles p_X as the system can offset with accompanying zeros z_X and only one additional pole p_1.

With 90° of phase margin, the closed-loop system responds to an input that rises instantly in a steplike fashion by raising its output s_O monotonically to s_O's targeted point S_O'. With 60° of phase margin, as the time-domain response of Fig. 6.48 illustrates, s_O first momentarily surpasses S_O', and then, s_O undershoots S_O' before finally settling within a few percentage points of S_O'. Worse yet, with 45° of phase margin, s_O overshoots S_O' one more time, for a total of three oscillating

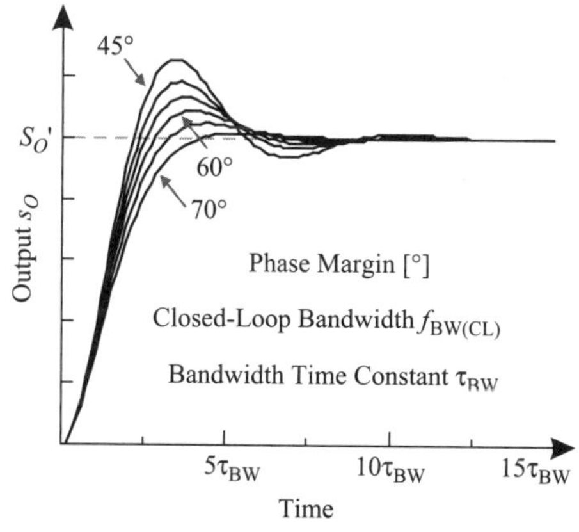

FIGURE 6.48 Compensated time-domain response to a stepped input.

rings, before finally converging on S_O'. In general, when bandwidth-limited to $f_{BW(CL)}$ with a time constant τ_{BW} that is equivalent to $1/2\pi f_{BW(CL)}$, s_O settles to S_O' after 3 to 10 time constants.

6.5.2 Compensation

Frequency compensation refers to the actions adopted to ensure that a feedback system meets the stability criterion. The fundamental aim is for the loop gain A_{LG} to reach the unity-gain frequency f_{0dB} with more than 45° of phase margin. This happens when A_{LG} arrives at f_{0dB} at a rate of 20 dB per decade, or more insightfully, under the influence of only one pole. To achieve this, engineers strive to establish in A_{LG} one dominant low-frequency pole.

Feedback loops with higher compensated unity-gain frequencies maintain desirable negative-feedback effects on gain, impedances, linearity, sensitivity, and other parameters through higher frequencies. In other words, higher bandwidth enables feedback systems to process higher-frequency signals. Unfortunately, as the ensuing discussion demonstrates, ensuring that A_{LG} reaches f_{0dB} under the effects of one pole normally reduces the operating bandwidth from its uncompensated state. The extent to which compensation reduces f_{0dB} and raises power consumption indicates the relative efficacy of the strategy adopted against alternate approaches.

Add Pole

Perhaps the easiest, though not always the best way of compensating an inverting feedback circuit is to *add* a pole p_A to A_{LG}. But for A_{LG} to reach its compensated unity-gain frequency $f_{0dB.A}$ at 20 dB per decade, the next pole in the system p_1 must reside at or above $f_{0dB.A}$, as Fig. 6.49 shows. As such, p_A must be low enough to ensure $f_{0dB.A}$ does not surpass p_1. This means that $f_{0dB.A}$ is at or below p_1, and therefore well below its uncompensated counterpart of f_{0dB}.

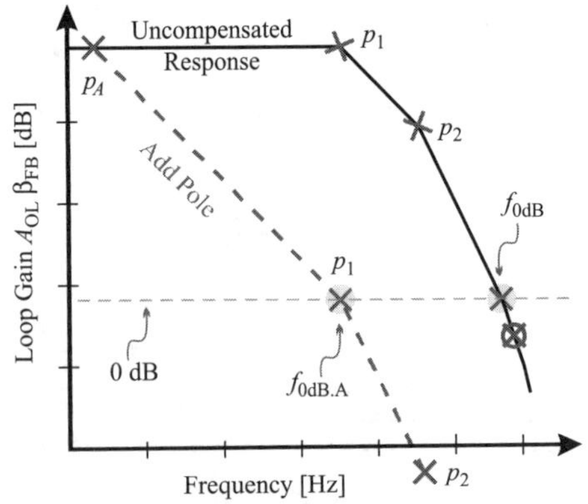

FIGURE 6.49 Compensated response after adding a pole.

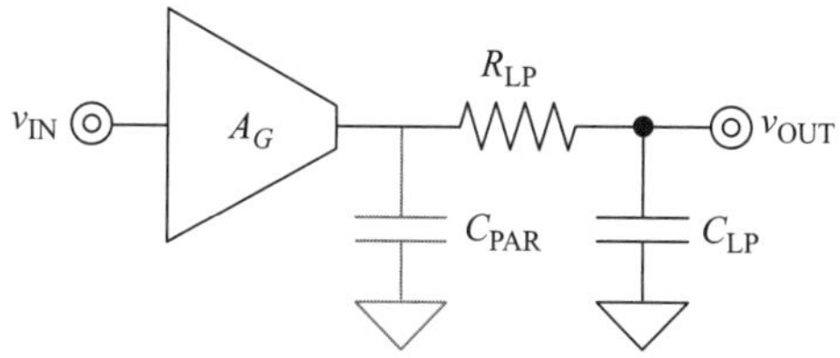

FIGURE 6.50 Inserting a low-pass filter into a transconductor gain stage.

Low-Pass Filter Adding a pole p_A amounts to inserting a low-pass filter into the loop. Perhaps the most straightforward means of doing this is to insert a decoupling resistor R_{LP} and a filter capacitor C_{LP} at the output of the transconductor gain stage A_G that incorporates the first pole p_1 of the uncompensated system, like Fig. 6.50 shows. For p_A to reside below p_1, C_{LP} must be greater than the parasitic capacitor C_{PAR} already present at A_G's output. This way, C_{LP} first shunts the parallel resistance that R_{LP} and A_G's output resistance R_{OA} create to establish p_A at $1/2\pi(R_{LP} + R_{OA})C_{LP}$:

$$\left.\frac{1}{sC_{LP}}\right|_{p_A \approx \frac{1}{2\pi(R_{LP}+R_{OA})C_{LP}}} \equiv R_{LP} + R_{OA} \tag{6.85}$$

With C_{LP} practically shorted, C_{PAR} then shunts $R_{OA} \parallel R_{LP}$ at p_1 or $1/2\pi(R_{OA} \parallel R_{LP})C_{PAR}$:

$$\left.\frac{1}{sC_{PAR}}\right|_{p_1 \approx \frac{1}{2\pi(R_{OA}\parallel R_{LP})C_{PAR}}} \equiv R_{OA} \parallel R_{LP} \tag{6.86}$$

Shift Pole

A better way of establishing a dominant low-frequency pole is to shift the lowest-frequency pole in the system to lower frequency, as p_1 in Fig. 6.51 demonstrates. This way, to reach the new compensated unity-gain frequency $f_{0dB.S}$ at 20 dB per decade, p_1 must be low enough to ensure $f_{0dB.S}$ is at or below the system's second pole p_2. The benefit of this approach over the previous one is higher bandwidth because $f_{0dB.S}$ is higher than the original p_1, where $f_{0dB.A}$ from adding a pole would have been. Luckily, shifting a moderate-frequency pole p_1 to lower frequencies amounts to adding capacitance to the node that introduced p_1 in the first place, which is easy. This is only possible, however, when the node is accessible, which is usually the case for integrated circuit (IC) engineers, but not always for system designers.

Split Poles

An improvement to pulling the lowest-frequency pole p_1 to lower frequency is to push the system's second pole p_2 to higher frequency.

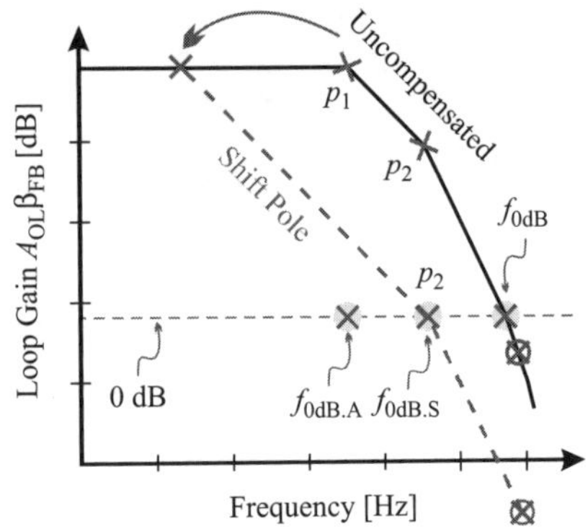

Figure 6.51 Compensated response after shifting the system's lowest-frequency pole.

This way, the new compensated unity-gain frequency $f_{0dB.M}$ in Fig. 6.52 can be at a higher frequency than p_2. Fortunately, adding the Miller capacitor C_M in Fig. 6.44 to the inverting stage that produces p_1 and p_2 achieves this functionality. As already discussed, the feedback circuit that C_M realizes presents high capacitance to the input and low resistance to the output, so the former shunts input resistances at low frequency and output capacitances shunt the latter at high frequency. In other words, C_M splits the input and output poles p_1 and p_2.

The beauty of this solution is that input and output resistances in Fig. 6.44 can be high at low frequency, which means pushing p_2 to higher frequencies does not sacrifice low-frequency loop gain. The drawback here is that C_M feeds forward an output-of-phase signal across the transconductor that introduces the right-half-plane zero z_M

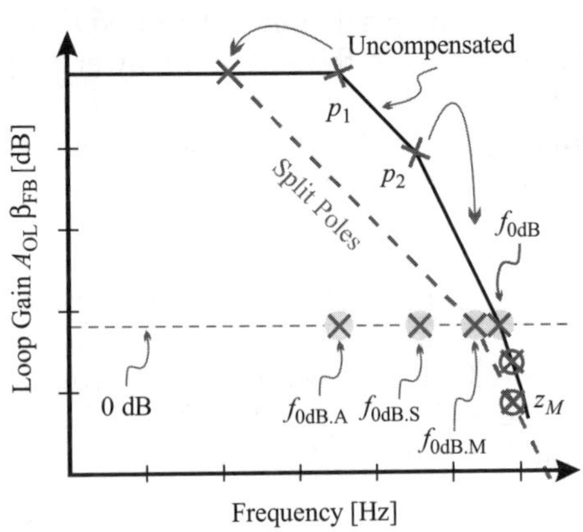

Figure 6.52 Compensated response after splitting the system's two lowest-frequency poles.

discussed in Chap. 4. Because z_M's effects of raising loop gain and lowering phase can only reduce phase margin, z_M must remain about a decade of frequency above $f_{0dB.M}$, or higher if possible.

Add In-Phase Zero

Cancelling the effect of the second pole p_2 with an in-phase zero z_{LHP} is even better because the new compensated unity-gain frequency $f_{0dB.Z}$ from Fig. 6.53 can now reside near the system's third pole p_3. Note, however, that high-frequency poles are typically the result of parasitic capacitors shunting $1/g_m$ resistances. This means that a circuit is liable to include several of these poles near p_3, or maybe even close to p_2. Thankfully, raising quiescent current in $1/g_m$ devices can shift these poles to higher frequency, but at the expense of additional power.

Current-Limiting Resistor One way to insert an in-phase zero is to add a current-limiting resistor to the capacitor added to shift the system's first pole p_1 to lower frequency, like R_Z and C_1 in Fig. 6.54 exemplify. This way, C_1 first shunts the output resistance R_{OA} of the transconductor A_G and R_Z to establish p_1 at $1/2\pi(R_Z + R_{OA})C_1$:

$$\left.\frac{1}{sC_1}\right|_{p_1\approx\frac{1}{2\pi(R_Z+R_{OA})C_1}} \equiv R_Z + R_{OA} \tag{6.87}$$

R_Z then limits how much current C_1 can sink when $1/sC_1$ falls below R_Z at current-limited zero z_R:

$$\left.\frac{1}{sC_1}\right|_{z_R\approx\frac{1}{2\pi R_Z C_1}} \equiv R_Z \tag{6.88}$$

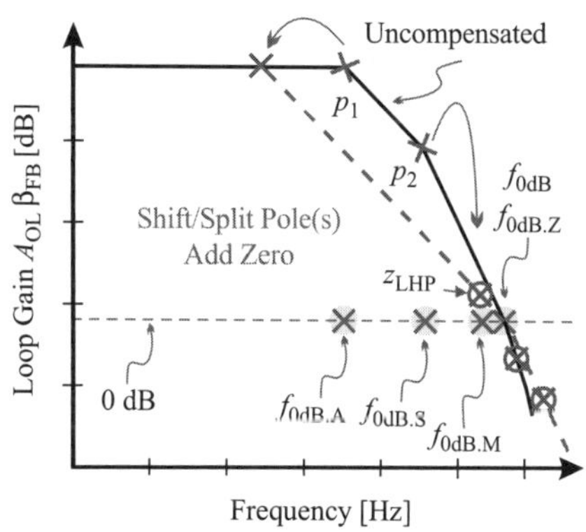

FIGURE 6.53 Compensated response after splitting poles and adding an in-phase zero.

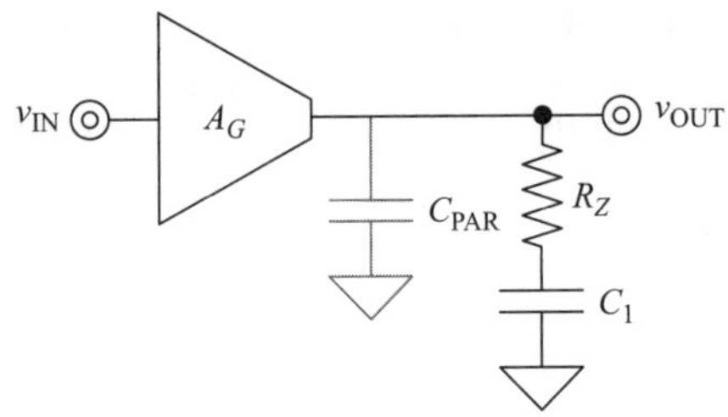

Figure 6.54 Adding a current-limiting resistor to the capacitor that shifts the first pole.

Bypass Capacitor A bypass capacitor can also feed forward in-phase signals across the noninverting gain stage A_G that incorporates the system's first pole p_1 to establish a zero at higher frequency, like C_{FF} in Fig. 6.55 shows. Here, the effects of the zero arise when C_{FF}'s current i_C surpasses A_G's output current i_O. To see where the zero z_C lands, first note that A_G's output capacitance C_1 shunts A_G's output resistance R_{OA} first to establish p_1. This means that output voltage v_{OUT} is close to zero at higher frequency, near z_C. Therefore, since z_C is the transitional frequency across which i_C crosses i_O, z_C reduces to $A_G/2\pi C_{FF}$:

$$i_C = (v_{IN} - v_{OUT})sC_{FF} \approx (v_{IN} - 0)sC_{FF}\Big|_{z_C \approx \frac{A_G}{2\pi C_{FF}}} \equiv i_O = v_{IN}A_G \tag{6.89}$$

Bypass Circuit A bypass circuit can similarly feed-forward signals across a noninverting amplifier that includes the system's first pole p_1 to establish an in-phase zero. In the case of Fig. 6.56, G_{FF} feeds forward current i_{FF} to the output of two cascaded transconductor gain stages G_1 and G_2. Since the first stage's output capacitance C_1 shunts R_1 first to establish p_1, the output current i_O of the two stages drops past p_1. As a result, feed-forward zero z_{FF} appears when i_O falls below i_{FF} at $G_1G_2/2\pi G_{FF}C_1$:

$$i_O = v_{IN}G_1\left(R_1 \parallel \frac{1}{sC_1}\right)G_2\Bigg|_{f \gg \frac{1}{2\pi R_1C_1}} \approx \frac{v_{IN}G_1G_2}{sC_1}\Bigg|_{z_{FF} = \frac{G_1G_2}{2\pi G_{FF}C_1}} \equiv i_{FF} = v_{IN}G_{FF} \tag{6.90}$$

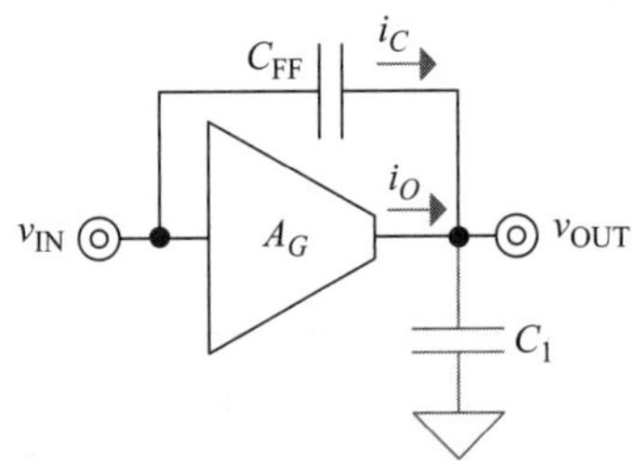

Figure 6.55 Inserting a bypass capacitor across a noninverting stage.

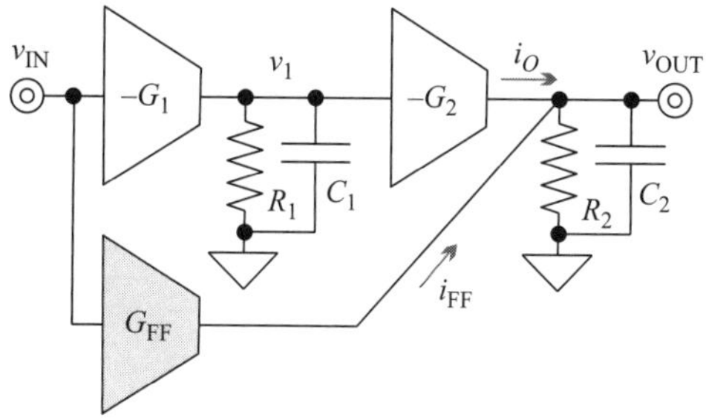

FIGURE 6.56 Inserting a bypass circuit across a noninverting stage.

In the transistor-level implementation of the bypass circuit in Fig. 6.57, Q_1 and M_2 realize G_1 and G_2, and Q_{FF} implements G_{FF}. Here, Q_{FF}'s loaded emitter resistance $2/g_{mFF}$ emitter-degenerates Q_1 lightly to set G_1 to roughly $g_{m1}/3$:

$$G_1 = \frac{i_1}{v_{in}} \approx \frac{g_{m1}}{1 + g_{m1}\left(\frac{2}{g_{mFF}}\right)} = \frac{g_{m1}}{3} \tag{6.91}$$

where the ratio of Q_1's and Q_{FF}'s transconductances g_{m1}/g_{mFF} is 1 because Q_1 and Q_2 match. In this way, Q_1 converts v_{IN} into a current i_1 that drops a voltage v_1 across R_1 to establish gain. M_2's transconductance g_{m2} then converts v_1 into an output current i_o that flows through M_2's r_{sd2} and bias transistor M_3's r_{ds3} to add more gain. Interestingly, Q_{FF} receives and buffers Q_1's i_1 to the output, so G_{FF} equals G_1 and the feed-forward zero z_{FF} that results therefore reduces to $g_{m2}/2\pi C_1$:

$$z_{FF} = \frac{G_1 G_2}{2\pi G_{FF} C_1} = \frac{G_1 g_{m2}}{2\pi G_1 C_1} = \frac{g_{m2}}{2\pi C_1} \tag{6.92}$$

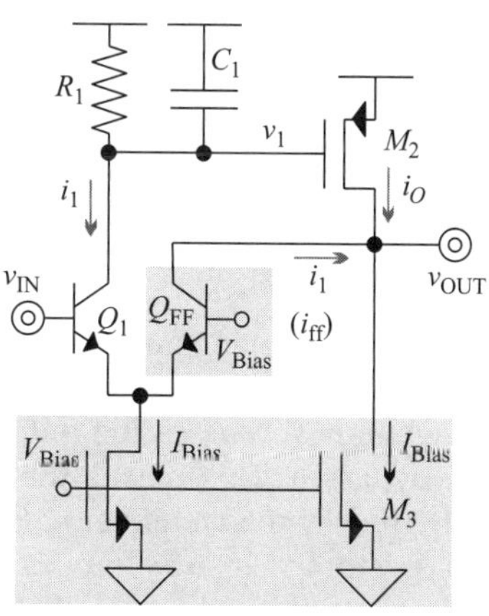

FIGURE 6.57 Transistor-level implementation of the bypass circuit.

Example 6.5 Under what conditions is the transistor-level realization of the regulated-cascode current mirror in Fig. 6.38 stable with at least 45° of phase margin?

Solution: Since source-degenerated cascode transistor M_C incorporates only one node, and as a result, only one pole, the inner loop that M_C establishes is stable. The outer loop includes two poles, one at M_C's gate and another at the source. Because M_A's r_{dsA} and M_B's r_{sdB} are high, r_{dsA} and r_{sdB} and M_C's gate capacitance C_{GC} establish a low-frequency pole p_1:

$$\left.\frac{1}{sC_{GC}}\right|_{p_1\approx\frac{1}{2\pi R_{OA}C_{GC}}} \equiv R_{OA} = r_{dsA} \parallel r_{sdB}$$

where R_{OA} is M_A's combined output resistance. M_A's gate capacitance C_{GA} then shunts M_C's source resistance R_{SC} at high frequency, at p_2, because R_{SC} nears $2/g_{mC}$ when load resistance R_{LOAD} is equivalent to r_o or lower, which is not uncommon:

$$\left.\frac{1}{sC_{GA}}\right|_{p_2\approx\frac{0.5g_{mC}}{2\pi C_{GA}}} \equiv r_{o2} \parallel R_{SC} \approx R_{SC} = \frac{R_{LOAD}+r_{dsC}}{1+g_{mC}r_{dsC}} \approx \frac{2}{g_{mC}}$$

Since the feedback factor is 1, forward open-loop gain $A_{I.OL}$ is roughly $g_{mA}(r_{dsA} \parallel r_{sdB})$, or $g_{mA}R_{OA}$, and unity-gain frequency f_{0dB} when assuming p_2 is equal to or greater than f_{0dB} is approximately $g_{mA}/2\pi C_{GC}$:

$$\left.A_{I.OL}\right|_{v_{out}=0} \equiv \frac{i_o}{i_e} = (-r_{o2})(-g_{mA})R_{OA}\left(\frac{g_{mC}}{1+g_{mC}r_{o2}}\right) \approx g_{mA}R_{OA}$$

and $$\left.f_{0dB}\right|_{One\ Pole} = GBW = A_{I.OL}\beta_{FB}p_{BW} \approx (g_{mA}R_{OA})(1)\left(\frac{1}{2\pi R_{OA}C_{GC}}\right) = \frac{g_{mA}}{2\pi C_{GC}}$$

The circuit is stable with 45° or more of phase margin when p_2 is nearly at f_{0dB} or higher, which happens when $0.5g_{mC}/C_{GA}$ is equal to or greater than g_{mA}/C_{GC}.

6.5.3 Out-of-Phase Zero

Miller capacitors are so efficient at splitting poles without sacrificing gain that many feedback loops often employ them for compensation. The downside is that they feed out-of-phase signals forward across the circuit to introduce right-half-plane zeros. Since phase starts dropping a decade before the zeros, engineers normally shift them to higher frequencies, or better yet, to the left-half plane to offset the effects of a pole. At times, designers even eliminate them.

Shift

The ultimate culprit behind the establishment of the out-of-phase zero z_M is the current i_C that the Miller capacitor C_M feeds to the output. The fact is that z_M arises when i_C surpasses the output current i_A of the amplifier A_G that C_M bypasses. This means that a lower i_C rises to exceed i_A at a higher frequency than a higher i_C. So, limiting C_M's current i_C with the series resistor in Fig. 6.58 impedes the process to shift z_M to higher frequency. To see this, first recall that z_M happens when i_C equals and cancels i_A, so A_G's output resistance R_{OA} drops zero volts at z_M.

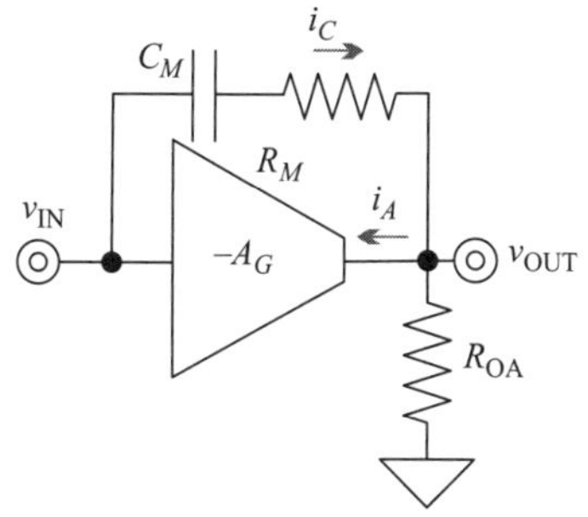

FIGURE 6.58 Current-limiting the Miller capacitor's current.

As a result, v_{out} is zero and R_M has the effect of reducing the equivalent resistance that defines the location of z_M to $(1/A_G) - R_M$, which is another way of saying that R_M raises z_M:

$$i_C = \frac{v_{IN} - v_O}{\frac{1}{sC_M} + R_M} = \left.\frac{v_{IN} - 0}{\frac{1}{sC_M} + R_M}\right|_{z_M = \frac{1}{2\pi C_M\left(\frac{1}{A_G} - R_M\right)}} \equiv i_A = v_{IN}A_G \tag{6.93}$$

What is more, equating R_M to $1/A_G$ shifts z_M to infinitely high frequency. Although matching R_M to $1/A_G$ perfectly across temperature and process is not practicable, reducing $1/A_G$ by 80 percent or more is, so z_M can be high enough not to matter.

Conversion

Interestingly, raising R_M above $1/A_G$ reverses the polarity of the zero. In other words, z_M becomes an in-phase zero when R_M is greater than $1/A_G$. For more insight, consider that elevating R_M to the point C_M's i_C is never high enough to surpass i_A basically removes the right-half-plane zero. Plus, raising R_M further limits how much current i_C draws from the input, which is the effect of offsetting the input pole with an in-phase current-limiting zero. The net result of all this is that R_M can shift the right-half-plane zero to the left-half plane.

Elimination

Eliminating the out-of-phase zero z_M that the Miller capacitor C_M introduces when C_M's current i_C bypasses the amplifier A_V across which C_M connects amounts to blocking the feed-forward component of i_C. Since splitting the input and output poles of a gain stage is the result of feedback, blocking feed-forward signals does not negate the benefits of C_M in compensating a circuit. The question is how to allow

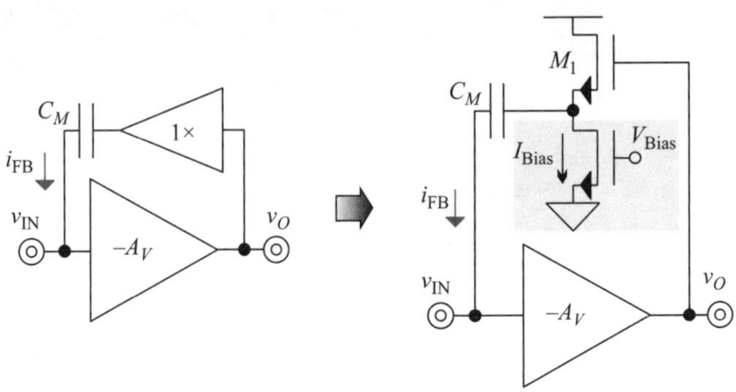

FIGURE 6.59 Blocking the feed-forward current with a voltage buffer.

feedback signals to reach the input without establishing a feed-forward path. In other words, C_M must still be able to convert the output voltage v_O into the feedback current i_{FB} that flows to the input.

Voltage Buffer One way to block feed-forward signals is to buffer the output voltage v_O, as Fig. 6.59 shows. While preserving the integrity of the voltage that C_M receives to establish feedback current i_{FB}, the buffer's low-impedance output shunts feed-forward signals. One simple transistor-level embodiment of this buffer is a source/emitter follower, like M_1 in Fig. 6.59.

Current Buffer Another approach is to buffer the feedback current that C_M establishes, which is what the current buffer of Fig. 6.60 achieves. This way, the buffer's high output impedance blocks feed-forward currents. One simple transistor-level embodiment of this current buffer is a common-gate/base transistor, like M_1 in Fig. 6.60.

Miller-Multiplying Current Buffer Replacing the current buffer with a current amplifier A_I not only blocks feed-forward currents but also multiplies the capacitor current that produces Miller effects. This way,

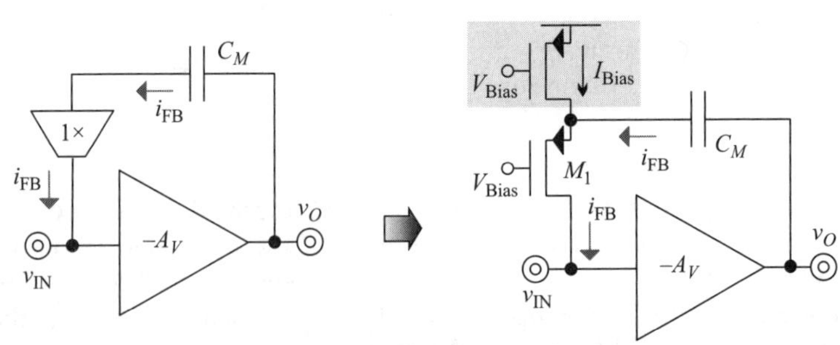

FIGURE 6.60 Blocking the feed-forward current with a current buffer.

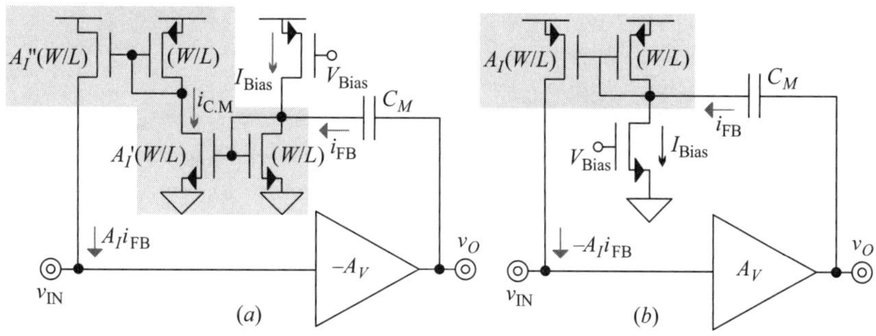

FIGURE 6.61 Blocking feed-forward currents with Miller-multiplying buffers.

A_I raises feedback factor β_{FB}, and as a result, the loop gain A_{LG} that β_{FB} helps establish, so input capacitance rises and output resistance falls by the same A_I factor to roughly $A_I A_V C_M$ and $1/A_I A_G$. In the implementation of Fig. 6.61*a*, for example, two current mirrors realize A_I because each mirror's gate–drain translation inverts C_M's feedback signal. As a result, the noninverting β_{FB} and the inverting amplifier $-A_V$ invert the loop once to implement negative feedback. Similarly, one current mirror in the feedback path inverts the loop that a noninverting amplifier completes, like A_V in Fig. 6.61*b* illustrates. Generally, feedback capacitors in series with one or two amplifying current mirrors can further split the input and output poles of not only inverting but also noninverting amplifiers.

6.5.4 Embedded Loops

Considering that the objective of a feedback loop is to translate the input to the output accurately, β_{FB} is normally both accurate and wide-bandwidth. This means forward open-loop A_{OL} usually incorporates the dominant low-frequency pole p_1 that drops the loop gain A_{LG} to the unity-gain frequency f_{0dB} at a rate of 20 dB per decade with 90° of phase shift. As such, no more than one additional pole can reside at the compensated unity-gain frequency f_{0dB} of the system to keep phase from shifting past 135°.

In light of these conditions, the closed-loop bandwidths of loops embedded in β_{FB} and A_{OL} are typically higher than f_{0dB}. This simplifies the analysis because the closed-loop gains of inner loops are, for the purposes of the outer loop, constant across frequency. The only exception to this practice is when one embedded loop establishes the dominant low-frequency pole of the system. Irrespective of the circumstances, though, secondary closed-loop poles of inner loops are almost always well above f_{0dB}. The moral in all this is that keeping track of closed-loop gains and bandwidths of embedded loops is important.

6.6 Design

6.6.1 Design Concepts

Ultimately, a feedback signal s_{FB} mirrors the input signal s_I with which it mixes. The effect of negative feedback is therefore to oppose the influence of external forces on this regulation action. The purpose of the feedback factor β_{FB} is to translate s_{FB} to the output s_O, because s_{FB} is not always in the same form or at the same level as the targeted s_O. The system therefore regulates s_O to the same extent that β_{FB} is accurate. Accuracy, however, only results when the loop has sufficient gain, that is, when the combined gain across β_{FB} and the forward open-loop gain A_{OL}, or loop gain A_{LG} for short, is considerably greater than 1. And for both small- and large-signal components of s_O to be a feedback-factor translation of s_I, both mixer and β_{FB} must process small- and large-signal components.

Generally, shunt feedback shunts the input or output terminal of the feedback network with an impedance that is roughly A_{LG} times lower than the open-loop impedance. Series feedback, on the other hand, introduces a series impedance that is A_{LG} times greater than the open-loop counterpart. This means that, with enough loop gain, the impedances of shunt-mixed inputs and shunt-mixed outputs are low and those of their series-mixed and series-sampled counterparts are high.

Differential pairs and transistor transconductances series-mix input voltages only when their opposing input terminals also carry loop signals. They can also sample output voltages, but only when their opposing input terminals are not in the loop. T connections can shunt-mix input currents and collectors, emitters, drains, and sources can sample output currents. Emitters and sources, however, sample currents only when their corresponding bases and gates carry loop signals; otherwise, they sample voltages. Note, by the way, that a sampler monitors a voltage if s_{FB} becomes zero after grounding the output terminal.

Open-loop impedances must account for all impedances in the network. Accordingly, opening a feedback loop amounts to shorting a voltage source or removing a current source in the feedback path. Inner loops, however, should not open, so voltage and current sources in embedded loops must remain intact. In the end, designing feedback systems hinges on linearizing and approximating loop-gain translations of nonlinear transistors, circuits, and intertwined loops.

6.6.2 Architectural Design

Design always starts with the targeted application, and more specifically, with the performance it demands of the system. Although the process is never unique, inputs and outputs are usually the first to consider. The following illustrates an example of this approach.

First, determine the input and output signals to be mixed and sampled from the required input–output translation. If output current must be an accurate translation of input current, the feedback network should implement a current amplifier. In other words, the loop should shunt-mix and series-sample currents.

From the mixed input s_I, determine the most suitable input impedance. Low input resistances, for example, absorb most of an incoming current, so T connections that feed error currents to emitter and source terminals are good candidates. In other words, the input of the forward open-loop amplifier A_{OL} can be a common-base/gate current buffer. High input resistances, on the other hand, drop most of an incoming voltage, so bases or gates are good A_{OL} inputs. Through this process, recall that T connections and differential pairs can mix both small- and large-signal components of the input, whereas base–emitter and gate–source terminals can only mix small-signal voltages.

Similarly, determine the best output impedance from the sampled output s_O. An output voltage, for instance, should vary little with current, so output resistance should be low. As a result, emitters and sources are good candidates for A_{OL}'s output. In other words, A_{OL} can drive the output with emitter and source followers. In contrast, output currents should not change with voltage, so output resistance should be high. Here, the collectors and drains of common-emitter and common-source transistors are good candidates for A_{OL}'s output.

The feedback factor β_{FB} is next. In this regard, voltage dividers are good for translating voltages to voltages because, when resistors match, small- and large-signal translations can be accurate. Common-base/gate current buffers and current mirrors are similarly good candidates for current amplifiers. Resistors are also good for both transconductance and transimpedance translations.

The next step in the process is to build forward gain in A_{OL} from the mixing input to the sampling output with the single-transistor primitives and analog building blocks from Chaps. 4 and 5. With gain stages in place, biasing and frequency compensation follow. Bear in mind that consulting with colleagues and conducting a literature survey can reveal circuits and techniques that can help leverage the experience of others in the field.

6.6.3 Frequency Compensation

The general approach for compensating a feedback system is to establish a dominant low-frequency pole in the forward open-loop gain A_{OL}. To ensure that phase margin does not drop below 45°, the second dominant pole should reside at or above the system's unity-gain frequency f_{0dB}. Because phase starts shifting a decade of frequency before a pole arises, all other poles should be at least one decade above f_{0dB}. Unfortunately, the phase shift of many parasitic poles can compound to such a pitch that phase can start avalanching near or just past f_{0dB}. This can be a problem when considering that f_{0dB} shifts with temperature and process and while the system powers up. Moving these parasitic poles to higher frequencies and adding in-phase zeros to this region help mitigate this snowball effect. Cancelling the effects of the second pole with an in-phase zero can also raise f_{0dB}, but only to the extent that mismatch between the two allow, because extending f_{0dB} closer to several parasitic poles can ultimately compromise phase margin. In this regard, out-of-phase zeros can only reduce phase margin, so like parasitic poles, they should reside at least one decade above f_{0dB}.

6.7 Summary

A feedback system regulates the feedback signal with which the loop mixes the input. This means that an accurate feedback-factor translation similarly controls the sampled output, which is why negative feedback is so ubiquitous in the universe and popular in microelectronics. Some of the positive side effects of feedback include lower sensitivity, higher linearity, wider bandwidth, and higher or lower impedances, depending on how the loop mixes signals and other circuits sample the loop. Shunt feedback, for example, lowers impedances and series feedback raises them.

Interestingly, the emitter/source-degenerated transistor first analyzed in Chap. 4 is an example of a transconductance feedback amplifier. What is more, the diode-connected transistor, the Miller capacitor, and the inverting operational-amplifier configuration are likewise classic examples of transimpedance feedback circuits. Perhaps the least surprising of the circuits explored in this chapter is the noninverting operational-amplifier configuration, which is a voltage-amplifier example.

For a feedback system to remain stable, the loop gain must reach the system's unity-gain frequency under the combined effects of only one pole, at 20 dB per decade of frequency. To realize this, designers normally establish one dominant low-frequency pole in the forward gain and ensure that all other poles reside at frequencies at or above the system's closed-loop bandwidth. Splitting poles with Miller capacitors and introducing in-phase zeros with current-limiting resistors and bypass capacitors and circuits help in this respect. The out-of-phase zeros that Miller capacitors produce, however, do not, so engineers typically shift them to high frequencies, convert them to in-phase zeros, or eliminate them.

In the end, insight is the key to design. So understanding how devices, transistors, circuits, and the feedback systems that they implement operate is imperative. The driving objective of Chaps. 3 through 6 is this, to develop the insight and intuition necessary to design analog ICs.

The purpose of the chapters that follow is to illustrate how to apply these concepts to popular analog systems like the linear regulator.

6.8 Review Questions

1. What are the basic functional components of a feedback system?

2. What is the most appealing and useful feature of a negative-feedback loop?

3. What is the effect of feedback on sensitivity?

4. What is the effect of series sampling on output impedance?

5. What type of signal does a feedback loop sample when shunt-sampling the output?

6. What is the effect of series mixing on input impedance?

7. What type of signal does a feedback loop mix when shunt-mixing the input?

8. What is the effect of feedback on a zero in the forward-gain path when loop gain is substantially higher than 1?

9. What is the effect of feedback on a high-pass filter in the forward-gain path?

10. What is the root cause of gain distortion in amplified signals?

11. What is the effect of a source resistance at the input when driving a transconductance feedback amplifier?

12. What is the effect of a current feedback amplifier's output resistance when driving a load?

13. Which transistor terminals can mix voltages and under what conditions?

14. Which transistor terminals can sample voltages and under what conditions?

15. Which transistor terminals can sample currents and under what conditions?

16. How can current mixers be distinguished identified?

17. What test conditions apply when deriving forward open-loop transconductance gains?

18. What test conditions apply when deriving forward open-loop transimpedance gains?

19. What test conditions apply when deriving the feedback factor of voltage feedback amplifiers?

20. What test conditions apply when deriving the output resistance of current feedback amplifiers?

21. Derive the open-loop parameters of the following circuit.

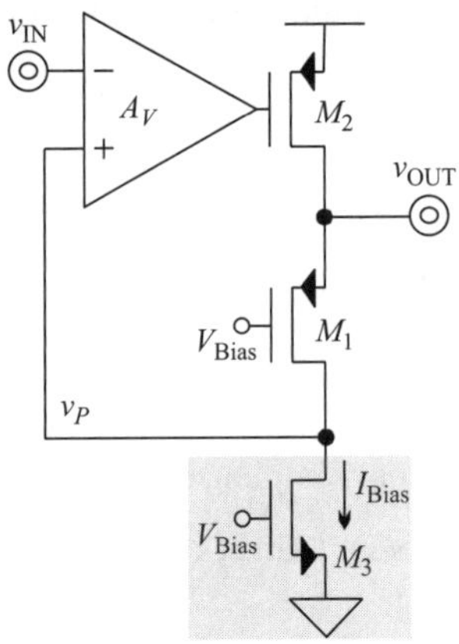

22. Derive the open-loop parameters and overall voltage gain of the following circuit.

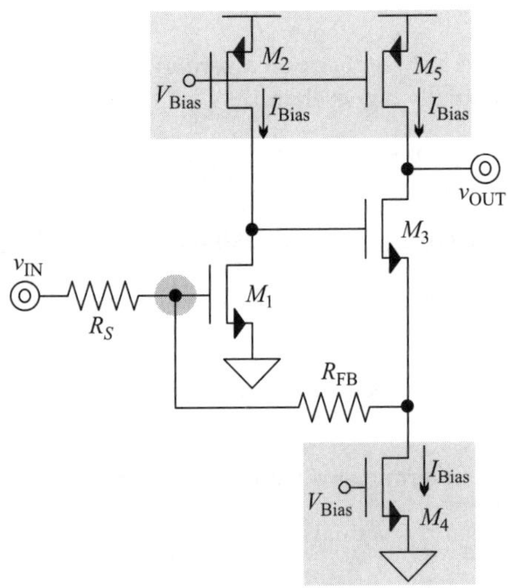

23. State the stability criterion.

24. How can an IC engineer shift a pole to low frequency?

25. How can an IC engineer split the input and output poles of an inverting gain stage?

26. How can an IC engineer introduce in-phase zeros?

27. How can an IC engineer shift to high frequency the right-half-plane zero that an out-of-phase feed-forward capacitor introduces?

28. How can an IC engineer eliminate the right-half-plane zero that an out-of-phase feed-forward capacitor introduces?

29. True or False: Feedback factors are normally accurate and wide-bandwidth.

CHAPTER 7

Bias Currents and Reference Circuits

Most, if not all, analog components in a system incorporate or require some form of bias circuit. The fundamental reason for this is simple: ensuring that transistors operate in their desired state, be it in triode or otherwise, across temperature, supply, and manufacturing corners requires a reliable bias reference. What is more, the accuracy of analog systems ultimately hinges on a reference. This is why sensor-interface circuits, analog-to-digital and digital-to-analog converters, thermostats, phase-locked loops, linear and switching power supplies, and energy-harnessing microelectronics, to name a few examples, include bias currents and reference circuits.

Not surprisingly, the design of almost all analog blocks starts with the circuits that bias them. The aim of this chapter is to introduce and describe the underlying concepts and circuits that govern the development of bias currents and references. But because references rely on bias currents, bias-current generators are first in the chapter. In that light, the text discusses issues like temperature dependence and compensation, startup, frequency compensation, and supply-noise suppression. After this, bandgap-voltage references follow. And throughout, the chapter shows how to substitute bipolar-junction transistors (BJTs) with PN-junction diodes to realize solutions that are compatible with complementary metal–oxide–semiconductor (CMOS) technologies.

7.1 Voltage Primitives

Engineers normally extract currents from predictable voltages. In this respect, a *forward-biased diode* voltage v_D at roughly 0.6–0.7 V is a good candidate for two fundamental reasons. The first one is that the diode's initial tolerance at room temperature is on the order of ±2 percent. The second is that the diode's logarithmic relationship to current suppresses variations in current to produce barely noticeable effects in voltage.

Or more specifically, because diode current i_D climbs exponentially with a linear rise in v_D:

$$i_D = I_S\left(e^{\frac{v_D}{V_t}} - 1\right) \approx I_S e^{\frac{v_D}{V_t}} \tag{7.1}$$

v_D is only a logarithmic translation of i_D:

$$v_D \approx V_t \ln\left(\frac{i_D}{I_S}\right) = \frac{KT}{q}\ln\left(\frac{i_D}{I_S}\right) \tag{7.2}$$

For similar reasons, another potential candidate is the *breakdown voltage* V_{BD} of a diode. In this case, initial tolerance is slightly higher at ±2–±4 percent. Similar to v_D, though, avalanche and zener tunneling suppress variations in current to yield logarithmic-like benefits. The drawback to V_{BD} is its magnitude. Typical base–emitter PN diode junctions in BJTs, to cite an example, enter breakdown at 6.8–7 V, which is normally too high for consumer electronics whose supply voltages range from 1–5 V.

The gate–source voltage v_{GS} of a metal–oxide–semiconductor field-effect transistor (MOSFET) is another option. When operating in strong inversion and saturation, MOS current i_{DS} rises quadratically with a linearly increasing v_{GS},

$$i_{DS} \approx 0.5\left(\frac{W}{L}\right)K'(v_{GS} - v_T)^2 \tag{7.3}$$

which means v_{GS} decomposes into threshold voltage v_T and saturation voltage $V_{DS(SAT)}$, the latter of which is a square-root function of i_{DS}, width–length ratio W/L, and transconductance parameter K':

$$v_{GS} = v_T + \sqrt{\frac{2i_{DS}}{K'(W/L)}} \equiv v_T + V_{DS(SAT)} \tag{7.4}$$

So, since the initial tolerance of v_T is approximately ±50–±100 mV about 0.4–0.6 V and that of K' is ±20 percent, v_{GS}'s tolerance can be about ±5 to ±10 percent. Plus, although not as effective as v_D's logarithm, the square root in v_{GS} suppresses i_{DS} variations. So as a whole, tolerance and current-induced variations in v_{GS} are both higher than in v_D. And even though a logarithm also suppresses i_{DS} variations in v_{GS} when operating in subthreshold because i_{DS} is an exponential function of v_{GS}, the effect of v_T variations in v_{GS} is nevertheless substantial:

$$i_{DS(SUB)} \approx \left(\frac{W}{L}\right)I_{ST}e^{\frac{v_{GS}-v_T}{nV_t}} \tag{7.5}$$

where n is a process constant that ranges from 1.5 to 3.

In all, diode voltages vary less than gate–source voltages, and breakdown voltages are too high for many consumer applications. But since resistors convert these voltages to currents, tolerance and accuracy also depend on on-chip resistance. Polysilicon resistors are popular in this respect because their initial tolerance is usually lower at ±20 percent than those of their diffused counterparts. As a result, currents derived from forward-biased PN-junctions and converted with polysilicon resistors are usually the most precise. Currents with better than ±20-percent accuracy, however, require resistor adjustments after the fabrication process. Luckily, deriving voltages from diodes excludes resistor tolerance, as subsequent sections discuss, so they can be more accurate, but even better when trimmed after the fabrication process.

Fortunately, diodes are readily available in most process technologies. BJTs, however, which incorporate PN junctions across their base–emitter terminals, enjoy the added benefit of a third terminal, so BJTs can ultimately realize more compact circuits. In fact, when deriving bias currents and references from diodes, the only difference between bipolar and CMOS technologies is the flexibility that a collector terminal offers. Therefore, to cater to most process technologies, this chapter develops both BJT- and diode-derived circuits.

7.2 PTAT Current

In practice, almost all electrical characteristics drift with temperature. In circuit design, cancelling the variation is not always as important as predicting it. The *thermal voltage* V_t is good in this respect, because V_t reduces to zero at 0 Kelvin and climbs linearly with temperature T:

$$V_t = \frac{KT}{q} \tag{7.6}$$

where K is the Boltzmann constant at 1.38×10^{-23}, q is charge at 1.60×10^{-19}, and V_t is 25.9 mV at 300 K or 27°C. In other words, V_t is *proportional to absolute temperature*, or PTAT for short. V_t is appealing because its PTAT behavior is predictable, reliable, and linear across a wide range of temperatures. And biasing diodes, BJTs, and MOSFETs in subthreshold with a PTAT current I_{PTAT} can cancel the V_t effects already present in their small-signal transconductances g_m, which reduce to I_{PTAT}/V_t.

Luckily, V_t appears often in semiconductor theory. In fact, because the diode voltage v_D is a logarithmic function of its current i_D, the voltage difference Δv_D between two diodes reduces to V_t and the logarithm of the ratio of their respective current densities i_{D1}/A_{D1} and i_{D2}/A_{D2}:

$$\Delta v_D = v_{D1} - v_{D2} \approx V_t \ln\left(\frac{i_{D1} I_{S2}}{I_{S1} i_{D2}}\right) = V_t \ln\left(\frac{i_{D1} A_{D2}}{A_{D1} i_{D2}}\right) = v_{\text{PTAT}} \tag{7.7}$$

where reverse-saturation current I_S is proportional to the cross-sectional area A_D of the PN junction that carries i_D. If areas and currents match and their combined ratio is greater than 1, Δv_D reduces further to V_t and a nonzero constant. Or more generally, the voltage difference of two matched diodes is PTAT.

Because subthreshold currents in MOSFETs are exponential functions of their gate–source voltages with a similar dependence on V_t, subthreshold MOSFETs enjoy similar features. As such, the difference between two gate–source voltages Δv_{GS} reduces to V_t and the logarithm of the ratio of their respective current densities $i_{DS1}/(W/L)_1$ and $i_{DS2}/(W/L)_2$:

$$\Delta v_{GS} = v_{GS1} - v_{GS2} \approx nV_t \ln\left[\frac{i_{DS1}\left(\frac{W}{L}\right)_2}{\left(\frac{W}{L}\right)_1 i_{DS2}}\right] = v_{PTAT} \tag{7.8}$$

As before, if the currents and the width–length ratios of their channels match and their combined ratio is greater than 1, Δv_{GS} reduces further to V_t and a nonzero constant. Or stated differently, the difference of two matched gate–source voltages in subthreshold is PTAT.

While diode voltages remain logarithmic functions of current across a wide range of currents, gate–source voltages do not. Unfortunately, high tolerance in threshold voltages exacerbates the problem because including margin constrains the range further. And since modeling the behavior of MOSFETs when they transition from subthreshold to strong inversion is often involved, simulations are normally less accurate in subthreshold. All this is yet another reason why diodes are usually more reliable reference candidates than MOSFETs.

Ultimately, PTAT currents i_{PTAT} are resistor-derived translations of V_t, and more specifically, directly proportional to V_t and inversely proportional to a resistance R_P. This means a temperature drift in the resistance of the translating resistor R_P alters the overall drift in i_{PTAT}. So resistance should drift as little as possible.

7.2.1 Cross-Coupled Quad

Since V_t results from the difference of two matched diode voltages, the difference of four matched diode voltages should similarly produce V_t. The four cross-coupled BJTs in Fig. 7.1 exploit this trait to generate a PTAT current i_{PTAT}. Here, Q_1, Q_2, Q_3, and Q_4 impress the difference of their base–emitter voltages Δv_{BE} across resistor R_P to establish a resistor current i_R that is equivalent to

$$\begin{aligned} i_R &= \frac{\Delta v_{BE}}{R_P} = \frac{v_{BE1} + v_{BE4} - v_{BE3} - v_{BE2}}{R_P} \\ &= \frac{V_t}{R_P}\ln\left(\frac{i_1 i_4 A_3 A_2}{A_1 A_4 i_3 i_2}\right) \approx \frac{V_t}{R_P}\ln(C_2 D_3) = i_{PTAT} \end{aligned} \tag{7.9}$$

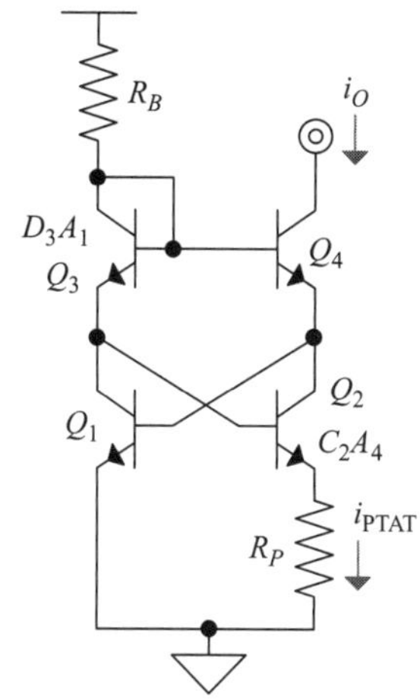

FIGURE 7.1 Cross-coupled quad PTAT generator.

Since Q_3's emitter area A_3 is D_3 times larger than Q_1's A_1, Q_2's A_2 is C_2 times larger than Q_4's A_4, and the same current flows through Q_1 and Q_3 and similarly through Q_2 and Q_4, currents and areas in the logarithm reduce to a constant C_2D_3, so i_R becomes PTAT. This, of course, assumes R_P drifts little with temperature.

The underlying feature of this circuit resides in the cancelling symmetry of the cross-coupled connection. Because the base–emitter voltages of Q_1 and Q_3 match and subtract in the diode loop they in part realize, supply-voltage variations, which manifest as changes in the current that bias resistor R_B channels into Q_1 and Q_3, cancel. This means the circuit is largely insensitive to the supply. Plus, since Q_1 and Q_2 emitter-degenerate Q_3 and Q_4 and Q_3 and Q_4 impress voltages at the collectors of Q_1 and Q_2 that for the most part match, the circuit hardly suffers from base-width modulation errors.

The *cross-coupled quad*, however, is not free of drawbacks. For one, Q_4's and Q_1's base currents do not match, and Q_2 adds a base current to R_P. As a result, Q_4's output current i_O is not equal to Q_2's collector current i_{C2}, and i_{C2} is less than R_P's i_{PTAT}. In other words, the circuit suffers from base-current beta β_0 errors. What is more, the circuit requires a supply that is higher than two diode voltages to establish a bias current through R_B, and therefore, through Q_3 and Q_1. *Headroom*, as a result, which refers to the minimum supply voltage that a circuit can tolerate, is not low.

7.2.2 Latched Cell

Fortunately, replacing the cross-coupled connection with a current mirror simplifies the base–emitter loop to the extent that headroom requirements ease. In the circuit of Fig. 7.2, for example, two BJTs Q_{P1}

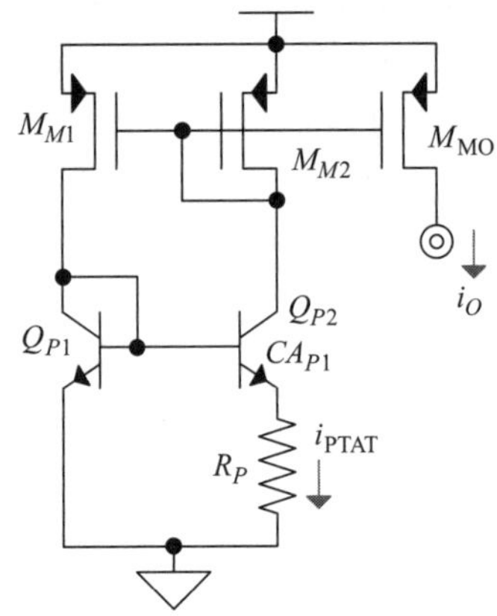

FIGURE 7.2 Latched BJT PTAT cell.

and Q_{P2} impress the difference of their base–emitter voltages Δv_{BE} across resistor R_P to establish a resistor current i_R that is equivalent to

$$i_R = \frac{\Delta v_{BE}}{R_P} = \frac{V_t}{R_P}\ln\left(\frac{i_{P1}A_{P2}}{A_{P1}i_{P2}}\right) \approx \frac{V_t}{R_P}\ln(C) = i_{PTAT} \tag{7.10}$$

Since Q_{P2}'s emitter area A_{P2} is C times larger than Q_{P1}'s A_{P1} and the mirror that M_{M1} and M_{M2} implement ensures that the currents through Q_{P1} and Q_{P2} are equal, currents and areas in the logarithm reduce to a constant C, which means i_R becomes PTAT. As before, this assumes R_P drifts little with temperature. The point with all this is that M_{M1} and Q_{P1} only require one source–drain voltage v_{SD} and one base–emitter voltage v_{BE} to operate, which is the fundamental headroom limit for analog integrated circuits.

One peculiarity here is that the circuit is bi-stable. More specifically, the mirror and the PTAT-generating transistors balance both when i_R is i_{PTAT} and when i_R is zero. Zero, of course, is an unacceptable state for a PTAT generator. Engineers therefore include startup features that ensure the circuit is never off, which is the subject of Section 7.5.

Another, though not unrelated idiosyncrasy in the *latched PTAT cell* is that it incorporates positive feedback. To see this, consider that Q_{P2} feeds the gates of M_{M2} and M_{M1}, which in turn drive the bases of Q_{P1} and Q_{P2} to close a feedback loop. The resulting loop is noninverting because M_{M1}'s and Q_{P2}'s gate–drain and base–collector terminals introduce two inversions. Interestingly, positive feedback is good here because, in the presence of external influences, the circuit's tendency is to latch to one of its stable states. Startup circuits, in fact, rely on this characteristic to start the cell. What is more, this feature is the reason this PTAT generator is largely insensitive to supply fluctuations.

This circuit, however, is not free of drawbacks. For one, Q_{P1}'s and Q_{P2}'s collector voltages and M_{M1}'s and M_{M2}'s drain voltages do not match across supply variations. The transistors therefore suffer from base-width and channel-length modulation errors. Plus, Q_{P1} and Q_{P2} draw two base currents away from Q_{P1}'s collector, which create an undesirable imbalance in collector currents in the BJTs. In other words, the circuit also suffers from base-current beta β_0 errors.

Example 7.1 What resistance in the latched BJT PTAT cell establishes 5 μA of PTAT current at room temperature when the ratio of emitter areas between the matched transistors is 8?

Solution

$$R_P = \frac{V_t}{i_{\text{PTAT}}}\ln(C) = \left(\frac{26m}{5\mu}\right)\ln(8) = 10.8\ \text{k}\Omega$$

MOS Variant

Because MOSFET currents in subthreshold are exponential functions of their corresponding gate–source voltages, replacing the BJTs in Fig. 7.2 with MOSFETs and ensuring the MOSFETs operate in subthreshold similarly establish a PTAT current. In the circuit of Fig. 7.3, for example, M_{P1} and M_{P2} impress the difference of their gate–source voltages Δv_{GS} across resistor R_P to define a resistor current i_R that is equivalent to

$$i_R = \frac{\Delta v_{GS}}{R_P} \approx \frac{nV_t}{R_P}\ln\left[\frac{i_{P1}\left(\frac{W}{L}\right)_{P2}}{\left(\frac{W}{L}\right)_{P1} i_{P2}}\right] = \frac{nV_t}{R_P}\ln(C) = i_{\text{PTAT}} \tag{7.11}$$

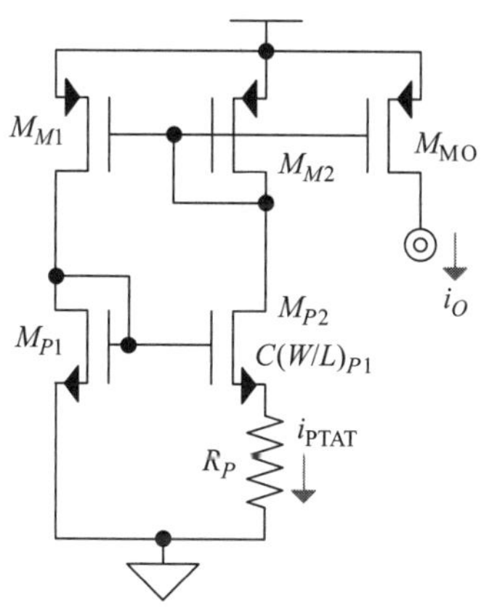

Figure 7.3 Latched MOS PTAT cell.

Since M_{P2}'s width–length ratio $(W/L)_{P2}$ is C times greater than M_{P1}'s $(W/L)_{P1}$ and the mirror that M_{M1} and M_{M2} implement ensures that the currents through M_{P1} and M_{P2} match, currents and aspect ratios in the logarithm reduce to a constant C, which means i_R becomes PTAT.

As in the original latched PTAT cell, R_P's drift across temperature alters the PTAT behavior of the output current, the mirror eases headroom requirements to the lowest possible limit, the circuit is bi-stable, positive feedback desensitizes the circuit to supply fluctuations, and the difference in drain voltages causes a channel-length modulation error. The only operational difference in this rendition is that MOS gates do not draw gate current, so base-current errors are absent. As mentioned earlier, though, with respect to their diode-based counterparts, MOSFETs remain in subthreshold for a shorter range of currents, and their tolerance is higher.

Error-Compensated BJT Cell

The circuit in Fig. 7.4 modifies the mirror from Figs. 7.2 and 7.3 to correct the errors that base-width and channel-length modulation and base currents produce. To understand how the circuit works, first recall that M_{M2}'s diode connection in Figs. 7.2 and 7.3 establishes the negative feedback that allows the circuit to mirror current. In this light, buffering M_{M2}'s gate-to-drain connection with a noninverting circuit achieves similar results. The purpose of bias transistors Q_{B1} and M_{B2} in Fig. 7.4 is just this, to implement the feedback connection that prompts M_{M1} to mirror M_{M2}'s current. The loop is inverting because Q_{P1}'s and Q_{B1}'s base–collector and M_{M2}'s gate–drain terminals invert loop signals three times. Note, by the way, that M_{M1} together with Q_{B1} and M_{B2} still establish the positive feedback that latches the circuit into one of its stable states.

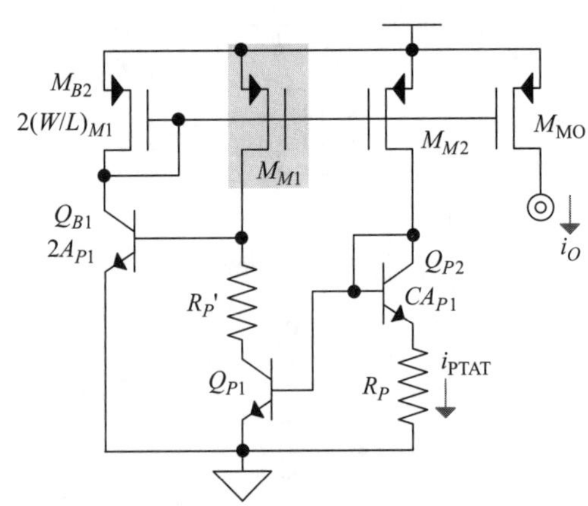

FIGURE 7.4 Error-compensated BJT PTAT cell.

The reason for changing the feedback connection around M_{M2} is to fix the deficiencies of the original BJT cell. For one, because Q_{B1} conducts twice the current that Q_{P1} and Q_{P2} each carry, Q_{B1} draws as much base current from Q_{P1}'s collector as Q_{P1} and Q_{P2} together pull from Q_{P2}'s collector, so base currents drawn from Q_{P1}'s and Q_{P2}'s collectors match and balance. Plus, R_P' and R_P and Q_{B1}, Q_{P1}, and Q_{P2} match to establish similar voltages across the collector–emitter terminals of Q_{P1} and Q_{P2} and the drain–source terminals of M_{M1} and M_{M2}, so base-width and channel-length modulation errors also disappear. In other words, the error-compensated PTAT cell enjoys all the benefits of the original and none of the shortcomings. The only drawback is the interaction and management of the negative- and positive-feedback loops that result, which is the subject of another section.

Diode Variant

Figure 7.5 shows yet another variant of the PTAT cell. As in others, MOS mirror M_{M1}–M_{M2} ensures that gate-coupled transistors M_{B1} and M_{B2} carry the same current. This way, M_{B1}'s and M_{B2}'s gate–source voltages match, so M_{B1} senses D_{P1}'s voltage and M_{B2} impresses D_{P1}'s voltage across R_P and D_{P2}. The net result is the voltage across R_P is the difference between two matched diode voltages, which means R_P's current i_R is PTAT.

As in the original cell, R_P's drift across temperature alters the PTAT behavior of the output current, the circuit is bi-stable, positive feedback desensitizes the circuit to supply fluctuations, and the difference in the mirror's drain voltages causes a channel-length modulation error. And as in the MOS counterpart, base-current errors are absent. The disadvantage here is the supply must surpass D_{P1}'s diode voltage v_{DP1}, M_{B1}'s gate–source voltage

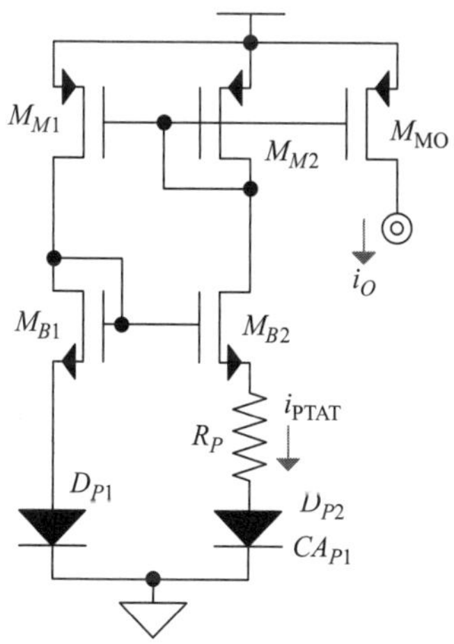

FIGURE 7.5 Latched diode PTAT cell.

v_{GSB1}, and M_{M1}'s source–drain voltage v_{SDM1} for the circuit to operate, so headroom is higher than in the previous two cases. The advantage of this latter circuit over the previous MOS variant, however, is that diodes produce a PTAT behavior across a wider range of currents, and the initial tolerance that results is lower and comparable to BJTs.

Error-Compensated Diode Cell

The circuit in Fig. 7.6 is a more general case of the diode variant in Fig. 7.5. The reason for this analogy is because, like M_{B1} and M_{B2} in Fig. 7.5, the differential amplifier A_G in Fig. 7.6 impresses D_{P1}'s voltage across R_P and D_{P2}. In fact, A_G not only series-mixes D_{P1}'s voltage with the combined voltage across R_P and D_{P2} but also establishes the negative-feedback path that diode-connects M_{M2}. In other words, A_G's input terminal voltages equal and M_{M1} mirrors M_{M2}'s current to ensure diode currents match.

The main advantage of this circuit is that the drain voltages of the mirror now match, so channel-length modulation error disappears. A more subtle merit is that A_G decouples diode voltages from the mirroring gates, so they no longer combine to limit headroom. In this case, either D_{P1}'s diode voltage and A_G's input common-mode range or M_{M2}'s source–gate voltage and A_G's output-voltage swing limit how low the supply voltage can drop before the circuit stops working. In other words, the error-compensated diode cell enjoys all the benefits of the original diode cell and, possibly, none of the shortcomings. As in the compensated BJT cell, the only drawback is the interaction and management of the inverting and noninverting feedback loops around M_{M2} and M_{M1}, which is the subject of Section 7.6.

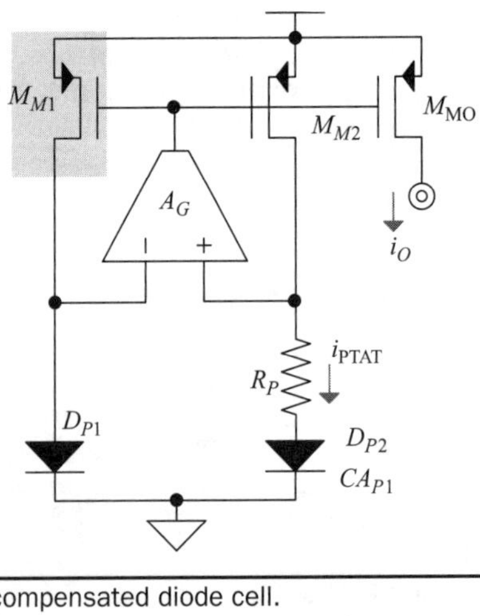

FIGURE 7.6 Error-compensated diode cell.

7.3 CTAT Current

Because currents that fall linearly with temperature oppose the behavior of those that are PTAT, engineers refer to this falling trait as *complementary to absolute temperature* (CTAT). The fundamental purpose of these CTAT currents is to cancel or offset the PTAT tendencies of other parameters, including thermal voltage V_t and PTAT currents. In this respect, the diode voltage v_D is again useful because v_D falls with temperature at roughly –2.2 mV/°C. Although v_D is not as linear across temperature when v_D falls as V_t is when V_t rises, v_D's first-order linear component is sufficiently stronger than higher-order components to be dominant.

As a side note, and for reference, MOS threshold voltages v_T and transconductance parameters K' usually exhibit CTAT drifts. When compared to v_D, however, extracting v_T and K' requires more circuit complexity, which raises risk, power consumption, and silicon area. Plus, initial tolerance in MOS devices is higher, so v_D is more reliable.

Similar to the PTAT case, generating a CTAT current amounts to impressing a diode voltage v_D across a resistor R_C. In this way, R_C's current i_R is the ratio of v_D and R_C:

$$i_R = \frac{v_D}{R_C} = \frac{V_t}{R_C}\ln\left(\frac{i_D}{I_S}\right) = i_{\text{CTAT}} \tag{7.12}$$

v_D, like before, decomposes into a logarithm of the ratio of diode current i_D and reverse-saturation current I_S. Unlike earlier, however, i_R's voltage is not the difference of two diode voltages, so I_S in the logarithm of v_D includes the temperature effects of mobility, diffusion, and other parameters that, together with V_t, produce a diode voltage that falls, for the most part, linearly with temperature. In fact, i_D is usually a PTAT bias current I_{PTAT} partially because a PTAT current can offset some of the nonlinear effects that I_S in the logarithm creates. More importantly, though, systems normally include a PTAT-current generator, so I_{PTAT} is already available. Regardless, i_R is CTAT to the extent that R_C's drift across temperature allows it.

7.3.1 Current-Sampled BJT

One way of generating a CTAT current is by sampling the current that flows through a resistor attached to the base–emitter terminals of a biased BJT. In the case of Fig. 7.7, for example, M_S is a voltage-following buffer that diode-connects Q_C to establish a diode voltage v_D across Q_C's base–emitter terminals, and as a result, across the resistor R_C between those terminals. This way, R_C's current i_R is proportional to v_D, and if R_C drifts little with temperature, i_R is therefore mostly CTAT.

The key to this circuit is how M_S samples Q_C's current. More to the point, M_S decouples Q_C's collector and base to steer and feed R_C's current i_R to a receiving mirror M_M–M_{MO} that outputs a replica of i_R.

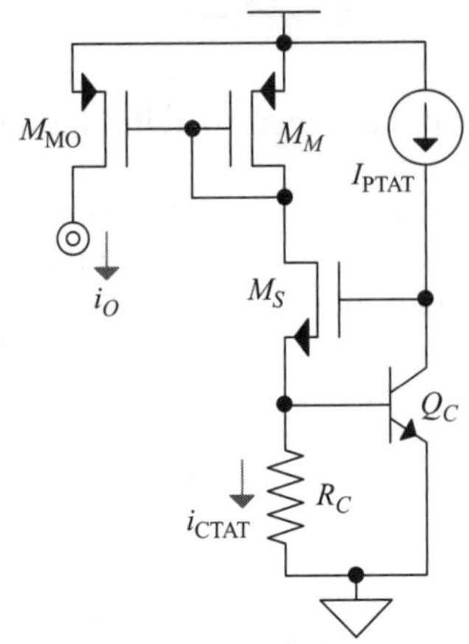

FIGURE 7.7 Current-sampled BJT CTAT current generator.

The bias current I_{PTAT} that feeds Q_C and M_S's gate is PTAT for the reasons mentioned earlier. One reason is that I_{PTAT} is probably already available, and the other is that I_{PTAT} can offset some of the nonlinear effects that I_S in the logarithm of v_D produces. One drawback to this circuit is headroom because the supply must accommodate M_M's source–gate voltage v_{SG}, M_S's drain–source voltage v_{DS}, and Q_C's base–emitter voltage v_{BE}, which is one more v_{SG} or v_{BE} than the fundamental lower limit for analog integrated circuits.

7.3.2 Voltage-Sampled Diode

In the absence of BJTs, a more general way of impressing a diode voltage across a resistor is with the feedback action of a differential amplifier. In the case of Fig. 7.8, for example, differential amplifier A_G and mirroring

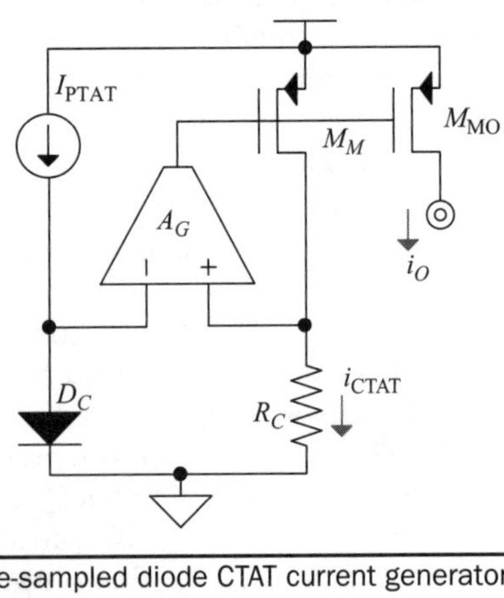

FIGURE 7.8 Voltage-sampled diode CTAT current generator.

transistor M_M establish an inverting feedback loop that series-mixes D_C's diode voltage v_D with the voltage across resistor R_C. Here, M_M decouples A_G's output from R_C to sample all of i_R. This way, A_G ensures that R_C's voltage mirrors v_D and M_M's output transistor M_{MO} outputs a replica of R_C's i_R.

Like before, PTAT current I_{PTAT} biases diode D_C for two reasons. One is that a PTAT-current generator is more than likely already available in the system. The other is that a PTAT current offsets some of the nonlinear effects that I_S produces in D_C's v_D. A_G, by the way, decouples D_C's v_D from M_M's source–gate voltage, so A_G's input common-mode range and output-voltage swing ultimately determine the minimum headroom limit of the circuit. Also notice that i_R and output current i_O are, like before, mostly CTAT, but again, only to the extent that R_C's drift across temperature allows.

7.4 Temperature Compensation

Since all electrical parameters in microelectronic devices ultimately drift with temperature, temperature independence only results when the dependence of one or more parameters offsets and cancels those of others. From this perspective, any two components whose drifts match and oppose can combine to produce a result that is largely independent of temperature. But when also aiming for accuracy, only predictable and reliable voltages or currents can produce a precise result.

Perhaps the most reproducible and reliable signal available is the thermal voltage V_t, which the voltage difference of two matched diodes Δv_D extracts. With roughly ±2 percent of initial tolerance, the diode voltage v_D is probably the second most precise, and after that at ±2–±4 percent, the diode's breakdown voltage V_{BD}, except V_{BD} is normally too high at 6–7 V to be useful in consumer electronics. MOS gate–source voltages are next with roughly ±5–±10 percent of initial variance. Luckily, the temperature drifts of the top two candidates in this list oppose, so combining appropriate proportions of them can yield temperature independence.

Since V_t is linear and v_D is not, V_t can only cancel the linear component of v_D. In other words, when V_t and v_D balance, nonlinearities in v_D surface in the compensated result. To see these nonlinearities, consider the Taylor-series expansion of the logarithm in v_D, which reveals that v_D reduces to the *bandgap voltage* V_{BG} of silicon, a first-order component of temperature T, and higher-order terms that a $T \ln T$ component captures:

$$v_D \approx V_{BG} - \left(\frac{V_{BG} - v_{D(ROOM)}}{T_{ROOM}}\right)T^1 - (\eta - x)V_t \ln\left(\frac{T}{T_{ROOM}}\right) \tag{7.13}$$

where V_{BG} is about 1.2 V, T_{ROOM} is room temperature, $v_{D(ROOM)}$ is v_D at T_{ROOM}, η is a process constant that is close to 4, and x is the order of the current flowing through the diode that establishes v_D, so if the current is PTAT and therefore proportional to T^1, x from T^x is 1.

From this relationship, first note that v_D is V_{BG} at 0 K. Secondly, the coefficient of v_D's first-order component T^1 is negative, and its slope is basically v_D's drop from V_{BG} to $v_{D(ROOM)}$ across 0 K through T_{ROOM}. Also notice that the coefficient of the $T \ln T$ term is negative, so v_D tends to fall faster with temperature at higher temperatures. Plus, a PTAT current reduces the impact of the $T \ln T$ term because x subtracts 1 from η's 4.

To generate a first-order temperature-independent current i_{REF}, a V_t-derived PTAT current i_{PTAT} must cancel the negative first-order term of a v_D-derived CTAT current i_{CTAT}. For this, appropriate balancing proportions of i_{CTAT} and i_{PTAT} should add to generate i_{REF}:

$$i_{REF} = i_{CTAT} + i_{PTAT} = \frac{v_D}{R_C} + \frac{\Delta v_D}{R_P} = \frac{v_D}{R_C} + \frac{V_t}{R_P}\ln(C)$$

$$\equiv \frac{V_{BG}}{R_C} - \left[\frac{(\eta - 1)V_t}{R_C}\right]\ln\left(\frac{T}{T_{ROOM}}\right) \approx \frac{V_{BG}}{R_C} \tag{7.14}$$

where a PTAT current flows through a diode to generate v_D. i_{REF} therefore reduces to the constant that V_{BG}/R_C sets and the nonlinearity that the $T \ln T$ term produces. To see this graphically, consider how i_{PTAT} in Fig. 7.9 cancels i_{CTAT}'s linear component to produce a sum i_{REF} that is mostly, but not perfectly, flat across temperature. In fact, $T \ln T$'s nonlinear tendency to fall causes the convex curvature in i_{REF}.

Since temperature independence from 0 K is not necessary, designers center the reference across the range that their system requires. And since i_{REF} from Fig. 7.9 rises less than it falls from T_{MIN} to T_{MAX}, more i_{PTAT} is necessary to center the new i_{REF}', so the resulting i_{REF}' is slightly higher than i_{REF}. In other words, the magnitude of the optimum reference not only changes with temperature range but is also slightly higher than the bandgap-derived counterpart V_{BG}/R_C.

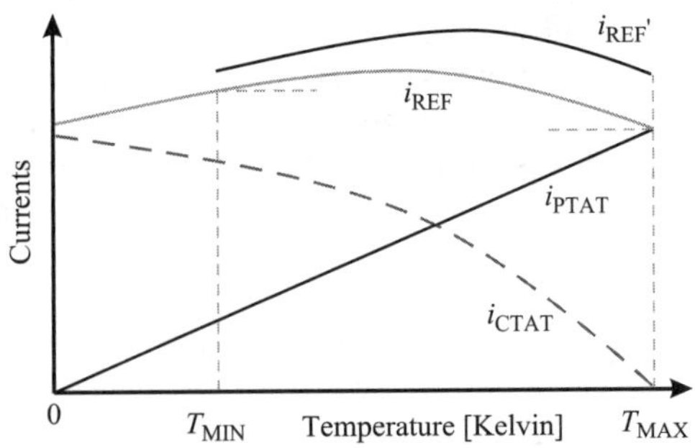

FIGURE 7.9 Temperature compensation.

But since i_{REF}' still nears V_{BG}/R_C and V_{BG} is roughly 1.2 V, i_{PTAT} at room temperature $i_{PTAT(ROOM)}$ should be roughly the bandgap-derived current V_{BG}/R_C minus the diode-derived current i_{CTAT} at room temperature $v_{D(ROOM)}/R_C$:

$$i_{PTAT(ROOM)} = \frac{V_{t(ROOM)}}{R_P}\ln(C) \approx \frac{V_{BG}}{R_C} - \frac{v_{D(ROOM)}}{R_C} \tag{7.15}$$

where $V_{t(ROOM)}$ is V_t at T_{ROOM}. This means R_C/R_P should be approximately the difference of V_{BG} and $v_{D(ROOM)}$ over $V_{t(ROOM)}$ ln C:

$$\frac{R_C}{R_P} \approx \frac{V_{BG} - v_{D(ROOM)}}{V_{t(ROOM)}\ln(C)} \tag{7.16}$$

where $v_{D(ROOM)}$ is normally between 0.6 and 0.7 V. As a result, R_C should be about 10 times greater than R_P when C is 8.

Example 7.2 What rough PTAT- and CTAT-generating resistances produce a temperature-compensated 5-μA reference current when the diode voltage at room temperature is 0.62 V and the area ratio of the matched diodes is 8?

Solution

Since

$$\frac{R_C}{R_P} \approx \frac{V_{BG} - v_{D(ROOM)}}{V_{t(ROOM)}\ln(C)} \approx \frac{1.2 - 0.62}{(26\text{m})\ln(8)} = 10.73$$

and

$$R_C \approx \frac{V_{BG}}{i_{REF}} = \frac{1.2}{5\mu} = 240\ \text{k}\Omega$$

$$R_P \equiv R_C\left(\frac{R_P}{R_C}\right) = (240\text{k})\left(\frac{1}{10.73}\right) = 22.4\ \text{k}\Omega$$

With these approximations in hand, simulate and adjust R_P or R_C until i_{REF} flattens across temperature.

7.4.1 Error-Compensated BJT Current Reference

Integrating CTAT currents into a BJT circuit amounts to adding resistors between the base and emitter terminals of BJTs. In the case of Fig. 7.10, for instance, whose predecessor is the error-compensated BJT PTAT cell of Fig. 7.4, resistors R_{C1} and R_{C2} establish CTAT currents that add to the PTAT components that Q_{P1} and Q_{P2} already generate. As such, the feedback loop that establishes M_{M2}'s current receives and mirrors the combined temperature-compensated sum, which M_{MO} outputs in the form of reference current i_{REF}:

$$i_{REF} \approx \frac{v_{BE}}{R_C} + \frac{\Delta v_{BE}}{R_P} = i_{CTAT} + i_{PTAT} \tag{7.17}$$

where Q_{B1}'s v_{BEB1} and Q_{P1}'s v_{BEP1} match and equal v_{BE} and R_{C1} and R_{C2} similarly match and equal R_C. The key feature here is that, other than the addition of R_{C1} and R_{C2}, no other circuit modifications are necessary.

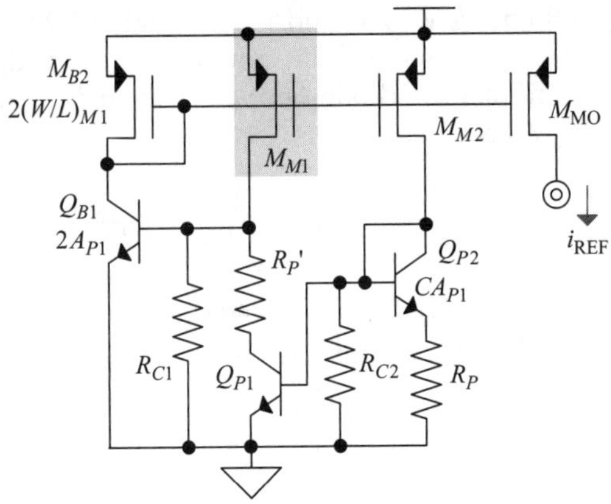

FIGURE 7.10 Error-compensated BJT current reference.

7.4.2 Diode-Derived Current Reference

Integrating CTAT currents into diode circuits is similar. Converting the diode-derived PTAT cell of Fig. 7.5 to incorporate CTAT currents, for example, amounts to attaching resistors R_{C1} and R_{C2} to the MOS source terminals that sample and reflect D_{P1}'s diode voltage v_{D1}, as Fig. 7.11 shows. This way, with v_{D1} across R_{C1} and R_{C2}, R_{C1} and R_{C2} conduct CTAT currents. And because M_{B2}'s source voltage remains v_{D1}, the voltage across R_P is still the voltage difference of the diodes, which means R_P's current is PTAT. So without further modifications to the circuit, R_{C1}'s and R_{C2}'s matched CTAT currents add to D_{P1}'s and

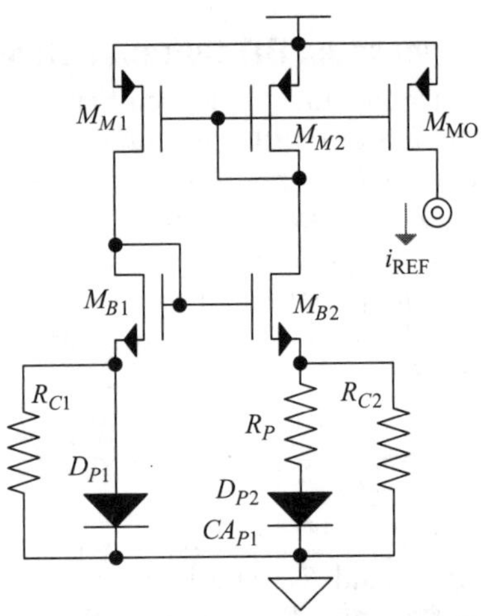

FIGURE 7.11 Diode-derived current reference.

D_{P2}'s PTAT counterparts. M_{B1} and M_{B2} then buffer and feed the combined temperature-compensated sum to the MOS mirror, which M_{MO} outputs in the form of i_{REF}:

$$i_{REF} \approx \frac{v_D}{R_C} + \frac{\Delta v_D}{R_P} = i_{CTAT} + i_{PTAT} \tag{7.18}$$

7.4.3 Error-Compensated Diode-Derived Current Reference

In the error-compensated diode PTAT cell of Fig. 7.6, attaching resistors R_{C1} and R_{C2} to the amplifier's input terminals that sample and reflect D_{P1}'s diode voltage v_{D1}, as Fig. 7.12 demonstrates, establishes CTAT currents in R_{C1} and R_{C2}. And because the feedback circuit that A_G implements ensures that the voltage at its noninverting input is v_{D1}, the voltage across R_P is still the voltage difference of the two diodes, so R_P's current is PTAT. As a result, R_{C1}'s and R_{C2}'s matched CTAT currents add to D_{P1}'s and D_{P2}'s PTAT counterparts. This way, without further modifications to the circuit, A_G and M_{M2} generate the source–gate voltage necessary to sustain the temperature-compensated sum of the currents, which M_{MO} then outputs in the form of i_{REF}.

Although not obvious at first, one side benefit to this circuit is low headroom. Here, A_G can tap R_{C1} and R_{C2} at lower points, as the dashed gray lines in Fig. 7.12 illustrate, to series-mix a fraction of D_{P1}'s diode voltage v_{DP1}. This way, in ensuring that the tapped voltages match, the feedback loop also impresses v_{DP1} across R_{C2} and R_P and D_{P2}, so the functionality of the circuit remains intact. The advantage of this connection is that A_G's input common-mode range need only include low voltages, so A_G's headroom limit can be low.

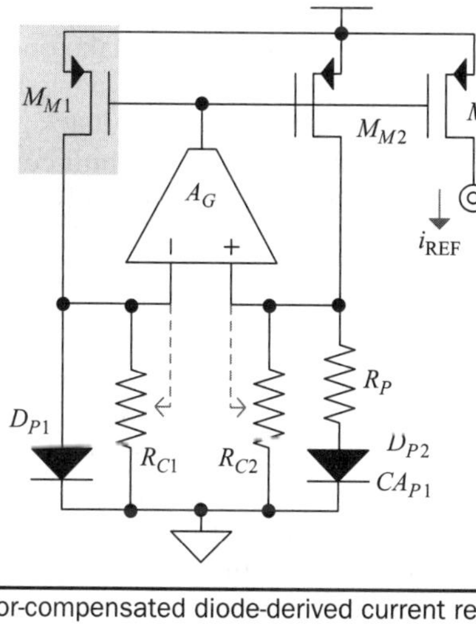

FIGURE 7.12 Error-compensated diode-derived current reference.

7.5 Startup

Basic PTAT cells incorporate a noninverting feedback loop that latches the circuit into one of two possible stable states. One is the PTAT state, when transistors carry the PTAT current that the voltage difference of two matched PN junctions establishes. The other is when all transistors are off, which is, of course, highly undesirable. Ensuring that the circuit latches to the PTAT state amounts to ensuring that the circuit is never off. One popular way of eliminating the possibility of an off state is injecting a small current into a diode-connected transistor in the positive-feedback loop. This way, the transistor can never shut, and with only one state left, the current prompts the positive-feedback path to regenerate and latch to the PTAT state.

Basic PTAT cells use a diode-connected transistor to command other matched transistors to mirror its current. The diode connection, as mentioned earlier and discussed in more detail in Chap. 6, realizes an inverting feedback loop. This means the latched cell incorporates two feedback loops, one positive and the other negative, and is stable in one of two possible states. Unfortunately, extending the diode connection to include other stages can not only compromise the stability of the inverting feedback loop but also create other stable states for the circuit. Adding more loops to the circuit, of course, only exacerbates the problem. For these reasons, engineers employ simpler and fewer loops whenever possible.

When more than two stable states exist, injecting currents into multiple points in the circuit and slowing selected loops with RC filters help. Still, starting a circuit with several stable states can be cumbersome. So much so, in fact, that resorting to an iterative and empirical trial-and-error process of repeated computer simulations is not uncommon in the industry. By far, the best solution is to eliminate and simplify feedback loops.

Fundamentally, startup circuits usually take one of two general forms. One is to push or pull currents continuously to or from diode-connected transistors in the noninverting feedback path. The other is for the circuit to sense its state and push or pull currents to or from similar diode-connected devices on demand, only when needed.

7.5.1 Continuous Startup

Since the PTAT cell is more than likely the primary current generator in the system, currents are not readily available for the startup function. Crude currents are therefore the only options left. Unfortunately, introducing a current whose tolerance can be ±50 percent compromises the accuracy of the PTAT current. So to keep errors from mounting, startup currents are usually about an order of magnitude lower than other currents in the circuit across temperature and process corners.

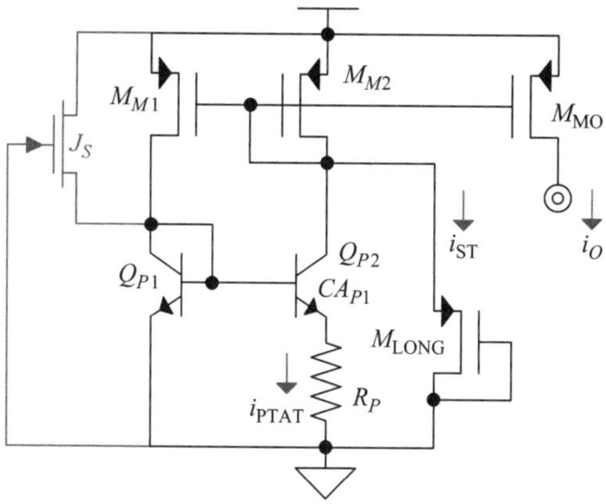

FIGURE 7.13 Continuous startup in the latched BJT PTAT cell.

One way of establishing a crude current is to insert a resistor across the supplies. In the case of PTAT cells, the resistor connects to the supplies via one of the diode-connected transistors in the noninverting feedback loop. For the current to be low, though, resistance must be high, so lowly doped semiconductor layers, such as diffused well and base regions, are good candidates. Pinching these resistors with oppositely doped layers, which is the recipe for building junction field-effect transistors (JFETs), produces even higher resistances. Photolithography in modern CMOS technologies is so fine, though, that long- and narrow-channel MOS transistors can exhibit higher resistances with less silicon area. And since the mobility of holes is lower than that of electrons, P-type devices are popular for this purpose.

In the case of Fig. 7.13, two diode-connected transistors—a minimum-width long-channel PFET M_{LONG} and the mirror's input device M_{M2}—connect across the supplies to establish a startup current i_{ST}. M_{LONG}'s channel is so narrow and long that i_{ST} is low, but still high enough to trigger the noninverting feedback loop into action. Because Q_{P1} is also in the loop, connecting a long and narrow diode-connected PFET or pinched JFET from the supply to Q_{P1} can similarly start the circuit, if again, the startup current is high enough to start the loop.

Irrespective of the means, a *continuous startup current* i_{ST} normally creates an imbalance in the circuit that produces an error in the PTAT current. Although inserting a balancing startup current into the other side of the PTAT cell is possible, the mismatch between the two is normally much greater than between other matched devices in the circuit. The best way to mitigate the impact of i_{ST} is to keep i_{ST} away from the matching PN junctions and their corresponding mirror transistors.

In the error-compensated BJT PTAT cell of Fig. 7.14, for example, M_{LONG} draws i_{ST} from the intermediate buffer stage that implements M_{M2}'s diode connection. This way, M_{M1} only conducts Q_{P1}'s current

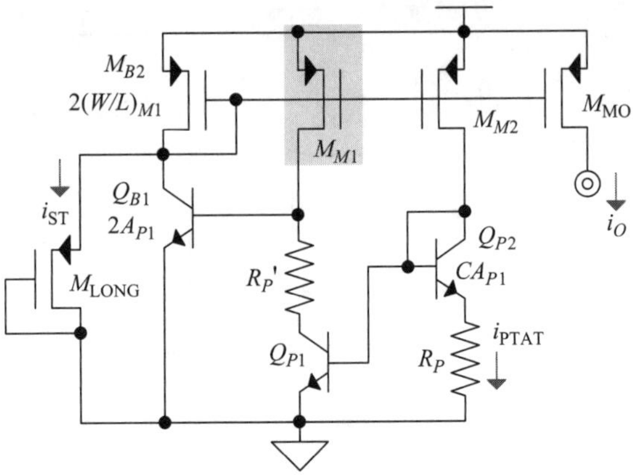

Figure 7.14 Continuous startup in the error-compensated BJT PTAT cell.

and M_{M2} only carries that of Q_{P2}, so the resulting currents match. But since M_{B2}'s current mirrors those of M_{M1} and M_{M2} and M_{LONG} draws i_{ST} from M_{B2}, Q_{B1} carries less current than it should under ideal conditions. As a result, Q_{B1} neither draws as much base current nor establishes as much base–emitter voltage to match those of Q_{P1} and Q_{P2} perfectly. Fortunately, Q_{B1}'s β_0 and exponential relationship suppress the variance. Still, ensuring that i_{ST} is low is important to render the errors that i_{ST} spawns in PTAT current i_{PTAT} negligible. Plus, i_{ST} must be well below M_{B2}'s i_{PTAT} to avoid starving Q_{B1} of current.

Example 7.3 What P-channel metal–oxide–semiconductor (PMOS) width and length dimensions for a diode-connected PFET can generate 250 nA at room temperature when impressing 2.5 V across the device and its threshold voltage is 0.5 V, its transconductance parameter is 25 μA/V², the lowest possible length is 0.5 μm, and the lowest possible width is 2 μm?

Solution When diode connected, a MOSFET is in saturation, so

$$i_{SD} \approx 0.5\left(\frac{W}{L}\right)K_P'(v_{SG} - |v_{TP}|)^2$$

which with a minimum width for low current means channel length should be about 400 μm to conduct 250 nA:

$$L \approx \left(\frac{0.5W}{i_{SD}}\right)K_P'(v_{SG} - |v_{TP}|)^2 = \left[\frac{0.5(2\mu)}{(250n)}\right](25\mu)[(2.5)-(0.5)]^2 = 400\ \mu\text{m}$$

7.5.2 On-Demand Startup

One way to eliminate startup error is to insert a startup device that only engages on demand. To push or pull a startup current i_{ST} only when needed, though, the startup circuit must somehow sense the operating state of the PTAT cell. Luckily, PTAT current i_{PTAT} and the voltages across diode-connected transistors are both zero in the off

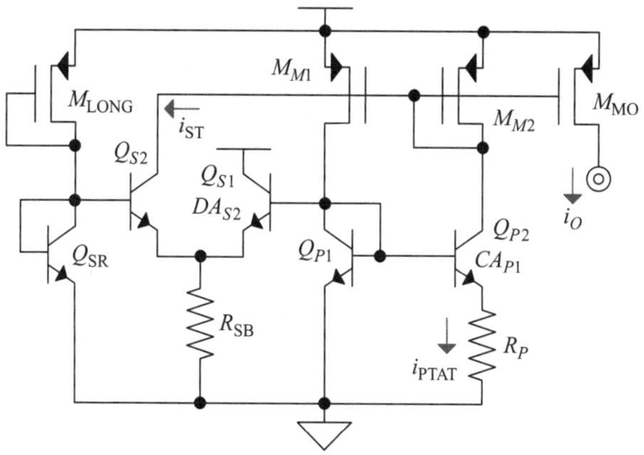

Figure 7.15 On-demand voltage-mode startup in the latched BJT PTAT cell.

state, so they can indicate the state of the cell. So if the circuit is off, a startup device can steer i_{ST} into a diode-connected transistor in the noninverting feedback loop.

Voltage Mode

Comparing a diode voltage v_D in the PTAT cell against a reference is one way of sensing the state of the circuit. In the latched PTAT cell of Fig. 7.15, for example, a long and narrow diode-connected PFET M_{LONG} and the diode in Q_{SR} connect across the supplies to establish both current flow and a reference diode voltage v_{DSR}. The differential pair that Q_{S1} and Q_{S2} realize then compares Q_{P1}'s diode voltage v_{DP1} against Q_{SR}'s v_{DSR}, and if v_{DP1} is well below v_{DSR}, near zero, Q_{S2} sinks startup current i_{ST} from the diode-connected transistor M_{M2} in the mirror. If on the other hand v_{DP1} nears v_{DSR}, most of the tail current that resistor R_{SB} establishes flows through Q_{S1} because Q_{S1}'s emitter area is D times larger than that of Q_{S2}. Notice, by the way, that the voltage across R_{SB} is the difference between Q_{SR}'s and Q_{S2}'s base–emitter voltages, so Q_{S1}–Q_{S2}'s tail current can be low. And although i_{ST} is not exactly zero in practice when the cell is in the PTAT state, i_{ST} is low enough to create a nearly negligible error in PTAT current i_{PTAT}.

Current Mode

Perhaps a more direct approach is to compare PTAT current i_{PTAT} against a crude current reference. In the latched PTAT cell of Fig. 7.16, for example, M_{MB} mirrors i_{PTAT} into a long and narrow diode-connected PFET M_{LONG}. This way, M_{LONG} pulls startup transistor M_S's gate to zero when i_{PTAT} is zero. As a result, the source–gate voltages of the diode-connected transistor in the mirror M_{M2} and M_S establish a startup current i_{ST}. But since M_{LONG}'s resistance is high, M_{MB}'s i_{PTAT} pulls M_S's gate voltage to shut M_S whenever the circuit is on, in the PTAT state.

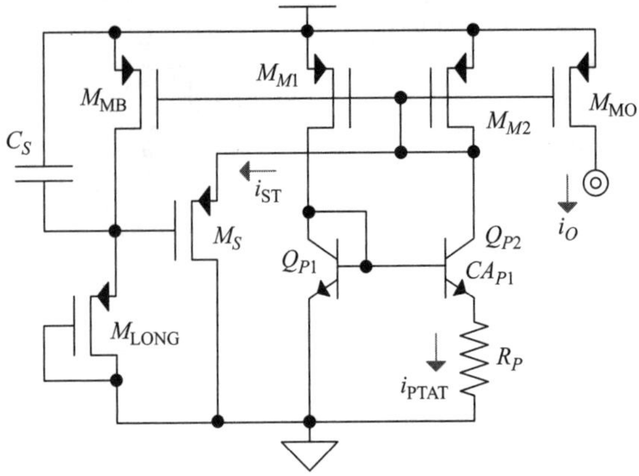

FIGURE 7.16 On-demand current-mode startup in the latched BJT PTAT cell.

One serious drawback to this circuit is noise sensitivity. The culprit for this vulnerability is the high-impedance node that results from connecting M_{LONG} to M_{MB}'s r_{sdB}. The problem is that noise energy into M_S's high-impedance gate can easily engage M_S, and as a result, disrupt the operation of the PTAT cell and the entire system it supports. The purpose of capacitor C_S is to keep M_S's gate from changing abruptly, to shunt noise energy away from M_S's gate.

7.6 Frequency Compensation

Latched PTAT cells normally incorporate two feedback loops. One is the noninverting loop that the matched PN junctions and the mirror establish. The other one is the inverting loop that diode-connects the mirroring transistor in the mirror. When this last loop is a short circuit, the diode connection is stable because the loop has no more than one node, and therefore, no more than one pole. As a result, once the startup circuit ensures that the noninverting feedback loop latches the cell to the PTAT state, the circuit stabilizes.

Inserting buffer and gain stages into the loop that diode-connects the mirroring device, however, compromises its stability. The error-compensated BJT cell, just to cite an example, introduces an additional node into the loop that both loops share. This not only adds a pole to the inverting loop but also integrates the noninverting loop into the inverting counterpart. In other words, the negative-feedback path includes positive-feedback components.

A way to see the interaction of these two loops is to consider the node they share. From this perspective, the loops are in parallel, and the loop with the highest gain across its path dominates how that node mixes signals. This means that stabilizing the inverting loop amounts to ensuring its loop gain A_{LG-} is greater than its noninverting

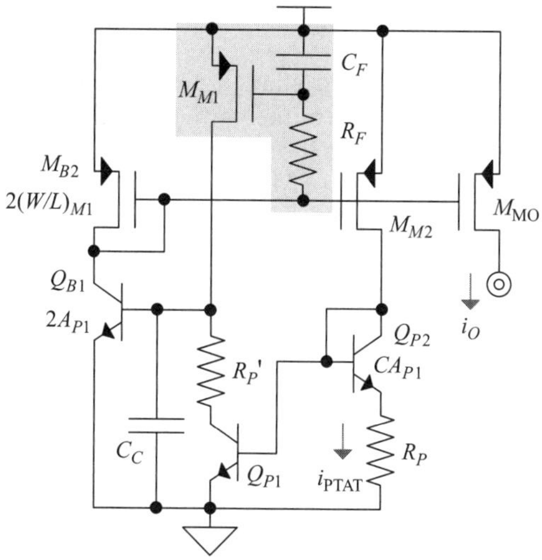

FIGURE 7.17 Error-compensated and stabilized BJT PTAT cell.

counterpart A_{LG+} through and slightly past A_{LG-}'s unity-gain frequency f_{0dB}. Luckily, the diode-connecting loop usually has more stages, so A_{LG-} can be higher than A_{LG+}. To help this, although not necessarily by a significant amount, R_P can reside near the BJT or diode that either favors A_{LG-} or suppresses A_{LG+}. Filtering the gain of the positive path also suppresses A_{LG+}.

In the error-compensated BJT cell of Fig. 7.17, for example, Q_{B1}, M_{B2}, M_{M2}, Q_{P2}, and Q_{P1} close the inverting loop; Q_{B1}, M_{B2}, and M_{M1} close the noninverting counterpart; and both loops share Q_{B1}'s base, which is the only high-impedance node in both loops. Capacitor C_C ensures Q_{B1}'s base is the dominant low-frequency pole for the inverting loop. The purpose of the low-pass filter that R_F and C_F realize is to reduce the gain of the noninverting path. Plus, R_P resides below Q_{P2}, instead of below Q_{P1}, to boost and favor the gain of the inverting path. This way, R_P raises the voltage that Q_{P1} receives.

With this in mind, the inverting loop's gain A_{LG-} is the gain that an incoming current i_i produces at Q_{B1}'s base to the current that Q_{P1} outputs i_{p1}. To quantify this gain, consider that i_i drops a voltage v_{bB1} across Q_{B1}'s $r_{\pi B1}$, Q_{P1}'s $r_{oP1} + R_P'$, M_{M1}'s r_{sdM1}, and C_C's $1/sC_C$. Q_{B1}'s g_{mB1} then converts v_{bB1} into a current that M_{B2}'s $1/g_{mB2}$ resistance translates into M_{M2}'s gate voltage. M_{M2}'s g_{mM2} converts this gate voltage into a current that drops a base voltage across Q_{P2}'s $1/g_{mP2}$ and R_P to drive Q_{P1}'s g_{mP1} and output i_{p1}:

$$A_{LG-} \equiv \frac{i_{p1}}{i_i} \approx \left[r_{\pi B1} \| (r_{oP1} + R_P') \| r_{sdM1} \| \left(\frac{1}{sC_C} \right) \right] g_{mB1} \left(\frac{1}{g_{mB2}} \right) g_{mM2} \left(\frac{1}{g_{mP2}} + R_P \right) g_{mP1} \tag{7.19}$$

where $1/g_{mB2}$ is so much lower than R_F that $1/g_{mB2}$ shunts the effects of R_F and C_F.

The noninverting gain A_{LG+} starts at the same point, at Q_{B1}'s base, and traverses across the same initial path, through Q_{B1} and M_{B2}, so the first few terms of A_{LG+} mirror those of A_{LG-}. The difference is that R_F and C_F feed a filtered version of M_{M2}'s gate to M_{M1}, whose g_{mM1} generates A_{LG+}'s output current i_{m1}:

$$A_{LG+} \equiv \frac{i_{m1}}{i_i} \approx \left[r_{\pi B1} \| (r_{oP1} + R_P') \| r_{sdM1} \| \left(\frac{1}{sC_C}\right) \right] g_{mB1} \left(\frac{1}{g_{mB2}}\right)\left(\frac{1}{1+R_F C_F s}\right) g_{mM1} \tag{7.20}$$

Not surprisingly, the product of the first two of three uncancelled terms in the ratio of A_{LG-} to A_{LG+}, which describes to what extent A_{LG-} is greater than A_{LG+}, is higher than 1, and the last term rises with frequency:

$$\frac{A_{LG-}}{A_{LG+}} \approx \frac{g_{mM2}}{g_{mM1}}\left(\frac{1}{g_{mP2}} + R_P\right) g_{mP1}(1+R_F C_F s) = \left(\frac{1}{g_{mP2}} + R_P\right) g_{mP1}(1+R_F C_F s) \tag{7.21}$$

where g_{mM1} and g_{mP1} equal g_{mM2} and g_{mP2}, respectively, because they correspond to matched transistor pairs. In other words, in boosting A_{LG-} with R_P and suppressing A_{LG+} with R_F and C_F, the circuit favors A_{LG-} over A_{LG+}.

7.7 Suppressing Supply Noise

Modern chips nowadays integrate numerous system components on the same die, from bias currents and references and the amplifiers, regulators, and data converters they support to memory and digital-signal processors (DSPs). In waking and reacting to events and commands, many of these blocks draw transient power. Unfortunately, the combined effect of so many synchronous and asynchronous loads on the same power supply is noise. What is more, because bias currents and references feed almost every analog block in a system, the effects of infecting them with this noise can be crippling. This is why *power-supply rejection* (PSR) is so important in integrated circuits today.

Perhaps the most effective way to eliminate this type of noise at the transistor level is to cancel it. In differential pairs, for example, in-phase noise that appears in both input terminals cancels and disappears from the output that the difference generates. This feature extends to transistors because noise that is common to both base and emitter or gate and source terminals also cancels and disappears from the difference that their transconductances output as currents. Inserting *common-mode capacitors* that couple noise between the input terminals of sensitive transistors is therefore the most fundamental way to battle supply and ground noise. Since negative feedback opposes the effects of disturbances in the system, extending the bandwidth of inverting feedback loops in bias currents and references is another way of suppressing this type of noise.

7.8 Bandgap Current Reference

7.8.1 BJT-Derived Bandgap Current Reference

The BJT-derived current reference in Fig. 7.18 incorporates all the concepts developed and discussed in this chapter. Operationally, Q_{P1}, Q_{B1}, and M_{B2} diode-connect M_{M2}, and together with M_{M1} and M_{MO}, they mirror currents in the circuit. Q_{P1} and Q_{P2} loop around R_P to impress the voltage difference of two matched PN junctions, which is V_t ln C and therefore PTAT, across R_P. Because R_{C1} and R_{C2} drop the matched diode voltages in Q_{B1} and Q_{P1}, their matched currents are CTAT. As a result, M_{B2}, M_{M1}, and M_{MO} mirror the temperature-compensated sum of R_P's and R_C's currents in the form of reference i_{REF}. Because cancelling the temperature-dependent components of a diode voltage reduces it to the bandgap voltage V_{BG}, some engineers call this type of current generator a *bandgap reference*.

Long and narrow FET M_{LONG} pulls startup current i_{ST} from diode-connected M_{B2} to ensure that M_{B2} and the transistors M_{B2} controls are never off. This way, Q_{B1}, M_{B2}, and M_{M1}'s noninverting feedback loop latches the circuit to the on state. And because R_P, R_P', and Q_{B1}'s and Q_{P2}'s bases–emitter junctions establish similar voltages at the drains

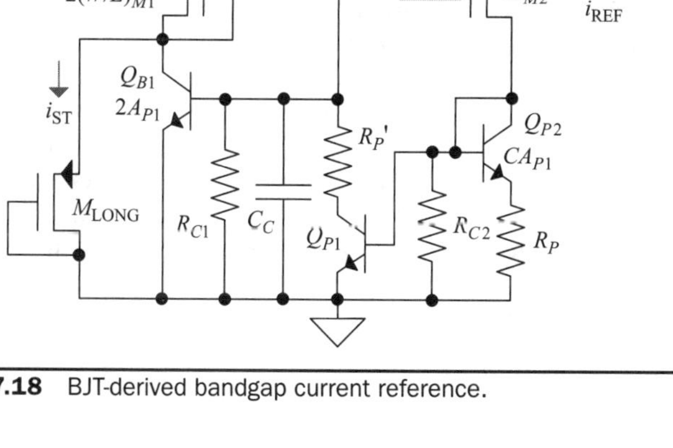

FIGURE 7.18 BJT-derived bandgap current reference.

and collectors of M_{M1}, M_{M2}, Q_{P1}, and Q_{P2}, R_P's PTAT current does not suffer from a channel-length or base-width modulation error. Note, by the way, that M_{LONG} does not pull i_{ST} from the drains of M_{M1} or M_{M2}, so i_{ST} does not introduce an offset between Q_{P1}'s and Q_{P2}'s currents. And since Q_{B1} conducts twice as much current as Q_{P1} and Q_{P2} alone carry, Q_{B1} pulls as much base current from Q_{P1}'s collector as Q_{P1} and Q_{P2} pull from Q_{P2}'s collector.

C_C adds capacitance to the only high-impedance node in M_{M2}'s diode-connecting path to establish the dominant low-frequency pole for the loop. But to extend the bandwidth of the loop so it can suppress as much noise as possible, C_C is only high enough to ensure that no other pole appears before the loop's unity-gain frequency f_{0dB}. R_P is under the BJT that appears in this loop, but not in the positive counterpart, to favor and boost this loop's inverting feedback gain. For similar ends, R_F and C_F filter and attenuate the positive gain around M_{M1}.

The reason for connecting C_F to the supply instead of ground is to couple supply noise present at M_{M1}'s source to M_{M1}'s gate. Similarly, M_{MO} derives its gate voltage from M_{M1}'s gate because C_F couples supply noise to M_{MO}'s gate. This way, M_{MO}'s g_{mMO} cancels common-mode power-supply ripples to reduce the noise content in M_{MO}'s i_{REF}.

7.8.2 Diode-Derived Bandgap Current Reference

Because diodes do not enjoy the benefit of a third terminal, the diode-derived current reference in Fig. 7.19 includes more devices than the BJT embodiment of Fig. 7.18. Operationally, amplifier A_G diode-connects M_{M2}, and together with M_{M1}, M_{MO}, and M_{MB}, they mirror currents in the circuit. A_G series-mixes the voltages across R_{C1}'–R_{C1}'' and R_{C2}'–R_{C2}'' to ensure that their voltages match. Accordingly, A_G impresses D_{P1}'s diode voltage v_D across R_P and D_{P2}, so the voltage difference between the two matched diodes, which is $V_t \ln C$ and therefore PTAT, appears across R_P. And because v_D drops across $R_{C1}' + R_{C1}''$ and $R_{C2}' + R_{C2}''$, their currents are CTAT. The end result is that M_{M1}, M_{MB}, and M_{MO} mirror the temperature-compensated sum of R_P's and R_C's currents in the form of reference i_{REF}.

M_S's gate essentially compares M_{MB}'s i_{REF} against the current that a long and narrow FET M_{LONG} establishes to determine whether or not the circuit is on or off. If i_{REF} is high, the circuit is on, so M_S's gate voltage rises to shut M_S. Otherwise, when i_{REF} is zero, M_{LONG} pulls M_S's gate to ground, so M_S responds by pulling a startup current i_{ST} from the diode-connected device in the mirror to start and latch the circuit to the on state. Note that, because the voltages across R_{C1}'–R_{C1}'' and R_{C2}'–R_{C2}'' are equal at v_D, M_{M1}'s and M_{M2}'s drain voltages match, so M_{M1} and M_{M2} do not suffer from a channel-length

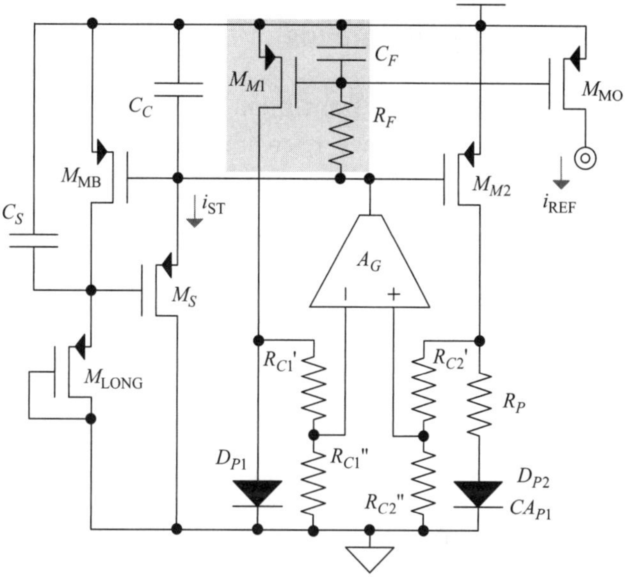

FIGURE 7.19 Diode-derived bandgap current reference.

modulation error. And since A_G's input voltages can be well below a diode voltage, the headroom limit that A_G's input common-mode range imposes can be low.

C_C adds capacitance to the only high-impedance node in M_{M2}'s diode-connecting path to establish the dominant low-frequency pole for the loop. But to extend the bandwidth of the loop so it can suppress as much noise as possible, C_C is only high enough to ensure that no other pole appears before the loop's unity-gain frequency f_{0dB}. R_P is on this loop's diode D_{p2} to boost its inverting gain. R_F and C_F, on the other hand, filter and attenuate the positive gain around M_{M1} and A_G.

The reason for connecting C_F to the power supply instead of ground is to couple supply noise in M_{M1}'s source to M_{M1}'s gate. Similarly, C_S connects to the supply to couple noise to M_S's gate, which keeps M_S from inadvertently engaging in the presence of supply noise. M_{MO} derives its gate voltage from M_{M1}'s gate because both C_F and C_C via R_F couple supply noise to M_{MO}'s gate, and of the two, C_F is more than likely higher than C_C because C_C should only be high enough to stabilize the system. Notice, by the way, that connecting C_S to ground can invoke startup assertions because supply noise that C_C and C_F couple to M_S's source would not appear at M_S's gate, the voltage difference of which could prompt M_S to sink i_{ST}. Note that these noise-induced startup events can propagate through the electronic network to destabilize the system.

7.9 Bandgap Voltage Reference

7.9.1 Current-to-Voltage Translation

Instead of currents, a voltage reference normally sums voltages whose drifts across temperature cancel. Most circuits add a voltage that is CTAT to one that is PTAT to produce a result v_{REF} that, when balanced, is largely independent of temperature:

$$v_{\text{REF}} = v_{\text{CTAT}} + v_{\text{PTAT}} = v_D + i_{\text{PTAT}} R_{\text{PTAT}} = v_D + \left(\frac{\Delta v_D}{R_P}\right) R_{\text{PTAT}}$$

$$= v_D + \left(\frac{R_{\text{PTAT}}}{R_P}\right) V_t \ln(C) \approx V_{\text{BG}} \tag{7.22}$$

Realizing this in a circuit amounts to establishing and stacking a PTAT voltage v_{PTAT} to a diode, whose voltage v_D is inherently CTAT and therefore equivalent to v_{CTAT}. Luckily, PTAT cells impress the voltage difference Δv_D of two matched diodes across a resistor R_P to generate and mirror a PTAT current i_{PTAT} that is equivalent to $(V_t \ln C)/R_P$. As a result, feeding i_{PTAT} into a resistor R_{PTAT} that matches R_P drops a PTAT voltage v_{PTAT} that, when connected in series with v_D, produces the desired temperature-compensated sum v_{REF}. Since cancelling the drift components in v_D leaves bandgap voltage V_{BG}, v_{REF} normally reduces to V_{BG}, which is how these circuits earn the appellation of bandgap references.

Interestingly, the mirror that M_{B2}, M_{M1}, and M_{M2} realize in the error-compensated BJT PTAT cell illustrated in Figs. 7.4, 7.14, and 7.17 ensures that the circuit conducts no more and no less than four PTAT currents $4i_{\text{PTAT}}$. This is convenient because inserting R_{PTAT} between the cell and ground, like Fig. 7.20 demonstrates, not only produces v_{PTAT} across R_{PTAT} at $4i_{\text{PTAT}}R_{\text{PTAT}}$ but also stacks Q_{P1}'s base–emitter voltage v_{BEP1} above v_{PTAT} to impress the desired v_{REF} at Q_{P1}'s base. Most bandgap circuits likewise feed the currents of a PTAT generator into a resistor that is in series with a diode. This particular embodiment, like many others, also derives a PTAT current from the mirroring transistors to bias other blocks in the system.

7.9.2 Output Regulation

Highly integrated systems normally include high-power blocks that generate and couple substantial noise into the substrate. Switching regulators and power amplifiers (PAs), for example, carry so much power that they inject vast amounts of noise energy into the substrate. The problem with substrate noise is that it ripples across the die and eventually couples into v_{REF}, which then feeds and infects the rest of the system. So when noise is excessive, engineers resort to

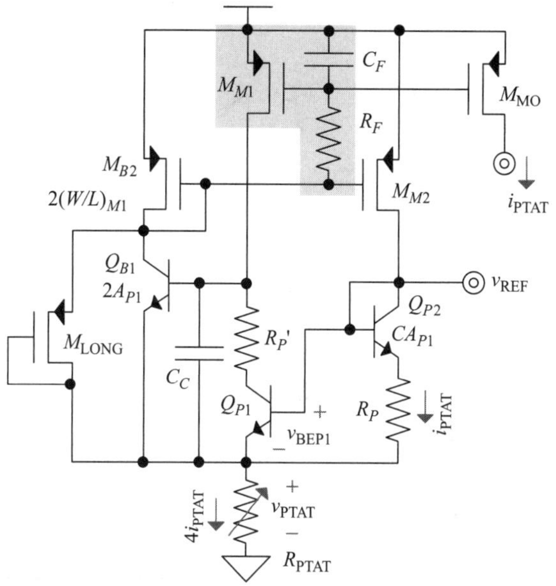

Figure 7.20 BJT-derived bandgap voltage reference.

shunt feedback for help. And for this, they normally modify the inverting feedback loop in the mirror of the PTAT cell so that it can sample and regulate v_{REF}, and in the process, oppose the deleterious influence of external forces on v_{REF}. This way, with shunt feedback, v_{REF}'s output impedance can be low.

BJT-Derived Bandgap Voltage Reference

The most popular way of implementing the mirror in the PTAT cell is with two resistors and an amplifier that series-mixes the resistors' voltages to ensure that the resistors conduct the same current, like Fig. 7.21 illustrates. In the case of BJTs, looping the amplifier back to the base of the matched pair is convenient. Amplifier A_V in this case adjusts Q_{P1} and Q_{P2}'s base v_{REF} this way to ensure that load resistors R_{L1} and R_{L2} drop equivalent voltages. As a result, R_{L1} and R_{L2} feed mirrored currents into Q_{P1} and Q_{P2}, which in turn, impress the voltage difference of their matched base–emitter diodes across R_P to generate a PTAT current i_{PTAT}. Steering Q_{P1}'s and Q_{P2}'s currents into a resistor R_{PTAT} that matches R_P then establishes the PTAT voltage v_{PTAT} with which Q_{P1}'s base–emitter diode voltage v_{BEP1} combines to set v_{REF}. In this way, the feedback action of A_V opposes the effects of noise on v_{REF}. And like before, R_P' matches R_P to drop similar voltages across the collector–emitter terminals of Q_{P1} and Q_{P2}.

Since the voltages across R_{L1} and R_{L2} can be low, the gate-coupled pair M_{D1}–M_{D2} in Fig. 7.22 can realize the differential input of the feedback amplifier A_V in Fig. 7.21 to impose a low headroom limit.

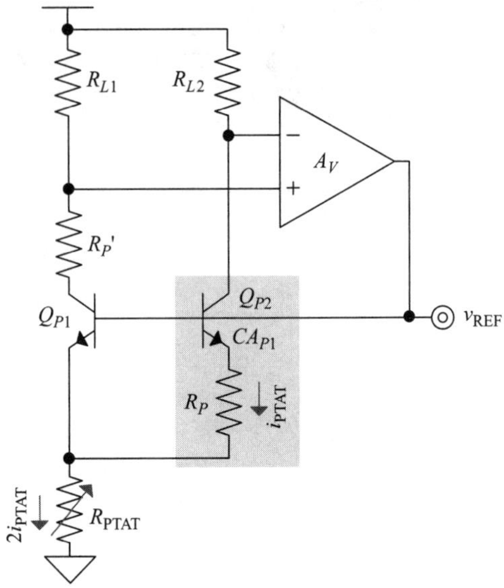

FIGURE 7.21 Regulated BJT-derived bandgap system.

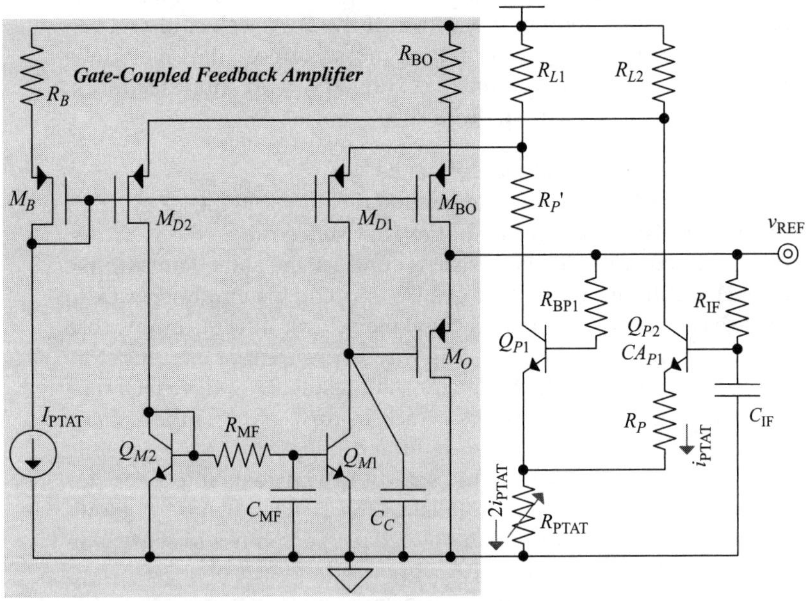

FIGURE 7.22 Regulated BJT-derived bandgap voltage reference.

To avoid the startup and stability complications of mixing yet another feedback loop into the system, a PTAT current I_{PTAT} from a simpler PTAT cell biases and starts this circuit. M_{D1}, M_{D2}, and M_{BO} together with their source-degenerating resistors R_{L1}, R_{L2}, and R_{BO} therefore mirror M_B and R_B's I_{PTAT} to bias the amplifier. Accordingly, M_{D1} and M_{D2} buffer small-signal currents from Q_{P1} and Q_{P2} into mirroring transistors Q_{M1} and Q_{M2} to establish a gain that voltage follower M_O buffers and drives back to v_{REF}.

Q_{P1} and Q_{P2} receive and process v_{REF} like an emitter-coupled differential pair would, except R_P emitter-degenerates Q_{P2}, so Q_{P2}'s transconductance gain is lower than that of Q_{P1}. The reason for degenerating Q_{P2} this way is that Q_{P2}, M_{D2}, Q_{M2}, Q_{M1}, and M_O invert feedback signals twice to loop a noninverting feedback path. To similarly favor the gain of Q_{P1}, M_{D1}, Q_{M1}, and M_O's inverting loop, R_{MF} and C_{MF} and R_{IF} and C_{IF} filter and attenuate positive-feedback signals. But since Q_{P2}'s base current drops a voltage across R_{IF}, the purpose of R_{BP1} is to match that voltage and avoid the offset R_{IF} alone would have otherwise generated.

Because M_O's gate not only appears in both loops but is also the only high-impedance node in the circuit, compensation capacitor C_C ensures that this gate establishes the dominant low-frequency pole in the system. To extend bandwidth and therefore suppress as much coupled noise as possible, C_C should only be as high as absolutely necessary. With respect to noise, C_{MF}, C_C, and C_{IF} connect to ground to couple noise that is present at and near Q_{M1}'s and Q_{P2}'s emitters and M_O's source to their corresponding bases and gate. Because common-mode signals cancel, coupling noise this way mitigates the impact of noise in the circuit.

Diode-Derived Bandgap Voltage Reference

The diode-derived bandgap voltage reference resembles its BJT counterpart. The only difference between the two is the absence of a third terminal in the diode, whose ultimate attribute in the BJT is to add flexibility. Like its predecessors from Figs. 7.6, 7.12, and 7.19, the amplifier A_G in the regulated diode-derived bandgap voltage reference of Fig. 7.23 diode-connects M_{M2} to ensure that M_{M1} mirrors the PTAT current i_{PTAT} that the voltage difference of two matched diodes D_{P1} and D_{P2} produces in R_P. With the mirror in place, steering D_{P1} and D_{P2}'s combined current into a resistor R_{PTAT} that matches R_P establishes a PTAT voltage v_{PTAT} that is equivalent to $2i_{PTAT}R_{PTAT}$. And stacking this voltage with D_{P1}'s diode voltage v_{DP1} produces the temperature-compensated sum desired in v_{REF}.

To favor the diode-connected loop over the noninverting counterpart, R_P is on the diode that boosts A_G and M_{M2}'s combined gain. R_F and C_F filter and therefore attenuate A_G and M_{M1}'s positive-feedback gain. Compensation capacitor C_C, like before, ensures that M_{M2}'s gate, which is the only high-impedance node in this circuit, produces

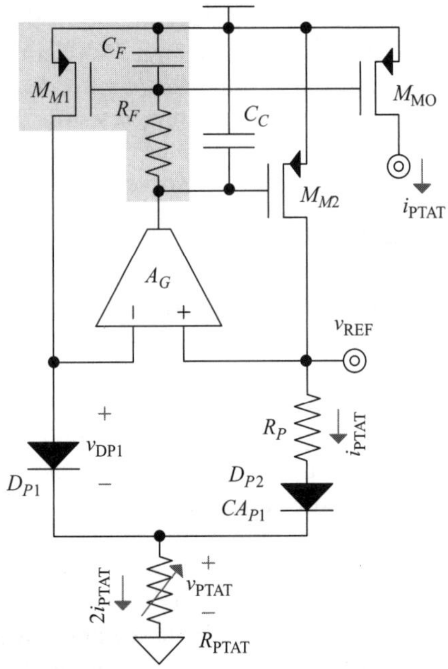

FIGURE 7.23 Regulated diode-derived bandgap voltage reference.

the dominant low-frequency pole in the system. To extend bandwidth and therefore suppress as much coupled noise as possible, C_C should only be as high as absolutely necessary. With respect to noise, C_F and C_C connect to the supply to couple noise present at M_{M1}'s and M_{M2}'s sources to their corresponding gates. And because C_F is more than likely higher than C_C, M_{MO} derives its gate voltage from M_{M1}. This way, M_{MO}'s output current i_{PTAT} is less sensitive to power-supply noise.

7.10 Accuracy

When resistors, diodes, and mirroring currents match and resistor ratio R_{PTAT}/R_P amplifies PTAT voltage $V_t \ln C$ to the level that linear components in diode voltage v_D cancel, reference voltage v_{REF} centers near bandgap voltage V_{BG} at around 1.2 V. This means the PTAT voltage v_{PTAT} that results is roughly $V_{BG} - v_{D(ROOM)}$, or 0.55 V at room temperature T_{ROOM} when v_D at T_{ROOM} is 0.65 V. Still, v_D's nonlinear $T \ln T$ component remains and manifests in v_{REF} to create the convex curvature that Fig. 7.24 illustrates. When compensated, v_{REF} peaks and flattens nearly halfway across the temperature range at T_{MID}. Interestingly, this halfway point does not usually correspond to T_{ROOM}. In fact, most conventional and expanded commercial ranges extend from 0 to 85°C or –40 to 125°C, so T_{MID} is normally 42.5°C. As a result, the slope and temperature coefficient of v_{REF} is normally zero at T_{MID}, but not at T_{ROOM}.

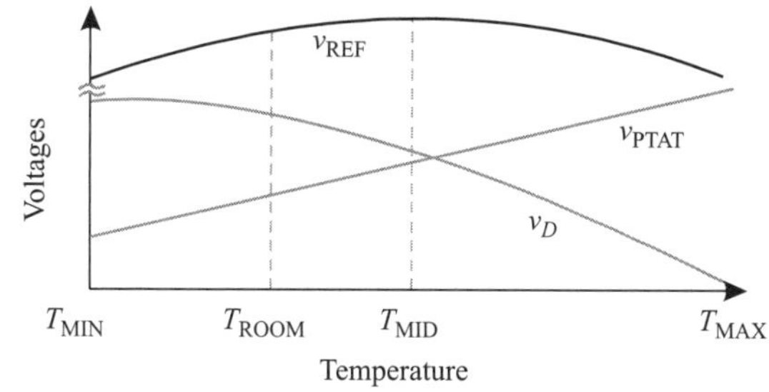

FIGURE 7.24 Trimmed first-order bandgap voltage reference.

When centered and trimmed, v_{REF} can vary about 1.2 V between 3 and 15 mV across 125°C and fabrication corners, which corresponds to 20 and 100 ppm/°C of three-sigma variation. Unluckily, tolerance in the diode voltage v_D and resistors; mismatches in diodes, mirror currents, and resistors; and package stresses produce a variance in v_{REF} that is practically impossible to predict in the design process. As a result, R_{PTAT} is normally adjustable and trimmed at T_{ROOM} after the fabrication process. This is unfortunate because modifying R_{PTAT} this way requires test time, the cost of which when multiplied by billions of microchips is substantial.

Reducing this trim range and the test time that trimming requires amounts to decreasing the overall tolerance of the reference. This is why matching the diodes, resistors, and mirroring transistors in the circuit is so important, especially when considering the nonlinearities that offsets can produce in v_{PTAT} and v_D. Compact, modular, cross-coupled, and common-centroid layouts are therefore the norm in reference circuits. The area ratio C of the PTAT-generating diodes or BJTs is normally 8 for this reason, because surrounding one device with eight others in a three-by-three array like Fig. 7.25 shows yields a layout that is compact, modular, and with a common center of mass. Keeping the space immediately above matched devices free of metallic paths is also good practice because metal traces can produce localized stress fields in the presence of package stresses. Although not obvious at first, also note that, like tolerance in R_P, a drift in R_P's resistance across temperature alters the PTAT characteristics of the current that establishes the diode voltage v_D, the effect of which is to add variance to v_D.

Reducing the trimmed variation of the reference to 5–20 ppm/°C is possible with *curvature correction*. Unfortunately, the improvement is normally incremental because trimming an additional component to adjust the nonlinear term that corrects curvature and trimming at more than one temperature are prohibitively expensive in a mass-production environment. So compensating the $T \ln T$ nonlinearity for

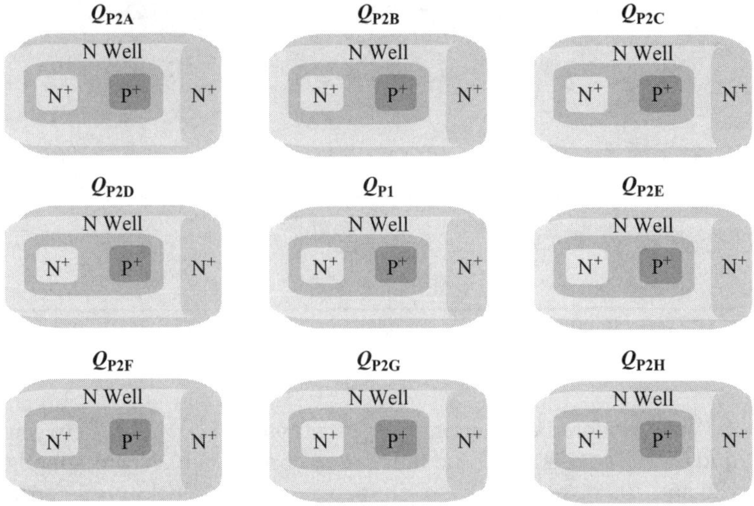

FIGURE 7.25 Popular layout arrangement of a PTAT-producing pair of devices.

a nominal improvement in accuracy seldom justifies the additional risk and overhead required to design and characterize the circuit.

Because diode voltages suffer from lower tolerance than gate–source voltages and BJTs match better than MOSFETs, all-bipolar bandgap references typically outperform their MOS counterparts. And when BJTs are not available, diode-derived CMOS references are the next best option. Since higher quiescent currents extend bandwidth and raise signal-to-noise ratio (SNR), higher-power references also suppress and overpower coupled noise more than their lower-power siblings.

Ultimately, the temperature dependence of mismatch-generated offsets in a circuit alters the temperature characteristics of the reference. In fact, temperature profiles are unique to circuit topology, physical layout, and process technology. This means that every design behaves differently, and therefore, the value that produces the least overall trimmed variation across temperature and process changes with design, layout, and technology. In other words, every new design requires characterization, which is why reusing a proven design whenever possible is oftentimes the most prudent course of action, because it saves time and money.

7.11 Summary

Of available voltages in most commercially available semiconductor technologies, diode voltages exhibit the lowest tolerance and produce the most reliable results across temperature, current density, and fabrication lots. Plus, since BJTs match better than MOSFETs and their third terminal offers more design options than diodes, BJT-derived circuits are typically more accurate and compact than their diode-derived counterparts. But when BJTs are unavailable, although not to the same extent, diodes can also provide the components needed to build reliable bias currents and reference circuits.

Interestingly, most references extract and use the thermal voltage V_t, which is PTAT, to cancel the linear component of a diode voltage, which is CTAT. In this respect, the latched PTAT cell is popular because it extracts V_t from two matched diodes to generate a current that is largely independent of the supply and linear across temperature, and even more so when base currents and collector and drain voltages match. PTAT cells, however, incorporate positive feedback, which not only produces two stable states but also compromises the stability of the mirror's current- and voltage-correcting feedback network. Circuits, as a result, usually include startup and frequency-compensating components that help start and stabilize the reference.

Noise is another important consideration. Common-mode capacitors are useful in this respect because they couple noise in emitters and sources to their corresponding bases and gates, and vice versa. This way, supply-noise effects cancel and disappear from transistor currents. Unfortunately, substrate- and system-coupled noise can also be a problem, especially in highly integrated systems. Under these circumstances, extending the loop bandwidth of the reference is good, because with a faster response, the circuit can oppose the undesirable effects of disturbances more quickly.

With reliable currents and voltages in hand, design engineers can bias their systems. In the case of linear regulators, a current biases the circuit and a voltage establishes the reference about which the feedback loop regulates its output. But before diving into the transistor-level realization of the regulator, a better understanding of the feedback requirements of the system is necessary. The purpose of Chap. 8 is therefore to explore how the load affects the small-signal dynamics of the feedback loop in a linear regulator, and under those conditions, how ripples in the supply leak to the output to produce noise in the output.

7.12 Review Questions

1. Bias currents and reference circuits normally derive their functionality from which readily available voltages?

2. Of these, which are the most predictable and precise across temperature, current density, and fabrication lots?

3. What is the meaning of PTAT?

4. Which PTAT voltage is popular in bias currents and reference circuits?

5. What is(are) the features of the cross-coupled quad?

6. What fundamental characteristic does the MOS variant of the latched PTAT cell exploit to generate a PTAT current?

7. From which systemic error(s) do basic PTAT cells suffer?

8. How do most diode-derived PTAT cells correct this(these) error(s)?

9. Cite examples of CTAT voltages in typical semiconductor technologies.

10. Which voltage components do reference circuits typically combine for temperature independence?

11. What relationship best describes nonlinearities in the diode voltage across temperature?

12. How do startup circuits ensure PTAT cells latch to the PTAT state?

13. How do most current-mode on-demand startup circuits sense the state of the PTAT cell?

14. When is the stability of the diode-connecting loop in the mirror of a PTAT cell not a concern?

15. In the case of possible feedback instabilities, which techniques help stabilize the system?

16. How can reference circuits suppress supply and ground noise?

17. How do bandgap circuits use PTAT cells?

18. How can a bandgap voltage reference shunt coupled noise?

19. What determines the trimming range of a bandgap voltage reference?

20. Why is the area ratio between the matched diodes that generate the PTAT voltage usually 8?

21. Why is curvature correction not as pervasive in industry as its first-order predecessor?

CHAPTER 8
Small-Signal Response

The fundamental aim of a supply circuit is to regulate the output about a user-defined target. *Linear regulators* employ negative feedback for this purpose, to oppose the effects of load and supply variations, load dumps, and supply and substrate noise on the output. Generally, an error amplifier, like A_E in Fig. 8.1, series-mixes a reference voltage v_{REF} with a feedback signal v_{FB} to generate an error difference that the loop amplifies and returns to the mixer as v_{FB}. With enough loop gain, v_{FB} mirrors v_{REF}, and the feedback factor β_{FB} translates v_{FB}'s impression of v_{REF} to the output voltage v_O. As a result, variations in v_O appear in v_{FB}, and the feedback action of A_E amplifies v_{FB}'s error with respect to v_{REF} to oppose and cancel the fluctuations that caused the error in the first place.

Unfortunately, loops cannot respond instantly to variations in v_O, so a filter capacitor C_{Filter} is usually at the output. The purpose of C_{Filter} is to keep v_O from drifting excessively while the system reacts. Feedback loops also incorporate a large pass device S_O because the systems they support normally supply substantial power to their loads. This is why engineers call them linear regulators, because the pass device is in series with the input supply v_{IN} and the load.

Although seemingly simple, the loop dynamics of a linear regulator can be difficult to manage. For one, the load-current range of modern consumer electronics is vast, from close to 1–10 μA when idling to 1–500 mA when fully engaged. The problem with this range is the variation it causes in the gain and output resistance of the pass device, which temperature and fabrication corners, capacitor tolerance, and aging extend. And arbitrarily reducing bandwidth to accommodate the gain and output-pole fluctuations that result slows the system, and in consequence, widens v_O's drift in response to fast load dumps.

The input v_{IN} does not help the situation. For one, v_{IN} normally supplies several blocks, so the combined effect of synchronous and asynchronous load currents on v_{IN} is wide-bandwidth noise. Plus, v_{IN}

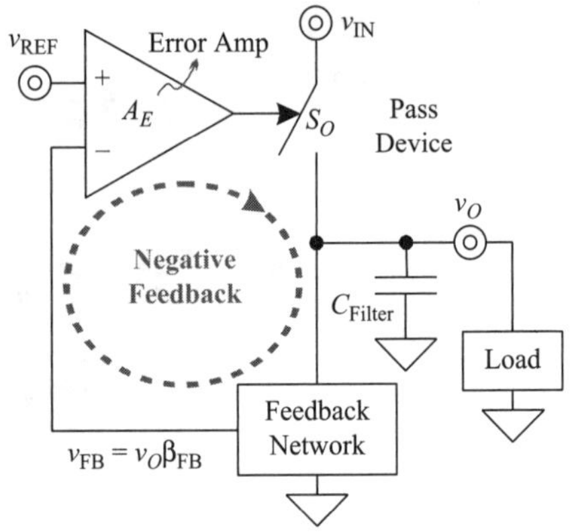

FIGURE 8.1 Loaded linear regulator.

is normally the output of a switching regulator whose switching action injects additional noise energy into v_{IN}. In fact, the purpose of a linear regulator is to suppress this noise. With this in mind, this chapter shows how to manage and design the loop dynamics of the system to both stabilize the feedback system and quell noise in the supply.

8.1 Equivalent Small-Signal Circuit

Developing an equivalent small-signal circuit is useful when determining the small-signal dynamics of a system. In the case of linear regulators, Fig. 8.2 illustrates how the error amplifier A_E, pass device S_O, output filter C_{Filter}, load, and feedback network β_{FB} map into the small-signal domain. Here, A_E's input capacitance C_{IE} loads the feedback voltage v_{fb}

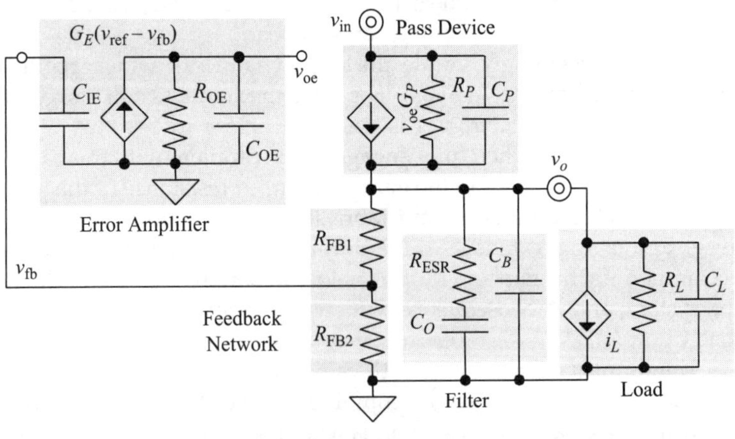

FIGURE 8.2 Equivalent small-signal circuit.

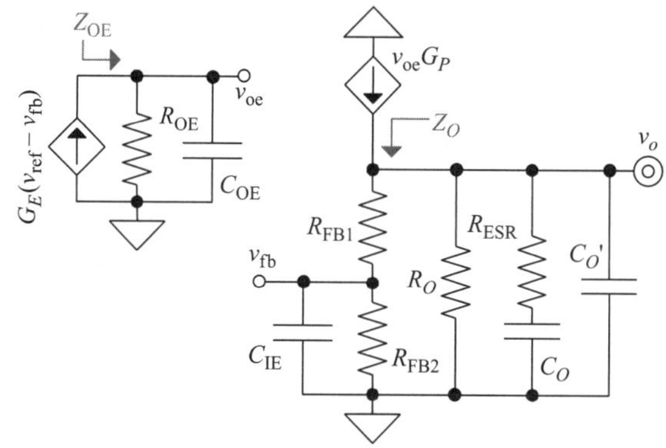

FIGURE 8.3 Simplified small-signal circuit of the feedback loop.

with which A_E mixes the reference v_{ref}. Going forward, A_E's transconductance G_E and output resistance R_{OE} establish a gain that A_E's output capacitance C_{OE} ultimately shunts. S_O's transconductance G_P then converts A_E's output voltage v_{oe} into a current that drives the output v_o. S_O also includes a resistance R_P and a capacitance C_P that connect the input v_{in} to v_o. To close the loop, the feedback factor β_{FB} usually resorts to a well-matched voltage divider of two resistors R_{FB1} and R_{FB2}.

Since load dumps can be vast, from maybe 1 to 200 mA, the filter normally includes an output capacitor C_O that can be as high as several microfarads. Unfortunately, the equivalent series resistance R_{ESR} of such a large device can also be substantial, so users sometimes shunt the undesirable voltage drops that appear across R_{ESR} with low-ESR bypass capacitors C_B. The load, of course, includes its own capacitance C_L and resistance R_L and the current i_L that produces load dumps in the system.

Since current variations produce hardly noticeable effects in the input source, v_{in} is practically a ground for small signals traversing the feedback loop. Likewise, as Fig. 8.3 further illustrates, i_L disappears from the equivalent small-signal circuit because loop signals do not usually affect i_L, which is another way of saying i_L is an independent source with respect to the feedback loop. Luckily, these modifications short R_P and R_L, so to simplify the circuit, equivalent output resistance R_O in Fig. 8.3 combines them. Similarly, output shunt capacitance C_O' incorporates C_P, C_L, and C_B because they, too, connect to each other.

8.2 Uncompensated Response

8.2.1 Relative Capacitances and Resistances

Knowing the relative resistances and capacitances in the network helps determine when the components assert their influence on gain across frequency. Low capacitances, for example, shunt the effects of resistances at high frequency, and at even higher frequency when resistances are low. In the case of linear regulators, C_O's series resistance R_{ESR} is normally the lowest resistance in the circuit at maybe 50–500 mΩ, depending on the type of capacitor used. In practice, R_{ESR} is typically on the high end of the range for *tantalum capacitors* and on the low end for *multilayer ceramic chip capacitors*.

Although load resistance R_L is difficult to predict, the resistance of the pass device R_P can be next between 0.1 and 50 kΩ, depending on the type of transistor used. The output resistances of N-type followers, for example, can be as low as 100 Ω. Collector and drain resistances are higher, but not much higher than 50 kΩ because base-width and channel-length modulation in lateral bipolar-junction transistors (BJTs) with lightly doped bases and short-channel metal–oxide–semiconductor (MOS) field-effect transistors (FETs) are normally significant. Irrespective of R_L, the combined resistance of R_L and R_P in R_O is therefore moderate to low, but still substantially higher than R_{ESR}.

Feedback resistors R_{FB1} and R_{FB2} are next. The fundamental purpose of these resistors is to voltage-divide the output to define an output voltage that is an accurate translation of the reference v_{REF}. Unfortunately, quiescent current flows through R_{FB1} and R_{FB2}, so they also dissipate power. To keep this power low, engineers normally choose high resistances on the order of 100–500 kΩ, and in some cases higher. With a 1.8-V output, for example, 500 kΩ of combined resistance draws 3.6 μA. R_{OE} is similarly high or higher because R_{OE} sets A_E's gain.

With respect to capacitances, C_{IE}, C_{OE}, C_P, and C_L are usually parasitic and C_O and C_B are intentional, which means the values of the latter set are higher than those of the former. C_{IE} is probably the lowest in this list because the input transistors of the error amplifier are normally small. While C_P can be close to negligible in BJTs and N-channel MOSFETs, C_P can be higher than C_{IE} in P-channel MOSFETs because the area of the drain–bulk junction is typically large. Since C_{OE} is R_{OE}'s loading capacitance, C_{OE} usually includes one gate–source or base–emitter capacitance, which can be on par with C_{IE} or higher. When C_{OE} establishes the dominant pole in the loop, C_{OE} might even include the gate–source or base–emitter capacitance of the large pass device, in which case C_{OE} is considerably higher than C_{IE}, but still lower than C_O and C_B.

The fundamental reason why designers add filter capacitors to the output of a regulator in the first place is because C_L is unpredictable and therefore unreliable, and even when predictable, more than likely insufficient. In fact, the reason for including C_O is because C_L is typically low. And since the purpose of C_B is to shunt C_O's R_{ESR}, C_B is normally one or two orders of magnitude lower than C_O. In other words, C_O is substantially higher than C_B, C_B is likewise higher than C_L and C_{OE}, C_L is higher than both C_P and C_{IE}, and C_{OE} is between C_P and C_{IE} and C_L.

8.2.2 Loop Gain

The stability of the feedback loop hinges on how the gain across the loop reaches the *unity-gain frequency* f_{0dB}. In fact, the *stability criterion* dictates that the *loop gain* A_{LG} should reach f_{0dB} with less than 180° of phase shift, which corresponds to the combined shift of two poles. This is why determining the gain across the loop is imperative.

Loop gain A_{LG} is, as the name implies, the gain across the loop, which is nothing but the product of the *forward open-loop gain* A_{OL} from the mixed difference between v_{ref} and v_{fb} to v_o, and the *feedback-factor* β_{FB} is the gain translation from v_o to v_{fb}. With respect to A_{OL}, A_E's G_E converts the difference between v_{ref} and v_{fb} into a current that flows into the equivalent impedance Z_{OE} that R_{OE} and C_{OE} present. G_P then translates the voltage across Z_{OE} into a current that drops v_o across the equivalent output impedance Z_O that R_{FB1}, R_{FB2}, C_{IE}, R_O, R_{ESR}, C_O, and C_O' establish to set A_{OL} to

$$A_{LG} = A_{OL}\beta_{FB} = \left(\frac{v_o}{v_{ref} - v_{fb}}\right)\left(\frac{v_{fb}}{v_o}\right) = (G_E Z_{OE} G_P Z_O)\left(\frac{Z_{FB2}}{R_{FB1} + Z_{FB2}}\right) \quad (8.1)$$

R_{FB1}, R_{FB2}, and C_{IE} then voltage-divide v_o to v_{fb} to set β_{FB}, and with it, A_{LG}.

Low-Frequency Gain

Luckily, capacitor impedances are so high at low frequency that capacitors practically disappear from the circuit. This means, Z_{OE} reduces to R_{OE}, and Z_O to the parallel combination of R_O and the combined resistance of R_{FB1} and R_{FB2}. And because R_P in R_O is lower than both R_L in R_O and the series combination of R_{FB1} and R_{FB2}, low-frequency gain A_{LG0} normally simplifies to

$$\begin{aligned} A_{LG0} &= G_E R_{OE} G_P [(R_{FB1} + R_{FB2}) \parallel R_O]\left(\frac{R_{FB2}}{R_{FB1} + R_{FB2}}\right) \\ &\approx G_E R_{OE} G_P R_P \left(\frac{R_{FB2}}{R_{FB1} + R_{FB2}}\right) \end{aligned} \quad (8.2)$$

Frequency Response

As the operating frequency climbs, capacitors shunt their parallel resistances to alter the gain characteristics of the circuit. Capacitors that short also change the impedances across other capacitors in the network, so the interaction of capacitors also affects gain. Luckily, C_{OE} and its parallel resistance R_{OE} do not connect to other capacitors, or for that matter, to other resistors, so C_{OE}'s influence on loop gain A_{LG} is largely independent of other capacitors. As a result, C_{OE} shunts no more and no less than R_{OE} to establish an *error-amplifier pole* at p_E:

$$\left.\frac{1}{sC_{OE}}\right|_{p_E = \frac{1}{2\pi R_{OE} C_{OE}}} \equiv R_{OE} \quad (8.3)$$

Unfortunately, the rest of the capacitors in the circuit do not enjoy such independence. Other than deriving abstract relationships that exclude circuit insight, the best way to analyze a system of interconnected capacitors is to keep track of their shunting effects across frequency. Knowing how resistances and capacitances compare can be revealing in this respect. At low frequencies, for example, low capacitances disappear, and only high capacitances shunt high resistances. High capacitances then short low impedances at higher frequencies, and at even higher frequencies, low capacitances further shunt whatever impedances remain.

In the case of linear regulators, C_O and R_O are the highest interconnected capacitance and resistance in the network. As such, C_O, together with C_B, C_L, and C_P in C_O' because R_{ESR} is negligibly low at low frequency, shunt R_O's R_L and R_P, R_{FB1}, and R_{FB2} to establish *output pole* p_O:

$$\left.\frac{1}{s(C_O+C_O')}\right|_{p_O=\frac{1}{2\pi[(R_{FB1}+R_{FB2})||R_L||R_P](C_O+C_B+C_L+C_P)}} \equiv (R_{FB1}+R_{FB2})\,||\,R_O \tag{8.4}$$

R_L and R_P are normally lower than R_{FB1} and R_{FB2}, and C_P is likewise lower than C_O, C_B, and C_L, so R_{FB1}, R_{FB2}, and C_P hardly affect p_O.

Since C_O is so much higher than C_B, C_L, and C_P, C_O's impedance falls first to such an extent that R_{ESR} limits the impedance into C_O. In other words, R_{ESR} current-limits C_O to offset the effects of p_O with an *in-phase ESR zero* z_{ESR}:

$$\left.\frac{1}{sC_O}\right|_{z_{ESR}=\frac{1}{2\pi R_{ESR}C_O}} \equiv R_{ESR} \tag{8.5}$$

C_B, C_L, and C_P in C_O' are together so much higher than C_{IE} that C_O' shunts R_{ESR} next past *bypass pole* p_B:

$$\left.\frac{1}{sC_O'}\right|_{p_B=\frac{1}{2\pi R_{ESR}(C_B+C_L+C_P)}} \equiv R_{ESR} \tag{8.6}$$

And C_B and C_L are typically so much higher than C_P that C_P has less impact on p_B. At this point, the only capacitor left is C_{IE}, and because C_O' at these frequencies practically shorts v_o to ground, C_{IE} shunts the parallel resistance that R_{FB1} and R_{FB2} establish past *feedback pole* p_{FB}:

$$\left.\frac{1}{sC_{IE}}\right|_{p_{FB}=\frac{1}{2\pi(R_{FB2}||R_{FB1})C_{IE}}} \equiv R_{FB2}\,||\,R_{FB1} \tag{8.7}$$

Of these effects, error-amplifier and output poles p_E and p_O appear at low-to-moderate frequencies, as Fig. 8.4 shows, because p_E incorporates

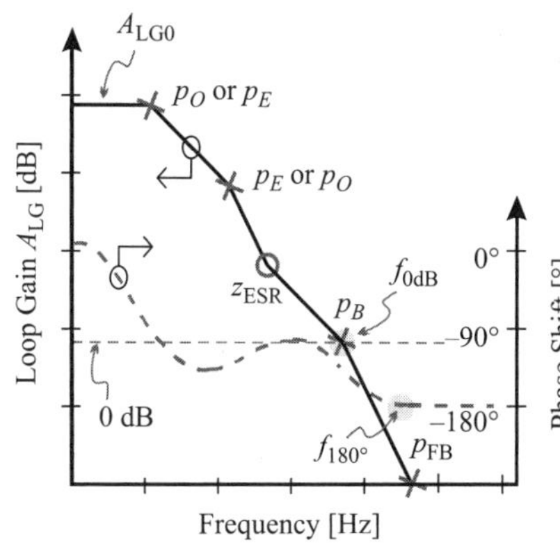

Figure 8.4 Uncompensated loop-gain response.

the highest resistance R_{OE} in the network and p_O includes the highest capacitance C_O. Which of these two appears first, though, is up to the designer and the application and technologies that constrain the system. ESR zero z_{ESR} is next, and bypass pole p_B is after that. In fact, because R_{ESR} sets both z_{ESR} and p_B, and C_O is greater than C_B, C_L, and C_P in C_O', p_B always trails z_{ESR} by the fraction of capacitances that set them:

$$\frac{p_B}{z_{ESR}} = \frac{2\pi R_{ESR} C_O}{2\pi R_{ESR} C_O'} = \frac{C_O}{C_O'} = \frac{C_O}{C_B + C_L + C_P} \tag{8.8}$$

In all, A_{LG} exhibits the combined effects of four poles and one zero:

$$A_{LG} \approx \frac{A_{LG0}\left(1+\frac{s}{2\pi z_{ESR}}\right)}{\left(1+\frac{s}{2\pi p_E}\right)\left(1+\frac{s}{2\pi p_O}\right)\left(1+\frac{s}{2\pi p_B}\right)\left(1+\frac{s}{2\pi p_{FB}}\right)} \tag{8.9}$$

where p_E and p_O are at low-to-moderate frequencies and p_B lags z_{ESR}. This means that reaching the unity-gain frequency f_{0dB} with the combined effect of less than two poles is not always a given, especially when R_{ESR} can be so low that the phase-saving effect of z_{ESR} can disappear.

Validation To validate how the interactions between C_O, C_B, C_L, and C_P affect A_{LG}, consider the output impedance Z_O into which G_P drives current to establish v_o when C_{IE} is negligibly lower than other capacitances in Z_O. Deriving Z_O directly from this network yields

$$Z_O = \frac{R_O'(1+sR_{ESR}C_O)}{s^2 R_O' R_{ESR} C_O C_O' + s(R_O' + R_{ESR})C_O + sR_O'C_O' + 1} \tag{8.10}$$

where R_O' combines the effects of R_{FB1}, R_{FB2}, and R_L. From this result, first notice that Z_O's numerator includes ESR zero z_{ESR}. To solve the mystery of the second-order polynomial in the denominator, consider that at low frequencies, the s^2 term, which includes the product of two capacitances, is considerably lower than both the s^1 term and 1. So, neglecting this s^2 term, realizing that R_{ESR} is substantially lower than R_O', and combining the remaining s^1 terms reveal output pole p_O. Because s^2 and s^1 overwhelm 1 at high frequencies, neglecting the 1 and factoring s^1, R_O', and C_O when R_O' overwhelms R_{ESR} similarly discloses bypass pole p_B. In other words, this derivation of Z_O demonstrates that the insight used to derive z_{ESR}, p_O, and p_B earlier is valid.

Example 8.1 Determine the location of relevant poles and zeros when output, bypass, load, and pass-transistor capacitances are 4.7, 0.3, 0.2, and 0.05 μF and pass-transistor, load, feedback, and ESR resistances are 100, 10,000, 500,000, and 0.1 Ω, respectively.

Solution

$$p_O = \frac{1}{2\pi[(R_{FB1}+R_{FB2})\,\|\,R_L\,\|\,R_P](C_O+C_B+C_L+C_P)}$$

$$= \frac{1}{2\pi[(500\,\text{k}\Omega)\,\|\,(10\,\text{k}\Omega)\,\|\,(100\,\Omega)](4.7+0.3+0.2+0.05\,\mu\text{F})} \approx 300\,\text{Hz}$$

$$z_{ESR} = \frac{1}{2\pi R_{ESR}C_O} = \frac{1}{2\pi(0.1\,\Omega)(4.7\,\mu\text{F})} \approx 340\,\text{kHz}$$

and $$p_B = \frac{1}{2\pi R_{ESR}(C_B+C_L+C_P)} = \frac{1}{2\pi(0.1\,\Omega)(0.3+0.2+0.05\,\mu\text{F})} \approx 2.9\,\text{MHz}$$

Compare these poles and zeros with the simulation results from Fig. 8.20.

8.3 Frequency Compensation

At first glance, ESR zero z_{ESR} seems to save phase. Unfortunately, the location of z_{ESR} in practice is unpredictable, and therefore unreliable. For starters, ESR resistance R_{ESR} varies considerably across temperature and process, by maybe ±50 to ±100 percent. What is more, characterization data is generally poor, so simulation models are correspondingly deficient. As a result, z_{ESR} may or may not lie within the bandwidth of the regulator. Adding series resistance to C_O intentionally to keep z_{ESR} near the unity-gain frequency f_{0dB} is unfortunately not a practicable option because load-dump response and *power-supply rejection* (PSR), as the subsequent section shows, deteriorate with R_{ESR}. Bypass pole p_B is, of course, no better, except C_B, C_P, and C_L are sometimes so low with respect to C_O that p_B's deleterious effects on loop gain A_{LG} at f_{0dB} may disappear.

Generally, high-power *breadboard* and *printed-circuit-board* (PCB) implementations often exhibit z_{ESR}–p_B pairs within the bandwidth of the regulator because high-power loads suffer most from wide load dumps.

This means that output capacitances are high and ESRs are, as a result, substantial, which also means local *point-of-load* (PoL) low-ESR bypass capacitances are high. In contrast, lower power levels and shallower load dumps in low-power *system-in-package* (SiP) and *system-on-chip* (SoC) applications demand less capacitance. Plus, linear regulators in highly integrated systems are close to their loads, so the need for local PoL bypass capacitors is less acute. Still, even if z_{ESR} and p_B were absent, output and error-amplifier poles p_O and p_E can nevertheless produce unstable conditions for the feedback system, even when excluding the feedback pole p_{FB} and other parasitic poles in the error amplifier A_E, which often prove difficult to place at high frequencies when constrained to a low-power budget.

The point of all this is that ensuring the feedback system remains stable requires careful scrutiny, and more than likely, proactive action. In this regard, to maintain stable conditions, the general approach is to establish a dominant low-frequency pole and place the secondary pole near the unity-gain frequency f_{0dB}. In the case of linear regulators, these two poles are the output and error-amplifier poles p_O and p_E. After deciding which of these should be dominant, the objective of compensation is to comprehend all possible variations in the system and ensure that the phase shift across the loop is never greater than maybe 140° at f_{0dB} across temperature and fabrication corners. Further ensuring that the frequency $f_{180°}$ where phase shift reaches 180° is nearly a decade away from f_{0dB} helps avoid transitional instabilities when the system first powers up or restarts after a dormant period.

8.3.1 Output Compensation

Output compensation in linear regulators refers to systems whose dominant low-frequency pole is at the output. This normally happens in high-power systems when load dumps are significant enough to call for high output capacitances. Since in the absence of other poles and zeros gain drops past the dominant pole as much as frequency climbs, by 20 dB per decade, the *gain–bandwidth product* $A_{LG0}p_O$, or GBW for short, is constant across frequency past the output pole p_O. In other words, if no other poles or zeros intervene, low-frequency gain A_{LG0} drops as much gain as frequency rises above p_O to reach the unity-gain frequency f_{0dB} at $p_O A_{LG0}$, or equivalently, at GBW frequency f_{GBW}:

$$f_{GBW} \equiv GBW = A_{LG0}p_O$$

$$= \frac{G_E R_{OE} G_P[(R_{FB1}+R_{FB2}) \parallel R_O]\beta_{FB}}{2\pi[(R_{FB1}+R_{FB2}) \parallel R_O](C_O+C_B+C_L+C_P)} = \frac{A_E G_P \beta_{FB}}{2\pi(C_O+C_B+C_L+C_P)} \tag{8.11}$$

Interestingly, of the parameters that set f_{GBW} in output-compensated systems, the transconductance G_P of the power transistor is normally the only one that is sensitive to load current i_L. Noting this dependence is important because i_L in modern systems spans several decades of current, from a few microamps to several hundred milliamps, so G_P's sensitivity to i_L subjects f_{GBW} and, as a result, f_{0dB} to considerable variations. The significance of this fluctuation in f_{0dB} surfaces when considering bandwidth and power dissipation. In this light, even though the system's worst-case bandwidth is at the lowest f_{0dB}, all parasitic poles in the system must still reside above the highest possible f_{0dB}. Shortening this range ultimately saves power because all parasitic poles are as high as quiescent currents in transistors allow.

Pushing parasitic poles in the error amplifier A_E to high frequencies is especially challenging. The reason for this is that high loop gain in negative-feedback systems, which is in large part the result of gain across A_E, suppresses the adverse effects of external forces on the output. Unfortunately, gain is usually the product of moderate transconductance current and high resistance. This is regrettable because that same resistance also establishes a moderate- to low-frequency pole. And in the case of output-compensated regulators, that pole, which corresponds to p_E, is parasitic. In other words, pushing p_E to high frequency for the sake of stability sacrifices loop gain. This is why, to keep loop gains around 50–60 dB, worst-case unity-gain frequencies in linear regulators normally hover between 10 and 500 kHz.

PMOS

In the case where the pass device is a P-channel MOSFET, G_P follows the square-law behavior of an MOS transconductance $g_{m.MOS}$:

$$g_{m.MOS} = \sqrt{2 i_L K_P'(W/L)_O} \tag{8.12}$$

where K_P' is the transconductance parameter and $(W/L)_O$ is the width–length ratio of the pass device. Luckily, G_P's square root suppresses the effects of i_L's variations in f_{GBW}. So in the absence of z_{ESR} and p_B, A_{LG} drops past p_O at 20 dB per decade of frequency like Fig. 8.5 shows, to f_{0dB} at f_{GBW}. Variations in i_L then shift f_{0dB}, but only by the square root of the variation. Channel-length modulation λ_L similarly sensitizes the output resistance of the pass device r_{ds} to i_L:

$$r_{sd} = \frac{1}{i_L \lambda_L} \tag{8.13}$$

Although this sensitivity in R_P translates to variations in R_O, and in consequence, in A_{LG} and p_O, f_{GBW} still captures the combined effects of R_O in both A_{LG} and p_O.

Unluckily, R_{ESR} may or may not be significant, so ESR zero z_{ESR} and bypass pole p_B may or may not interfere before A_{LG} reaches f_{0dB}. This means that z_{ESR} can extend f_{0dB} to the extent p_B allows, by a factor of p_B/z_{ESR}, which as noted earlier is equivalent to C_O/C_O' or $C_O/(C_B+C_L+C_P)$.

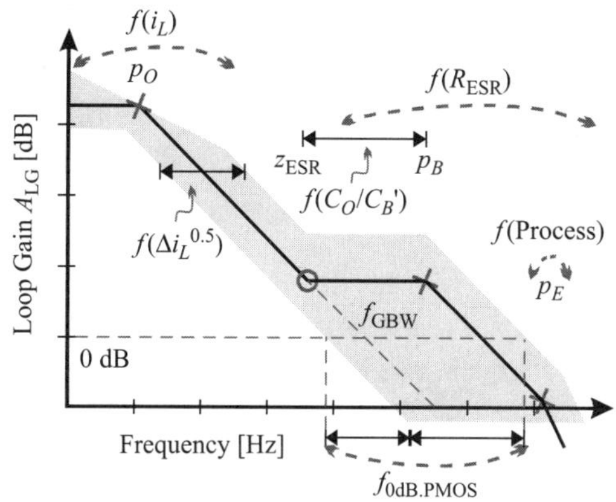

FIGURE 8.5 Output-compensated response when the pass device is a PMOSFET.

In other words, i_L in f_{GBW} shifts f_{0dB} by the square root of the variation and z_{ESR} and p_B can extend that by another factor:

$$\frac{f_{0dB(MAX).PMOS}}{f_{0dB(MIN).PMOS}} = \left(\frac{f_{GBW(MAX)}}{f_{GBW(MIN)}}\right)\left(\frac{p_B}{z_{ESR}}\right) = \left(\frac{A_E G_{P(MAX)} \beta_{FB}}{A_E G_{P(MIN)} \beta_{FB}}\right)\left(\frac{p_B}{z_{ESR}}\right)$$
$$\approx \left(\sqrt{\frac{i_{L(MAX)}}{i_{L(MIN)}}}\right)\left(\frac{C_O}{C_B + C_L + C_P}\right) \tag{8.14}$$

For reference, consider that the maximum–minimum f_{0dB} ratio for a 1–100-mA load with 1 μF and 0.1 μF of output and bypass capacitances is 100 to 1. This range essentially specifies that, for a targeted $f_{0dB(MIN)}$, parasitic poles in the system must nevertheless lie at or above $f_{0dB(MAX)}$.

PNP

G_P in PNP BJTs is unfortunately more sensitive to i_L than in MOSFETs because transconductance $g_{m.BJT}$ in BJTs is a linear translation of i_L:

$$g_{m.BJT} = \frac{i_L}{V_t} \tag{8.15}$$

This means that variations in i_L appear uninhibited in G_P and, as a result, also in f_{GBW} and f_{0dB}, as Fig. 8.6 shows. As in FETs, base-width modulation V_A also sensitizes the output resistance of the pass device r_o to i_L:

$$r_o = \frac{V_A}{i_L} \tag{8.16}$$

This sensitivity in R_P appears in R_O, and in consequence, in A_{LG} and p_O. Interestingly, R_O's dependence on i_L cancels that of G_P in A_{LG}'s

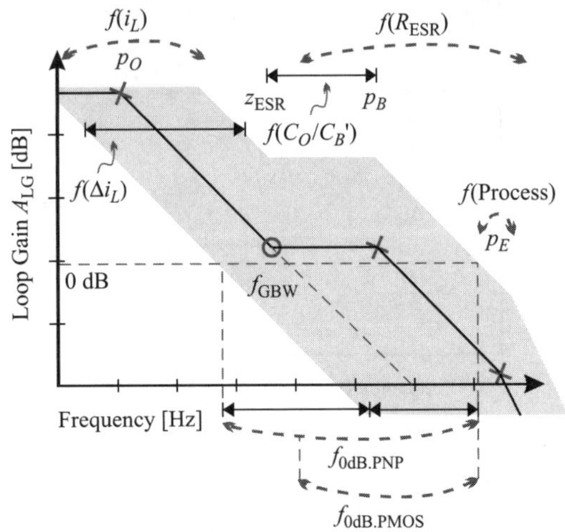

Figure 8.6 Output-compensated response when the pass device is a PNP BJT.

$G_P R_O$, so A_{LG} does not change with i_L. p_O, however, still changes with R_O, so f_{GBW}, which captures the combined effects of A_{LG} and p_O, is nonetheless sensitive to i_L.

Like before, z_{ESR} and p_B may or may not appear below f_{0dB}, so z_{ESR} can further extend f_{0dB} to the extent that p_B allows, by p_B/z_{ESR}, or equivalently, by $C_O/(C_B + C_L + C_P)$:

$$\frac{f_{0dB(MAX).PNP}}{f_{0dB(MIN).PNP}} = \left(\frac{f_{GBW(MAX)}}{f_{GBW(MIN)}}\right)\left(\frac{p_B}{z_{ESR}}\right) = \left(\frac{A_E G_{P(MAX)} \beta_{FB}}{A_E G_{P(MIN)} \beta_{FB}}\right)\left(\frac{p_B}{z_{ESR}}\right)$$
$$\approx \left(\frac{i_{L(MAX)}}{i_{L(MIN)}}\right)\left(\frac{C_O}{C_B + C_L + C_P}\right) \tag{8.17}$$

Here, the maximum–minimum f_{0dB} ratio for the 1–100-mA example with 1 μF and 0.1 μF of output and bypass capacitances is 1000 to 1, which is 10 times greater than that of the MOS counterpart. A vast variation like this is unfortunate because, for a specified minimum bandwidth, the circuit needs sufficient power to push parasitic poles above the highest possible bandwidth.

Worst Cases

The bandwidth of the system is at its minimum point $f_{0dB(MIN)}$ when gain–bandwidth frequency f_{GBW} is low and ESR zero z_{ESR} is high. In other words, $f_{0dB(MIN)}$ results when error-amplifier gain A_E, G_P's load current i_L, feedback factor β_{FB}, and ESR resistance R_{ESR} are low and C_O, C_B, C_L, and C_P are high. Meeting the stability criterion under these conditions is less likely when A_E's pole p_E is low. The situation worsens when temperature and fabrication corners raise R_{OE}, C_{OE}, C_O, and C_B and lower R_{ESR}. Simulations must therefore account for this worst-case condition at the temperature setting and process corner that produces the highest R_{OE}, C_{OE}, C_O, and C_B and the lowest R_{ESR}.

Bandwidth similarly rises to the opposite extreme $f_{0dB(MAX)}$ when f_{GBW} is high and z_{ESR} is low, which occurs when A_E, G_P's i_L, β_{FB}, and R_{ESR} are high and C_O, C_B, C_L, and C_P are low. Meeting the stability criterion under these new conditions is least likely when low-frequency loop gain A_{LG0} is high and p_E is low. And the situation worsens when temperature and fabrication corners strengthen transistors with higher transconductances; raise R_{OE}, C_{OE}, and R_{ESR}; and lower C_O and C_B. Simulations must therefore account for this other worst-case condition at the temperature setting and process corner that produces the highest gain, R_{OE}, C_{OE}, and R_{ESR} and lowest C_O and C_B.

8.3.2 Internal Compensation

Internal compensation in linear regulators refers to systems whose dominant low-frequency pole is inside the feedback loop and away from the output. This means that the output pole p_O is secondary and therefore parasitic. Engineers normally resort to internal compensation when designing linear regulators for low-power systems because load dumps are usually shallow and output and bypass capacitances can, as a result, be low. And since gain across the loop is ultimately the result of high resistance, the internal pole is usually in the error amplifier A_E in the form of p_E.

Unfortunately, ensuring that the output pole remains above the unity-gain frequency f_{0dB} is not easy. Load dumps, for one, even when shallow, are considerable, so output capacitance C_O is still substantial at hundreds of picofarads or even nanofarads. The second reason is supply noise. The problem is that negative feedback can only suppress supply noise up to f_{0dB}. Unfortunately, the input supply v_{IN} can ripple at 0.1–10 MHz, at or above typical f_{0dB}'s of 10–500 kHz. When confronted with megahertz ripples, past f_{0dB}, C_O is by and large the only component that can shunt them, as the next section shows.

R_P's sensitivity to load current i_L exacerbates the problem because i_L in low-power applications spans several decades of current, from 1–10 μA to 1–10 mA. As a result, because R_P in both MOSFETs and BJTs are linear translations of i_L, p_O rises linearly with i_L across the same number of decades. This means that the lowest possible p_O, which limits how high f_{0dB} can rise, is very low. This explains why f_{0dB} in practice hardly ever rises above 500 kHz in internally compensated systems.

Thankfully, f_{0dB} is normally independent of i_L. To see this, first consider that gain drops past the dominant pole in the absence of other poles and zeros as much as frequency climbs, at 20 dB per decade. As a result, the gain–bandwidth product GBW or $A_{LG0}p_E$ is constant across frequency past p_E. In other words, if no other poles or zeros intervene, low-frequency gain A_{LG0} drops as much gain as frequency rises above p_E to reach the unity-gain frequency f_{0dB} at $p_E A_{LG0}$, or equivalently, at GBW frequency f_{GBW}:

$$f_{GBW} = A_{LG0} p_E = \frac{G_E R_{OE} G_P [(R_{FB1} + R_{FB2}) \parallel R_O] \beta_{FB}}{2\pi R_{OE} C_{OE}}$$
$$\approx \frac{G_E G_P R_O \beta_{FB}}{2\pi C_{OE}} \approx \frac{G_E A_P \beta_{FB}}{2\pi C_{OE}} \tag{8.18}$$

where A_P is the gain across the pass device $G_P R_O$, R_O is the parallel combination of R_P and R_L, and R_P is normally so low that R_P shunts R_L, R_{FB1}, and R_{FB2}. Although maybe not obvious at first, for reasons to be addressed momentarily, A_P/C_{OE} in internally compensated systems is typically insensitive to i_L.

N-Type Follower

Other than reducing output capacitance, the only way to keep p_O at high frequency is by reducing output resistance. And what better way to do this than by using an N-type follower for the pass device. This way, R_P reduces to $1/g_m$ and p_O to $g_m/2\pi(C_O + C_B + C_L + C_P)$. Plus, the gain is nearly 1 because the effective load R_{LOAD} degenerates the input voltage of the pass device to produce a transconductance current that recreates the input voltage across the same degenerating load R_{LOAD}:

$$A_{P.Follower} = G_{M(DEG)} R_{LOAD} = \left(\frac{g_m}{1 + g_m R_{LOAD}}\right) R_{LOAD} \approx 1 \tag{8.19}$$

where R_{FB1}, R_{FB2}, and R_L and R_P in R_O comprise R_{LOAD}. Luckily, A_P's independence from i_L, as Fig. 8.7 illustrates, carries to f_{GBW}, and as a result, also to f_{0dB}.

P-Type Miller Transistor

Another way of reducing output impedance is by looping a Miller capacitor C_M across the pass device when the power device is a P-type transistor. This way, Miller capacitance $(1 + g_m R_{LOAD})C_M$ dominates C_{OE} to control p_E, and C_M shunts at higher frequency to reduce the output impedance of the regulator to $1/g_m$, where again, g_m is the transconductance of the pass device. R_P essentially becomes $1/g_m$ past p_E, so as with followers, C_O, C_B, C_L, and C_P shunt $1/g_m$ at the same p_O, at $g_m/2\pi(C_O + C_B + C_L + C_P)$. And since the gain across the power transistor is $g_m R_{LOAD}$, $g_m R_{LOAD}$ in A_P and C_{OE} practically cancel in A_P/C_{OE}:

$$\frac{A_{P.Miller}}{C_{OE}} = \frac{g_m R_{LOAD}}{(1 + g_m R_{LOAD})C_M} \approx \frac{1}{C_M} \tag{8.20}$$

In other words, variations that i_L causes in A_P cancel those in p_E in Fig. 8.8 to render f_{GBW}, and in consequence, f_{0dB} insensitive to i_L.

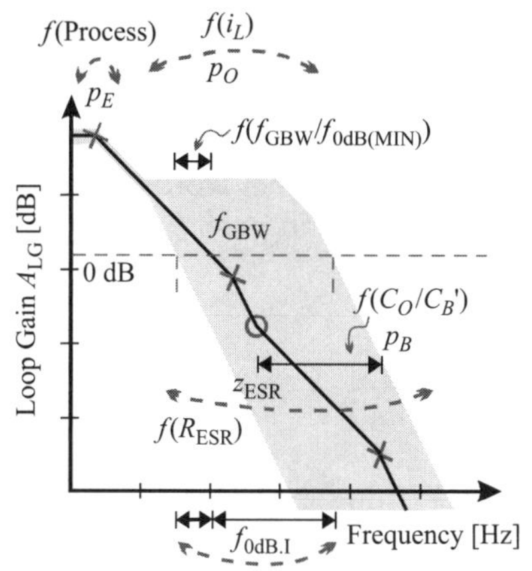

Figure 8.7 Internally compensated response when the pass device is an N-type follower.

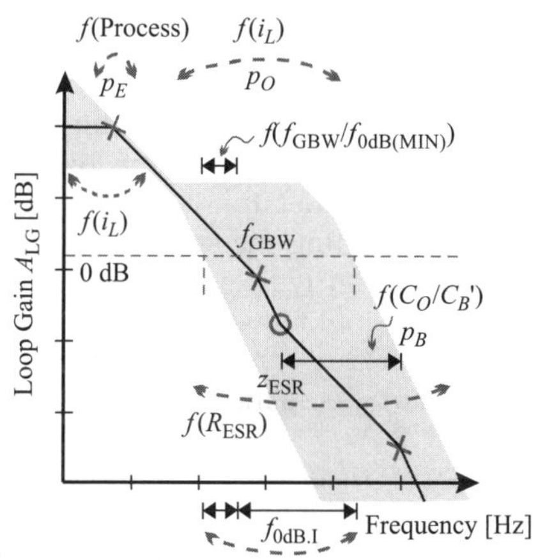

Figure 8.8 Internally compensated response when the pass device is a Miller-compensated P-type transistor.

Worst-Case Variance

Unfortunately, R_P's sensitivity to i_L manifests in p_O. Under worst-case conditions, in fact, p_O can be so low that, to maintain a practical bandwidth, designers allow p_O to fall below f_{0dB}. But to keep phase shift below 180°, p_O cannot fall further than a decade below f_{0dB}. Still, letting p_O cross f_{0dB} effectively extends f_{0dB}'s lower frequency range. Since A_{LG} drops 40 dB per decade, or by a factor of 100 for every 10 factor of rise in frequency, $f_{0dB(MIN)}$ should be no lower than half a decade or $10^{½}$ below f_{GBW}, which happens when p_O is 10 times lower than f_{GBW}.

But this is not all because R_{ESR} in some systems may still be high enough to pull ESR zero z_{ESR} below f_{0dB}. This means that z_{ESR} can extend f_{0dB} to the extent p_O allows, and under worst-case conditions when p_O disappears to high frequency, to the extent p_B allows by a factor of p_B/z_{ESR}, or $C_O/(C_B + C_L + C_P)$. In other words, while $p_{O(MIN)}$ can pull f_{0dB} below f_{GBW} by $10^{½}$, z_{ESR} and p_B can push f_{0dB} to higher frequency by a factor of $C_O/(C_B + C_L + C_P)$:

$$\frac{f_{0dB(MAX)}}{f_{0dB(MIN)}} = \left(\frac{f_{GBW}}{f_{0dB(MIN)}}\right)\left(\frac{p_B}{z_{ESR}}\right)$$

$$= \left(\frac{f_{GBW}}{f_{0dB(MIN)}}\right)\left(\frac{C_O}{C_B + C_L + C_P}\right) < \left(\sqrt{10}\right)\left(\frac{C_O}{C_B + C_L + C_P}\right) \tag{8.21}$$

Fortunately, R_{ESR}, C_B, C_L, and C_P can be so low in highly integrated low-power systems that the resulting range reduces to more manageable levels. Still, p_O remains a concern because p_O sets an upper boundary for f_{0dB}.

Worst Cases

The bandwidth of the system is at its minimum point $f_{0dB(MIN)}$ when error-amplifier pole p_E and output pole p_O are low and ESR zero z_{ESR} is high, which results when load current i_L and R_{ESR} are low and C_O, C_B, C_L, and C_P are high. Meeting the stability criterion under these conditions is less likely when temperature and fabrication corners raise R_{OE}, C_{OE}, R_P, C_O, and C_B and lower R_{ESR}. Simulations must therefore account for this worst-case condition at the process corner and temperature setting that produces the highest R_{OE}, C_{OE}, R_P, C_O, and C_B and the lowest R_{ESR}.

Bandwidth similarly rises to the opposite extreme $f_{0dB(MAX)}$ when p_E is high and z_{ESR} is low, which results when R_{ESR} is high. Meeting the stability criterion under these conditions is less likely when p_O is low, which results when i_L is low and C_O, C_B, C_L, and C_P are high. Meeting the stability criterion under these conditions is less likely when temperature and fabrication corners lower R_{OE} and C_{OE} and raise C_O, C_B, and R_{ESR}. Simulations must therefore account for this worst-case condition at the temperature setting and process corner that produces the lowest R_{OE} and C_{OE} and the highest C_O, C_B, and R_{ESR}.

8.4 Power-Supply Rejection

Power-supply rejection, or as it is also commonly known in the industry, *power-supply ripple rejection*, or *ripple rejection* for short, refers broadly to the ability of a circuit to oppose the effects of small-signal fluctuations in the input supply on the output. PSR is therefore the complement of supply injection, or equivalently, the reciprocal of *supply gain* A_{IN}. A_{IN} here is the small-signal variation in the output voltage ∂v_O, or v_o, that results in response to small-signal changes in the input supply ∂v_{IN}, or v_{in}:

$$\text{PSR} \equiv \frac{1}{A_{IN}} = \frac{\partial v_{IN}}{\partial v_O} \equiv \frac{v_{in}}{v_o} \tag{8.22}$$

Similarly, but with respect to steady-state signals, *line regulation* (LNR) is essentially supply gain A_{IN} at low frequency. In other words, LNR refers to steady-state variations in v_O, or ΔV_O, in response to steady-state changes in v_{IN}, or ΔV_{IN}. But because LNR applies to the linear region of the regulator, v_O displays linear translations of ΔV_{IN} like v_o manifests those of v_{in}. This means that LNR is basically A_{IN} at low frequency, which is also equivalent to PSR's complement at low frequency:

$$\text{LNR} \equiv \frac{\Delta V_O}{\Delta V_{IN}} \approx \frac{\partial v_O}{\partial v_{IN}} = A_{IN0} = \frac{1}{\text{PSR}_0} \tag{8.23}$$

where A_{IN0} and PSR_0 are A_{IN} and PSR at low frequency.

Regrettably, "power-supply ripple rejection" produces the same acronym as "*power-supply rejection ratio*," the latter of which designers most often apply to amplifiers, not regulators. Although the two terms relate and are similar in many respects, they are not the same. The latter, for example, refers to how much a circuit favors changes in the input signal to be processed over variations in the supply, so PSRR is the ratio of forward input–output voltage gain A_V to supply–output gain A_{DD}:

$$\text{PSRR} \equiv \frac{A_V}{A_{DD}} = A_V\text{PSR} \tag{8.24}$$

Power-supply ripple rejection, on the other hand, refers to PSR, which again, is the complement of supply–output gain, which in this case is $1/A_{DD}$, and in the context of linear regulators, is $1/A_{IN}$. In other words, PSRR is the product of forward gain A_V and PSR.

Interestingly, applying a supply ripple to an operational amplifier in unity-gain configuration resembles the linear-regulator setup. In fact, this is a rather useful means of extrapolating the PSRR performance of an operational amplifier because A_{DD} transfers supply ripple v_{dd} to v_o and loop gain A_V suppresses it. As a result, v_o exhibits the effects of both A_{DD} and A_V:

$$v_o = v_{dd}A_{DD} + v_{id}A_V = v_{dd}A_{DD} - v_o\beta_{FB}A_V \tag{8.25}$$

where feedback factor β_{FB} is 1 in unity gain and A_V amplifies the amplifier's differential voltage v_{ID}, which is $v_{REF} - v_O$, where v_{REF} excludes small-signal components, so v_{id} reduces to $-v_o$ and

$$\frac{v_{dd}}{v_o} = \left.\frac{A_{DD}}{1 + A_V\beta_{FB}}\right|_{\beta_{FB}=1} = \left.\frac{A_{DD}}{1 + A_V}\right|_{A_V>>1} \approx \frac{A_{DD}}{A_V} = \frac{1}{\text{PSRR}} \tag{8.26}$$

This approximation for PSRR, however, is only valid as long as forward gain A_V is considerably higher than 1. In other words, PSRR is v_o/v_{dd} up to the unity-gain frequency f_{0dB}, which is where PSRR differs from PSR, because v_o/v_{dd} is accurate for PSR past f_{0dB}. But not to digress further, and to avoid confusion, this book only uses the acronym PSR to refer to ripple-rejection performance, that is, to power-supply rejection.

8.4.1 Voltage-Divider Model

Since PSR is the complement of input-supply gain A_{IN}, extracting PSR performance amounts to deciphering the small-signal gain from the input supply v_{in} to the output v_o. With this in mind, recognizing that the pass device S_O, the feedback network β_{FB}, the output filter C_{Filter}, and the load comprise a voltage divider from v_{in} to v_o is helpful. Since S_O's output resistance and capacitance R_P and C_P from Fig. 8.9 connect v_{in} to v_o, together they constitute the top impedance of the network Z_T. Similarly, β_{FB}'s R_{FB1} and R_{FB2} and C_{Filter}'s C_O, R_{ESR}, and C_B and C_L in C_B' compose the bottom impedance Z_B. In this light, since PSR is high when A_{IN} is low, good PSR performance results when S_O drops most of v_{in}, or stated differently, when β_{FB}, C_{Filter}, and the load shunt as much supply noise energy as possible, and S_O couples as little as possible.

The transconductance of the pass device G_P affects A_{IN} in two ways. First, G_P is the medium through which supply noise that feeds

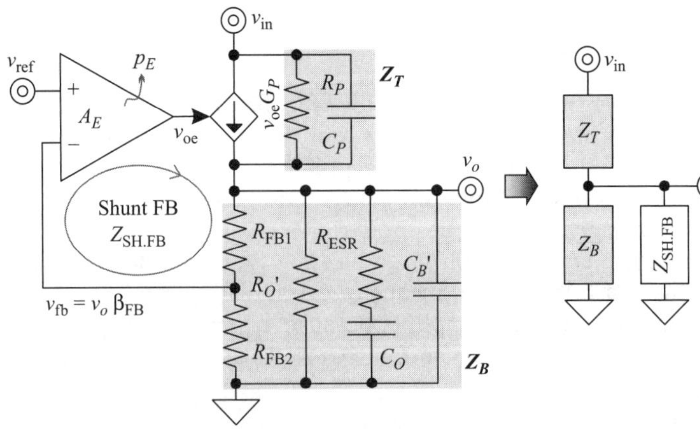

FIGURE 8.9 Voltage-divider model for PSR analysis.

through the error amplifier A_E reaches v_O. How this noise affects v_O depends on the nature of the pass device, which is the subject of the following subsection. Neglecting feed-through effects for now, G_P is also the element that closes the loop in the feedback network, which means G_P opposes the effects of supply ripple on the output. In this case, since R_{FB1} and R_{FB2} shunt-sample v_o, G_P presents the shunting impedance $Z_{SH.FB}$ that negative feedback establishes when sampling the output voltage.

Without resorting to feedback theory, $Z_{SH.FB}$ is fundamentally G_P's output impedance v_o/i_{GP}, where i_{GP} is a $\beta_{FB}A_EG_P$ translation of v_o. As such, $Z_{SH.FB}$ reduces to $1/\beta_{FB}A_EG_P$, or equivalently, to the reciprocal of the equivalent feedback transconductance G_{FB}:

$$Z_{SH.FB} \equiv \frac{v_o}{i_{GP}} = \frac{v_o}{v_o\beta_{FB}A_EG_P} \equiv \frac{1}{G_{FB}} = \frac{Z_T \parallel Z_B}{\beta_{FB}A_EG_P(Z_T \parallel Z_B)} = \frac{Z_{O.OL}}{\beta_{FB}A_{OL}} = \frac{Z_{O.OL}}{A_{LG}} \tag{8.27}$$

For reference, multiplying the top and bottom of this relation by the open-loop output impedance of the network $Z_T \parallel Z_B$, or $Z_{O.OL}$, reveals that the denominator $\beta_{FB}A_EG_PZ_{O.OL}$ is the gain across the loop A_{LG}, and the resulting ratio is the shunting impedance that feedback theory predicts. But not to deviate the discussion from PSR, the effect of $Z_{SH.FB}$ is to shunt noise energy away from v_o, so $Z_{SH.FB}$ reduces A_{IN} and raises PSR.

8.4.2 Feed-Through Noise

Common-Mode Concept

As mentioned earlier, the transconductor G_P in the pass device can feed supply noise in the output v_{OE} of the error amplifier A_E to the output v_O. To what extent noise feeds through ultimately depends on the power transistor. In the case of N-type devices, for example, the voltage followers that the transistors implement reproduce the supply noise they receive in v_{OE} across their loads, so as Fig. 8.10 illustrates, v_O displays the same noise that v_{OE} carries. A_E should therefore

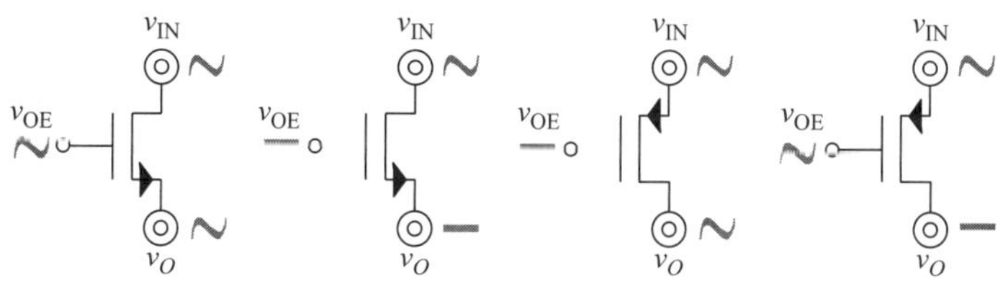

FIGURE 8.10 Illustration of the common-mode concept in N- and P-type transistors.

suppress as much supply noise as possible when driving N-type pass devices.

Interestingly, eliminating supply noise in v_{OE} is not always the best solution. The reason for this is that the source and emitter terminals of P-type transistors connect directly to the input supply v_{IN}. So in the absence of supply noise at their gates and bases, as Fig. 8.10 shows, their source–gate and emitter–base voltages ripple with v_{IN}. And because small-signal drain and collector currents are transconductance translations of source–gate and emitter–base voltages, P-type transistors inject noise energy into v_O when v_{OE} is free of noise. Recreating the supply noise in v_{OE}, on the other hand, removes the small-signal difference between sources and gates and emitters and bases. So to eliminate feed-through noise, A_E should replicate in v_{OE} ripples in the supply when driving P-type power devices.

The underlying theme in this discussion is the concept of *common mode*. Because transistors, like differential amplifiers, amplify the small-signal difference between their input voltages, noise that is common to both input terminals cancels. As a result, suppressing noise amounts to reproducing the small-signal conditions present at the source or emitter on the gate or base, and vice versa. In the case of linear regulators, A_E should therefore reproduce and drive whatever noise the source or emitter of the power transistor already incorporates.

Current Mirrors

As one might expect, current mirrors clone supply currents. To perceive the significance of this statement, consider the small-signal translations of N- and P-type current mirrors, which Fig. 8.11 displays. Because the input supply v_{IN} is across the I_{Bias} source and the diode-connected input transistor of the mirror, supply ripple v_{in} establishes a small-signal current i_{in} through their resistances R_{EQ} and $1/g_m$:

$$i_{in} = \frac{v_{in}}{R_{EQ} + \frac{1}{g_m}} \approx \frac{v_{in}}{R_{EQ}} \tag{8.28}$$

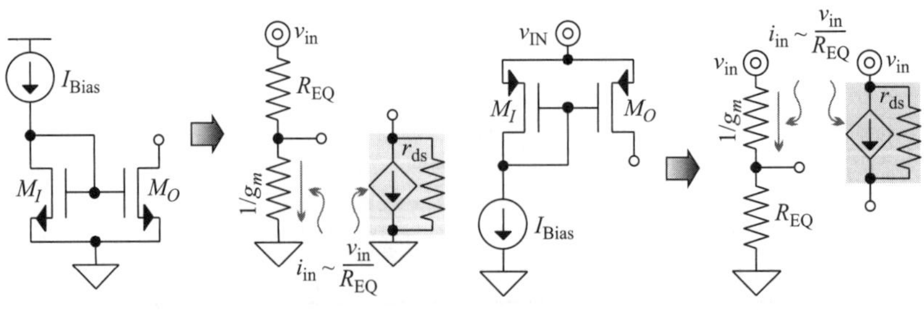

FIGURE 8.11 Small-signal translations of N- and P-type current mirrors.

but $1/g_m$ is so much lower than R_{EQ} that i_{in} reduces to v_{in}/R_{EQ}. The transconductance of the output transistor mirrors this supply-ripple current i_{in} to the output. Notice, however, that while ground-referenced N-type mirrors sink i_{in}, their supply-referenced P-type counterparts source i_{in}, which means one can probably cancel the effects of the other, and of other similar ripples in a circuit.

N-Type Mirror in the Error Amplifier

Within the context of the error amplifier A_E, N-type mirrors sink supply ripple from A_E's output v_{OE}. To see the impact of this action, consider the differential stage that the P-type input pair M_{I1}–M_{I2} and N-type mirror M_{M3}–M_{M4} in Fig. 8.12 realize. Because the effect of negative feedback in a linear regulator is to impress one of A_E's inputs onto the other, v_{I1} and v_{I2} equal, and their small-signal contributions are therefore nil. In fact, since shunt-impedance $Z_{SH.FB}$ already models the effects of negative feedback on the overall system, including those effects here would be redundant. So with no small signals at their gates, M_{I1} and M_{I2} buffer the supply-ripple current that v_{in} injects through the tail-current source M_T into the N-type mirror.

As in the basic mirrors of Fig. 8.11, M_T and M_{I1} and M_T and M_{I2} present equivalent resistances R_{EQ} that connect the power supply to the $1/g_{m3}$ resistance of the diode-connected transistor M_{M3} and to the resistance r_{ds4}, or R_M, of the mirroring output M_{M4}. And since R_{EQ} is considerably higher than M_{M3}'s $1/g_{m3}$, the ripple current i_{in} that v_{in} feeds into the mirror is nearly v_{in}/R_{EQ}. As a result, with respect to the output v_{oe}, R_{EQ} feeds supply ripple and M_{M4} sinks ripple. In other words, M_{M4} opposes the effects of R_{EQ}. To quantify their combined impact on v_{oe}, first consider their individual effects and then combine them using superposition. In this light, R_{EQ} and R_M when M_{M4}'s transconductance is absent voltage-divides v_{in} to v_{oe},

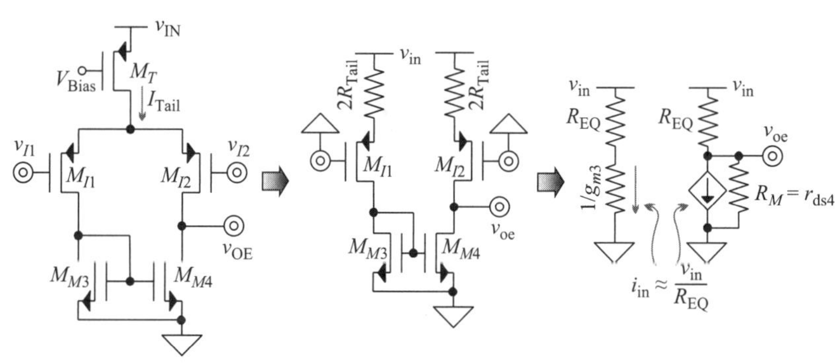

FIGURE 8.12 P-type differential pair with an N-type mirror and its small-signal translation.

and M_{M4} when v_{in} is zero pulls i_{in} from the parallel combination of R_{EQ} and R_M:

$$A_{IN(N.MIR)} = \frac{v_{oe}}{v_{in}} = \frac{\left(\frac{v_{in} R_M}{R_{EQ} + R_M}\right) - [i_{in}(R_{EQ} \parallel R_M)]}{v_{in}}$$
$$\approx \frac{\left(\frac{v_{in} R_M}{R_{EQ} + R_M}\right) - \left[\left(\frac{v_{in}}{R_{EQ}}\right)(R_{EQ} \parallel R_M)\right]}{v_{in}} = 0 \quad (8.29)$$

Interestingly, the symmetry of the circuit ensures that the contributions of R_{EQ} and M_{M4} are equal and opposite, so their ripple contributions cancel in v_{oe}. This of course assumes $1/g_{m3}$ is considerably lower than R_{EQ}, which is normally the case. Still, reversing this assumption nevertheless shows that M_{M4} sinks $v_{in}/(R_{EQ} + 1/g_{m3})$ to cancel some of the ripple that R_{EQ} injects:

$$A_{IN(N.MIR)} \equiv \frac{v_{oe}}{v_{in}} = \frac{\left(\frac{v_{in} R_M}{R_{EQ} + R_M}\right) - [i_{in}(R_{EQ} \parallel R_M)]}{v_{in}}$$
$$\approx \left(\frac{R_M}{R_{EQ} + R_M}\right) - \left(\frac{R_{EQ} \parallel R_M}{R_{EQ} + \frac{1}{g_{mM3}}}\right) \quad (8.30)$$

This cancelling effect extends to all circuits that present symmetrical connections from the input supply to an N-type mirror. The N-type mirror M_{M3}–M_{M4} in the folded-cascode amplifier of Fig. 8.13,

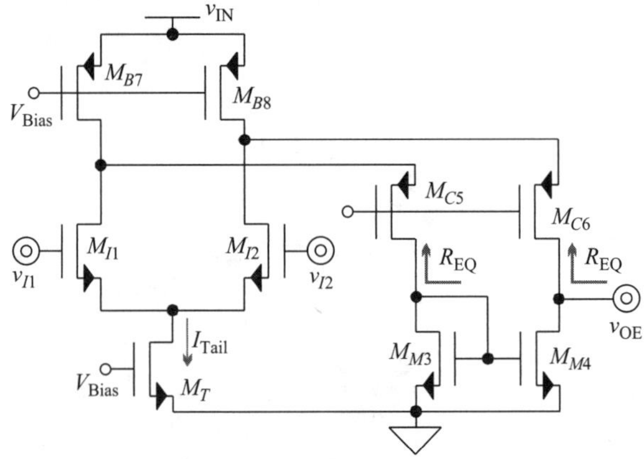

FIGURE 8.13 Folded-cascode amplifier with an N-type input pair and an N-type mirror load.

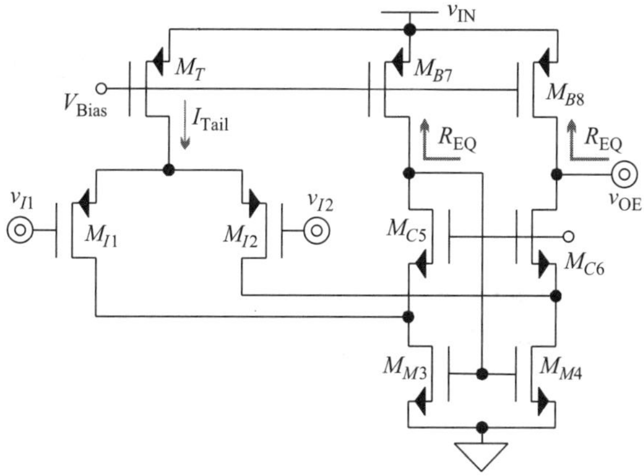

FIGURE 8.14 Folded-cascode amplifier with a P-type input pair and an N-type mirror load.

for example, similarly sinks and cancels ripple current from the input supply. In this case, N-type input pair M_{I1}–M_{I2} outputs differential currents that cascode transistors M_{C5} and M_{C6} buffer and fold into the N-type mirror. From the perspective of the mirror, M_{C5} and M_{C6} establish symmetrical connections to the input supply v_{IN}, so M_{C5} and M_{C6} present equivalent resistances R_{EQ} to supply noise v_{in}. As a result, the circuit reduces to the same equivalent circuit illustrated in Fig. 8.12, which means the N-type mirror removes the supply ripple that R_{EQ} injects to produce a noise-free output.

Any variation that employs an N-type mirror and presents symmetrical connections from the supply to the mirror produces the same result. The low-voltage cascode topology of Fig. 8.14 is yet another example. Here, the P-type differential pair M_{I1}–M_{I2} produces differential currents that cascode transistors M_{C5} and M_{C6} fold and buffer into N-type mirror M_{M3}–M_{M4}, where as a current buffer, M_{C5} diode-connects M_{M3}. This way, mirroring transistors M_{M3} and M_{M4} receive bias currents from M_{B7} and M_{B8} and differential currents from M_{I1} and M_{I2}. From the perspective of the mirror, M_{B7} and M_{B8} present symmetrical resistances R_{EQ} to the supply. M_{I1} and M_{I2} also feed equivalent ripple currents into the mirror, so M_{M3} receives M_{I1}'s ripple and M_{M4} mirrors what M_{M3} collects to cancel M_{I2}'s matching contribution. And like before, the N-type mirror cancels the ripple that M_{B8}'s R_{EQ} injects to produce a noise-free output.

Example 8.2 Determine the supply gain of an error amplifier that incorporates an N-type current mirror when transistor transconductances are 50 μS and resistances are 500 kΩ.

Solution

$$A_{IN} \equiv \frac{v_{oe}}{v_{in}} \approx \left(\frac{R_M}{R_{EQ}+R_M}\right) - \left(\frac{R_{EQ} \parallel R_M}{R_{EQ}+\dfrac{1}{g_{mM3}}}\right) \approx \left(\frac{0.5\,\text{M}\Omega}{0.5\,\text{M}\Omega+0.5\,\text{M}\Omega}\right) - \left(\frac{0.5\,\text{M}\Omega \parallel 0.5\,\text{M}\Omega}{0.5\,\text{M}\Omega+\dfrac{1}{50\,\mu\text{S}}}\right)$$

$$= 0.5 - \left(\frac{0.25\,\text{M}\Omega}{0.5\,\text{M}\Omega+\dfrac{1}{50\,\mu\text{S}}}\right) \approx 19\,\text{mV/V} = -34\,\text{dB}$$

P-Type Mirror in the Error Amplifier

In the context of the error amplifier A_E, P-type mirrors source power supply ripple into A_E's output v_{OE}. To see the impact of this action, consider the differential stage that the N-type input pair M_{I1}–M_{I2} and P-type mirror M_{M3}–M_{M4} in Fig. 8.15 realize. Because the effect of negative feedback in a linear regulator is to impress one of A_E's inputs onto the other, v_{I1} and v_{I2} equal, and their small-signal contributions are therefore nil. And since shunt-impedance $Z_{SH.FB}$ already models the effects of negative feedback on the overall system, including those effects here would be redundant. So with no small signals at their gates, M_{I1} and M_{I2} establish symmetrical connections from the P-type mirror to ground.

As in the basic mirrors of Fig. 8.11, M_T and M_{I1} and M_T and M_{I2} present equivalent resistances R_{EQ} that connect ground to the $1/g_{m3}$ resistance of the diode-connected transistor M_{M3} and to the resistance r_{ds4}, or R_M, of the mirroring output M_{M4}. And since R_{EQ} is considerably higher than M_{M3}'s $1/g_{m3}$, the ripple current i_{in} that v_{in} establishes through the mirror is nearly v_{in}/R_{EQ}. As a result, with respect to the output v_{oe}, both R_M and M_{M4}'s mirroring transconductance g_{m4} feed supply ripple. In other words, R_M and g_{m4} reinforce one another to produce noise in v_{oe}. To quantify their combined impact on v_{oe}, first consider their individual effects and then combine them using superposition. In this light, R_M and R_{EQ} when g_{m4} is absent voltage-divide v_{in}

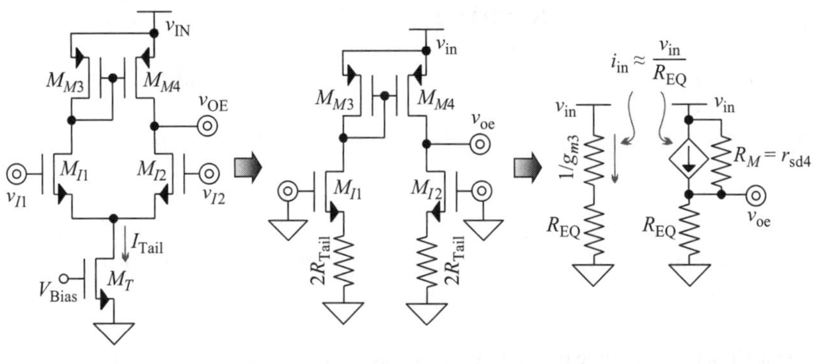

FIGURE 8.15 N-type differential pair with a P-type mirror and its small-signal translation.

to v_{oe}, and g_{m4} when v_{in} is zero pushes i_{in} into the parallel combination of R_{EQ} and R_M:

$$A_{IN(P.MIR)} = \frac{v_{oe}}{v_{in}} = \frac{\left(\frac{v_{in}R_{EQ}}{R_{EQ}+R_M}\right)+[i_{in}(R_{EQ}\parallel R_M)]}{v_{in}} \approx \frac{\left(\frac{v_{in}R_{EQ}}{R_{EQ}+R_M}\right)+\left[\left(\frac{v_{in}}{R_{EQ}}\right)(R_{EQ}\parallel R_M)\right]}{v_{in}} = 1 \tag{8.31}$$

Interestingly, the symmetry of the circuit ensures that the contributions of R_M and g_{m4} not only match but also reinforce one another to reproduce the input ripple v_{in} in v_{oe}. This of course assumes $1/g_{m3}$ is considerably lower than R_{EQ}, which is normally the case. Still, reversing this assumption nevertheless shows that g_{m4} sources $v_{in}/(R_{EQ}+1/g_{m3})$ to help R_M reconstruct the noise in v_{oe}:

$$A_{IN(P.MIR)} \equiv \frac{v_{oe}}{v_{in}} = \frac{\left(\frac{v_{in}R_{EQ}}{R_{EQ}+R_M}\right)+[i_{in}(R_{EQ}\parallel R_M)]}{v_{in}} \approx \left(\frac{R_{EQ}}{R_{EQ}+R_M}\right)+\left(\frac{R_{EQ}\parallel R_M}{R_{EQ}+\frac{1}{g_{mM3}}}\right) \tag{8.32}$$

This in-phase and constructive effect extends to all circuits that present symmetrical connections from a P-type mirror to ground. The P-type mirror M_{M3}–M_{M4} in the folded-cascode amplifier of Fig. 8.16, for example, similarly supplies and reinforces ripple

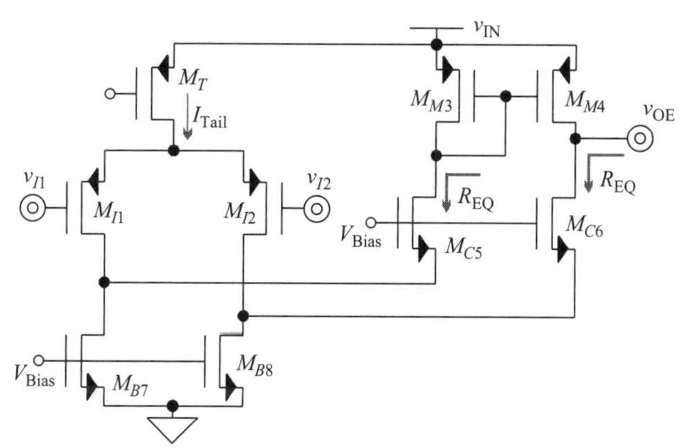

FIGURE 8.16 Folded-cascode amplifier with a P-type input pair and a P-type mirror load.

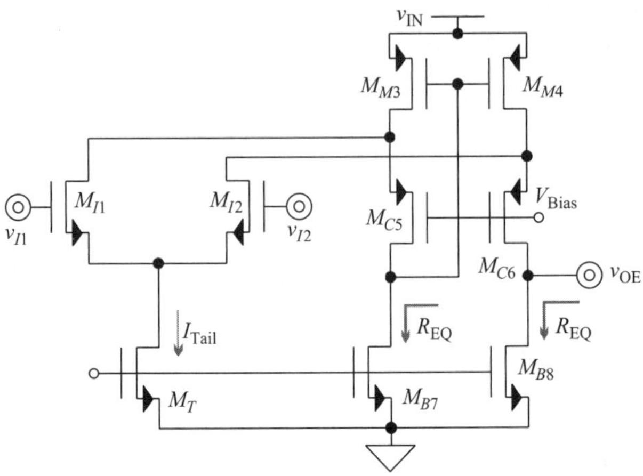

FIGURE 8.17 Folded-cascode amplifier with an N-type input pair and a P-type mirror load.

current into the output. In this case, P-type input pair M_{I1}–M_{I2} outputs differential currents that cascode transistors M_{C5} and M_{C6} buffer and fold into the P-type mirror. From the perspective of the mirror, M_{C5} and M_{C6} establish symmetrical connections to ground, so M_{C5} and M_{C6} present equivalent resistances R_{EQ} to ground. As a result, the circuit reduces to the same equivalent circuit illustrated in Fig. 8.15, which means the transconductance and resistance of the P-type mirror reinforce one another to reproduce supply noise in the output.

Any variation that employs a P-type mirror and presents symmetrical connections to ground produces the same result. The low-voltage cascode topology of Fig. 8.17 is another example. Here, the N-type differential pair M_{I1}–M_{I2} produces differential currents that cascode transistors M_{C5} and M_{C6} fold and buffer into P-type mirror M_{M3}–M_{M4}, where as a current buffer, M_{C5} diode-connects M_{M3}. This way, mirroring transistors M_{M3} and M_{M4} source M_{B7}'s and M_{B8}'s bias currents and M_{I1}'s and M_{I2}'s differential currents. From the perspective of the mirror, M_{B7} and M_{B8} present symmetrical resistances R_{EQ} to ground. M_{I1} and M_{I2} also sink equivalent ripple currents from the mirror, so M_{M3} sources M_{I1}'s ripple and M_{M4} mirrors what M_{M3} supplies to cancel M_{I2}'s matching contribution. And like before, the transconductance and resistance of the P-type mirror reinforce one another to reproduce supply noise in the output.

Example 8.3 Determine the supply gain of an error amplifier that incorporates a P-type current mirror when transistor transconductances are 50 μS and resistances are 500 kΩ.

Solution

$$A_{IN} \equiv \frac{v_{oe}}{v_{in}} \approx \left(\frac{R_{EQ}}{R_{EQ}+R_M}\right)+\left(\frac{R_{EQ} \parallel R_M}{R_{EQ}+\frac{1}{g_{mM3}}}\right) \approx \left(\frac{0.5\,\mathrm{M\Omega}}{0.5\,\mathrm{M\Omega}+0.5\,\mathrm{M\Omega}}\right)+\left(\frac{0.5\,\mathrm{M\Omega} \parallel 0.5\,\mathrm{M\Omega}}{0.5\,\mathrm{M\Omega}+\frac{1}{50\,\mu\mathrm{S}}}\right)$$

$$= 0.5+\left(\frac{0.25\,\mathrm{M\Omega}}{0.5\,\mathrm{M\Omega}+\frac{1}{50\,\mu\mathrm{S}}}\right) \approx 0.98\,\mathrm{V/V} = -0.17\,\mathrm{dB}$$

Conclusions

Several conclusions merit emphasis here. One is that transistors cancel common-mode noise, so while noise-free gates and bases keep N-type devices from propagating supply noise, P-type transistors achieve similar results with in-phase supply noise in their gate and base voltages. Second, N-type mirrors sink supply ripple and P-type mirrors source supply ripple. And when presented with symmetrical loads, N-type mirrors cancel supply ripple and P-type mirrors reproduce supply ripple. So combining all these assertions teaches that error amplifiers in linear regulators should employ N-type mirrors when driving N-type power transistors and P-type mirrors otherwise. This way, the transconductance of the power transistor G_P does not feed supply noise to the output of the regulator.

8.4.3 Miller Capacitor

A Miller capacitor across a P-type power device, like C_M and M_O illustrate in Fig. 8.18, alters the dynamics of the connection between the supply and the output. Here, C_M and the output resistance of the error amplifier R_{OE} realize a voltage divider that diode-connects M_O. So at high frequencies, when C_M is practically a short circuit with respect to R_{OE}, M_O's $1/g_{mO}$ diode-connected resistance connects the input supply v_{in} to v_o. Unfortunately, reducing the resistance between v_{in} and v_o this way couples high-frequency supply noise to the output.

As a whole, M_O's transconductance g_{mO}, or G_P, establishes an impedance Z_{GP} that together with M_O's output resistance r_{sdO}, or R_P,

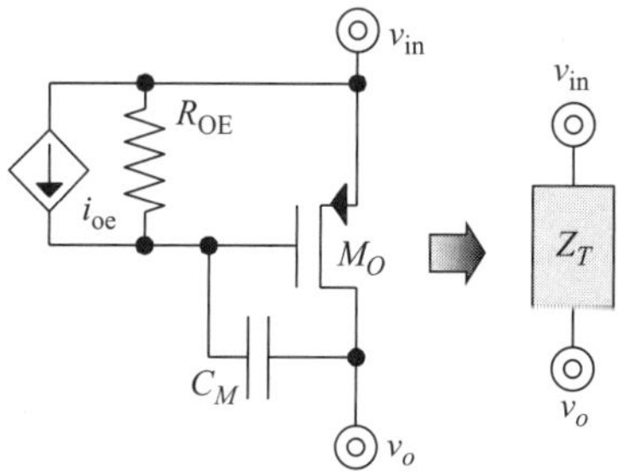

Figure 8.18 Miller capacitor across a P-type power transistor and their combined impedance.

and drain–bulk capacitance C_{DB}, or C_P, connect v_{in} to v_o in the form of top impedance Z_T:

$$Z_T = Z_{GO} \parallel R_P \parallel \frac{1}{sC_P} = Z_{GP} \parallel r_{sdO} \parallel \frac{1}{sC_{DB}} \quad (8.33)$$

Although C_M and R_{OE} diode-connect M_O, C_M's impedance Z_{CM}, or $1/sC_M$, and C_P's $1/sC_P$ are so high at low frequency that Z_T reduces to M_O's R_P. As operating frequency rises, however, M_O's diode impedance Z_{GP} falls, and when Z_{GP} falls below R_P, Z_T begins to fall with Z_{GP}. To see when this happens, consider that C_M and R_{OE} voltage-divide v_o to M_O's gate, so g_m in the $1/g_m$ resistance of a diode connection is here a voltage-divided ratio of G_P:

$$Z_{GP} = \frac{1}{\left(\frac{G_P R_{OE}}{R_{OE} + Z_{CM}}\right)} \approx \left.\frac{R_{OE} + \frac{1}{sC_M}}{G_P R_{OE}}\right|_{\frac{1}{sC_M} >> R_{OE}} \approx \left.\frac{1}{sC_M G_P R_{OE}}\right|_{p_M \approx \frac{1}{2\pi R_{OE}(G_P R_P C_M)}} \equiv R_P \quad (8.34)$$

But since C_M's $1/sC_M$ is still high, $1/sC_M$ overwhelms R_{OE}, so Z_{GP} falls below R_P past the pole that effective capacitance $G_P R_P C_M$ establishes with R_{OE}, which is the same Miller pole p_M discussed in Chaps. 4 and 6. Seen in a different light, Z_T loads the error amplifier A_E, so the gain across A_E falls with Z_T past p_M.

Z_T continues to fall with Z_{GP} until C_M shorts with respect to R_{OE}, at which point Z_{GP} and, as a result, Z_T flatten to diode-connected resistance $1/G_P$. This transition happens when $1/sC_M$ crosses R_{OE}, at zero z_{GP}, which is equivalent to $1/2\pi R_{OE} C_M$:

$$\left.\frac{1}{sC_M}\right|_{z_{GP} = \frac{1}{2\pi R_{OE} C_M}} \equiv R_{OE} \quad (8.35)$$

Because C_M is normally intentional and C_P is parasitic, $1/sC_P$ shunts $1/G_P$ at higher frequency, at pole p_{CP}, which is equivalent to $G_P/2\pi C_P$:

$$\left.\frac{1}{sC_P}\right|_{p_{CP} = \frac{G_P}{2\pi C_P}} \equiv \frac{1}{G_P} \quad (8.36)$$

In all, Z_T starts as R_P at low frequency, falls past Miller pole p_M, flattens past G_P's zero z_{GP}, and falls again past C_P's pole p_{CP}:

$$Z_T = \frac{R_P\left(1 + \frac{s}{2\pi z_{GP}}\right)}{\left(1 + \frac{s}{2\pi p_M}\right)\left(1 + \frac{s}{2\pi p_{CP}}\right)} \quad (8.37)$$

8.4.4 Analysis

How much supply ripple v_{in} reaches the output v_O hinges on the components that connect to v_O. This is why the output stage of a circuit is so important, because it voltage-divides v_{in} to v_o. In this respect, the role of the power device S_O is to drop as much, or to state it differently, couple as little of v_{in} as possible. The job of shunt feedback, on the other hand, is to shunt v_{in}. But since negative feedback loses its effects past its unity-gain bandwidth f_{0dB}, the purpose of the output filter C_{Filter} is to continue shunting v_{in} past f_{0dB}. Unfortunately, C_{Filter} is neither ideal nor sufficiently high to be as effective near f_{0dB} as feedback is well below f_{0dB}. This is why supply gain A_{IN} usually peaks and PSR troughs near f_{0dB} in linear regulators.

Still, S_O's possible diode-connection impedance Z_{GP}, output resistance R_P, and output capacitance C_P in Fig. 8.19 should drop most of v_{in}. Shunt-feedback impedance $Z_{SH.FB}$ and the combined impedance of C_O, R_{ESR}, and C_B in C_{Filter} should be low enough to do the opposite, to shunt energy away from R_{FB1} and R_{FB2} in the feedback factor β_{FB} and load resistance R_L in output resistance R_O'. Load capacitance C_L helps in this regard, so C_L and C_B combine in C_B'. Overall, Z_{GP}, R_P, and C_P constitute the top impedance Z_T and $Z_{SH.FB}$, R_O', C_O, R_{ESR}, and C_B' comprise the combined bottom impedance Z_B' that voltage-divide v_{in} to v_o to set A_{IN} and 1/PSR to

$$A_{IN} \equiv \frac{1}{PSR} = \frac{v_o}{v_{in}} = \frac{Z_B'}{Z_T + Z_B'} \tag{8.38}$$

What remains to be seen is how these impedances relate and change across frequency. For this, recall that C_O is usually the highest capacitance and R_{OE}, R_{FB1}, and R_{FB2} the highest resistances

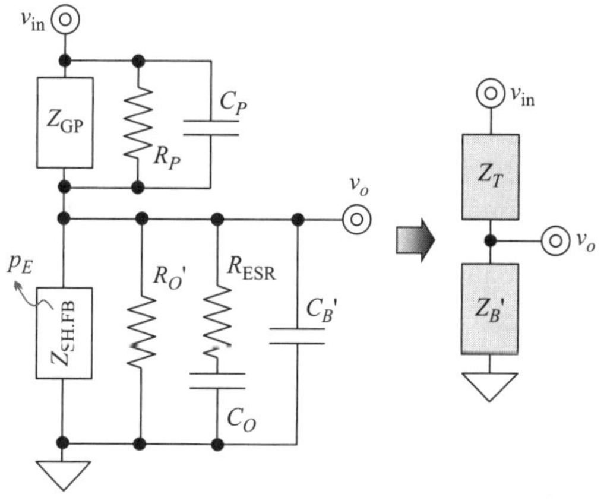

FIGURE 8.19 Voltage-divider model of the linear regulator.

in the network. C_B is the next highest capacitance and R_P and R_L the next highest resistances. C_L and C_P are the lowest capacitances and R_{ESR} and $Z_{SH.FB}$'s low-frequency equivalent $R_{SH.FB}$ are the lowest resistances.

Low Frequency

At low frequency, the effects of all capacitors in the network disappear. G_P's possible diode connection is therefore absent, which means Z_{GP} disappears, and the feedback translation of v_o to G_P's current through β_{FB}, A_E's low-frequency gain A_{E0}, and G_P sets $Z_{SH.FB}$'s equivalent resistance $R_{SH.FB}$ to $1/G_{FB}$ or $1/\beta_{FB}A_{E0}G_P$. In fact, feedback gain reduces $R_{SH.FB}$ to such a degree that $R_{SH.FB}$ not only shunts R_O' in Z_B' but also becomes negligible with respect to R_P in $Z_T + Z_B'$, so low-frequency supply gain A_{IN0} reduces to $R_{SH.FB}/R_P$:

$$A_{IN0} = \frac{1}{PSR_0} = \frac{R_O' \parallel R_{SH.FB}}{R_P + (R_O' \parallel R_{SH.FB})} \approx \frac{R_{SH.FB}}{R_P} = \frac{1}{\beta_{FB}A_{E0}G_P R_P} \approx \frac{1}{A_{LG0}} \approx LNR \tag{8.39}$$

Interestingly, $R_{SH.FB}/R_P$ is roughly equivalent to the reciprocal of the low-frequency gain across the loop A_{LG0}. In other words, loop gain determines both line regulation LNR with A_{IN0} and low-frequency power-supply rejection PSR_0 with $1/A_{IN0}$.

Example 8.4 Determine line regulation and power-supply rejection at low frequency when the feedback factor is 0.5 V/V, the gain of the error amplifier is 440 V/V, and the transconductance and resistance of the pass device are 50 mA/V and 100 Ω.

Solution

$$LNR = A_{IN0} = \frac{1}{PSR_0} \approx \frac{R_{SH.FB}}{R_P} = \frac{1}{\beta_{FB}A_{E0}G_P R_P}$$
$$\approx \frac{1}{(0.5\,V/V)(440\,V/V)(50\,mA/V)(100\,\Omega)} \approx 0.9\,mV/V = -61\,dB$$

Compare this LNR and PSR_0 with the simulation results from Fig. 8.20.

Output Compensation

Variations across frequency are the result of capacitors shunting the effects of resistances across their terminals. In output-compensated regulators, C_O is high and error-amplifier pole p_E is at high frequency, so the first variation corresponds to C_O and C_B and C_L in C_B' shunting R_L, R_{FB1}, and R_{FB2} in R_O'. This occurs near output pole p_O, which happens when C_O, C_B, C_L, and C_P shunt R_P, R_L, R_{FB1}, and R_{FB2}. But $R_{SH.FB}$ is so low at these frequencies that $R_{SH.FB}$ dominates in Z_B':

$$Z_B' = R_{SH.FB} \parallel R_O' \parallel \left.\frac{1}{s(C_O + C_B')}\right|_{f>p_O} \approx R_{SH.FB} \parallel \frac{1}{s(C_O + C_B')} \approx R_{SH.FB} \tag{8.40}$$

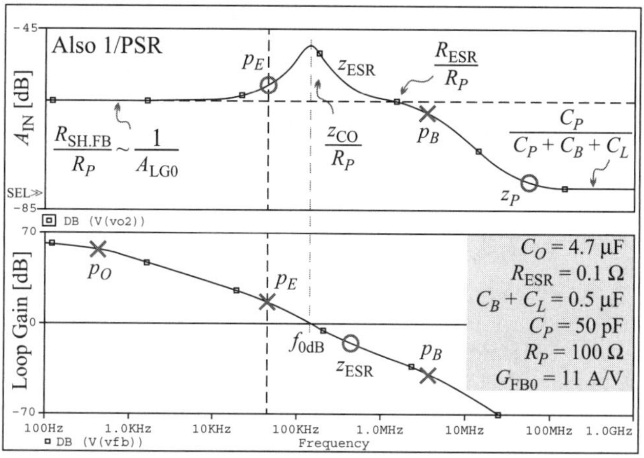

Figure 8.20 The loop gain and corresponding supply gain of an output-compensated regulator.

and R_P dominates in Z_T and in $Z_T + Z_B'$, so A_{IN0} remains $1/A_{LG0}$, as Fig. 8.20 demonstrates.

The next transition can be at p_E because, for the sake of gain, R_{OE} in the error amplifier A_E is usually so high that A_E's C_{OE} can shunt R_{OE} near f_{0dB}. The effect of p_E in the system is to reduce A_E in the feedback translation that sets shunt impedance $Z_{SH.FB}$:

$$Z_{SH.FB} = \frac{1}{\beta_{FB} A_E G_P} \approx \frac{\left(1 + \frac{s}{2\pi p_E}\right)}{\beta_{FB} A_{E0} G_P} \equiv R_{SH.FB}\left(1 + \frac{s}{2\pi p_E}\right) \tag{8.41}$$

This means that $Z_{SH.FB}$ rises from $R_{SH.FB}$ past p_E, and since $Z_{SH.FB}$ still dominates in Z_B' and in R_P in Z_T and in $Z_T + Z_B'$, A_{IN}'s $Z_{SH.FB}/R_P$ climbs with $Z_{SH.FB}$ past p_E.

$Z_{SH.FB}$ rises so much, in fact, that C_O, C_B, and C_L shunt $Z_{SH.FB}$ past the system's bandwidth f_{0dB}. To see this, recall that $Z_{SH.FB}$ is the loop-gain translation of Z_B and Z_T, where Z_B's C_O, C_B, C_L, and C_P in the loop gain A_{LG} shunt Z_T's R_P past p_O:

$$Z_{SH.FB} = \frac{Z_B \| Z_T}{A_{LG}} \approx \left.\frac{Z_B \| R_P}{A_{LG}}\right|_{f>p_O} \approx \left.\frac{Z_B}{A_{LG}}\right|_{f_{0dB}} = Z_B$$

$$\approx \frac{1}{s(C_O + C_B + C_L + C_P)} \approx \frac{1}{s(C_O + C_B + C_L)} \tag{8.42}$$

so $Z_{SH.FB}$ reaches Z_B at f_{0dB}, when A_{LG} drops to 1, where C_P is negligibly low with respect to C_O, C_B, and C_L. The impact of this in A_{IN} is that C_O, C_B, and C_L's falling impedance Z_{CO} dominates Z_B and Z_B' past f_{0dB}, and since Z_{CO} is lower than R_P past p_O, R_P overwhelms Z_B' in $Z_T + Z_B'$ to reduce A_{IN} to Z_{CO}/R_P at and past f_{0dB}:

$$A_{IN(0dB)} = \left.\frac{Z_B'}{Z_T + Z_B'}\right|_{f > f_{0dB}} \approx \frac{Z_{CO}}{R_P + Z_{CO}} \approx \frac{Z_{CO}}{R_P} = \frac{1}{s(C_O + C_B + C_L)R_P} << 1 \tag{8.43}$$

As a result, A_{IN} peaks to Z_{CO}/R_P at f_{0dB} and falls after that. And since Z_{CO} is considerably lower than R_P at f_{0dB}, A_{IN} at f_{0dB} is well below 0 dB, which highlights the utility of output capacitance in Z_{CO} near f_{0dB}.

As the highest capacitance, R_{ESR} current-limits C_O next past ESR zero z_{ESR}:

$$\left.\frac{1}{sC_O}\right|_{z_{ESR} \approx \frac{1}{2\pi R_{ESR} C_O}} \equiv R_{ESR} \tag{8.44}$$

At these frequencies, Z_B' reduces to R_{ESR} and $Z_T + Z_B'$ remains R_P because R_{ESR} is usually much lower than R_P. As a result, A_{IN} at moderate frequencies $A_{IN(MF)}$ becomes R_{ESR}/R_P:

$$A_{IN(MF)} = \left.\frac{Z_B'}{Z_T + Z_B'}\right|_{f > z_{ESR}} \approx \frac{R_{ESR}}{R_P + R_{ESR}} \approx \frac{R_{ESR}}{R_P} \tag{8.45}$$

This is an important observation because it underlines the significance of R_{ESR} in PSR. It shows that R_P and R_{ESR} set and limit PSR past f_{0dB}, when the benefits of output capacitance and shunt feedback vanish.

Past z_{ESR}, C_B and C_L in C_B' shunt R_{ESR} near p_B, which happens when C_B and C_L shunt R_{ESR}:

$$\left.\frac{1}{sC_B'}\right|_{f = \frac{1}{2\pi R_{ESR}(C_B + C_L)} \approx p_B} \equiv R_{ESR} \tag{8.46}$$

As a result, C_B and C_L's falling impedance shunts R_{ESR} in Z_B', and R_P still dominates in $Z_T + Z_B'$ to pull A_{IN} below $A_{IN(MF)}$. C_P's impedance, however, eventually shunts R_P in Z_T and in $Z_T + Z_B'$ to arrest A_{IN}'s drop with a zero z_P:

$$\left.\frac{1}{sC_P}\right|_{z_P \approx \frac{1}{2\pi R_P C_P}} \equiv R_P \tag{8.47}$$

At this point, C_P's Z_{CP} dominates Z_T, and C_B and C_L's combined impedance $Z_{CB'}$ dominates Z_B', so A_{IN}'s high-frequency gain $A_{IN(HF)}$ reduces to the voltage-divided ratio that C_P, C_B, and C_L define:

$$A_{IN(HF)} = \left.\frac{Z_B'}{Z_T + Z_B'}\right|_{f>z_P} \approx \frac{Z_{CB'}}{Z_{CP} + Z_{CB'}} = \frac{C_P}{C_P + C_B'} = \frac{C_P}{C_P + C_B + C_L} \tag{8.48}$$

In other words, C_P limits the extent that C_B and C_L help when shunting R_{ESR}.

In review, the system relies on shunt feedback for high PSR at low frequencies. Feedback gain, however, diminishes when A_E falls past p_E, so PSR similarly drops after p_E. Near the system's bandwidth f_{0dB}, where the effect of feedback disappears, C_O overtakes shunt feedback in suppressing supply ripple v_{in}. PSR bottoms at f_{0dB}, when C_O, C_B, and C_L alone shunt the noise that R_P couples into v_o. Unfortunately, R_{ESR} limits C_O's ability to shunt v_{in} after that. Although C_B and C_L shunt R_{ESR} at higher frequency, C_P eventually starts coupling more supply noise v_{in} to v_o. This means that only C_P, C_B, and C_L remain to determine PSR at high frequencies.

Example 8.5 Determine power-supply rejection at the system's bandwidth and past the ESR zero when bandwidth is 100 kHz; output, bypass, load, and pass-transistor capacitances are 4.7, 0.3, 0.2, and 0.05 μF; and pass-transistor, load, feedback, and ESR resistances are 100, 10,000, 500,000, and 0.1 Ω, respectively.

Solution

$$\text{PSR}_{0dB} = \frac{1}{A_{IN(0dB)}} \approx \frac{R_P}{Z_{CO}} = s(C_O + C_B + C_L)R_P$$

$$\approx 2\pi(100\text{ kHz})[(4.7 + 0.3 + 0.2\ \mu\text{F})](100\ \Omega) \approx 330\text{ V/V} \approx 50\text{ dB}$$

and $$\text{PSR} = \left.\frac{1}{A_{IN}}\right|_{f>z_{ESR}} \approx \frac{R_P}{R_{ESR}} = \frac{100\ \Omega}{0.1\ \Omega} = 1\text{ kV/V} = 60\text{ dB}$$

Compare these PSR numbers with the simulation results from Fig. 8.20.

Internal Compensation

Variations across frequency are the result of capacitors shunting the effects of resistances across their terminals. In internally compensated regulators, error-amplifier pole p_E is the dominant low-frequency pole, so the first variation corresponds to A_E's C_{OE} shunting R_{OE}. Since the effect of p_E is to reduce A_E in $Z_{SH.FB}$'s feedback translation, $Z_{SH.FB}$ rises from $R_{SH.FB}$ past p_E, and since $Z_{SH.FB}$ still dominates in Z_B' and R_P in Z_T and in $Z_T + Z_B'$, A_{IN}'s $Z_{SH.FB}/R_P$ climbs with $Z_{SH.FB}$ past p_E, as Fig. 8.21 demonstrates.

$Z_{SH.FB}$ rises to the extent that $Z_{SH.FB}$ surpasses R_P in $Z_T + Z_B'$ near the system's unity-gain frequency f_{0dB}, when loop gain A_{LG} falls below 1:

$$Z_{SH.FB} = \frac{Z_B \parallel Z_T}{A_{LG}} \approx \left.\frac{R_P}{A_{LG}}\right|_{f_{0dB}} = R_P \tag{8.49}$$

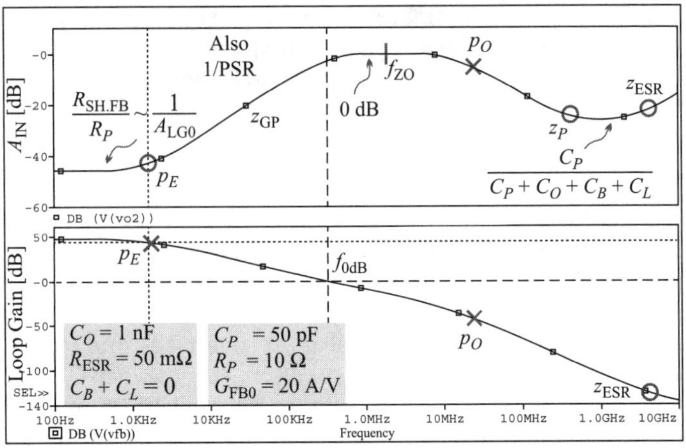

FIGURE 8.21 The loop gain and corresponding supply gain of an internally compensated regulator.

where Z_T's R_P is lower than C_O, C_B, C_L, C_P, R_{FB1}, R_{FB2}, and R_L's combined bottom impedance Z_B. This means that $Z_{SH.FB}$ not only dominates R_O' in Z_B' but also overwhelms R_P in $Z_T + Z_B'$, so supply gain past f_{0dB} reduces to $Z_{SH.FB}/Z_{SH.FB}$, which is 1:

$$A_{IN(0dB)} = \frac{Z_B'}{Z_T + Z_B'} = \frac{Z_{SH.FB}}{R_P + Z_{SH.FB}}\bigg|_{f>f_{0dB}} \approx \frac{Z_{SH.FB}}{Z_{SH.FB}} = 1 \tag{8.50}$$

$Z_{SH.FB}$ rises so much, in fact, that $Z_{SH.FB}$ surpasses R_O' in Z_B' past cross-over frequency f_{ZO}, so R_O' still dominates in both Z_B' and $Z_T + Z_B'$, which means A_{IN} at moderate frequencies remains 1:

$$A_{IN(MF)} = \frac{Z_B'}{Z_T + Z_B'} = \frac{R_O'}{R_P + R_O'}\bigg|_{f>f_{ZO}} \approx \frac{R_O'}{R_O'} = 1 \tag{8.51}$$

The next transition corresponds to C_O and C_B and C_L from C_B' shunting R_L, R_{FB1}, and R_{FB2} in R_O':

$$\frac{1}{s(C_O + C_B')}\bigg|_{f=\frac{1}{2\pi[R_L \| (R_{FB1}+R_{FB2})](C_O+C_B+C_L)} \approx p_O} \equiv R_O' \tag{8.52}$$

This happens near output pole p_O, when C_O, C_B, C_L, and C_P shunt R_P, R_L, R_{FB1}, and R_{FB2}. At this point, when the benefits of shunt feedback vanish, the system relies on C_O, C_B, and C_L for PSR. C_P's impedance shunts R_P in Z_T next at zero z_P. As a result, C_P's Z_{CP} dominates Z_T, and C_O, C_B, and C_L's combined impedance Z_{CO} controls Z_B', so A_{IN}'s high-frequency gain $A_{IN(HF)}$ reduces to the voltage-divided ratio that C_P, C_O, C_B, and C_L establish:

$$A_{IN(HF)} = \left.\frac{Z_B'}{Z_T + Z_B'}\right|_{f>z_P} \approx \frac{Z_{CO}}{Z_{CP} + Z_{CO}} = \frac{C_P}{C_P + C_O + C_B'} = \frac{C_P}{C_P + C_O + C_B + C_L} \tag{8.53}$$

In other words, C_P limits the extent to which C_O, C_B, and C_L help when shunting R_O'. And if C_O is much higher than C_B and C_L, C_O's R_{ESR} current-limits C_O past ESR zero z_{ESR} to halt Z_B's descent, which allows C_P's coupling effect to raise A_{IN}. This usually happens at considerably higher frequencies in internally compensated regulators because C_O is normally a low-ESR capacitor.

In all, the system relies on shunt feedback for high PSR at low frequencies. Feedback gain, however, diminishes when A_E falls past p_E, so PSR similarly drops after p_E. At and past the system's bandwidth f_{0dB}, shunt feedback is no longer effective, so the circuit couples nearly all the noise present in the supply to the output. Fortunately, C_O, C_B, and C_L shunt this ripple, but only to the extent that C_P's coupling effect allows.

Miller-Compensated Systems When a Miller capacitor C_M loops a P-type power transistor, Z_T also incorporates a diode-connected impedance Z_{GP}. As C_M's diode connection shorts, Z_{GP} decreases, and when Z_{GP} drops below Z_T's R_P past Miller pole p_M, Z_T falls with Z_{GP}. Then, when C_M shorts with respect to R_{OE} past zero z_{GP}, at $1/2\pi R_{OE}C_M$, Z_T flattens to Z_{GP}'s $1/G_P$. Z_T falls again when C_P shunts $1/G_P$ past p_{CP}, at $G_P/2\pi C_P$. In other words, Z_T falls past p_M, flattens past z_{GP}, and falls again past p_{CP}.

In fact, the reason why A_E drops past p_E is C_M's shunting effect on A_E's load past p_M. This means that p_E is p_M and A_E not only drops with Z_{GP} past p_M but also stops falling with Z_{GP} past diode-connected zero z_{GP}. Interestingly, the gain across the power transistor A_P also suffers a pole at z_{GP} when C_M shunts the transistor's output resistance. As a result, z_{GP}'s effects in A_E and A_P cancel and disappear from the overall loop gain A_{LG}. To see the combined effects of C_M's feedback connection on PSR, recall that $Z_{SH.FB}$ is ultimately the loop-gain translation of the open-loop impedance, which in this case is $(Z_T \| Z_B)/A_{LG}$. In this light, since Z_B' reduces to $Z_{SH.FB}$ and $Z_T + Z_B'$ to Z_T at low frequencies, $A_{IN(LF)}$ becomes $Z_{SH.FB}/Z_T$:

$$A_{IN(LF)} = \left.\frac{Z_B'}{Z_T + Z_B'}\right|_{f<f_{0dB}} \approx \frac{Z_{SH.FB}}{Z_T} = \left.\left(\frac{Z_T \| Z_B}{A_{LG}}\right)\left(\frac{1}{Z_T}\right)\right|_{f>p_E} \approx \left(\frac{Z_T}{A_{LG}}\right)\left(\frac{1}{Z_T}\right) = \frac{1}{A_{LG}} \tag{8.54}$$

And since Z_T's diode connection reduces both Z_T and $Z_{SH.FB}$'s $Z_T \| Z_B$ past p_E, the effects of Z_T cancel in $A_{IN(LF)}$ and A_{LG} alone determines the fate of $A_{IN(LF)}$. As a result, A_{IN} rises with A_{LG}'s p_E without displaying the effects of z_{GP}, as Fig. 8.21 shows.

Beyond z_{GP}, the shunting effects of negative feedback diminish near f_{0dB} through and past crossover frequency f_{ZO}, when $Z_{SH.FB}$ rises above R_O'. C_O, C_B, and C_L then shunt R_{FB1}, R_{FB2}, and R_L in R_O' near p_O. C_P shunts Z_T's $1/G_P$ next past p_{CP}, which corresponds to z_P in Fig. 8.21. And R_{ESR} eventually current-limits C_O past z_{ESR}. Note that all these effects in A_{IN} mirror those already described and illustrated for internally compensated systems. The only difference here is that C_P shunts $1/G_P$, not R_P, so z_P is higher in Miller-compensated circuits than in other internally compensated circuits.

Interestingly, had C_M looped a ground-referenced N-type transistor instead of the P-type power device, $Z_T \| Z_B$ would have dropped with Z_B to reduce $A_{IN(LF)}$ to $Z_B/A_{LG}Z_T$, or $Z_B/A_{LG}R_P$. If this were possible in a linear regulator, the Miller effects of C_M on Z_B and A_{LG} would cancel, and $A_{IN(LF)}$ would not rise with A_{LG} past p_E until Z_B stopped falling at z_{GP}. In other words, PSR bandwidth would extend to z_{GP}, instead of p_M. All this simply corroborates the fact that, while C_M's diode connection to ground shunts supply ripple, C_M's diode connection to the power supply does the opposite, couples supply noise.

8.4.5 Conclusions

Several important conclusions are worth recounting at this point. One is that PSR is a manifestation of power-supply gain A_{IN}, where low gain is, of course, desirable. And although LNR is a large-signal phenomenon, it is nonetheless a linear event, so A_{IN} at low frequency is a good approximation of LNR. Since the role of negative feedback is to oppose the effects of external forces, A_{IN}, LNR, and PSR all benefit from high loop gain. In fact, PSR at low frequency and LNR reduce to the low-frequency gain across the loop, which means high loop gain is beneficial.

Since PSR ultimately refers to ripple at the output, the voltage divider that the output stage of the linear regulator establishes determines PSR performance. In this respect, the power transistor couples supply ripple to the output. Luckily, negative feedback shunts this noise up to the system's bandwidth. Past this point, the system relies on the output filter for shunting supply noise. As a result, PSR is higher when the coupling effects of C_P and R_P in the power transistor are lower, which happens when C_P is low and R_P is high. PSR is similarly high when the loop gain and C_O, R_{ESR}, and C_B in the filter shunt the most noise, that is, when loop gain, C_O, and C_B are high and R_{ESR} is low. In other words, high loop gain, high output capacitance, and low equivalent series resistances in output capacitors are important attributes in linear regulators.

The favorable effects of shunt feedback hinge on the loop-gain translation of the output voltage to the current that the power transistor delivers. From this respect, extending the bandwidth of this translation is beneficial for PSR. This is why internally compensated regulators suffer from lower PSR than their output-compensated counterparts, because they limit the bandwidth of this translation. Miller-compensated circuits are no better in this respect.

All these conclusions assume the error amplifier does not feed supply noise through the pass device. This is a reasonable assumption because the amplifier can either reproduce the ripple or altogether eliminate it. In the case of P-type power transistors, for example, a P-type mirror load in the error amplifier reproduces supply noise so that noise is common to sources and gates and emitters and bases. This way, the transistor's transconductance cancels their effect to generate a noise-free current. And of course, N-type followers do not propagate noise when the error amplifiers that drive them output signals that are free of noise. All this means that designers can design the error amplifier to ensure that the transconductance of the pass device does not feed noise to the output.

Perhaps the most important lesson in this section is how to decipher PSR performance. Extracting the voltage divider network at the output of the system and including the shunting impedance that negative feedback presents are key in this respect. After that, unraveling PSR amounts to keeping track of how impedances in the network change across frequency. In all this, bear in mind that feedback effects vanish near the system's bandwidth and Miller capacitors diode-connect transistors to either the positive supply to couple supply noise to the output or to ground to further shunt supply noise.

8.5 Comparison of Compensation Strategies

The overriding advantages of output-compensated regulators are high output capacitance C_O and high error-amplifier bandwidth p_E. High C_O for one, enhances the regulator's ability to suppress fast and wide high-power load dumps. High p_E, on the other hand, extends the benefits of shunt feedback on PSR to higher frequencies. The fundamental challenge here is keeping p_E above the system's bandwidth f_{0dB} to avoid unstable feedback conditions. One reason for this difficulty is that high gain results from running moderate transconductance currents through high resistances, and capacitors shunt high resistances at low frequencies. Another reason is the power pass transistor is normally large, so it can load the output resistance of the error amplifier with so much capacitance that the resulting pole ends at low frequency.

The driving benefit of internally compensated circuits is lower output capacitance C_O, which conforms well with the integration demands of many consumer products today. Unfortunately, lower capacitance suffers higher voltage excursions when confronted with wide load dumps. So much so, in fact, that engineers ensure that C_O is nevertheless substantial. Plus, load current changes output resistance

	Output Compensated	**Internally Compensated**
Dominant Pole	Output Pole p_O	Error-Amplifier Pole p_E
C_O	Higher	Lower
Integration	Off-chip or in-package C_O	On-chip or in-package C_O
Load-dump variation	Lower	Higher
Stability requirement	Low p_O and high p_E	Low p_E and high p_O
Worst-case stability	High p_O, low p_E, and no R_{ESR}	High p_E, low p_O, and no R_{ESR}
PSR or $1/A_{IN}$	Wider bandwidth	Lower bandwidth
Typical application	**Higher power**	**Lower power**

Table 8.1 Comparison of Compensation Strategies

to such an extent that keeping the output pole p_O above the system's bandwidth f_{0dB} across operating conditions and temperature and fabrication corners is challenging. This is why internally compensated circuits normally supply less power than their output-compensated counterparts, because they cannot sustain wide load dumps. Table 8.1 summarizes these comparative conclusions for reference.

When aiming to extend battery life, lowering quiescent power is paramount. Unfortunately, reducing currents in an analog circuit raises resistances, which in turn, pulls parasitic poles to lower frequencies. And because parasitic poles compromise the stability of the feedback loop, lower poles also pull the bandwidth of the system to lower frequency. To aggravate matters, the possible presence of an ESR zero and a bypass pole extends f_{0dB} to such an extent that the worst-case f_{0dB} must be lower by the same amount. This means that, to reach unity gain at such a low frequency, the loop gain must also be low. This explains why low-frequency loop gains in linear regulators normally hover between 50 and 60 dB and bandwidths between 0.05 and 1 MHz.

8.6 Summary

The variations of load current i_L and the equivalent series resistance R_{ESR} of the output capacitor are probably the two most significant obstacles to overcome when stabilizing a linear regulator. In a battery-operated environment, for example, the former varies by up to five decades, from maybe 1 μA during light loading conditions to 100 mA when fully engaged. As a result, the output pole p_O also shifts by up to five decades of frequency, because p_O is inversely proportional to the resistance of the pass device and therefore proportional to $\sqrt{i_L}$ in the case of N-type voltage followers or i_L in the case of P-type transconductors. To aggravate matters, although not to the same degree, R_{ESR} varies considerably across temperature and process, from maybe 1 to 500 mΩ. This means that R_{ESR} can pull or push the ESR zero z_{ESR} and the bypass pole p_B below or above the loop's unity-gain bandwidth f_{0dB}, which is equivalent to the system's bandwidth. In other words, R_{ESR} can extend f_{0dB} by one to two decades, depending on the ratio that output, bypass, load, and output-transistor capacitances establish for p_B and z_{ESR}. In all, f_{0dB} can shift across five to seven decades of frequency.

Luckily, load dependence in p_O is not the only factor that shifts f_{0dB}. What matters in output-compensated regulators is the combined effect of both gain and bandwidth on f_{0dB}. In this regard, since the transconductance g_m of MOSFETs rises with the square root of i_L, g_m's gain–bandwidth effect on f_{0dB}, wherein the effects of output resistance cancel, is to climb with the square root of i_L. This translates to a square-root suppression of i_L variations. Unfortunately, BJTs do not enjoy this benefit because their transconductances are linear translations of i_L. f_{0dB} in internally compensated circuits is no better, because p_O variations in light of feedback stability requirements set an upper boundary for f_{0dB}. Here again, the $1/g_m$ resistance of MOSFETs falls with the square root of i_L, so MOS transistors similarly quell i_L variations in p_O and BJTs do not.

The difficulties compound when considering that high output capacitance and high loop gain are highly desirable. Output capacitance, for one, subdues transient voltage excursions in the output, and feedback gain opposes the effects of large- and small-signal fluctuations in the load and supply. Unfortunately, high output capacitance pulls p_O to lower frequency and the high resistance that high loop gain requires establishes a low- to moderate-frequency pole p_E. So with no zeros to save phase, reaching f_{0dB} before phase-shifting 180° across the loop means loop gain should be low and secondary and parasitic poles should be high. And pushing p_O or p_E to high frequency with a limited power budget and when loaded with C_O or the high capacitance that a large power device presents is challenging.

PSR also manifests the effects of negative feedback and output capacitance. Loop gain, for example, determines the extent to which shunt feedback suppresses supply noise in the output. This shunting effect is actually the result of the loop-gain translation from the output voltage to the current that the output transistor delivers. As a result, PSR drops when the gain across the error amplifier falls, which first happens past pole p_E. PSR then bottoms when loop gain vanishes, near and past the system's bandwidth f_{0dB}, which is where the shunting effect of output capacitance becomes significant. Unfortunately, ESR in the capacitor limits the extent to which capacitance shunts noise. All this means that PSR hinges on high loop gain, wide error-amplifier bandwidth, high output capacitance, and low ESR.

The application ultimately determines what is most important in PSR. Deriving energy from, for example, 2.7–4.2-V lithium-ion batteries exposes linear regulators to 1.5-V steady-state variations in the supply. This type of application demands good line regulation, which results from high PSR at low frequency. When drawing power from a 1-MHz switching dc–dc converter, on the other hand, PSR at 1 MHz may be more critical. Extending the bandwidth of the error amplifier, raising output capacitance, and reducing ESR improve PSR in this region. Digital-signal processors and power amplifiers, to cite two more examples, can also load the input supply to produce even higher-frequency noise. Low ESR and low coupling capacitance across the power device are crucial at these higher frequencies.

Transistor-level realizations of the linear regulator must in the end accommodate and conform to the small-signal constraints that feedback stability and supply fluctuations impose. Topological choices and their physical-layout translations are vital in this regard, because seemingly minor missteps can easily undermine the system. The purpose of Chap. 9 is to explore how transistors, resistors, capacitors, and the analog building blocks they comprise can combine to produce high loop gain, wide bandwidth, and widely stable feedback conditions, which combined suppress the effects of load and supply fluctuations on the output.

8.7 Review Questions

1. What capacitors can load the output of a linear regulator?
2. Which of these usually has the highest capacitance?
3. What resistances load the output of a linear regulator?
4. Which of these is normally the highest resistance?
5. Which components establish the output pole?
6. Which components, if any, establish an in-phase zero?
7. Which components establish the bypass pole?
8. What other poles does the feedback loop incorporate?
9. What is the impact of load-current variations on the unity-gain frequency f_{0dB} of output-compensated regulators that use a PMOS power transistor?
10. What is the impact of a widely variable R_{ESR} on f_{0dB}?
11. What conditions compromise the feedback stability of output-compensated circuits the most when load current is high?
12. What is the most challenging aspect of compensating internally compensated regulators?

13. What is the impact of load-current variations on the unity-gain frequency f_{0dB} of Miller-compensated regulators?

14. What type of output devices in internally compensated circuits produce the most variations in the output pole?

15. Which components couple noise in the supply to the output of regulators?

16. Which components shunt power-supply noise in the output?

17. What type of base signals keep NPN power transistors from propagating supply noise to the output of regulators?

18. What type of gate signals keep PMOS power transistors from injecting supply noise to the output of regulators?

19. How can error amplifiers reproduce supply noise in their outputs?

20. What type of current mirror negates the effects of noise in the positive supply?

21. What are the effects of a Miller capacitor around a power P-type transistor on the impedance that connects the supply to the output?

22. What suppresses supply ripple the most at low frequency?

23. What parameter raises PSR the most at low frequency?

24. What pole first attenuates the effect of feedback on PSR?

25. At which frequencies do the effects of feedback vanish?

26. What suppresses supply ripple the most near and past the system's bandwidth?

27. What limits PSR performance near and past the system's bandwidth?

28. Which compensation strategy extends the bandwidth of feedback-aided suppression the most?

CHAPTER 9

Integrated Circuits

The purpose of linear regulators is to produce and maintain an independent, stable, noise-free, and predictable output voltage that is capable of delivering substantial current to the load. For this, regulators normally resort to negative feedback. Accordingly, a typical system like Fig. 9.1 shows shunt-samples of the output v_O and series-mixes an accurate reference v_{REF} with a feedback signal v_{FB} derived from v_O. This way, the differential amplifier A_E and a power switch S_O amplify the difference between v_{REF} and v_{FB}, and a network of resistors R_{FB1} and R_{FB2} feed the amplified error back to the input. With enough loop gain, the feedback system reduces the difference between v_{REF} and v_{FB}. As a result, v_{FB} mirrors v_{REF}, R_{FB1} and R_{FB2} translate the mirrored impression of v_{REF} in v_{FB} to v_O, and negative feedback opposes the effects of load and supply fluctuations on v_{FB} and, by translation, on v_O.

9.1 Design Process

The design process of a typical linear regulator, as with most other electronic systems, begins with its target specifications, and more particularly, with its load and the power transistor that drives it. To start, the targeted steady-state output voltage prescribes the *feedback factor* β_{FB} that the feedback resistors R_{FB1} and R_{FB2} must realize to translate the mirrored impression of the reference v_{REF} on v_{FB} to v_O. Steady-state parameters like input supply voltage v_{IN}, load current i_L, and *dropout voltage* V_{DO} then set the basic operating requirements of the power device S_O. After that, fluctuations in i_L and v_{IN} determine how quickly S_O must react to suppress their effects on v_O in the presence of a nonideal filter.

Because the operating specifications often require S_O to carry substantial current, S_O is large and its input control terminal is, as a result, highly capacitive. A low-impedance buffer circuit A_B must therefore drive S_O with sufficient current and voltage swing to quickly charge and discharge a large parasitic capacitance. For this, A_B demands circuit-specific attributes from the error amplifier A_E, other than the ones the application already imposes, like low *input-referred offset*, good *load* and *line regulation*, high *signal-to-noise ratio*, high *power-supply rejection*, and so on.

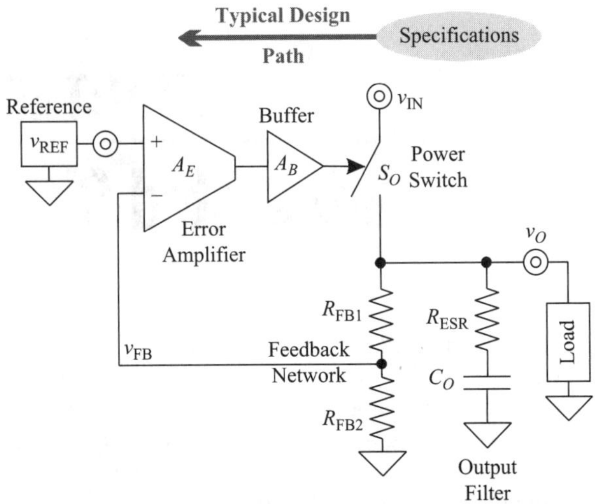

FIGURE 9.1 Linear-regulator system.

With the design of a linear regulator in mind, the underlying objective of this chapter is to bridge and convert the system-level objectives outlined in Chaps. 1 and 2 into practical integrated circuits (ICs). The initial approach is to compartmentalize the design into its constituent components: the power transistor S_O, the buffer A_B, and the error amplifier A_E, and design them according to the specifications of the system and their interdependent requirements. The design process starts, as is usually the case, at the output with S_O and ends with A_E because the demands of sustaining a specified load normally ripple back through the system. To cater to practical loads, each section in the chapter discusses how circuit- and layout-design choices ultimately affect efficiency and regulation performance.

9.2 Power Transistors

The design goal of the power switch S_O is to supply output current i_O to the load without dissipating much power in the process. Minimizing the fraction of input power P_{IN} lost to S_O as P_{SO} amounts to reducing the voltage across S_O, between v_{IN} and v_O, and losing little ground current i_{GND}:

$$\eta_C \equiv \frac{P_O}{P_{IN}} = \frac{P_O}{P_{SO}+P_O} = \frac{P_O}{(v_{IN}-v_O)i_O + v_{IN}i_{GND} + P_O} \leq \frac{P_O}{V_{DO}i_O + v_{IN}i_{GND} + P_O} \quad (9.1)$$

where η_C is power-conversion efficiency and dropout voltage V_{DO} is the minimum voltage that S_O can drop. S_O must achieve this objective while presenting minimal capacitance to its driving buffer A_B because this capacitance slows the system, the effects of which compromise stability, noise performance, and load-dump response. As a result, S_O must operate with a low input supply v_{IN}, survive a low-dropout voltage V_{DO}, lose little ground current i_{GND}, and be only large enough to sustain the highest load current under worst-case temperature and fabrication corners and extreme operating conditions.

9.2.1 Alternatives

The power device S_O is a transistor because a feedback loop need only modulate its input terminal to control how much current S_O derives from v_{IN} to supply v_O. Although a *junction field-effect transistor* (JFET) can accomplish this task, shutting a JFET to source little to no current in the presence of light loads is not easy. *Bipolar-junction transistors* (BJTs) and *metal–oxide–semiconductor field-effect transistors* (MOSFETs), on the other hand, are normally-off devices that can conduct considerable current under typical supply rails. Because raising or lowering the control voltage of N-type BJTs and MOSFETs produces the same general effects, which oppose those of their P-type counterparts, the following subsections split the discussion into N- and P-type devices, rather than BJTs and MOSFETs.

N-Type Power Transistor

In the context of linear regulators, the collectors and drains of NPN BJTs and N-channel MOSFETs connect to v_{IN} and the emitters and sources to v_O, as Fig. 9.2 illustrates. This configuration carries an intrinsic advantage in that the output is already low impedance at roughly $1/g_m$, which means power transistors Q_{NO} and M_{NO} respond quickly to load dumps. And of the two, because collector currents rise exponentially with base–emitter voltages v_{BE} and drain currents quadratically with gate–source voltages v_{GS}, the $1/g_m$ resistance of the BJT is lower, which is good for high bandwidth.

The main disadvantage with N-type transistors is *high-dropout* (HDO) voltages. To be more specific, the minimum dropout voltage required to conduct maximum output current $i_{O(MAX)}$ in an NPN BJT is

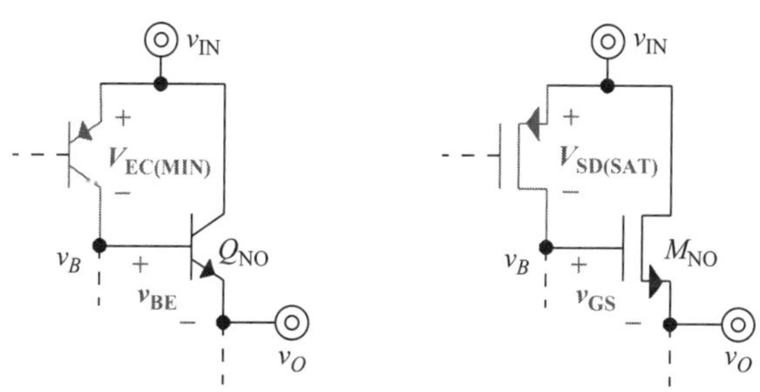

FIGURE 9.2 NPN and N-channel power transistors.

its v_{BE} and the minimum emitter–collector voltage $V_{EC(MIN)}$ of the transistor that drives its base. In the case of a MOSFET, the circuit similarly drops out of regulation when v_{IN} and v_O are apart by less than the saturation voltage $V_{SD(SAT)}$ of the driving transistor and the v_{GS} of the power MOSFET. In other words, v_{IN} must exceed v_O by 0.2–0.3 V and a diode's or MOSFET threshold's 0.4–0.8 V for the circuit to supply the load, which corresponds to dropout voltages V_{DO} of 0.6–1.5 V. Luckily, their fast response time can at times outweigh the efficiency disadvantage that results from a higher V_{DO}. Note, by the way, that the base current of the BJT ultimately flows to the output via the emitter, so the circuit does not lose any current to ground as i_{GND}.

P-Type Power Transistor

The current-sourcing terminals attached to v_O in P-type configurations, as Fig. 9.3 shows, are collectors and drains. This means that their output resistances are higher at r_o and r_{ds} than those of their N-type counterparts. Here, however, v_{IN} need only rise one $V_{EC(MIN)}$ or $V_{SD(SAT)}$ above v_O to sustain $i_{O(MAX)}$ without collapsing the power device Q_{PO} or M_{PO} into its ohmic region, which is another way of saying V_{DO} is lower at 0.2–0.4 V than with N-type transistors. The higher efficiency that results from such a *low-dropout* (LDO) range is often more beneficial to a system than low impedance, especially when drawing power from small batteries, which are easy to exhaust. Unlike the NPN follower, though, the base of a PNP transistor diverts current away from v_O to ground. This is unfortunate because this ground current i_{GND} can be substantial at $i_{O(MAX)}/\beta_{PO}$, especially under high temperatures and with weak fabrication lots, when β_{PO} is low.

Tradeoffs

As with most designs, choosing the power device that is most suitable for a given application is full of tradeoffs. The prevalence and importance of high efficiency in both battery-powered microelectronics for battery life and high-power applications for thermal management, for example, accentuate the value of low V_{DO} and low i_{GND}. In this respect, as Table 9.1 shows, P-type transistors offer the lowest V_{DO} and, because gate

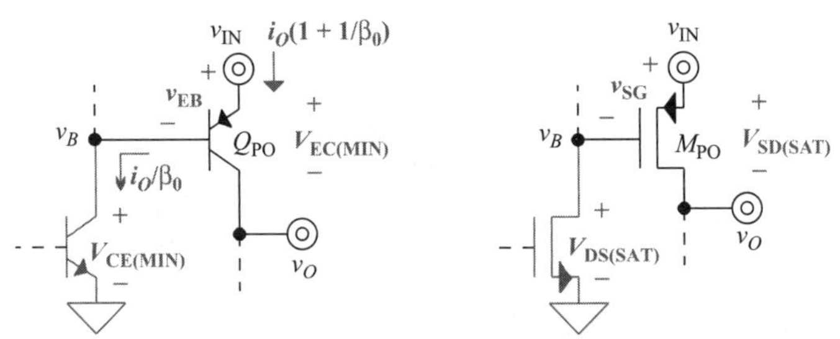

Figure 9.3 PNP and P-channel power transistors.

Parameter	N-Type Transistors BJT	N-Type Transistors MOSFET	P-Type Transistors BJT	P-Type Transistors MOSFET
V_{DO}	$V_{EC(MIN)} + v_{BE}$	$V_{SD(SAT)} + v_{GS}$	$V_{EC(MIN)}$	$V_{SD(SAT)}$
Drive	$i_{O(MAX)}/\beta_N$	$v_B - v_O$	$-i_{O(MAX)}/\beta_P$	$v_{IN} - v_B$
$i_{O(MAX)}$	Highest	Low	High	Moderate
i_{GND}	0 A	0 A	$i_{O(MAX)}/\beta_P$	0 A
R_O	$1/g_{m(BJT)}$	$1/g_{m(MOS)}$	r_o	r_{sd}

Table 9.1 Comparison of Power Transistors

currents are nil in steady state, MOS devices feature the lowest i_{GND}. This is why P-channel metal–oxide–semiconductor (PMOS) transistors are so popular in the industry today, because V_{DO} is low and i_{GND} is zero. Still, response time can be critical when load dumps are vast and output capacitance is low. In these latter cases, the low-impedance and therefore fast-response features of N-type power devices can trump the efficiency benefits of their P-type counterparts.

Because of their exponential behavior, BJTs typically source more current $i_{O(MAX)}$ than MOSFETs under similar silicon-area and input-supply constraints. Optimized vertical PNP transistors, however, are seldom available and their lateral counterparts, which are available in standard single-well CMOS technologies, are slower and less powerful. Plus, establishing a gate-drive voltage is often more practicable than driving a base current. Between MOSFETs, N-channel transistors source less current than their P-type counterparts because v_O limits its gate drive in NFETs. Again, the speed associated with N-channel metal–oxide–semiconductor (NMOS) followers partially offsets their high V_{DO} and low $i_{O(MAX)}$ performance.

Example 9.1 Use the process parameters outlined in Table 9.2 to design a power transistor that supplies up to 50 mA to a 1-V output when supplied from a 0.9–1.6-V nickel–metal–hydride (NiMH) battery with less than 200 mV of dropout and no ground current.

Circuit Parameter	Specification	Process Parameter	Value
v_{IN}	0.9–1.6 V	$\lvert V_{TPO}\rvert$	0.6 ± 0.15 V
v_O	1 V	K_P'	40 μA/V² ± 20%
i_O	≤ 50 mA	L	≥ 0.35 μm
V_{DO}	≤ 200 mV		
i_{GND}	0 A		

Table 9.2 Target Specifications and Process Parameters of the Power Transistor

Solution

1. For zero i_{GND} and a low V_{DO}, use a PMOSFET.
2. $V_{DO} \le R_{DS(MAX)} i_{O(MAX)} \approx \dfrac{i_{O(MAX)}}{K_{P(MIN)}'\left(\dfrac{W}{L}\right)(v_{IN(MIN)} - |v_{TP(MAX)}|)} \le 100\text{ mV}$

   or $$\frac{W}{L} \ge \frac{50\text{m}}{(32\mu)(200\text{m})(0.9 - 0.75)} = 52{,}083$$

   so W can be 18.3 mm and L can remain minimum at 0.35 μm.
3. $V_{DS(SAT)} \le \sqrt{\dfrac{2i_{O(MAX)}}{K_{P(MIN)}'\left(\dfrac{W}{L}\right)}} = \sqrt{\dfrac{2(50\text{m})}{(32\mu)\left(\dfrac{18{,}300}{0.35}\right)}} \approx 244\text{ mV}$

   so while V_{DO} in the dropout region is about 100 mV, the onset of dropout is when v_{IN} is within roughly 250 mV of v_O.

9.2.2 Layout

Reliability and performance drive the physical layout considerations of the power transistor. High-power conditions, for example, if designed to reduce silicon area, can stress devices to the outer boundaries of their *safe-operating areas* (SOAs), beyond which power transistors begin to break. As a result, the physical design of the device must avoid occurrences of *electron migration*, *hot spots* on the IC that lead to *secondary-breakdown* events, *substrate de-biasing* that leads to further de-biasing and *thermal runaway*, and other adverse effects. Typical high-power considerations therefore account for *current-density limits* and *current-crowding effects*. At the cost of additional power, engineers sometimes resort to resistors to ballast and distribute currents that would otherwise crowd and overstress a device. In other cases, low-dropout demands cause the size of the power device to be so large that reaching current-density limits is unlikely.

With respect to dropout, the aim is to reduce parasitic resistances in the power-conducting path, because they not only dissipate power and increase dropout voltage but also generate heat. In practice, ICs introduce parasitic ground-current paths whose effects exacerbate under high-power conditions, for large surface-area devices, and at high temperature. Leaking this ground current further lowers power efficiency, propagates noise through the common substrate, and increases the propensity for latch-up conditions. In attenuating these unfavorable effects, the general layout approach is to raise the impedance through these parasitic paths, and for the current that leaks through, to provide low-impedance paths to the supplies, all of which amounts to steering current away from the common substrate.

In light of the plethora of process technologies available and their possible mask-set compositions, several flavors of a particular transistor type may be available. The preponderance of standard *junction-isolated* bipolar, CMOS, and stripped-down biCMOS technologies, however, reduce layout approaches to a few general strategies: vertical and lateral BJTs and substrate and welled MOSFETs. CMOS technologies that include a P-base diffusion and standard P-substrate N-epitaxy bipolar processes, for example, offer vertical NPN and lateral PNP structures. Fully complementary bipolar processes, on the other hand, feature vertical NPN and PNP BJTs as well as their lateral counterparts. In the same vein, N-well CMOS technologies include substrate N-channel and welled P-channel transistors, and their P-well complements offer the opposite, substrate P-channel and welled N-channel devices. Dual-well technologies, of course, offer both, welled N- and P-channel devices as well as substrate NMOS transistors. Given the commonality of these variants, the intent of the following subsections is to highlight the driving physical considerations associated with these four basic layout strategies by citing and discussing common structures in view of dropout and ground-current considerations.

Vertical BJT

The vertical NPN BJT in Fig. 9.4 is often available in standard bipolar and basic biCMOS technologies. Here, electrons that carry i_O flow vertically from the emitter at v_O down to the intrinsic collector that the N^+ buried layer establishes, and then sideways and up to the extrinsic collector terminal to which v_{IN} attaches. To reduce resistance and the dropout voltage that series resistance in this path produces, the emitter contact is close to that of the collector. This way, the physical distance that carriers traverse from v_O to v_{IN} is minimal. A deep N^+ collector plug and an N^+ buried layer further lower this resistance.

When viewed from the top, increasing the surface area of the emitters or the number of emitter strips raises the amount of current the device can carry when the base–emitter junction is forward biased. And inserting thin base strips between the emitters lowers the series resistance of the base. Reducing this base resistance accelerates the large-signal response of the transistor because completely shutting or

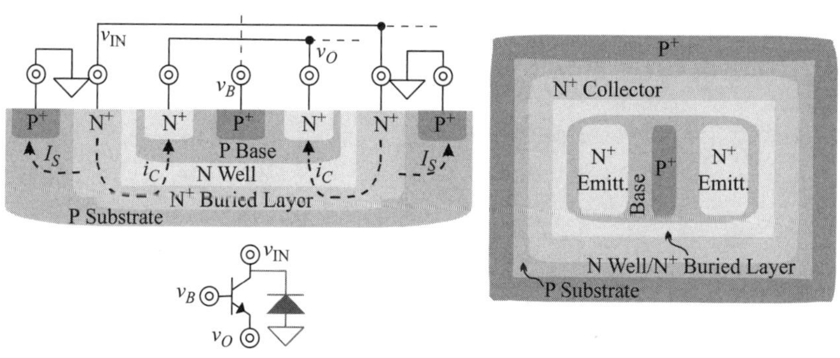

FIGURE 9.4 Profile and top views of a vertical NPN BJT in a P-type substrate.

fully engaging the device requires a *uniform* and substantial change in base–emitter voltage, which resistance and parasitic capacitance oppose in the form of momentary voltage drops. The deep N⁺ plug and corresponding collector contact surround and ring the entire device to reduce the overhead voltage that series resistance through the collector adds to dropout voltage and the device's minimum collector–emitter voltage $V_{CE(MIN)}$.

The vertical BJT structure shown in Fig. 9.4 only presents one reverse-biased diode from the collector attached to v_{IN} to the substrate that, other than conducting reverse-saturation current I_S, induces little additional substrate current. The P-base, N-collector, and P-substrate regions also comprise a substrate PNP transistor. Luckily, the effects of this parasitic vertical device are minimal because its emitter–base junction, which corresponds to the base–collector junction of the power NPN BJT, is typically reverse biased. Still, given the high-power nature of the structure, extending the N^+ buried layer to the edge of the N-well region is a good idea because overlapping the entire base with a highly doped N-type layer provides electrons with which parasitic minority-carrier holes can recombine. The end result is that less current reaches the substrate, which means the β_P of the substrate PNP BJT falls. A highly doped P-type guard ring around the entire device collects and prevents any I_S that leaks into the common substrate from propagating to the rest of the IC.

Lateral BJT

The holes that carry i_O in lateral PNP BJTs traverse laterally from the emitter to the collector, as Fig. 9.5 shows. Therefore, the periphery

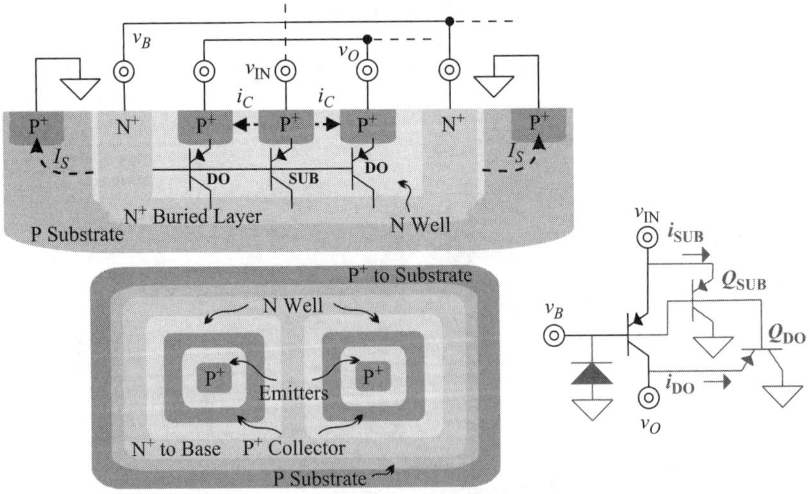

FIGURE 9.5 Profile and top views of a lateral PNP BJT in a P-type substrate.

and depth of the diffused emitter set the current-conducting area of the device, not the top surface area. So to collect as much current as possible, the collector tied to v_O usually rings the emitter attached to v_{IN}. Unfortunately, the P-type emitter and collector terminals, the N base, and the P substrate also constitute vertical PNP transistors Q_{DO} and Q_{SUB} that connect v_O and v_{IN} to ground. Because the collector–base junction of the intrinsic device is reverse biased, though, the parasitic device attached to v_O is normally off. Unfortunately, the emitter and base terminals of the parasitic BJT attached to v_{IN} share connections with those of the intrinsic BJT, so current i_{SUB} flows to ground through the substrate.

Decreasing the surface area of a unit emitter favors intrinsic lateral current i_C over parasitic vertical current i_{SUB}. This is why paralleling many small *emitter dots* raises the power rating of lateral transistors. An N^+ buried layer in the N well that houses the lateral BJT further reduces i_{SUB} by supplying electrons with which the parasitic vertical minority hole carriers can recombine. During dropout conditions, however, when the collector–base junction of the lateral transistor forward-biases, the parasitic vertical device attached to v_O also forward-biases, so Q_{DO} carries additional dropout current i_{DO} to ground. Although the N^+ buried layer also helps reduce the gain of this parasitic BJT and its resulting substrate current, the buried layer steers the current back to the lateral BJT's base, losing it as ground current i_{GND} into the buffer that drives the base. Generally, PNP BJTs are not ideal for battery-powered devices because their base currents lose i_O/β_P to ground, which under worst-case temperature and fabrication corners can be substantial.

Although parasitic current paths lower power efficiency, they also impede current crowding and the formation of hot spots on the IC, naturally increasing the ruggedness of the structure. Ringing shallow and deep N^+ base regions around each collector–emitter dot combination reduces base resistance and consequently improves response time when prompted to drive and sustain wide load dumps. This ring is especially important in the absence of a buried layer, as in the case of standard CMOS and basic biCMOS technologies. Like before, P^+ substrate contacts around the entire device collect and steer reverse-bias current I_S away from the rest of the IC.

Substrate MOSFET

Conventional MOSFETs, as Fig. 9.6 illustrates, conduct drain current i_D laterally. Therefore the separation between the source and drain, which constitutes the length of the induced channel, must be short and the width must be wide. As a result, channel resistance and, in consequence, dropout voltage are correspondingly low. From a top-view perspective, the length of each source/drain finger, which establishes channel width, should therefore be long, but not extraordinarily so. An excessively long finger unfortunately raises the series

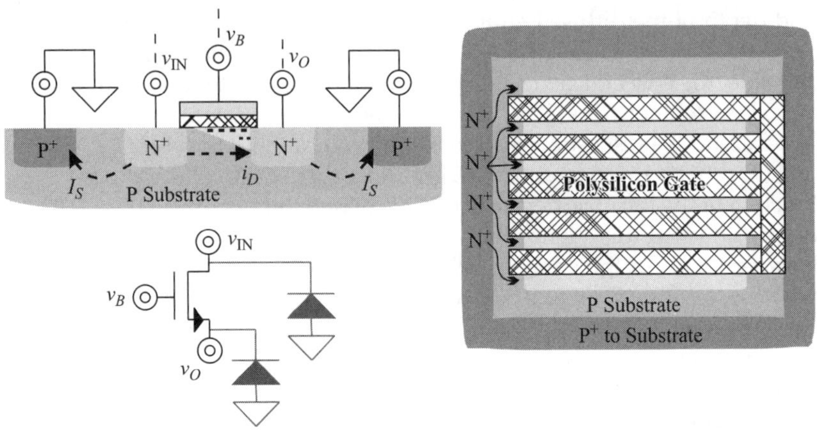

FIGURE 9.6 Profile and top views of a P-substrate NMOSFET.

resistance of the gate. This slows the large-signal response time of the device because gate resistance and capacitance keep the drain current that traverses across the extreme ends of the finger strip from reacting quickly.

Typical polysilicon gate layouts resemble a "fork" or a "tree" structure with multiple, but equally sized branches. Increasing the area of the gate, however, raises the parasitic capacitance that the transistor presents to the buffer that drives it. This of course slows the response of the regulator, which is why the base width of the fork or tree should not be excessive. The best way to raise the power rating of the device is to first extend the length of each finger to the point where series gate resistance is just shy of becoming problematic, and after that, multiply the number of fingers until the device can supply and sustain the power needs of the application. If many fingers are necessary, mirroring fingers on either side of the polysilicon "base" or "backbone," about the vertebra of the gate structure, minimizes the gate resistance of the entire device.

Fortunately, the only parasitic devices present in this structure are reverse-biased diodes to the substrate. Other than conducting reverse-saturation current I_S, these diodes conduct little else. In spite of this, given the high-power nature of the device, a P+ contact region that connects to the substrate normally rings this power transistor. This way, parasitic current I_S flows to ground directly instead of through the common substrate.

Welled MOSFET

A welled MOSFET is also a lateral device, as Fig. 9.7 shows so the series resistance through the device is lower if the distance between the source and diffusion contacts, which constitutes the length of

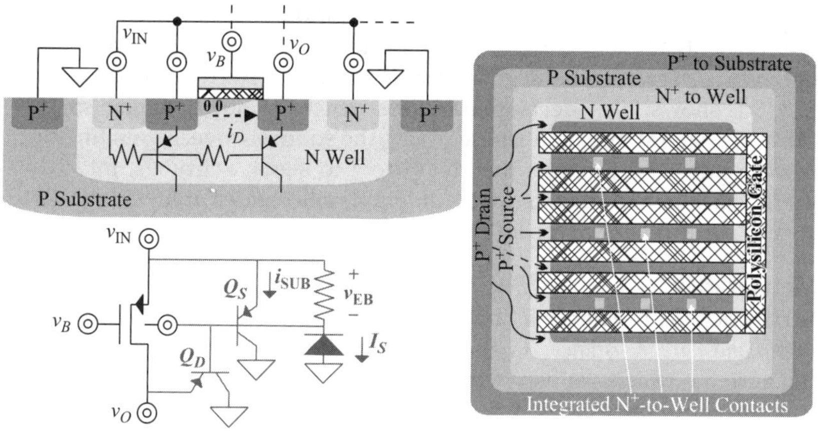

FIGURE 9.7 Profile and top views of an N-welled PMOSFET.

the channel, is short and the width of the *active gate* is wide. The polysilicon gate should also mimic a fork or a tree when viewed from the top to reduce gate resistance. The only difference between a substrate MOSFET and its welled counterpart is the back gate, because that of the former is the substrate and that of the latter is a well.

In spite of the seemingly benign difference, embedding a PMOS device in an N well presents, as in the case of lateral PNP BJTs, parasitic vertical PNP transistors from source and drain diffusions through the base that the well avails to the collecting substrate. Unlike the lateral PNP device, neither parasitic BJT is intrinsically forward biased, and the one attached to v_O is reverse biased. Unfortunately, because the power transistor conducts high currents and a highly resistive well connects the emitter and base terminals of the parasitic device Q_S attached to v_{IN}, the well is prone to drop a voltage that *de-biases* the base to the extent that Q_S steers current i_{SUB} into the substrate. When the transistor is small, this de-biasing voltage is not high enough to become a problem. The issue arises when the transistor is large, because sufficient reverse-saturation current I_S through reverse-biased junctions can flow through the resistive well to forward-bias the parasitic BJT and therefore induce it to conduct i_{SUB}.

As in the lateral BJT case, inserting a buried layer reduces the beta of the parasitic vertical BJT, except a buried layer is not typically available in standard CMOS technologies. The next strategy is to reduce the series resistance between the extrinsic and intrinsic base terminals of the PNP transistor, that is, between the well immediately around and below the P+ source diffusion that attaches to v_{IN} and its N+ contact. Conventionally, butting P+ sources against N+ well contacts lowers this series resistance. Unfortunately, inserting N+ well contacts next to every source finger in a high-power PMOS array requires considerable silicon area, which not only costs money but also presents higher overall capacitance to the driving buffer. Alternatively, integrating back-gate contacts into the source finger, as Fig. 9.7 shows, also decreases this resistance. Although doing so raises the lateral diffusion length across the source between gate fingers slightly and prompts the layout tool to signal warnings, integrated back-gate contacts save considerable silicon area at little to no cost in performance or reliability.

Metallization

Series metallic resistance R_M between the power device and the load and *bond-wire resistance* R_{BW} between the *lead frame* and *bond pad* increase the resistance between the power transistor and v_O, which raises the dropout voltage V_{DO} of the regulator to

$$V_{DO} = i_O R_{ON} = i_O(R_{SW} + R_M + R_{BW}) \tag{9.2}$$

where i_O is output current and R_{SW} is the "on" resistance of the power switch. Similarly, R_M and R_{BW} between v_{IN} and the input terminal of the device also raises R_{ON}, but more intrinsically in the case of P-type transistors because the critical dropout path is across the emitter–collector or source–drain terminals. The critical path of N-type transistors is through their base–emitter or gate–source terminals, so the connection to v_{IN} is less important than the connection to v_O. Still, irrespective of the type, power devices are physically large, and contacting their respective current-carrying terminals inevitably introduces series resistances that can in some instances overwhelm the resistance of the switch R_{SW}. For reference, consider that the total resistance of a 75-mΩ switch that connects v_{IN} to v_O via a 25-mΩ metallic link and two 25-mΩ bond wires is 150 mΩ, 50 percent of which rests outside the power transistor. Note, by the way, that lead-frame and board resistances are also parasitic, except the resistance of copper is usually much lower than those of the semiconductor and metals in the IC.

Several strategies can reduce this series on-chip resistance, and they all start and finish where the power device ends, at the silicon interface. The first step is to steer output current i_O laterally from the input bond pad where v_{IN} connects to one or several low-impedance planes above the power transistor, as the second-level plane connected to the source in Fig. 9.8 shows. Vertical *vias* then channel i_O down into the transistor. After traversing through the transistor, vias direct i_O up to one or several low-impedance planes that finally guide i_O laterally to the output pad to which v_O attaches. To decrease the vertical resistance, first-level metal covers the entire ohmic surface area of all emitter, collector, source, and drain regions.

Integrating the first-to-second-level vias into the first-level metal fingers decreases the first-to-second metal interface resistance. Some process technologies now allow vias to sit directly above and overlap

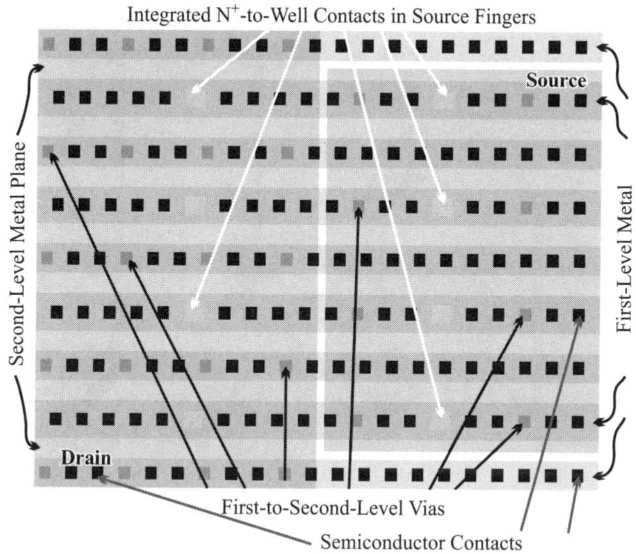

FIGURE 9.8 First-to-second-level metallization strategy for an N-welled PMOSFET.

the semiconductor–metal contacts, which means that integrating metallic vias into the fingers incurs little to no tradeoff with respect to semiconductor-to-metal resistance, because vias do not take the place of semiconductor contacts. Note that the PMOS source fingers include integrated N^+ back-gate/N-well contacts to reduce the N-well de-biasing effects discussed earlier, which is why the width of the source fingers is slightly wider than that of drain fingers. The ratio of the number of source-diffusion to back-gate contacts is an engineering judgment call. Inserting four or more diffusion contacts into each well contact is not a bad strategy because the source carries considerably more current than the well.

Note that a lateral metal track above the transistor collects or disperses charges as it crosses the IC. When sourcing current from the output terminal of a power transistor to the output pad v_O, for example, current climbs as the metal plane carrying the current draws more charges from the underlying semiconductor. If this lateral metal plane has a constant width, the current density through the plane is inconsistent, which results in current crowding and inefficient use of the area available. Increasing the width of the current-sourcing plane as it collects charges is ideal, as Fig. 9.9 shows, as is decreasing the width of the current-receiving plane as it loses charges to the semiconductor. Depending on the ultimate location of the bond pads and the number of metal layers available, current may grow and attenuate in a number of different directions, giving the top-view layout

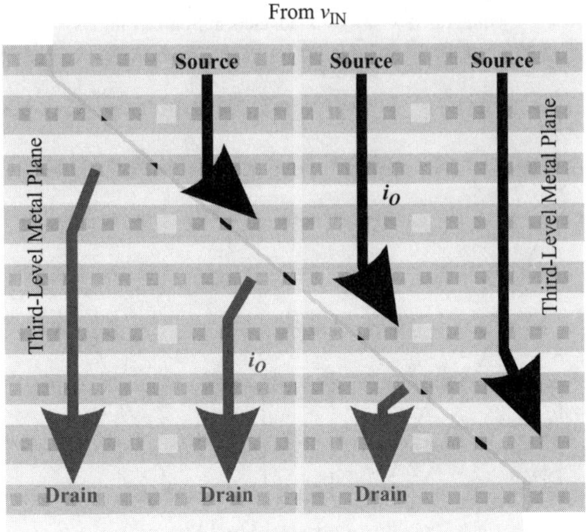

FIGURE 9.9 Top metallization strategy for an N-welled PMOSFET.

geometry of a power device a multitriangular look, but seldom a rectangular appearance, unless current levels are sufficiently low to incur little or no additional series voltage drop.

Ideally, the power device should sit close to the input and output bond pads, like in the corner of a large IC, to reduce the lateral resistance of the top-level metal planes that carry i_O. If placing the device immediately next to its bond pads is impossible, multiple metal planes in parallel with the top-level plane, which usually has the lowest sheet resistivity, should carry the current to the pads. If more than three levels of metal are available, they should parallel the efforts of the other planes with the objective of lowering series resistance and ballasting current flow to keep current densities low and consistent through the device. Similarly, using the pins in the lead frame that are closest to their respective bond pads and using multiple bond wires and bond pads reduces bond-wire resistance R_{BW}.

Series current sensors or parallel current replicas normally monitor the power device to protect it from *short-circuit*, or more generally, *overcurrent* conditions. *Thermal-shutdown* features can likewise, though indirectly, protect the device against similar events. A substantially smaller version of the power transistor in parallel is a popular way of sensing current because it does not add any series resistance to the power path. Placing this transistor sensor in the middle of the large power-transistor array is important to combat the mismatches that short-channel lengths and large spreads between devices cause. This strategy often creates a discontinuity in the layout structure because one of the sensor terminals is not common to the power transistor. The benefits of matching the sensor to the power device, however, far outweigh the routing idiosyncrasies that can viably result. Because the power device is the fundamental heat source in the regulator IC, a temperature sensor should similarly lie close to the power transistor, but not necessarily in the middle of the array, especially if the sensor is a different type of transistor.

9.3 Buffers

The objective of the buffer A_B is to drive the power switch S_O quickly and with minimal power. Because the large transistor loads the buffer with high parasitic capacitance C_{IP}, the buffer must not only exhibit low output impedance Z_{OB} but also sink and source sufficiently high slew-rate currents I_{SR} to quickly charge and discharge C_{IP}. A_B must also be able to swing its output voltage v_B wide enough to both shut the power device off and fully engage it. And it must do so even when the input supply v_{IN} is low, because the power lost across the power transistor drops when v_{IN} is closer to v_O. A_B should also load the error amplifier A_E minimally, with high resistance R_{IB} and low capacitance C_{IB}, so A_E's gain and bandwidth can both remain high.

9.3.1 Driving N-Type Power Transistors

NPN BJT

Among all the single-transistor configurations possible, emitter and source followers subscribe best to the high-bandwidth, low-output-impedance requirements of the buffer. Of these, since the voltage drop from v_{IN} to the output of the buffer v_B is in the critical dropout path of a power NPN BJT, P-type followers are well suited for the task, because their respective emitter–base and source–gate voltages from Fig. 9.10 do not affect the power transistor's dropout voltage V_{DO}.

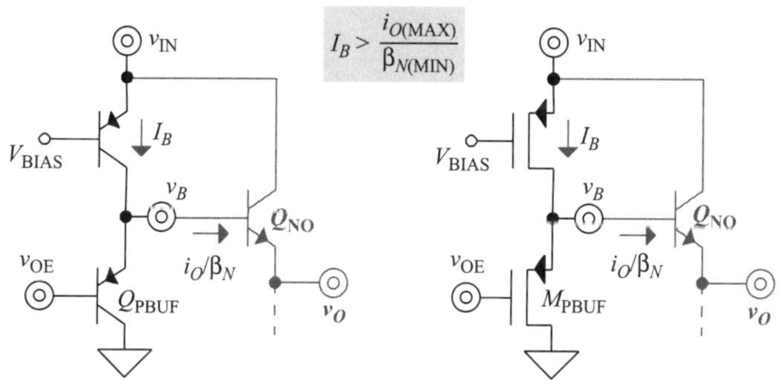

FIGURE 9.10 Driving NPN BJTs with P-type followers.

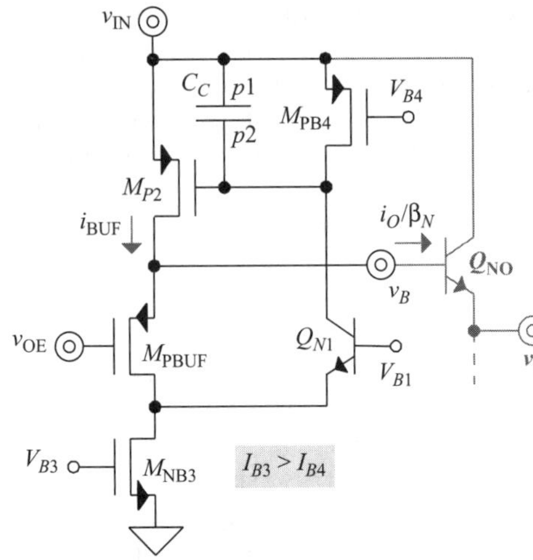

FIGURE 9.11 Driving an NPN BJT with an on-demand P-type follower.

Despite the seemingly attractive feature of simplicity, P-type followers suffer from poor power efficiency because the current source I_B that biases them must be sufficiently high to supply the base current necessary to drive the NPN power transistor. The requirement is worse when the load is heavy, temperature is low, and the device is weak—in other words, when base current is highest at $i_{O(MAX)}/\beta_{N(MIN)}$.

On-Demand Drive Designing a dynamically adaptive, on-demand bias current is probably the best approach for improving efficiency. Adding a negative feedback path from the follower buffer to the sourcing current, as Fig. 9.11 illustrates, achieves this. This way, the feedback loop supplies whatever sourcing current i_{BUF} is necessary to ensure that v_B changes with A_E's output v_{OE}. Here, the P-type follower M_{PBUF} series-mixes v_{OE} with v_B and shunt-samples v_B with a feedback factor that is equivalent to 1. This means the circuit's output resistance R_{OB} is the loop-gain translation of M_{PBUF}'s $1/g_{m.BUF}$ resistance:

$$R_{OB} \approx \frac{1}{g_{m.BUF}A_{LG}} \approx \frac{1}{g_{m.BUF}(r_{sd4}g_{m2})} \tag{9.3}$$

where M_{PBUF} and Q_{N1} buffer feedback current into M_{PB4}'s r_{sd4} and M_{P2}'s g_{m2} converts the voltage dropped across r_{sd4} into i_{BUF} to establish the gain across the loop A_{LG}. Note that replacing N- and P-type FETs with NPN and PNP BJTs, and vice versa, yields similar benefits, albeit at slightly different current and impedance levels.

M_{NB3} must sink more current than M_{PB4} sources to bias M_{PBUF}. Although the base of the power transistor connects to M_{PBUF}'s source, whose output resistance is low, parasitic capacitance C_{IP} is so high that it pulls the pole at v_B close to the one at the gate of sourcing transistor

M_{P2}, whose drain resistance is high. The purpose of compensation capacitor C_C is therefore to pull M_{P2}'s gate pole to lower frequency, and that way, ensure that it remains the dominant low-frequency pole for the feedback loop.

Capacitor C_C's first-level polysilicon plate connects to v_{IN} to shunt substrate noise coupled to that poly-one terminal to v_{IN}. The reason for this is that the *poly-one plate* is more prone to receive noise than the *poly-two plate,* because the former is closer to the substrate. And the capacitor connects to v_{IN} instead of ground to couple supply noise to M_{P2}'s gate. This way, C_C replicates supply noise at M_{P2}'s source onto M_{P2}'s gate, and M_{P2}'s transconductance cancels the noise that is common to both terminals. M_{P2}'s width–length ratio W/L and minimum gate drive $v_{SG2(MIN)}$, which depends on v_{IN} and Q_{N1}'s bias voltage V_{B1}, ultimately set how much current the circuit can drive into the base of the power transistor:

$$i_{BUF} \geq 0.5K_P'\left(\frac{W}{L}\right)(v_{SG2(MIN)} - |v_{TP}|)^2 \geq \frac{i_{O(MAX)}}{\beta_{N(MIN)}} \tag{9.4}$$

The key advantage here is that, under light loading conditions, the quiescent current of the circuit can be as low as a few microamps. In other words, i_{BUF} is high only when needed, which means the circuit loses little current otherwise. The tradeoff for this efficiency improvement is a slower response time because, for the sake of feedback stability, C_C slows the small-signal dynamics of the buffer.

NMOSFET

The power NMOS device has similar requirements to those of its NPN counterpart, except driving a base current is no longer necessary. Accordingly, the requirements of the P-type follower buffers shown in Fig. 9.10 are less stringent. Bias current I_B must now supply and sink the slew-rate current I_{SR} necessary to charge and discharge a high parasitic capacitance C_{IP}, which is not trivial, but usually easier than supplying a worst-case base current:

$$I_B \geq C_{IP}\left(\frac{\Delta v_C}{\Delta t_C}\right) \approx C_{IP}\Delta v_{B(MAX)} f_{BW(MAX)} \equiv I_{SR} \tag{9.5}$$

where Δt_C is the time required for the voltage across C_{IP} to change Δv_C in response to a load dump, and $\Delta v_{B(MAX)}$ and $f_{BW(MAX)}$, which corresponds to $1/\Delta t_{C(MIN)}$, represent the maximum Δv_C required and the minimum Δt_C allowed. Note that, with enough base or gate drive, a P-type follower can more easily sink slew-rate current I_{SR-}. And again, as with the NPN BJT, adding a feedback loop like the one Fig. 9.11 shows can reduce the quiescent current of the circuit by dynamically adjusting I_B in Fig. 9.10 to supply maximum current on demand, only when needed. Like before, though, efficiency improves at the expense of response time.

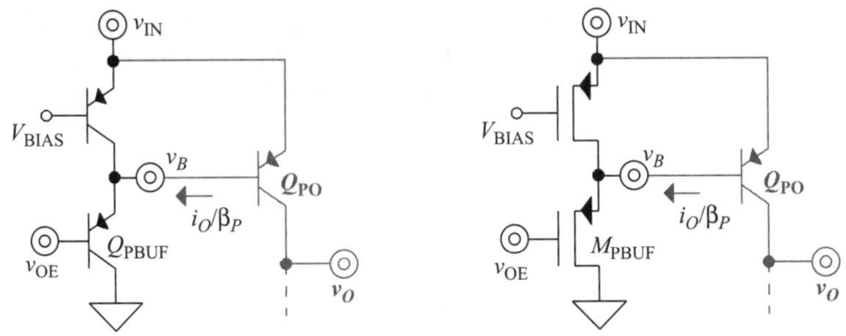

FIGURE 9.12 Driving PNP BJTs with P-type followers.

9.3.2 Driving P-Type Power Transistors

PNP BJT

As already mentioned, the electrical characteristics of emitter and source followers synergize with the requirements of the driving buffer A_B. Of possible configurations, though, P-type followers from Fig. 9.12 are often more practical for driving P-type power transistors because their outputs can rise close enough to v_{IN} to reliably shut the power device off. Raising v_B only within a base–emitter or gate–source voltage of v_{IN}, which is what N-type followers can do, is insufficiently high to ensure that a large and leaky power transistor is off across temperature and fabrication corners. This is especially troublesome in portable applications because leakage currents drain the battery. The other major steady-state requirement of the buffer here is to sink the base current of the PNP BJT. This current, of course, can be substantial when the load is heavy, temperature is low, and the transistor is weak, which is when base current peaks to $i_{O(MAX)}/\beta_{P(MIN)}$.

BJT buffers propagate base-current requirements back to the error amplifier A_E. In the case of a PNP follower buffer, for example, A_E must sink roughly $i_{O(MAX)}/\beta_{P(MIN)}{}^2$, which in the case of a 100-mA load when β_P is 15 A/A translates to 444 μA. Sinking this much current is challenging when A_E's quiescent-current budget is less than 10 μA. A PMOS follower is no better because establishing a source–gate voltage that is high enough to sink $i_{O(MAX)}/\beta_{P(MIN)}$ from the PNP BJT is not trivial when constrained to low input-supply voltages. In the case of the 100-mA load when β_P is 15 A/A, sinking 9.7 mA of base current requires gate-drive voltages of up to 1.5–3 V.

On-Demand Drive Like before, a feedback loop around the sinking P-type follower buffer can enlist the help of another transistor for additional on-demand current. Feedback transistor M_{N3} in the buffer driver of Fig. 9.13, for example, supplements PMOS follower M_{PBUF} with substantial sinking current only when needed. Here, M_{PBUF} only

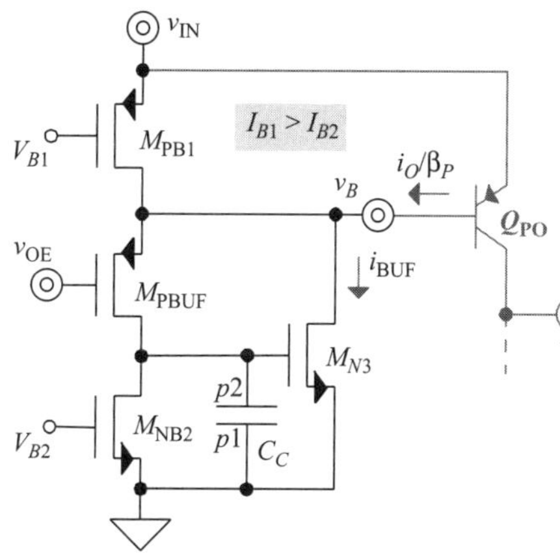

FIGURE 9.13 Driving a PNP BJT with an on-demand P-type follower.

sinks the bias current I_{B2} that M_{NB2} establishes, and M_{N3} sinks what remains of M_{PB1}'s bias current I_{B1} when the power PNP transistor Q_{PO} is off. As Q_{PO}'s base current climbs, and v_{OE} and subsequently v_B fall, M_{N3} sinks whatever current M_{PBUF}'s I_{B2} cannot. As before, compensation capacitor C_C ensures that the lowest-frequency pole in the loop is at the gate of M_{N3}. Connecting C_C's poly-one plate to ground couples ground-bounce noise in M_{N3}'s source to its gate. This way, M_{N3}'s transconductance can cancel it in the form of common-mode noise. In the end, as before, the quiescent current that negative feedback saves offsets the slower response that results.

Native NMOSFET Although not always an option, more sophisticated power biCMOS technologies offer the necessary mask sets and processing steps to isolate a low-threshold NMOS transistor. Figure 9.14, for example, illustrates how a deep N^+ ring and an overlapping N^+ buried layer can isolate a P-type region that an NFET can use as its bulk.

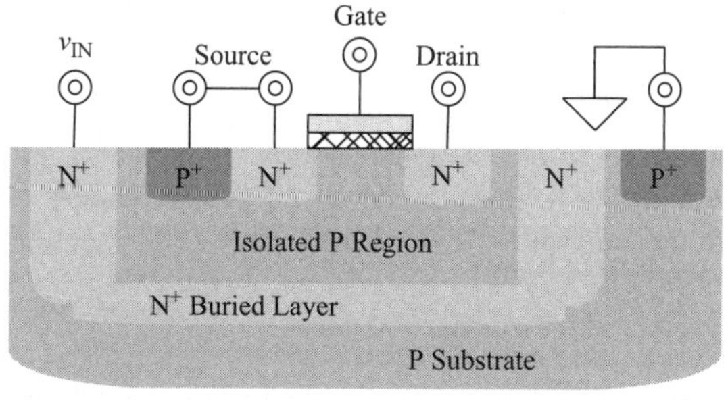

FIGURE 9.14 Isolated substrate NMOS transistor in a P-type substrate.

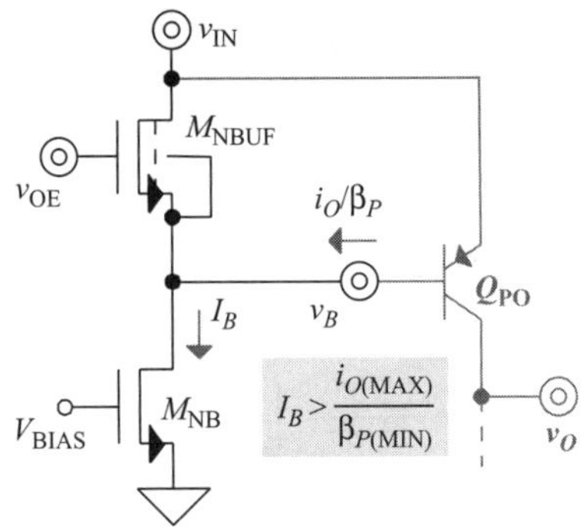

FIGURE 9.15 Driving a PNP BJT with a native NMOS follower.

Connecting this isolated bulk to the source of a low-threshold NFET eliminates the bulk effects that would otherwise render a low-threshold device unreliable. Luckily, substrate NMOS transistors that do not undergo the threshold-adjustment step in the fabrication process exhibit zero bulk-bias threshold voltages V_{TN0} that near zero. The gate–source voltages of these so-called *natural* or *native NFETs* reduces to a saturation voltage $V_{DS(SAT)}$ when their isolated bulks connect to their corresponding sources.

When configured as a follower, as Fig. 9.15 shows, the isolated natural NMOS transistor does not, on its own, overcome the need for sinking the high base current of a PNP power transistor. M_{NB}'s bias current I_B must therefore be high enough to sink and sustain Q_{PO}'s base current when the load is heavy, temperature is low, and the device is weak. In other words, I_B must sink no less than $i_{O(MAX)}/\beta_{P(MIN)}$.

Like before, negative feedback around the follower can help sink additional on-demand current. Transconductor G_{FB} and M_{NBUF} in Fig. 9.16, for example, series-mix V_{BIAS} and v_R and v_{OE} and v_B. This way, v_R mirrors V_{BIAS} to establish a bias current through R_B and M_{NBUF}, and changes in v_{OE} appear in v_B to drive Q_{PO}. The feedback action of M_{N1} is to sink whatever current M_{NBUF}'s bias current and Q_{PO}'s base current supply. Capacitor C_C sets the dominant low-frequency pole of the loop at the gate of M_{N1}, and its poly-one plate connection ensures that ground noise remains common mode across M_{N1}'s gate and source terminals. Because v_B must climb high enough to shut the PNP transistor off, the voltage across R_B should be low at around 100–300 mV. Incidentally, note that G_{FB} need not be anything more than an N-type differential pair loaded with a P-type bias current source M_{PB}, as Fig. 9.16 shows.

PMOSFET

Like PNP devices, P-type MOS transistors require high gate voltages to shut them, so the P-type followers in Fig. 9.12 also work in

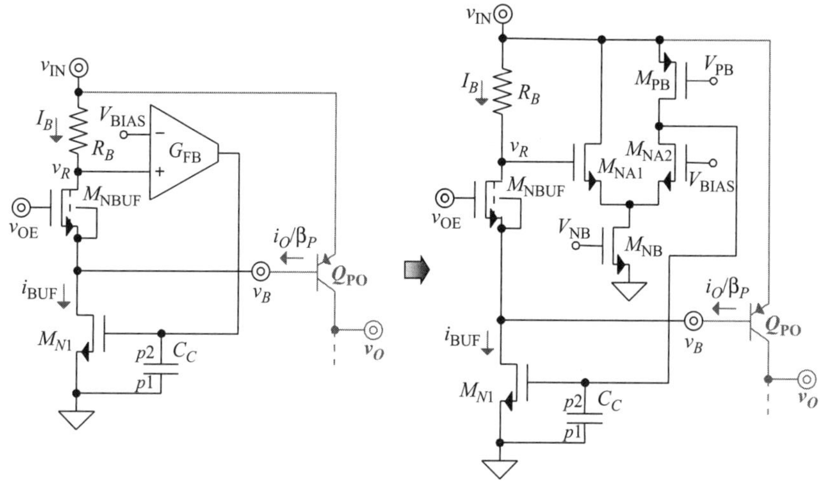

FIGURE 9.16 Driving a PNP BJT with an on-demand native N-type follower.

PMOS applications. The only problem with a P-type follower is poor dropout performance when the input supply v_{IN} is low. The reason for this is that the dropout voltage V_{DO} across a power PMOSFET M_{PO} is inversely proportional to its gate drive $v_{SG.O}$. And in the case of a P-type follower, the follower's source–gate or emitter–base voltage v_{SG} or v_{EB} limits how low M_{PO}'s gate voltage v_B can fall with respect to v_{IN}:

$$V_{DO} = i_O R_{DO} \propto \frac{1}{v_{SG.O}} = \left.\frac{1}{v_{IN} - v_B}\right|_{PBUF} = \frac{1}{v_{IN} - (v_{SG.BUF} + v_{OE})} \tag{9.6}$$

where R_{DO} is M_{PO}'s dropout resistance and v_{OE} drives the gate of the buffer. In this respect, the isolated native N-type follower buffer M_{NBUF} in Fig. 9.15 is more attractive than a P-type follower because it can not only shut the power PMOS transistor but can also extend its gate drive during dropout conditions by the source–gate or emitter–base voltage that a P-type follower would otherwise impose:

$$V_{DO} = i_O R_{DO} \propto \frac{1}{v_{SG.O}} = \left.\frac{1}{v_{IN} - v_B}\right|_{NBUF} = \frac{1}{v_{IN} - V_{DS(SAT)}} \tag{9.7}$$

where $V_{DS(SAT)}$ is the voltage across the sinking bias transistor M_{NB}.

On-Demand Drive As with NMOS power transistors, driving a base current is no longer an issue, but charging and discharging the high parasitic capacitance that the gate of a power PFET presents is. In this regard, follower buffers can supply substantial slew-rate currents in one direction, but not in the other, because a biasing current

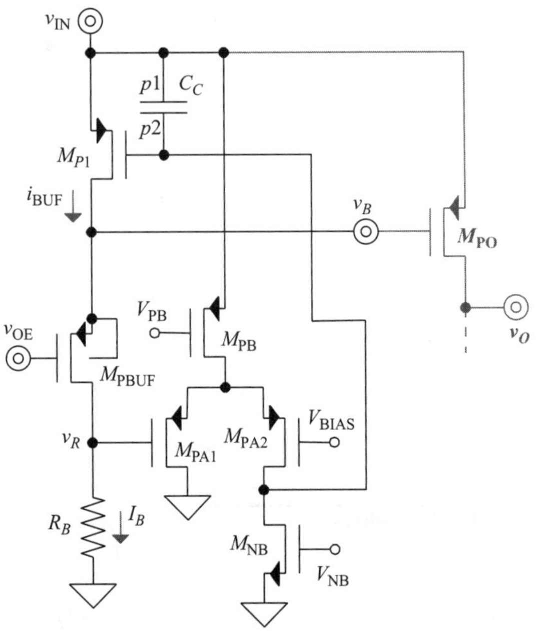

FIGURE 9.17 Driving a PMOSFET with an on-demand P-type follower.

I_B bounds them. Applying negative feedback, as before, can supplement the driving deficiencies of the follower at a fraction of the cost in quiescent current, as exemplified by the N-type follower M_{NBUF} embodiment of Fig. 9.16 and its P-type equivalent in Fig. 9.17. Here, M_{PA1} and M_{PA2} series-mix V_{BIAS} and v_R, and buffer follower M_{PBUF} series-mixes v_{OE} and v_B. As a result, v_R mirrors V_{BIAS} to bias M_{PBUF} with R_B's current, and changes in v_{OE} appear in v_B to drive the power transistor M_{PO}. V_{BIAS} should be low enough to ensure that the voltage across R_B does not push M_{PBUF} into triode during sinking slew-rate conditions, which would otherwise raise M_{PBUF}'s v_{SG} and v_B. And again, like before, compensation capacitor C_C slows the circuit for the sake of feedback stability.

Regenerative Drive The benefits of dynamically adapting the driving capabilities of the buffer with follower transistors in negative feedback may not always sufficiently offset their speed limitations to make them attractive solutions. This is especially true in *system-on-chip* (SoC) applications, when output capacitance C_O is low and load dumps are vast and wide. A regenerative circuit like Fig. 9.18 shows in place of a biasing transistor and its negative-feedback counterpart can accelerate and bootstrap its otherwise slew-rate-limited response without surrendering to the bandwidth limitations of negative feedback.

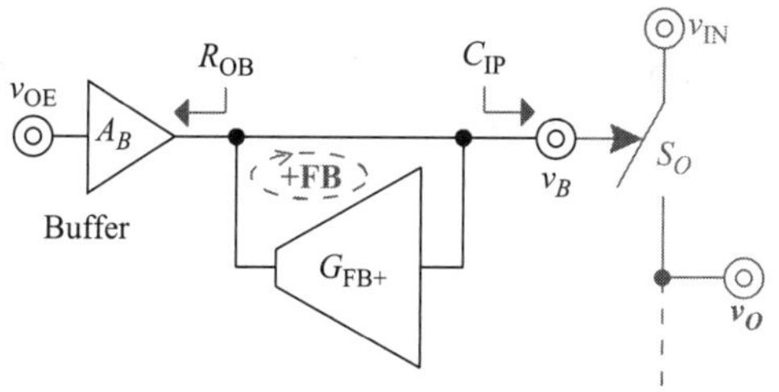

FIGURE 9.18 Driving a power transistor with a regenerative buffer.

The problems with positive feedback are possible oscillations and difficult-to-reverse latched events. This is why the gain across the loop A_{LG+} must remain below unity. This is not a problem at low frequencies because the feedback loop that controls the output of the regulator also shunt-samples v_B to reduce the output resistance of the buffer R_{OB} by the loop gain of the system A_{REG}:

$$A_{LG+} = G_{FB+}Z_{OB} = G_{FB+}\left(\frac{R_{OB}}{A_{REG}} \parallel \frac{1}{sC_{SW}}\right) < G_{FB+}\left(R_{OB} \parallel \frac{1}{sC_{SW}}\right) < 1 \tag{9.8}$$

At high frequency, however, at and past the unity-gain frequency f_{0dB} of the system, A_{REG} no longer helps. Unfortunately, A_{LG+} must still remain below 1 at and past f_{0dB}, so the gain that G_{FB+} and R_{OB} establish must similarly stay below 1. This means that R_{OB} must be low enough to damp the oscillations and latching tendencies of the regenerative loop.

Load Tracking When considering the biasing condition of the follower buffer, also recall that the bandwidth f_{0dB} of a PMOS regulator varies with output current i_O, and the pole at the output of the buffer p_{OB} is parasitic. The problem is that the output pole p_O and the unity-gain frequency f_{0dB} that p_O establishes shift with i_O, as discussed in Chap. 8 and as Fig. 9.19 now illustrates in the context of p_{OB}. In addition, R_{ESR}

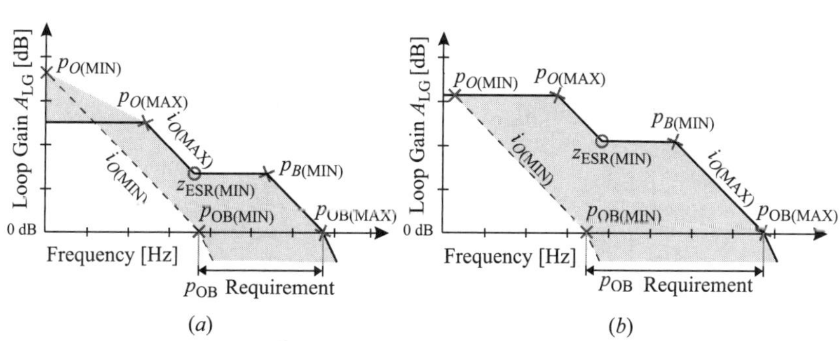

FIGURE 9.19 Buffer-pole requirements in output-compensated regulators for power (*a*) PMOS and (*b*) PNP transistors.

extremes extend this variation. As a result, p_{OB} must remain above f_{0dB} at all times. R_{OB} should therefore be so low that not even C_{IP} should be able to shunt R_{OB} at or below the highest possible f_{0dB}. This of course translates to quiescent current because higher currents produce lower resistances.

In the same spirit of dynamic adjustments, considering that high efficiency is paramount in battery-powered applications, integrating a steady-state component into the regenerative loop of Fig. 9.18 that raises the biasing current of the buffer with i_O would be ideal because p_{OB} can rise with p_O. The circuit illustrated in Fig. 9.20 does just this, employs positive feedback to both accelerate the transient response of v_B and adjust the steady-state biasing conditions of the follower buffer M_{NBUF}. Here, sensing circuit M_{PS}–Q_{P3}–Q_{P4} sources a current into a folding mirror M_{N1}–M_{N2} that is proportional to i_O. This establishes a bias current for M_{NBUF} that rises with i_O. This way, with the same no-load bias current, the regenerative loop can accelerate and adjust M_{NBUF}'s current to such an extent that a response time of 60 μs, for example, can drop to 2–5 μs.

Operationally, M_{PS} is a sense PMOS transistor that is substantially smaller than the power device M_{PO}. Accordingly, M_{PS} directs a diminutive fraction of i_O through current buffer Q_{P3} into current mirror M_{N1}–M_{N2}, whose output biases M_{NBUF} with a mirrored impression of i_O. Source current I_{B1} drives a small bias current into the mirror M_{N1}–M_{N2} through Q_{P3} that, in turn, sets M_{NBUF}'s bias current during zero-to-light loading conditions, when i_O is low and M_{PS} sources negligibly low currents. Notice that bypassing Q_{P3} and M_{N1}–M_{N2} and pulling an

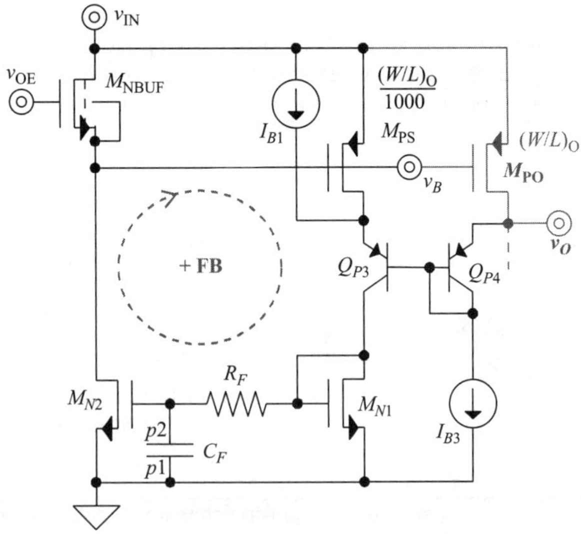

FIGURE 9.20 Driving a power PMOS transistor with a regenerative N-type buffer.

independent bias current directly from M_{NBUF}'s source to ground would starve Q_{P3} and M_{N1}–M_{N2} of current during light loads, when M_{PS} sources negligible current, which would unfortunately extend the response time of the loop. The underlying purpose of the sensing buffer Q_{P3}–Q_{P4} is to impress, to first order, M_{PO}'s drain voltage on M_{PS} to prevent M_{PS} from entering a dissimilar region of operation with respect to M_{PO}. This fix is important during dropout conditions, when M_{PO} enters the triode region and v_B drops close to ground. Without Q_{P3}–Q_{P4}, these circumstances would keep M_{PS} in saturation and induce it to source a disproportionately higher current into the mirror than a linear sensor should.

During a rising load dump, when i_O suddenly rises and A_E responds by lowering v_{OE}, v_B drops, and as it does, M_{PS} drives more current into the current mirror. As a result, M_{N2} pulls more current from v_B and raises the discharge rate of C_{IP} and the falling rate of v_B. Increasing the width–length ratio of M_{PS} and the mirror gain of M_{N1}–M_{N2} raises the transconductance gain $G_{\text{FB+}}$ of the positive feedback loop and amplifies its "speed-up" effects. Doing so, however, can compromise the feedback stability of the system. As a result, the width–length ratios of the sensing PFET M_{PS} and the mirroring transistor M_{N2} with respect to M_{PO} and M_{N1} should be high, but not high enough to induce long and large oscillations in any one of the worst-case operating, temperature, and fabrication corners of the system. This worst-case point can fall in one of many corners, when M_{PS} and other PFETs are strong, M_{NBUF} and other NFETs are weak, temperature is low and transconductances are correspondingly high, input voltage v_{IN} is high and gate drive is correspondingly high, i_O is high and C_O is low and p_O is correspondingly high, and R_{ESR} is high and z_{ESR} is correspondingly low.

A low-pass filter R_{FB}–C_{FB} in the positive-feedback path helps decouple the biasing and dynamic requirements of the loop. For maximum efficiency, M_{NBUF}'s current should be low when i_O is low and high otherwise, so the current gain that M_{PS} and M_{N1}–M_{N2} establishes should therefore accommodate this range. This gain, however, compromises the feedback stability of the loop. This is how the filter helps, because the filter attenuates the feedback gain to ensure that $A_{\text{LG+}}$ remains below 1 near and past f_{0dB}. The filter, of course, slows the speed-up benefits of the loop, but not to the extent that the accelerating effect vanishes.

Example 9.2 Refer to the specifications and process parameters outlined in Table 9.3 and design the regenerative load-tracking buffer illustrated in Fig. 9.20 to drive the power transistor designed in Example 9.1. The no- and full-load 50-mA bandwidths of the circuit should be higher than 0.1 and 1 MHz, respectively, with a no-load quiescent current of less than 4 μA and an output voltage that can swing from 0.2 V to within a P-type threshold voltage of the input supply. A 0.9–1.6-V NiMH battery supplies the buffer, but its practicable range is 1.1–1.6 V because the energy that remains in the battery at 0.9–1 V is impracticably low, and the headroom limit of MOSFETs with 0.6 ± 0.15-V thresholds in analog circuits is about 1–1.1 V across temperature and fabrication corners.

Circuit Parameter	Specification	PMOS Process Parameter	Value
v_{IN}	1.1–1.6 V	$\lvert V_{TPO}\rvert$ & V_{TNO}	0.6 ± 0.15 V
v_O	1 V	$V_{TNO.NAT}$	0 ± 0.15 V
i_O	≤ 50 mA	K_P'	40 μA/V² ± 20%
f_{0dB} at 0 A	≥ 100 kHz	K_N'	100 μA/V² ± 20%
p_{OB} at 50 mA	≥ 1 MHz	L	≥ 0.35 μm
I_Q at 0 A	≤ 4 μA	$\lambda_{L(MIN)}$	500 mV⁻¹
$v_{BUF(MAX)}$	$v_{IN} - \lvert V_{TPO(MIN)}\rvert$	L_{OV}	35 nm
$v_{BUF(MIN)}$	0.2 V	C_{OX}''	5 fF/μm²
		β_{PO}	50–150 A/A
		V_A	15 V

Table 9.3 Target Specifications and Process Parameters for the Buffer in Example 9.2

Architectural Design

1. Current buffer Q_{P3}: When v_O is 1 V and $v_{GS.N1}$ is about 1 V, Q_{P3} is in deep saturation, so shift Q_{P3}'s collector voltage with the level-shifting resistor R_{DC} in Fig. 9.21 to ensure that Q_{P3} remains in light saturation. This way,

$$v_{EC.P3} \approx v_O - (v_{GS.N1} - I_{B1}R_{DC}) \geq 300\text{ mV}$$

Component Design

2. Parasitic zero-load capacitance $C_{IP(0)}$ at v_B:

$$C_{IP(0)} \approx C_{SG.PO} \approx \frac{2}{3}C_{OX}''W_{PO}L_{PO} + C_{OX}''W_{PO}L_{OV}$$
$$\approx \frac{2}{3}(5\text{f})(18.3\text{k})(0.35) + (5\text{f})(18.3\text{k})(0.035) \approx 25\text{ pF}$$

3. Equivalent full-load capacitance $C_{IP(50mA)}$ at v_B:

$$C_{IP(50mA)} \approx C_{SG.PO} + C_{GD.PO(Miller)} \approx \frac{2}{3}C_{OX}''W_{PO}L_{PO} + C_{OX}''W_{PO}L_{OV}(g_{m.PO}r_{sd.PO})$$

$$\approx \frac{2}{3}C_{OX}''W_{PO}L_{PO} + C_{OX}''W_{PO}L_{OV}\left(\frac{\sqrt{2i_{O(MAX)}K_P'\left(\frac{W}{L}\right)_{PO}}}{i_{O(MAX)}\lambda_{L(MIN)}}\right)$$

$$\approx \frac{2}{3}(5\text{f})(18.3\text{k})(0.35) + (5\text{f})(18.3\text{k})(0.035)\left(\frac{\sqrt{2(50\text{m})(40\mu)\left(\frac{18.3\text{k}}{0.35}\right)}}{(50\text{m})(500\text{m})}\right) \approx 80\text{ pF}$$

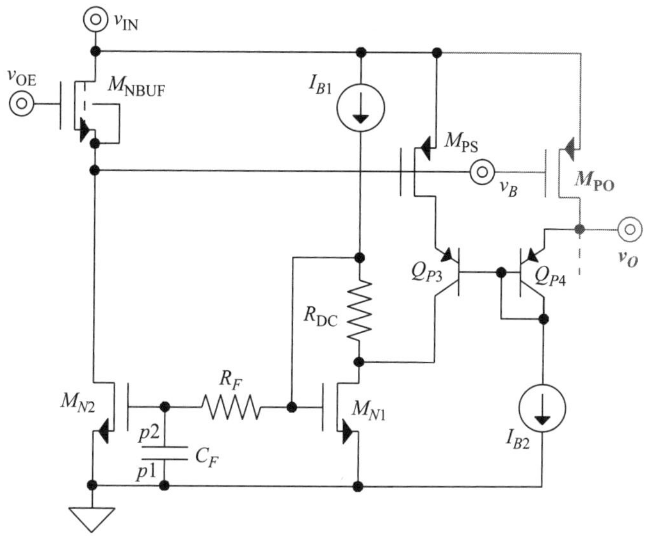

FIGURE 9.21 Regenerative load-tracking N-type follower buffer.

4. Full-load bandwidth $p_{OB(50mA)}$:

$$p_{OB(50mA)} \approx \frac{1}{2\pi C_{IP(50mA)} R_{OB}} \approx \frac{g_{m.NBUF}}{2\pi C_{IP(50mA)}}$$

$$= \frac{\sqrt{2 i_{NBUF(50mA)} K_N' \left(\frac{W}{L}\right)_{NBUF}}}{2\pi C_{IP(50mA)}} \geq 1\,\text{MHz}$$

or $$i_{NBUF(50mA)} \left(\frac{W}{L}\right)_{NBUF} \geq \frac{(2\pi C_{IP(50mA)} p_{OB(50mA)})^2}{2K_{N(MIN)}'} = \frac{[(2\pi)(80p)(1M)]^2}{2(80\mu)}$$

$$\approx 1.6\,\text{mA}\left(\frac{\text{m}}{\text{m}}\right)$$

so if W_{NBUF} is 12 μm and L_{NBUF} is as short as possible at 0.35 μm, $i_{NBUF(50mA)}$ should be 46 μA. Note that a smaller M_{NBUF} raises A_{LG+} and a larger M_{NBUF} raises the parasitic capacitance that M_{NBUF} presents to A_E. So with a one-to-one mirror ratio in M_{N1}–M_{N2}, which means $(W/L)_{N1} \equiv (W/L)_{N2}$,

$$\frac{i_O}{i_{PS}} = \frac{\left(\frac{W}{L}\right)_{PO}}{\left(\frac{W}{L}\right)_{PS}} = \frac{\left(\frac{18.3k}{0.35}\right)}{\left(\frac{W}{L}\right)_{PS}} \equiv \frac{50\,\text{mA}}{46\,\mu\text{A}}$$

or $(W/L)_{PS} \geq 48$, which means W_{PS} can be 17 μm and L_{PS} can be minimum at 0.35 μm.

5. Zero-load bandwidth $p_{OB(0)}$:

$$p_{OB(0)} \approx \frac{1}{2\pi C_{IP} R_{OB}} \approx \frac{g_{m.NBUF}}{2\pi C_{IP(0)}} = \frac{\sqrt{2 i_{NBUF(0)} K_N' \left(\frac{W}{L}\right)_{NBUF}}}{2\pi C_{IP(0)}} \geq 100\text{ kHz}$$

or $$i_{NBUF(0)} \geq \frac{(2\pi C_{IP(0)} p_{OB(0)})^2}{2K_{N(MIN)}' \left(\frac{W}{L}\right)_{NBUF}} = \frac{[2\pi(25\text{p})(100\text{k})]^2}{2(80\mu)\left(\frac{12}{0.35}\right)} \approx 0.05\ \mu\text{A}$$

so with some margin to overwhelm coupled noise, $i_{NBUF(0)}$ can be 0.5 μA.

6. $v_{B(MIN)}$: $$v_{B(MIN)} \approx V_{DS.N2(SAT)} \leq \sqrt{\frac{2 i_{NBUF(MAX)}}{K_{K(MIN)}' \left(\frac{W}{L}\right)_{N2}}} \leq 0.2\text{ V}$$

so $$\left(\frac{W}{L}\right)_{N2} \geq \frac{2 i_{NBUF(MAX)}}{V_{DS.N2(SAT)}{}^2 K_{N(MIN)}'} = \frac{2 i_{NBUF(50mA)}}{V_{DS.N2(SAT)}{}^2 K_{N(MIN)}'}$$

$$= \frac{2(46\mu)}{(0.2)^2(80\mu)} = 28.7$$

which means W_{N1} and W_{N2} can be 30 μm and L_{N1} and L_{N2} can be three times the minimum of 0.35 μm to reduce the effects of channel-length modulation λ_L.

7. Positive-feedback filter R_{FB}–C_{FB}: Set feedback-filter pole $f_{P.FB}$ at 100 kHz to ensure positive loop-gain A_{LG+} starts dropping at 100 kHz:

$$p_{FB+} \approx \frac{1}{2\pi C_F R_F} \leq 100\text{ kHz}$$

so if C_F is 5 pF, because higher capacitance requires more than 32 μm × 32 μm of silicon area when C_{OX}'' is 5 fF/μm²,

$$R_F \geq \frac{1}{2\pi C_F p_{FB+}} = \frac{1}{2\pi(5\text{p})(100\text{k})} \approx 320\text{ k}\Omega$$

8. To ensure that Q_{P3} does not enter deep saturation:

$$V_{EC.P3(MIN)} \approx v_O - (v_{GS.N1(MAX)} - I_{B1} R_{DC})$$

$$\approx v_O - v_{TN(MAX)} - \sqrt{\frac{2 i_{NBUF(MAX)}}{K_{N(MIN)}' \left(\frac{W}{L}\right)_{N1}}} + I_{B1} R_{DC} \geq 0.3\text{ V}$$

and since M_{N1} and M_{N2} comprise a one-to-one mirror and $i_{\text{NBUF(0)}}$ is 0.5 μA, I_{B1} should also be 0.5 μA and

$$R_{\text{DC}} \geq \frac{V_{\text{EC.P3(MIN)}} - v_O + v_{\text{TN(MAX)}} + \sqrt{\dfrac{2i_{\text{NBUF(MAX)}}}{K_{N(\text{MIN})}'\left(\dfrac{W}{L}\right)_{N1}}}}{I_{B1}}$$

$$\approx \frac{0.3 - 1 + 0.75 + \sqrt{\dfrac{2(46\mu)}{(80\mu)\left(\dfrac{30}{1.05}\right)}}}{0.5\mu} = 501\text{ k}\Omega$$

so R_{DC} can be 500 kΩ.

9. Bias current I_{B2}: Choosing i_{P4} to be about 0.5 μA,

$$I_{B2(\text{MIN})} \geq \frac{i_{P3(\text{MAX})}}{\beta_{P0(\text{MIN})}} + I_{P4} \approx \frac{46\mu}{50} + 0.5\mu = 1.4\,\mu\text{A}$$

so I_{B2} can be 2 μA to accommodate a ±30-percent variation in I_{B2} and still meet the 1.4 μA target.

Design Checks

10. Feedback stability: The loop gain $A_{\text{LG+}}$ from v_B back to v_B is zero when M_{PS} is off because M_{PS} opens the loop, so the circuit is stable when i_O is 0. When i_O is not zero, M_{PS} closes the loop and $A_{\text{LG+}}$ becomes

$$A_{\text{LG+}} \equiv \frac{v_b}{v_b} = \left(\frac{v_{g.\text{N1}}}{v_b}\right)\left(\frac{v_b}{v_{g.\text{N1}}}\right) \approx \left[g_{m.\text{PS}}\left(\frac{1}{g_{m.\text{N1}}}\right)\left(\frac{1}{1+R_F C_F s}\right)\right]$$
$$\left[g_{m.\text{N2}}\left(\frac{R_{\text{OB}}}{A_{\text{REG}}} \parallel \frac{1}{sC_{\text{IP}}}\right)\right] < g_{m.\text{PS}} R_{\text{OB}}$$

$$= \frac{g_{m.\text{PS}}}{g_{m.\text{NBUF}}} \approx \frac{\sqrt{2i_{\text{PS}}K_P'\left(\dfrac{W}{L}\right)_{\text{PS}}}}{\sqrt{2i_{\text{PS}}K_N'\left(\dfrac{W}{L}\right)_{\text{NBUF}}}} = \sqrt{\frac{K_P'\left(\dfrac{W}{L}\right)_{\text{PS}}}{K_N'\left(\dfrac{W}{L}\right)_{\text{NBUF}}}} = \sqrt{\frac{(40\mu)\left(\dfrac{17}{0.35}\right)}{(100\mu)\left(\dfrac{12}{0.35}\right)}} \approx 0.75 < 1$$

where R_{OB} is $1/g_{m.\text{NBUF}}$ and R_F–C_F and C_{IP} attenuate $A_{\text{LG+}}$, so $A_{\text{LG+}}$ is never closer to 1 than 0.75. Note that applying worst-case three-sigma (3σ) extremes for K_P' and K_N' is probabilistically unreasonable.

11. Small-signal gain across the buffer A_B: Decomposing M_{NBUF}'s g_m into its gate–source components like Fig. 9.22 shows reveals that the regenerative loop mixes M_{NBUF}'s v_{oe}-derived current i_i with M_{N2}'s feedback current i_{fb}, M_{PS}'s gate shunt-samples v_b, and the resistance into M_{NBUF}'s v_b-derived transconductance is $r_{\text{ds.NBUF}} \parallel (1/g_{m.\text{NBUF}})$, or roughly $1/g_{m.\text{NBUF}}$, so the open-loop

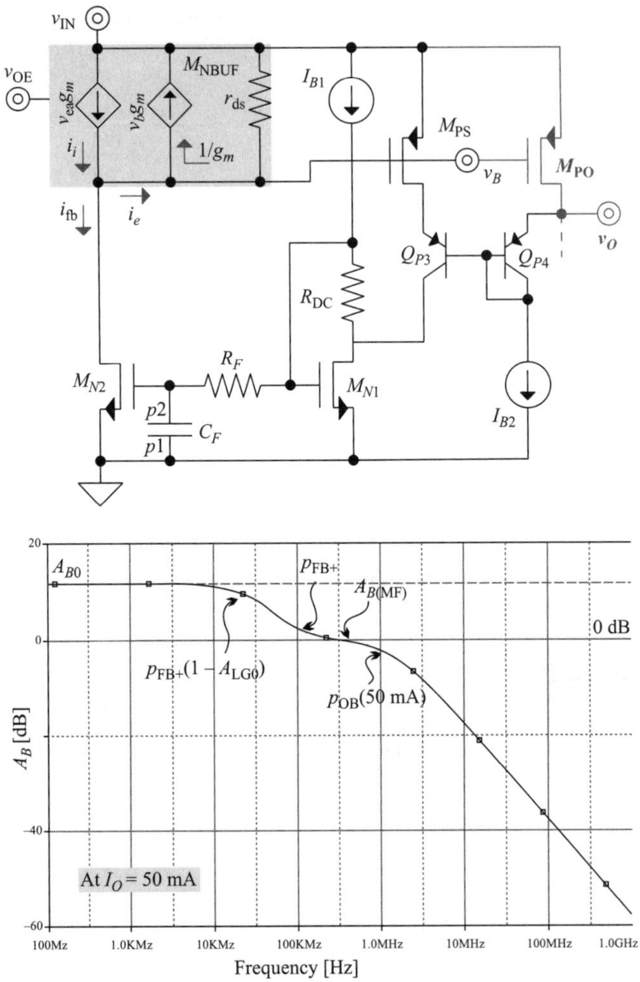

FIGURE 9.22 Decomposed follower and resulting frequency response across the regenerative buffer.

transresistance gain A_{ROL}, or v_b/i_e, is $1/g_{m.NBUF}$. As a result, A_B is v_{ea}'s translation into i_i times the closed-loop transresistance gain A_{RCL}, so A_B's low-frequency component A_{B0} is therefore

$$A_{B0} \equiv \frac{v_b}{v_{oe}} = \left(\frac{i_i}{v_{oe}}\right)\left(\frac{v_b}{i_i}\right) = \left(\frac{i_i}{v_{oe}}\right)A_{RCL0} = g_{m.NBUF}\left(\frac{A_{ROL0}}{1 - A_{LG0}}\right)$$

$$\approx g_{m.NBUF}\left[\frac{1}{g_{m.NBUF}(1-0.75)}\right] \approx 4$$

where the loop gain is noninverting and the denominator therefore includes a negative sign, and the low-frequency loop gain A_{LG0} is the gain established by feeding M_{PS}'s v_b-derived current into M_{N1}'s diode connection and receiving M_{N2}'s transconductance current back to M_{NBUF}'s R_{OB}, or $1/g_{m.NBUF}$:

$$A_{LG0} \approx g_{m.PS}\left(\frac{1}{g_{m.N1}}\right)g_{m.N2}R_{OB} = g_{m.PS}\left(\frac{1}{g_{m.N1}}\right)g_{m.N2}\left(\frac{1}{g_{m.NBUF}}\right)$$

$$= \frac{g_{m.PS}}{g_{m.NBUF}} \approx 0.75$$

which from before is 0.75. As a result, as the simulation results from Fig. 9.22 corroborate, the buffer's low-frequency gain is 4, or 12 dB, and feedback shifts p_{FB+} to $(1 - A_{LG0+})p_{FB+}$, or $0.25p_{FB+}$, and introduces a zero at p_{FB+} when the effects of feedback diminish:

$$A_{RCL} = \frac{A_{ROL}}{1-\left(\dfrac{A_{LG0}}{1+\dfrac{s}{2\pi p_{FB+}}}\right)} = \frac{\left(\dfrac{A_{ROL}}{1-A_{LG0}}\right)\left(1+\dfrac{s}{2\pi p_{FB+}}\right)}{\left[1+\dfrac{s}{2\pi(1-A_{LG0})p_{FB+}}\right]}$$

$$= \frac{A_{ROL0}\left(1+\dfrac{s}{2\pi p_{FB+}}\right)}{\left[1+\dfrac{s}{2\pi(1-A_{LG0})p_{FB+}}\right]}$$

Since A_{LG0} is less than 1 and R_F–C_F filters A_{LG} near $p_{OB(0)}$ and below $p_{OB(50mA)}$, A_{LG}'s effects in A_B vanish past p_{FB+} and $p_{OB(0)}$, so A_B reduces to M_{NBUF}'s follower gain of 1, or 0 dB:

$$A_{B(MF)} \equiv \frac{v_b}{v_{oe}} = \left(\frac{i_i}{v_{oe}}\right)\left(\frac{v_b}{i_i}\right)\Bigg|_{p_{FB+}<f<p_{OB(50mA)}} \approx g_{m.NBUF}\left(\frac{A_{ROL}}{1-A_{LG(MF)}}\right)$$

$$\approx g_{m.NBUF}A_{ROL} \approx \frac{g_{m.NBUF}}{g_{m.NBUF}} = 1$$

And since A_{LG} is zero when i_O is zero, when M_{PS} is off, A_B also reduces to M_{NBUF}'s follower gain of 1, or 0 dB, when the regulator is lightly loaded.

12. $v_{B(MAX)}$ when assuming $v_{OE(MAX)}$ is $v_{IN} - 0.3$ V:

$$v_{B(MAX)} > v_{OE(MAX)} - v_{GS.NBUF(0)} = v_{OE(MAX)} - \left[v_{TN.NAT(MAX)} + \sqrt{\frac{2i_{NBUF(0)}}{K_{N(MIN)}'\left(\frac{W}{L}\right)_{NBUF}}}\right]$$

$$\approx (v_{IN} - 0.3) - \left(0.15 + \sqrt{\frac{2(0.5\mu)}{(80\mu)\left(\dfrac{12}{0.35}\right)}}\right) \approx v_{IN} - 0.47\text{V} < v_{IN} - |v_{TP0(MIN)}|$$

which is below the specification by a small margin. Note that this relationship combines 3σ limits for $v_{TN.NAT}$ and v_{TP} linearly, which is not realistic, so exceeding the specification by a small margin when applying improbable worst-case conditions implies that the circuit probably meets the specification with an acceptable degree of confidence.

13. Zero-load quiescent current $i_{Q(0)}$:

$$i_{Q(0)} = i_{N2(0)} + i_{N1(0)} + i_{P4} \approx I_{B1} + I_{B1} + I_{B2} = 0.5\mu + 0.5\mu + 2\mu = 3\ \mu A$$

so a 30-percent variation in currents raises $i_{Q(0)}$ to 3.9 μA, which still meets the specified target. Full-load quiescent current $i_{Q(50mA)}$ is

$$i_{Q(50mA)} = i_{N2(50mA)} + i_{N1(50mA)} + i_{P4} \approx 46\mu + 46\mu + 2\mu = 94\ \mu A$$

which with a 30-percent variation can be 120 μA. Notice that 120 μA is substantially lower than $i_{O(MAX)}$'s 50 mA, so the impact of $i_{Q(50mA)}$'s 120 μA on full-load efficiency is negligible.

14. $v_{IN(MIN)}$ assuming the minimum voltage across I_{B1} is 150 mV:

$$v_{IN(MIN)} = v_{GS.N1(MAX)} + V_{B1(MIN)} = \left[v_{TN(MAX)} + \sqrt{\frac{2i_{NBUF(MAX)}}{K_{N(MIN)}'\left(\frac{W}{L}\right)_{N1}}} \right] + V_{B1(MIN)}$$

$$\approx 0.75 + \sqrt{\frac{2(46\mu)}{(80\mu)\left(\frac{30}{1.05}\right)}} + 0.15 \approx 1.1\ \text{V}$$

which meets the specified target.

9.3.3 Layout

Perhaps the most important layout consideration for the buffer A_B is its placement relative to the error amplifier A_E and the power transistor S_O because parasitic delays caused by physical separation lower bandwidth performance. The silicon distance between A_E's output and A_B's input should therefore be as short as possible, as should the separation between A_B's output and S_O's input. Similarly, to raise the bandwidth at every point in the circuit, the layout should be generally small and modular, so that its constituent paths are short, and as a result, low impedance.

Although A_B does not drive the level of currents that S_O sources, A_B still supplies considerably higher currents than A_E, especially during wide and heavy load dumps, which is why supply rails v_{IN} and ground should be relatively wide at 5–8 μm, or wider, and fairly ubiquitous in the IC. They should also spread evenly through the circuit, as in a *bus*, to avoid introducing unmatched series voltages that could otherwise emitter- or source-degenerate transistors unevenly.

Although matching devices in this circuit is not critical, reliable and uniform performance dictate that biasing transistors, current mirrors, and differential pairs should match.

9.4 Error Amplifiers

Power efficiency and accuracy are vital parameters in the design of the error amplifier A_E. To begin, as with the power transistor S_O and the buffer A_B, A_E must survive low input-supply v_{IN} conditions when the voltage across S_O is low and consume low quiescent current i_Q for high power efficiency and extended battery life. A_E must also respond quickly to dampen the adverse effects of fast load dumps and supply fluctuations on the regulated output v_O. Also, since steady-state accuracy, which refers to load and line regulation and initial accuracy, depends on the reference v_{REF} and A_E, A_E's gain must be high, although not excessively so to avoid compromising feedback stability. Systemic and random input-referred offset V_{OS} must similarly be low. Architecturally, A_E's output stage must also consider the current and voltage driving demands of A_B. So to summarize, $v_{IN(MIN)}$ should be low, i_Q should be low, bandwidth should be high, power-supply rejection should be high, gain should be high, offset should be low, and A_E's output v_{OE} should be compatible with the driving demands of A_B.

9.4.1 Headroom

Headroom limit $v_{IN(MIN)}$ defines the voltage beyond which further drops in v_{IN} push one or more high-gain transistors into their low-gain state, into triode. In this respect, all circuit paths between v_{IN} and ground require a minimum voltage below which the path no longer sustains the targeted current desired. Or stated differently, every node must fall below v_{IN} and rise above ground by minimum positive and negative headroom limits V_{ROOM+} and V_{ROOM-}. $v_{IN(MIN)}$ is therefore the worst-case sum of the headroom limits in a circuit:

$$v_{IN(MIN)} = \text{Max}\{V_{ROOM+} + V_{ROOM-}\} \tag{9.9}$$

Differential Pair

The reference v_{REF} can be a limiting factor in $v_{IN(MIN)}$. v_{REF} should, for example, be low enough below v_{IN} to bias a P-type differential pair M_{D1}–M_{D2} with a single-transistor P-type tail-current source M_T, as Fig. 9.23 shows. As such, v_{REF}'s positive headroom limit V_{ROOM+} to v_{IN} is M_{D1}–M_{D2}'s v_{SG} and M_T's $V_{SD(SAT)}$. This is a limiting constraint to v_{IN} because v_{REF} is usually the output of a 1.2-V bandgap voltage reference:

$$v_{IN(MIN)} = V_{ROOM+} + v_{REF} = V_{SD(SAT)} + v_{SG} + v_{REF} = 2V_{SD(SAT)} + |v_{TP}| + v_{REF} \tag{9.10}$$

where v_{SG} decomposes into threshold voltage $|v_{TP}|$ and another saturation voltage $V_{SD(SAT)}$. Considering that $|v_{TP}|$ can be 0.6 V and $V_{SD(SAT)}$

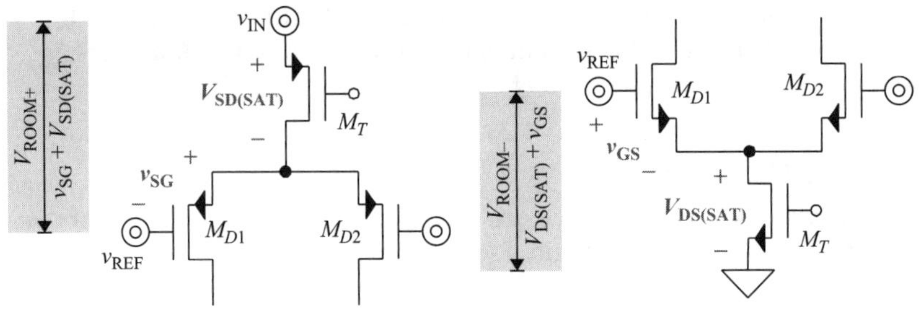

FIGURE 9.23 Inherent headroom limits of P- and N-type differential pairs.

can be 0.15 V, $v_{IN(MIN)}$ cannot be lower than 1.9 V, and perhaps not lower than 2.2 V under worst-case temperature and fabrication corners. Although requiring a lower v_{REF} is possible, the fraction that noise in v_{REF} represents is higher when v_{REF} is lower. In the end, many engineers avoid this undesirable tradeoff between $v_{IN(MIN)}$ and *signal-to-noise ratio* (SNR) by steering away from P-type differential input pairs.

The situation changes for N-type differential pairs. In this latter case, as Fig. 9.23 also illustrates, v_{REF} should be high enough above ground to bias an N-type differential pair M_{D1}–M_{D2} and its corresponding single-transistor N-type tail-current sink M_T. A_E's negative headroom limit V_{ROOM-} to ground at v_{REF} is therefore M_{D1}–M_{D2}'s v_{GS} and M_T's $V_{DS(SAT)}$:

$$v_{REF} > V_{ROOM-} = v_{GS} + V_{DS(SAT)} = v_{TN} + 2V_{DS(SAT)} \tag{9.11}$$

where v_{GS} decomposes into threshold voltage v_{TN} and another saturation voltage $V_{DS(SAT)}$. This is a limiting constraint to the input pair, and more generally, to A_E's *input common-mode range* (ICMR), but not $v_{IN(MIN)}$. Thankfully, this is not always a concern because v_{REF} is normally 1.2 V, v_{TN} can be 0.6 V, and $V_{DS(SAT)}$ can be 0.15 V, so between v_{REF} and V_{ROOM-}, the circuit includes 300 mV of margin to accommodate temperature and process variations. This margin and the fact that v_{REF} and V_{ROOM-} do not constrain $v_{IN(MIN)}$ are two compelling reasons for using N-type differential pairs in A_E.

Load Mirror

Another important consideration is the load of the differential pair. In the case that a basic P-type current mirror loads an N-type differential pair, like M_{M3}–M_{M4} does to Q_{D1}–Q_{D2} in Fig. 9.24, v_{IN} cannot fall below a limit that, among other components, includes v_{REF} and the mirror's v_{SG} diode connection:

$$v_{IN(MIN)} = V_{ROOM+} + v_{REF} = v_{SG} + V_{CE(MIN)} - v_{BE} + v_{REF} \tag{9.12}$$

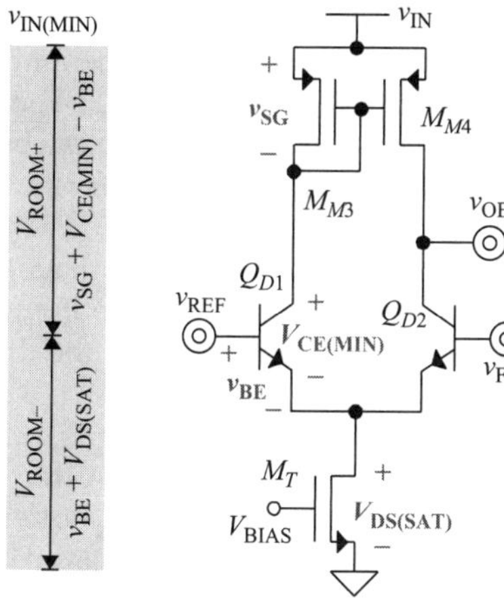

FIGURE 9.24 N-type differential pair loaded with a basic P-type mirror.

where the differential pair imposes $V_{CE(MIN)}$ and v_{BE} on $v_{IN(MIN)}$. Considering that v_{SG}'s $|v_{TP}|$ can be 0.6 V and $V_{SD(SAT)}$ can be 0.15 V, $V_{CE(MIN)}$ can be 0.2 V, v_{BE} can be 0.65 V, and v_{REF} is normally 1.2 V, $v_{IN(MIN)}$ cannot be lower than 1.55 V, and perhaps not lower than 1.8 V under worst-case temperature and fabrication corners. This is why basic P-type load mirrors are not ideal in modern battery-powered applications.

One way to reduce this limit is by reducing the overhead that the diode connection in the mirror imposes. Level-shifting the mirror's v_{SG} up with an NPN BJT's v_{BE}, for example, like Q_B in Fig. 9.25 exemplifies,

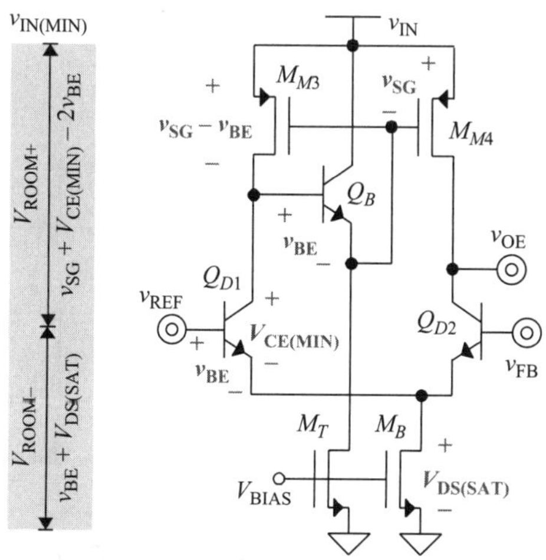

FIGURE 9.25 N-type differential pair loaded with a level-shifted P-type mirror.

reduces the voltage across the mirror to $v_{SG} - v_{BE}$. As a result, V_{ROOM+}, and in consequence, $v_{IN(MIN)}$ drop by the same amount. Considering that v_{SG}'s $|v_{TP}|$ can be 0.6 V, $V_{SD(SAT)}$ can be 0.15 V, or better yet, 0.3 V, and v_{BE} can be 0.65 V, M_{M3}'s source–drain drop can be 0.1–0.25 V. From an operational standpoint, Q_B is a voltage buffer biased with M_B's current that diode-connects M_{M3}. One additional design constraint in this setup is ensuring that M_{M3}'s source–drain voltage is positive across temperature and fabrication corners, because if v_{SG} is low and v_{BE} is high, as can be the case at low temperatures, the drop can fall below M_{M3}'s $V_{SD(SAT)}$, which would reduce A_E's gain.

Another way to reduce the mirror's drop is to fold the output of the N-type mirror into a P-type current buffer. In the circuit of Fig. 9.26, for example, cascode transistor M_{C5} buffers and folds M_{D1}'s output current to diode-connect M_{M3}. When biased properly with V_{BP}, M_{M3}'s drop can be $V_{SD(SAT)}$, which is a threshold voltage lower than that of the basic diode connection in Fig. 9.24. This reduction propagates to V_{ROOM+}, and as a result, to $v_{IN(MIN)}$:

$$v_{IN(MIN)} = V_{ROOM+} + v_{REF} = V_{SD(SAT)} + V_{DS(SAT)} - v_{GS} + v_{REF}$$
$$= V_{SD(SAT)} - v_{TN} + v_{REF} \qquad (9.13)$$

where $V_{DS(SAT)}$ in M_{D1}'s v_{GS} cancels M_{D1}'s $V_{DS(SAT)}$. Considering that v_{GS}'s v_{TN} can be 0.6 V and $V_{SD(SAT)}$ can be 0.15 V, and v_{REF} is normally 1.2 V, $v_{IN(MIN)}$ cannot be lower than 0.75 V, and perhaps not lower than 1 V under worst-case temperature and fabrication corners.

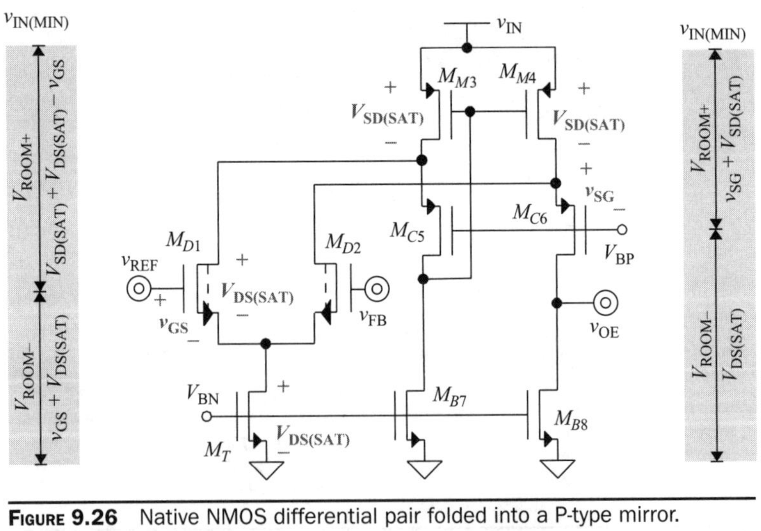

FIGURE 9.26 Native NMOS differential pair folded into a P-type mirror.

The headroom limit that v_{REF} imposes is so low, in fact, that other nodes in the circuit can be just as, or even more limiting. Here, the headroom requirement of the bias voltage V_{BP} can be more constraining because V_{BP} must fall below v_{IN} by M_{C5}'s v_{SG} and M_{M3}'s $V_{SD(SAT)}$ in V_{ROOM+}, and must still have enough room V_{ROOM-} to ground to accommodate a $V_{DS(SAT)}$:

$$v_{IN(MIN)} = V_{ROOM+} + V_{ROOM-} = (v_{SG} + V_{SD(SAT)}) + V_{SD(SAT)} = |v_{TP}| + 3V_{SD(SAT)} \tag{9.14}$$

Again, considering v_{SG}'s $|v_{TP}|$ can be 0.6 V and $V_{SD(SAT)}$ and $V_{DS(SAT)}$ can be 0.15 V, $v_{IN(MIN)}$ cannot be lower than 1.05 V, and perhaps not lower than 1.3 V under worst-case temperature and fabrication corners. Still, V_{BP}'s headroom limit is the lowest an analog circuit can tolerate when including cascode transistors.

9.4.2 Power-Supply Rejection

N-Type Power Transistors

Power-supply rejection (PSR) superimposes architectural requirements on the load mirror of the error amplifier A_E that primarily depend on the power transistor S_O. In the case of N-type followers, S_O impresses whatever noise is at its input on the output v_O. Since the buffer A_B reproduces what A_E generates in the first place, A_E should therefore output a noise-free signal. Thankfully, N-type mirror loads, as discussed in Chap. 8 and as now shown in Fig. 9.27, do just this, sink and cancel the supply ripple that impedances to the supply inject. Here, M_{M3} receives a supply ripple that matches M_{D2}'s counterpart, which M_{C6} buffers to the output v_{OE} in the folded-cascode circuits. M_{M4} then mirrors M_{M3}'s noise to cancel M_{D2}'s contribution. This way, with an N-type mirror load, A_E produces a noise-free output.

P-Type Power Transistors

As veritable complements, the input of P-type transistors must receive the same supply ripple v_{in} already present at their emitters and sources to keep the noise from propagating to the output of the regulator. The idea is to ensure that v_{in} is common to both emitter–base and source–gate terminals. That way, v_{in} disappears from the difference that S_O's transconductance processes to generate a noise-free output current i_O. And since A_B reproduces what A_E generates, A_E should reproduce v_{in} in v_{OE}. Luckily, P-type mirror loads, as discussed in Chap. 8 and as now shown in Fig. 9.28, do just this, source enough supply ripple to reproduce v_{in} in v_{OE}. Here, M_{M3}, and as result, M_{M4}'s g_{m4} source a supply ripple that reinforces and complements the ripple that M_{M4}'s r_{sd4} injects into v_{OE}, which M_{C6} buffers to v_{OE} in the folded-cascode circuit that incorporates an N-type differential pair. M_{M4}'s g_{m4}

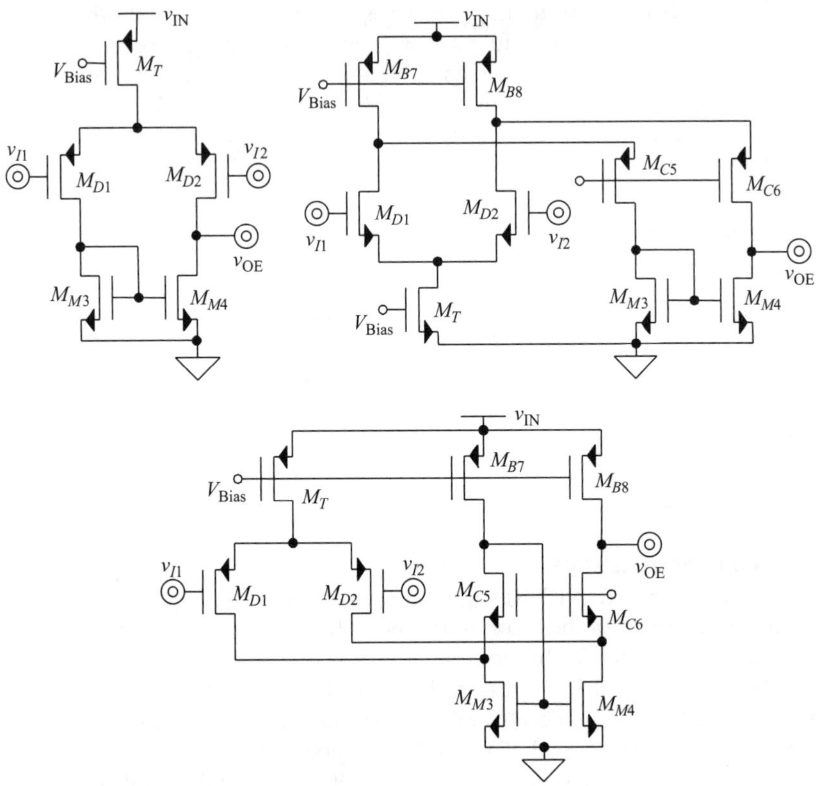

FIGURE 9.27 Amplifier configurations with N-type mirror loads.

then mirrors M_{M3}'s noise to reinforce r_{sd4}'s contribution. This way, with a P-type mirror load, A_E reproduces supply noise in v_{OE}.

Generalities

The canceling and summing effects of these ground- and supply-referenced N- and P-type mirrors hinge on the symmetry of the circuit and the mirroring features of the load, so keeping the design symmetrical and the mirror as ideal as possible are important. The overriding assumption in the previous discussion, in fact, is that the $1/g_m$ input resistance of the mirrors is substantially lower than the resistances driving them. This resistance actually distorts the mirror's input current with respect to the supply-injected current already present at the output, but since $1/g_m$ is low, the effect is minute. As such, power-supply rejection generally improves when the impedances to the rails v_{IN} and ground are high, so high Early-voltage BJTs and long-channel MOSFETs favor noise rejection.

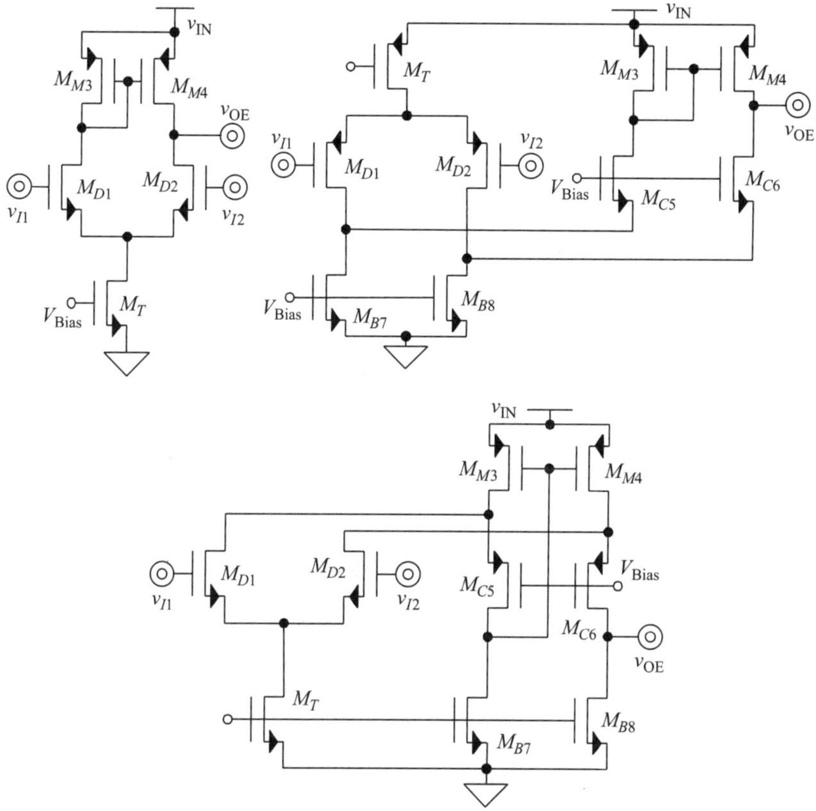

FIGURE 9.28 Amplifier configurations with P-type mirror loads.

9.4.3 Input-Referred Offset

Any input-referred offset V_{OS} in A_E affects the mirrored impression that the system's feedback loop produces in v_{FB} of the reference v_{REF}. This is troublesome because the feedback network that R_{FB1} and R_{FB2} realize in Fig. 9.1 propagates this offset to the output of the regulator. Unfortunately, this translation is usually a gain because R_{FB1}–R_{FB2}'s voltage divider amplifies v_{FB}'s impression of v_{REF} to v_O:

$$v_O = v_{FB}\left(\frac{R_{FB1} + R_{FB2}}{R_{FB1}}\right) \approx (v_{REF} + V_{OS})\left(\frac{R_{FB1} + R_{FB2}}{R_{FB1}}\right) \tag{9.15}$$

The end result is that the offset degrades the accuracy of the system.

When perfectly balanced, zero volts across the inputs of A_E produces currents that match and cancel to balance the circuit. Unfortunately, transistors do not match perfectly, so identical gate–source

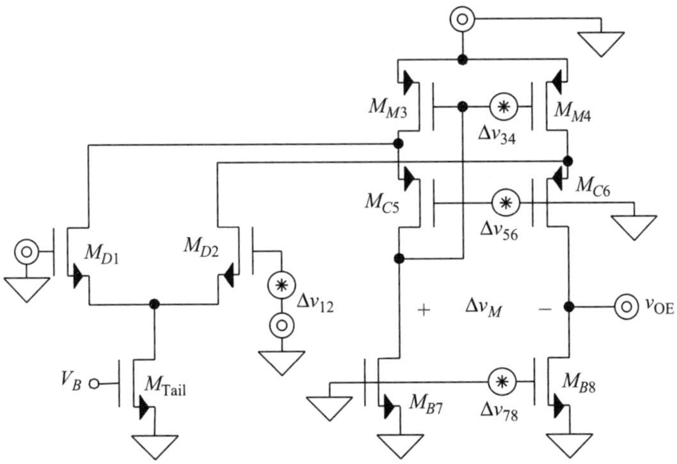

FIGURE 9.29 Random and systemic mismatches in the error amplifier.

and base–emitter voltages in the input pair M_{D1}–M_{D2}, mirror load M_{M3}–M_{M4}, cascode pair M_{C5}–M_{C6}, and bias pair M_{B7}–M_{B8} in Fig. 9.29 produce mismatched currents. Plus, a voltage difference across the mirror Δv_M also establishes an imbalance in currents. Normally, Δv_M's systemic component overwhelms its random counterpart, and transistor mismatches are all random. As a result, a finite, nonzero offset voltage V_{OS} that includes systemic and random components $V_{OS(S)}$ and V_{OS}^* across the inputs balances the circuit like zero volts would under perfectly matched conditions:

$$V_{OS} = V_{OS(S)} \pm V_{OS}^* \tag{9.16}$$

Random

Three-sigma base- and gate-referred mismatches between BJTs and FETs can be 1–3 mV and 5–15 mV, respectively, which are small enough to produce linear effects in the circuit like small signals would. A_E's small-signal equivalent circuit can therefore predict the effects of mismatches between transistors that should otherwise match. Since these offsets are random, summing them linearly is unrealistically pessimistic. Instead, the square-rooted sum of the squares of the components is, in statistical terms, a more practical prediction.

Ultimately, transistor transconductances translate and transpose offsets across the circuit. The bias pair's transconductance g_{m78} in Fig. 9.29, for example, multiplies its gate-referred offset Δv_{78} to produce an offset current that M_{C6} buffers and the input pair's transconductance g_{m12} reflects back to its input as $\Delta v_{78} g_{m78}/g_{m12}$. Similarly, the mirror's transconductance g_{m34} multiplies its gate-referred offset Δv_{34}

to produce an offset current that the input pair's g_{m12} reflects back to its input as $\Delta v_{34}g_{m34}/g_{m12}$. The situation is slightly different for the cascode pair's offset Δv_{56} because the mirror transistors and the differential pair degenerate how much voltage the cascode transconductances process. Still, their degenerated transconductance $G_{M56(DEG)}$ multiplies Δv_{56} and the input pair's g_{m12} reflects it back as $\Delta v_{56}G_{M56(DEG)}/g_{m12}$, where $G_{M56(DEG)}$ is much lower than g_{m56}. And since the offset in the input pair Δv_{12} is already across A_E's inputs, Δv_{12} appears in V_{OS}'s random component V_{OS}^* as is, without a gain translation:

$$V_{OS}^* = \sqrt{\Delta v_{12}^{\ 2} + \left(\frac{\Delta v_{34}g_{m34}}{g_{m12}}\right)^2 + \left(\frac{\Delta v_{56}G_{M56(DEG)}}{g_{m12}}\right)^2 + \left(\frac{\Delta v_{78}g_{m78}}{g_{m12}}\right)^2}$$

$$\approx \sqrt{\Delta v_{12}^{\ 2} + \left(\frac{\Delta v_{34}g_{m34}}{g_{m12}}\right)^2 + \left(\frac{\Delta v_{78}g_{m78}}{g_{m12}}\right)^2} \qquad (9.17)$$

Luckily, $G_{M56(DEG)}$ is usually so much lower than g_{m12}, g_{m34}, and g_{m78} that other offset components normally overwhelm Δv_{56}'s contribution in V_{OS}^*. In other words, nondegenerated transistors contribute the most offset. In this light, reducing the impact of nondegenerated transistors on the input lowers V_{OS}^*, which amounts to raising the pair-to-mirror and pair-to-bias transconductance ratios g_{m12}/g_{m34} and g_{m12}/g_{m78}. Of course, matching the input, mirror, and bias transistor pairs is more fundamental and is therefore an engineer's first line of attack.

Alternate means of reducing this offset are offset-calibration schemes like *dynamic-element matching, chopper techniques, trimming*, and so on. Unfortunately, these methods add complexity, power losses, switching noise, test time, and/or silicon area. One low-cost possibility is to shift the trimming point from v_{REF} to the regulator output v_O or v_{FB}. The problem with this approach is that the benefits of trimming shift from v_{REF} to v_O, so v_{REF} is no longer as accurate. This means that other components in the IC that rely on v_{REF} for accuracy similarly suffer as a result. This is acceptable in stand-alone applications when only the regulator relies on v_{REF}, but not on system-on-chip solutions.

Systemic

In the case of a mismatched voltage across the mirror Δv_M, A_E's voltage gain refers Δv_M back to A_E's input in the form of a systemic offset $V_{OS(S)}$. Accordingly, the gain that the differential pair's g_{m12} establishes across A_E's output resistance R_{OE} determines how much impact Δv_M has on $V_{OS(S)}$:

$$V_{OS(S)} = \frac{\Delta v_M}{A_E} = \frac{\Delta v_M}{g_{m12}R_{OE}} \qquad (9.18)$$

When matched within 50 mV, however, A_E is typically high enough to suppress Δv_M well below V_{OS}^*, to the point that $V_{OS(S)}$ becomes negligible in V_{OS}.

Load Regulation Even when matched, v_{OE} changes with output current i_O, so Δv_M includes a component that changes with i_O. To be more specific, the power transistor's transconductance G_P translates variations in i_O into voltage fluctuations that the buffer A_B must drive with v_B. And since the gain across the buffer is usually about 1, Δv_M incorporates the buffer variation Δv_B that results from fluctuations in i_O:

$$\Delta V_{OS(S)} = \frac{\Delta v_M(i_O)}{A_E} \approx \frac{\Delta v_B}{A_E} = \left|\frac{i_{O(MAX)}}{G_{P(MAX)}} - \frac{i_{O(MIN)}}{G_{P(MIN)}}\right|\left(\frac{1}{A_E}\right) \tag{9.19}$$

Interestingly, this variation in $V_{OS(S)}$ is how load regulation manifests in linear regulators, because R_{FB1} and R_{FB2} amplify $\Delta V_{OS(S)}$ to v_O by R_{FB1} and R_{FB2}'s voltage-divider ratio.

Typically, this load dependence is not a problem for conventional amplifiers because the gain across the circuit is high. Unfortunately, load-current and capacitor-ESR variations are so vast in linear regulators that feedback stability requirements limit the gain across the system to 50–60 dB. The fact that i_O changes so much and A_E's portion of the 50–60 dB can be 30–40 dB means that $\Delta V_{OS(S)}$ can be significant. In Example 9.1, for instance, since the PFET's G_P is the square-root translation of i_O, i_O climbs with a rising load faster than G_P does, so $i_{O(MAX)}/G_{P(MAX)}$ overwhelms $i_{O(MIN)}/G_{P(MIN)}$ in $\Delta V_{OS(S)}$. As a result, because $i_{O(MAX)}$ is 50 mA and $G_{P(MAX)}$ is $(2i_{O(MAX)}K_P'W_O/L_O)^{0.5}$, or 320 mS, $\Delta V_{OS(S)}$ is 5 mV when A_E is 30 dB or 32 V/V.

Load Tracking A way to mitigate the swinging effects of v_B on v_{OE} in response to large-signal variations in i_O is to shift v_B by an oppositely phased load-dependent voltage drop. The regenerative load-tracking buffer of Figs. 9.20 and 9.21 incorporates this feature when driving a P-type power transistor M_{PO}. To see how this happens, consider that the system responds to a rise in i_O by reducing v_{OE} and v_B to the extent that M_{PO}'s gate drive is high enough to sustain a higher i_O. Since M_{PS} mirrors M_{PO}'s i_O and M_{N1}–M_{N2} mirrors M_{PS}'s impression of i_O to the follower buffer M_{NBUF}, M_{NBUF}'s current i_{NBUF} rises with i_O, and in consequence, so does M_{NBUF}'s gate–source voltage $v_{GS.NBUF}$. The net result of this action is that a rise in $v_{GS.NBUF}$ opposes a fall in v_B, and vice versa, to produce smaller variations in v_{OE}, which is the sum of $v_{GS.NBUF}$ and v_B.

What is more, if $v_{GS.NBUF}$'s sensitivity to i_O is high enough to match that of v_B, load-regulation effects in $V_{OS(S)}$ and v_O disappear. To see how this can happen, first consider that variations in i_{NBUF} produce variations in $v_{GS.NBUF}$. Similarly, M_{PO}'s G_P translates variations in i_O into source–gate voltage variations, and as a result, into v_B variations. So if the current translation that M_{PS}–M_{PO} and M_{N2}–M_{N1} establish from i_O to i_{NBUF} in A_I is high enough to establish a current that, with M_{NBUF}'s lower transconductance g_m, produces the same voltage variation that i_O produces in v_B, A_I/g_m matches $1/G_P$ and variations in A_E's v_{OE} vanish:

$$\Delta v_{OE} = \Delta v_{GS.NBUF} - \Delta v_B \approx \left(\frac{i_{NBUF(MAX)}}{g_{m(MAX)}} - \frac{i_{NBUF(MIN)}}{g_{m(MIN)}}\right) - \left(\frac{i_{O(MAX)}}{G_{P(MAX)}} - \frac{i_{O(MIN)}}{G_{P(MIN)}}\right)$$

$$\approx \frac{i_{NBUF(MAX)}}{g_{m(MAX)}} - \frac{i_{O(MAX)}}{G_{P(MAX)}} = \frac{i_{O(MAX)}A_I}{g_{m(MAX)}} - \frac{i_{O(MAX)}}{G_{P(MAX)}} \approx \Delta i_{O(MAX)}\left(\frac{A_I}{g_{m(MAX)}} - \frac{1}{G_{P(MAX)}}\right) \tag{9.20}$$

where like before, g_m and G_P are the square-root translations of i_{NBUF} and i_O, and $i_{O(MIN)}$ is so much lower than $i_{O(MAX)}$ that $i_{O(MAX)}$ is roughly equivalent to i_O's maximum variation $\Delta i_{O(MAX)}$ and $i_{NBUF(MAX)}/g_{m(MAX)}$ and $i_{O(MAX)}/G_{P(MAX)}$ overwhelm $i_{NBUF(MIN)}/g_{m(MIN)}$ and $i_{O(MIN)}/G_{P(MIN)}$. Notice that matching $A_I/g_{m(MAX)}$ to $1/G_{P(MAX)}$ means the gain across the noninverting feedback loop of the buffer is relatively high, which compromises the feedback stability of the system. Even when the gains do not match, though, the regenerative loop still diminishes load-regulation effects in the regulator.

9.4.4 Layout

Matching Hierarchy

Perhaps the most important layout notes for the error amplifier A_E relate to random mismatches. Accordingly, categorizing and grouping transistors with respect to their matching requirements is the first step in the layout process. In this respect, the highest matching degree applies to *critical* devices, which should conform, if and when possible, to the following seemingly extreme guidelines:

a. Place transistors close to one another to minimize the total variation spread that results from two-dimensional gradients across the die;

b. Ensure that the transistor array is as modular, that is, as square as possible to reduce silicon area and the parameter spreads that result from nonlinear gradients across the die;

c. Orient transistors in the same direction to ensure that fabrication effects from deposition and other steps are consistent across matching transistors;

d. Employ a common-centroid layout configuration to balance the effects of linear gradients on matching transistors;

e. Cross-couple equally sized transistors to reduce the parameter spreads that result from nonlinear gradients across the die;

f. Place small dummy transistors around the peripheral transistors in the array to mitigate proximity-mismatch effects;

g. Avoid placing metal routes immediately above matching transistors to prevent packaged-induced stress fields from establishing mismatched piezo effects locally, and if routing across these devices is unavoidable, match the routing tracks across these devices so that all matched transistors experience similar effects;

h. Use a uniform sheet of one or two top-level metals above the entire array to spread and disperse the effects of packaged-induced stress fields.

Deciding which, when, and how to apply the aforementioned practices is subjective and dependent on the target specifications required of the regulator and the technology used to fabricate it. Notwithstanding, the second-highest matching degree usually backs off from the extreme measures just mentioned, which often demand considerably more silicon area. Adopting (a) through (c), for example, requires less area and still retains *good* matching performance. The last matching degree applies to *nominal* matching requirements, in which case (a) alone may be sufficient.

With respect to A_E, the input pair M_{D1}–M_{D2}, the load mirror M_{M3}–M_{M4}, and the bias transistors M_{B7}–M_{B8} from Figs. 9.23–9.29 must match best of all, so critical matching layout techniques apply. Although the matching performance of the tail current source M_T with respect to the corresponding bias transistors M_{B7}–M_{B8} in Figs. 9.26–9.29 does not affect random offset performance, it has an impact on quiescent current, unity-gain frequency, slew rate, and other performance parameters. These devices should therefore match well, but not to the same extent that critically matched transistors do, so good matching practices normally suffice. Although the matching performance of cascode devices M_{C5}–M_{C6} exerts little influence on random and systemic offsets and other performance parameters, completely neglecting them can raise their impact to noticeable levels, so matching them to first order is prudent.

Notes

Other important layout notes relate to bandwidth and noise, and to a lesser extent, power and thermal effects. To begin, introducing parasitic capacitance to A_E's high output resistance slows the response of the regulator. To avoid this, all transistors connected to v_{OE} should be as close as possible, which means A_E and A_B should be close or integrated into one block. For similar reasons, and to avoid noise injection from the substrate and other lines in the system, A_E's inputs should be as close as possible to their respective bond pads. In fact, the sense path from the output of the regulator to A_E's v_{FB} should be as short as possible and surrounded by low-impedance tracks to shunt and steer incoming noise away from A_E.

Series voltage drops in the supplies introduce systemic offset voltages in supposedly matched transistors, so power-supply buses of maybe 5–8 μm in width should surround A_E. Displacing differential pairs from power output devices mitigates the impact of thermal gradients that "hot spots" in the die induce. Power transistors in linear regulators, however, normally account for about 70 percent of the layout, so separating the differential pair from the power device is not so practical, especially when the thermal impedance across modern plastic packages is so low that temperature gradients almost disappear.

Example 9.3 Refer to the specifications and process parameters outlined in Table 9.4 to design a 6-μA error amplifier for the regenerative load-tracking buffer and PMOS power transistor designed in Examples 9.1 and 9.2. The complete internally compensated low-dropout regulator should generate a 1-V output with a unity-gain frequency of 1 MHz from a 0.9-V reference and a 0.9–1.6-V NiMH battery, the practical operating range of which is 1.1–1.6 V. The output of the amplifier should swing from 0.2 V to within 0.3 V of v_{IN} and incur no more than 25 mV of random uncorrelated input-referred offset and 10 mV of load-regulation effects in v_O across i_O's 10-μA–50-mA range.

Architectural Design

1. Differential input pair: Since v_{REF} is 0.9 V, use native NFETs to accommodate a tail-current transistor. Since $v_{TN.NAT(MIN)}$ is –0.15 V, connect the bulk to ground to use bulk effect to raise $v_{TN.NAT}$ to a positive value under worst-case conditions. This way, the source-coupled node of the differential pair remains below v_{REF}.

Circuit Parameter	Specification	Process Parameter	Value
v_{IN}	1.1–1.6 V	$\|V_{TPO}\|$ & V_{TNO}	0.6 ± 0.15 V
v_O	1 V	$V_{TNO.NAT}$	0 ± 0.15 V
v_{REF}	0.9 V	K_P'	40 μA/V² ± 20%
A_E	≈ 40 dB	K_N'	100 μA/V² ± 20%
i_Q	≤ 6 μA	L	≥ 0.35 μm
$v_{OE(MAX)}$	v_{IN} – 0.3 V	C_{OX}''	5 fF/μm²
$v_{OE(MIN)}$	0.2 V	$\lambda_{L(MIN)}$	500 mV⁻¹
f_{0dB}	1 MHz	$\lambda_{3L(MIN)}$	10 mV⁻¹
V_{OS}^*	≤ 25 mV	β_{PO}	50–150 A/A
$\Delta v_{O(LD)}$	≤ 10 mV	V_A	15 V

Table 9.4 Target Specifications and Process Parameters for the Error Amplifier in Example 9.3

2. Load mirror: Since the power device is a PFET, use a P-type load mirror to reproduce the supply ripple at the gate and ensure that the noise is common to both transistor inputs for higher power-supply rejection. Since $v_{IN(MIN)}$ is low at 1.1 V, use cascode transistors to fold the output of the differential pair into the P-type mirror, as Fig. 9.30 shows. Use a diode-connected PMOS M_{CB} to bias the cascode transistors.
3. Frequency compensation: Use a Miller capacitor C_M across A_B and M_{PO} to ensure that v_{OE} establishes the dominant low-frequency pole in the system.
4. Negative feedback: Since M_{PO}'s gate–drain gain inverts, connect the feedback signal v_{FB} to A_E's noninverting input, which is at the gate of M_{D1}.
5. Low random offset: Use large transistors for the differential pair and load mirror, so set their channel lengths L_{D1}, L_{D2}, L_{M3}, and L_{M4} to 10 times the minimum allowable length of 0.35 μm at 3.5 μm and use critical matching criteria to match the transistors. To ensure that A_E's gain is not too high, use minimum channel lengths for the bias pair M_{B7}–M_{B8}, so L_{B1} and L_{B2} can be 0.35 μm. But still, use critical matching criteria to match M_{B7} to M_{B8}.

Component Design

6. Feedback resistors R_{FB1} and R_{FB2}: Use resistor divider R_{FB1}–R_{FB2} to set v_O's steady-state target of 1 V from v_{REF}'s 0.9 V:

$$\frac{v_O}{v_{REF}} = \frac{R_{FB1} + R_{FB2}}{R_{FB2}} \equiv \frac{1}{0.9} = 1.11\ \text{V/V}$$

Set the current drawn by R_{FB1} and R_{FB2} to 3 μA to ensure that the resistor string can sink M_{PO}'s subthreshold current during worst-case conditions and extreme temperature and fabrication corners:

$$i_R = \frac{v_O}{R_{FB1} + R_{FB2}} \approx 3\ \mu\text{A}$$

so R_{FB2} can be 300 kΩ and R_{FB1} can be 33.3 kΩ.

7. Open-loop gain A_E:

$$A_{E0} \approx g_{m.D1} r_{ds.B8} \approx \frac{\sqrt{2 I_{D1} K_N' \left(\frac{W}{L}\right)_{D1}}}{I_{B8} \lambda_{L(MIN)}} = \frac{\sqrt{2\left(\frac{I_{Tail}}{2}\right) K_N' \left(\frac{W}{L}\right)_{D1}}}{I_{B8} \lambda_{L(MIN)}} \equiv 100$$

or

$$\frac{\sqrt{I_{Tail}\left(\frac{W}{L}\right)_{D1}}}{I_{B8}} \equiv \frac{A_{E0} \lambda_{L(MIN)}}{\sqrt{K_N'}} = \frac{(100)(0.1)}{\sqrt{100\mu}} \approx 1000$$

so choosing I_{Tail} to be 2 μA and M_{B8}'s I_{B8} to be 1 μA,

$$\left(\frac{W}{L}\right)_{D1} \approx \frac{(1000)^2 (1\mu)^2}{2\mu} = 0.5$$

so W_{D1} and W_{D2} can be $0.5L_{D1}$ and $0.5L_{D2}$ at 1.75 μm.

8. Maximum output $v_{OE(MAX)}$: To be able to shut M_{PO},

$$v_{OE(MAX)} = v_{IN} - v_{SD.M4} - V_{SD.C6(SAT)} \geq v_{IN} - 0.3\ \text{V}$$

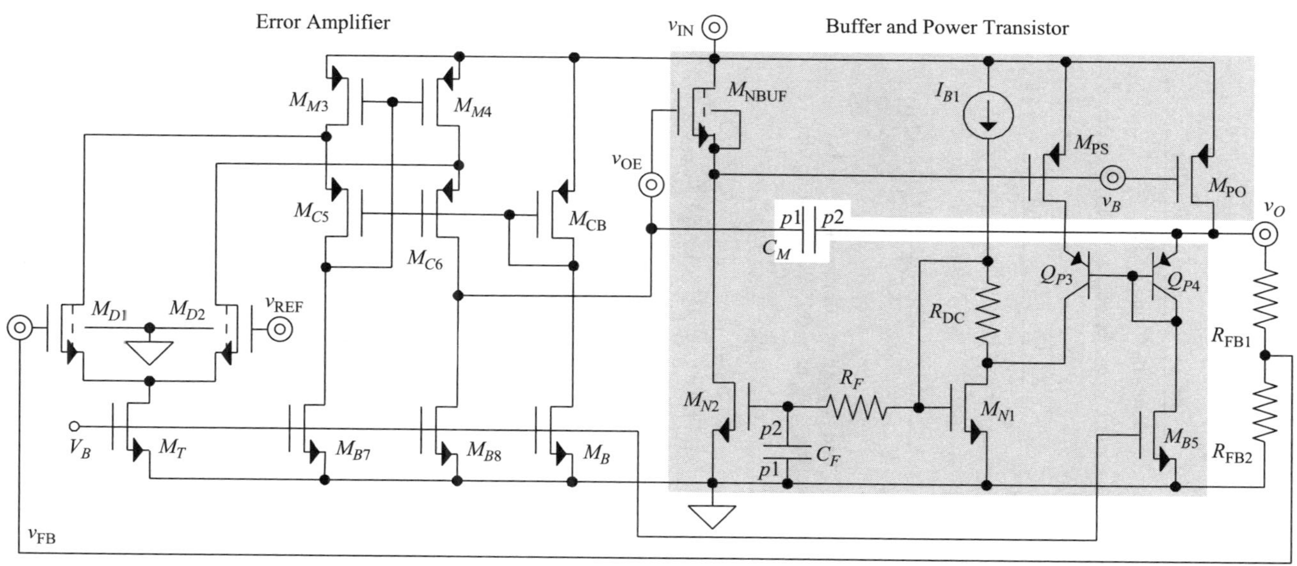

FIGURE 9.30 Low-dropout regulator design for Example 9.3.

so $V_{SD.M4(SAT)}$ and $V_{SD.C6(SAT)}$ can be less than 0.15 V:

$$V_{SD.M4(SAT)} = \sqrt{\frac{2I_{M4}}{K_K'\left(\frac{W}{L}\right)_{M4}}} \le \sqrt{\frac{2(I_{D2}+I_{B8})}{K_{K(MIN)}'\left(\frac{W}{L}\right)_{M4}}} \le 0.15 \text{ V}$$

or
$$\left(\frac{W}{L}\right)_{M4} \ge \frac{2(I_{D2}+I_{B8})}{V_{SD.M4(SAT)}{}^2 K_{P(MIN)}'} = \frac{2(1\mu+1\mu)}{(0.15)^2(32\mu)} = 5.55$$

which means W_{M3} and W_{M4} can be $6L_{M3}$ and $6L_{M4}$ at 21 μm,

and
$$V_{SD.C6(SAT)} = \sqrt{\frac{2I_{C6}}{K_P'\left(\frac{W}{L}\right)_{C6}}} \le \sqrt{\frac{2I_{B8}}{K_{P(MIN)}'\left(\frac{W}{L}\right)_{C6}}} \le 0.15 \text{ V}$$

or
$$\left(\frac{W}{L}\right)_{C6} \ge \frac{2I_{B8}}{V_{SD.C6(SAT)}{}^2 K_{P(MIN)}'} = \frac{2(1\mu)}{(0.15)^2(32\mu)} = 2.8$$

which means W_{C5} and W_{C6} can be 1.05 μm and L_{C5} and L_{C6} can be minimum at 0.35 μm. Choosing $v_{SD.M4}$ to be 0.15 V and noting that

$$v_{G.CB} = v_{IN} - v_{SG.CB} \equiv v_{IN} - v_{SD.M4} - v_{SG.C6}$$

means
$$v_{SG.CB} = |v_{TP}| + \sqrt{\frac{2I_{BP}}{K_P'\left(\frac{W}{L}\right)_{CB}}} \equiv v_{SD.M4} + v_{SG.C6}$$

$$= v_{SD.M4} + |v_{TP}| + \sqrt{\frac{2I_{C6}}{K_P'\left(\frac{W}{L}\right)_{C6}}}$$

so choosing I_{BP} to be 0.5 μA reduces $(W/L)_{CB}$ to

$$\left(\frac{W}{L}\right)_{CB} = \frac{2I_{BP}}{K_P'\left[v_{SD.M4} + \sqrt{\frac{2I_{C2}}{K_P'\left(\frac{W}{L}\right)_{C6}}}\right]^2} = \frac{2(0.5\mu)}{(40\mu)\left[0.15 + \sqrt{\frac{2(1\mu)}{(40\mu)\left(\frac{1.05}{0.35}\right)}}\right]^2} = 0.32$$

which means W_{CB} can be 1 μm and L_{CB} can be 3 μm.

9. Minimum output $v_{OE(MIN)}$: To be able to fully engage M_{PO},

$$v_{OE(MIN)} = V_{DS.B8(SAT)} \le \sqrt{\frac{2I_{B8}}{K_{N(MIN)}'\left(\frac{W}{L}\right)_{B8}}} \le 0.2 \text{ V}$$

or
$$\left(\frac{W}{L}\right)_{B8} \ge \frac{2I_{B8}}{K_{N(MIN)}' v_{OE(MIN)}{}^2} = \frac{2(1\mu)}{(80\mu)(0.2)^2} = 0.62$$

which means W_{B8} can be 2 μm and L_{B8} can be minimum at 0.35 μm.

10. Since the unity-gain frequency f_{0dB} is roughly equal to the gain–bandwidth product, and M_{PO}'s gain A_P and A_B Miller-multiply the effects of C_M on A_E's R_{OE},

$$f_{0dB} \approx A_E A_B A_P p_{OE} \approx \frac{A_E A_B A_P}{2\pi C_M (A_B A_P) R_{OE}} \approx \frac{(g_{m.D1} R_{OE}) A_B A_P}{2\pi C_M (A_B A_P) R_{OE}}$$

$$= \frac{g_{m.D1}}{2\pi C_M} \equiv 1\ \text{MHz}$$

or
$$C_M = \frac{g_{m.D1}}{2\pi f_{0dB}} = \frac{\sqrt{2 I_{D1} K_N' \left(\frac{W}{L}\right)_{D1}}}{2\pi f_{0dB}} = \frac{\sqrt{2(1\mu)(100\mu)\left(\frac{1.75}{3.5}\right)}}{2\pi(1\text{M})} = 1.59\ \text{pF}$$

which means C_M can be 1.6 pF.

Design Checks

11. Quiescent current i_Q:

$$i_Q = i_{M3} + i_{M4} + I_{BP} = 2\mu + 2\mu + 0.5\mu = 4.5\ \mu\text{A}$$

which with a 30-percent variation is 5.85 μA.

12. $v_{IN(MIN)}$ assuming the minimum voltage $V_{IBP(MIN)}$ across I_{BP} is 0.2 V:

$$v_{IN(MIN)} = \text{Max}\begin{cases} v_{SG.M3} + V_{DS.B7(SAT)} = |v_{TP(MAX)}| + V_{SD.M3(SAT)} + V_{DS.B7(SAT)} \\ v_{SG.CB} + v_{I_{BP}} = |v_{TP(MAX)}| + V_{SD.CB(SAT)} + V_{IBP(MIN)} \end{cases}$$

$$\approx \text{Max}\begin{cases} |v_{TP(MAX)}| + \sqrt{\dfrac{2I_{M3}}{K_{P(MIN)}'\left(\frac{W}{L}\right)_{M3}}} + \sqrt{\dfrac{2I_{B7}}{K_{P(MIN)}'\left(\frac{W}{L}\right)_{B7}}} \\ |v_{TP(MAX)}| + \sqrt{\dfrac{2I_{BP}}{K_{P(MIN)}'\left(\frac{W}{L}\right)_{CB}}} + V_{IBP(MIN)} \end{cases}$$

$$\approx \text{Max}\begin{Bmatrix} 0.75 + \sqrt{\dfrac{2(2\mu)}{(32\mu)\left(\frac{21}{3.5}\right)}} + \sqrt{\dfrac{2(1\mu)}{(80\mu)\left(\frac{2}{0.35}\right)}} \sim 0.96 \\ 0.75 + \sqrt{\dfrac{2(0.5\mu)}{(32\mu)\left(\frac{1}{3}\right)}} + 0.2 \approx 1.26 \end{Bmatrix} \approx 1.26\ \text{V}$$

Although combining three-sigma (3σ) parameters $K_{P(MIN)}'$, $K_{N(MIN)}'$, and $v_{TP(MAX)}$ is probabilistically unreasonable, applying $v_{TP(MAX)}$ alone also violates the $v_{IN(MIN)}$ specification:

$$v_{IN(MIN)} \approx |v_{TP(MAX)}| + \sqrt{\frac{2I_{BP}}{K_P'\left(\frac{W}{L}\right)_{CB}}} + V_{IBP(MIN)}$$

$$= 0.75 + \sqrt{\frac{2(0.5\mu)}{(40\mu)\left(\frac{1}{3}\right)}} + 0.2 \approx 1.22\text{ V}$$

So, reduce $v_{IN(MIN)}$ to 1.05 V by lowering M_{CB}'s v_{SG} by 150 mV, which can be achieved by lowering M_{C5}–M_{C6}'s $V_{SD(SAT)}$ by 75 mV and M_{M3}–M_{M4}'s $V_{SD(SAT)}$ by 50 mV, and pushing M_{M1}–M_{M2} into triode slightly by 25 mV because M_{M1}–M_{M2}'s r_{ds}'s have little impact on A_E:

$$\left(\frac{W}{L}\right)_{C5} \equiv \left(\frac{W}{L}\right)_{C6} \geq \frac{2I_{B8}}{V_{SD.C6(SAT)}{}^2 K_{P(MIN)}'} = \frac{2(1\mu)}{(0.075)^2(32\mu)} = 11$$

which means W_{C5} and W_{C6} can be 3.85 μm and L_{C5} and L_{C6} can be minimum at 0.35 μm,

and $$\left(\frac{W}{L}\right)_{M3} \equiv \left(\frac{W}{L}\right)_{M4} \geq \frac{2(I_{D2}+I_{B8})}{V_{SD.M4(SAT)}{}^2 K_{P(MIN)}'} = \frac{2(1\mu+1\mu)}{(0.1)^2(32\mu)} = 12.5$$

which means W_{M3} and W_{M4} can be $12.5L_{M3}$ and $12.5L_{M4}$ at 22 μm when reducing L_{M3} and L_{M4} to $5L_{MIN}$ at 1.75 μm to keep overall dimensions from growing excessively,

and $$\left(\frac{W}{L}\right)_{CB} = \frac{2I_{BP}}{K_P'\left[v_{SD.M4} + \sqrt{\frac{2I_{C6}}{K_P'\left(\frac{W}{L}\right)_{C6}}}\right]^2}$$

$$= \frac{2(0.5\mu)}{(40\mu)\left[0.075 + \sqrt{\frac{2(1\mu)}{(40\mu)\left(\frac{3.85}{0.35}\right)}}\right]^2} = 1.2$$

which means W_{CB} can be 1.2 μm and L_{CB} can be 1 μm.

13. Systemic offset $V_{OS(S)}$:

$$V_{OS(S)} = \frac{v_{DS.B8} - v_{DS.B7}}{A_E} = \frac{v_{OE} - (v_{IN} - v_{SG.M3})}{A_E}$$

where $$V_{OS(S)(MAX)} \leq \frac{v_{OE(MAX)} - \left\{v_{IN} - \left[|v_{TP(MAX)}| + \sqrt{\frac{2I_{M3}}{K_{P(MIN)}'\left(\frac{W}{L}\right)_{M3}}}\right]\right\}}{A_E}$$

$$= \frac{v_{IN} - 0.3 - \left[v_{IN} - 0.75 - \sqrt{\frac{2(2\mu)}{(32\mu)\left(\frac{22}{1.75}\right)}}\right]}{(100)} \approx 4.6\text{ mV}$$

and $$V_{OS(S)(MIN)} \le \frac{v_{OE(MIN)} - \left\{v_{IN(MAX)} - \left[|v_{TP(MIN)}| + \sqrt{\frac{2I_{M3}}{K_{P(MAX)}'\left(\frac{W}{L}\right)_{M3}}}\right]\right\}}{A_E}$$

$$= \frac{0.2 - \left[1.6 - 0.75 - \sqrt{\frac{2(2\mu)}{(48\mu)\left(\frac{22}{1.75}\right)}}\right]}{100} \approx -5.7\text{ mV}$$

$$-5.7\text{ mV} \le V_{OS(S)} \le 4.6\text{ mV}$$

so $$-5.7\text{ mV} \le V_{OS(S)} \le 4.6\text{ mV}$$

14. Load regulation in v_O:

$$\Delta v_{OE} = \Delta v_{GS.NBUF} - \Delta v_B \approx \Delta i_{O(MAX)}\left(\frac{A_I}{g_{m(MAX)}} - \frac{1}{G_{P(MAX)}}\right)$$

$$= \Delta i_{O(MAX)}\left[\left(\frac{(W/L)_{PS}}{(W/L)_{PO}}\right)\left(\frac{(W/L)_{N2}}{(W/L)_{N1}}\right)\left(\frac{1}{\sqrt{2i_{NBUF(MAX)}K_N'(W/L)_{NBUF}}}\right) - \frac{1}{\sqrt{2i_{O(MAX)}K_P'(W/L)_{PO}}}\right]$$

$$\approx 50\text{m}\left[\left(\frac{17}{18.3\text{k}}\right)(1)\left(\frac{1}{\sqrt{2(46\mu)(100\mu)(12/0.35)}}\right) - \frac{1}{\sqrt{2(50\text{m})(40\mu)(18.3\text{k}/0.35)}}\right]$$

$$= 50\text{m}\left[\left(\frac{17}{18.3\text{k}}\right)(1)(1.78\text{k}) - 2.2\right] \approx -27\text{ mV}$$

so $$\Delta V_{OS(S)} = \frac{\Delta v_{OE}}{A_E} \approx \frac{27\text{m}}{100} = 270\ \mu\text{V}$$

and $$\Delta v_{O(LD)} = \Delta V_{OS(S)}\left(\frac{v_O}{v_{REF}}\right) \approx (270\mu)\left(\frac{1}{0.9}\right) < 1\text{ mV}$$

15. Random offset V_{OS}^*: Assuming differential-pair, mirror, and bias transistors match well with 5 mV of 3σ input-referred offsets,

$$V_{OS}^* \approx \sqrt{\Delta v_{12}^2 + \left(\frac{\Delta v_{34} g_{m34}}{g_{m12}}\right)^2 + \left(\frac{\Delta v_{78} g_{m78}}{g_{m12}}\right)^2}$$

$$\approx \sqrt{\Delta v_{12}^2 + \Delta v_{34}^2 \left(\frac{2I_{M34} K_P'(W/L)_{M34}}{2I_{D12} K_N'(W/L)_{D12}}\right) + \Delta v_{12}^2 \left(\frac{2I_{D78} K_P'(W/L)_{D78}}{2I_{D12} K_N'(W/L)_{D12}}\right)}$$

$$\approx 5\text{m}\sqrt{1 + \left[\frac{2(2\mu)(40\mu)(21/1.75)}{2(1\mu)(100\mu)(1.75/3.5)}\right] + \left[\frac{2(1\mu)(100\mu)(2/0.35)}{2(1\mu)(100\mu)(1.75/3.5)}\right]} \approx 29\text{ mV}$$

which exceeds the specification by ±4 mV. Had it not been for the increase in $(W/L)_{M1}$ and $(W/L)_{M2}$ to accommodate $v_{IN(MIN)}$, V_{OS}^* would have been 24 mV. Decreasing V_{OS}^* amounts to raising g_{m12}, which increases A_E and f_{0dB} and compromises the stability of the system in light of A_B's finite bandwidth and M_{PO}'s output pole. Lowering g_{m34}, which raises $v_{IN(MIN)}$, also decreases V_{OS}^*. So choose to risk some yield loss and/or tighten v_{REF}'s accuracy to accommodate an increase in V_{OS}^*.

General Design Notes

A low-$v_{IN(MIN)}$ design usually compromises stability because, to realize a similar gain, more pole-producing nodes are necessary. And as the foregoing example shows, a low $v_{IN(MIN)}$ also sacrifices V_{OS}^*. Low i_Q, on the other hand, slows the response of the circuit because parasitic poles fall with reductions in i_Q. Ensuring that the circuit maintains performance across temperature and fabrication corners exacerbates the effects of these tradeoffs, forcing the designer to balance linear worst-case and probabilistic approaches to forecast probable risk. In the end, the optimum solution, as in operational amplifiers, targets *reasonable* specifications with emphasis on *key* application-specific parameters to avoid the sacrifices associated with overdesigned circuits.

9.5 Summary

Although not always the case, the design cycle of a linear regulator usually begins at the output and ends with the input, with the error amplifier. The load, for instance, determines the type of power transistor needed, the power device defines the parametric limits of the buffer driving it, and the buffer sets the limits of the preceding amplifier. As a whole, taking into account their interdependencies, the entire chain must cater to the increasingly stringent efficiency and accuracy demands of state-of-the-art portable and stationary systems, such as surviving low input-supply voltages, requiring low quiescent currents, and producing high loop gains, high bandwidths, low offsets, and so on.

For high power efficiency and extended battery life, power PMOS transistors not only accommodate lower input supplies for a given target output voltage, which is a feature that is inherent to P-type devices, but also demand less current. N-type devices, on the other hand, normally respond more quickly to load dumps, so they tend to be more accurate, and of those, BJTs source higher currents and withstand higher voltages. As a result, low-dropout regulators typically employ PFETs, but for applications that demand high accuracy, engineers often prefer N-type power transistors, and choose BJTs for high power. Irrespective of the device, the circuit must ultimately account for all parasitic components. In this regard, parasitic vertical and lateral PNP BJTs and their constituent diodes not only conduct ground current but also compromise the integrity and performance of the system.

Since power devices are large, they present unique and stringent requirements. A driving buffer must not only survive low input-supply voltages and use low quiescent currents but also comply with the wide voltage swings and high currents that large power transistors demand. To this end, emitter- and source-follower transistors are compact, fast, and efficient because they generate moderately low output impedances at low currents. Dynamically adjusting their bias points with local feedback loops can lower the buffer's power consumption during light loading conditions, when the quiescent current's toll on efficiency is greatest. While negative feedback is popular in this regard, a regenerative loop can boost the gain and accelerate the response of the buffer, albeit at the risk it poses to the feedback stability of the system. To avoid problems, the gain across the positive feedback path should fall below 1, especially near and above the system's bandwidth where the benefits of negative feedback disappear.

Ultimately, several key specifications converge on the error amplifier, such as steady-state accuracy, power-supply rejection, frequency response, and loop gain, which determines regulation performance. From a design perspective, N-type differential pairs normally favor signal-to-noise ratio because their biasing reference voltages are higher, which means endemic switching noise constitutes a lower fraction of the reference. While folding loads synergize with low input-supply voltages, P-type load mirrors appeal to P-type power devices and N-type mirrors to their N-type counterparts for high power-supply rejection. Since input-referred random offsets depend almost entirely on nondegenerated transistors in the error amplifier, matching their respective physical layouts is critical.

As shown in this chapter, design in its very nature is the art of tradeoff because improving one parameter normally degrades another. Raising power efficiency by reducing the input supply voltage, for example, first constrains the solution to P-type power transistors. And in further decreasing the input supply, input-referred offset and stability also suffer because less flexibility exists in the transconductor ratios of critical offset-setting transistors and mirror loads. Plus, adding transistors to compensate these deficiencies introduces pole-producing nodes. Reducing quiescent currents also costs the system performance in the form of speed and input-referred offsets. Worst-case operating conditions and extreme temperature and fabrication corners further correlate parameters in a way that iteration in the design process is almost impossible to avoid. Managing and assessing the effects of these probabilistic variations on the circuit is an intrinsic part of the design process, without which product development and practical solutions could not succeed in the marketplace.

In many ways, the IC design practices and tradeoffs discussed in this chapter represent the climax of this book, because they employ the device, circuit, and feedback tools developed in Chaps. 3–6 to achieve the system and small-signal design objectives discussed in Chaps. 1, 2, and 8. While the text may not cover the entire expanding field of analog microelectronics, the design considerations presented here exemplify and describe, to an extent, the core of analog IC design, from device physics, circuit techniques, and positive- and negative-feedback theory to system integration and design. The components discussed to this point, however, do not yet constitute a complete linear regulator solution, not without considering the system as a whole, possible performance-specific enhancers, protection against extreme operating conditions, and characterization, which Chaps. 10 and 11 address.

9.6 Review Questions

1. Which type(s) of power transistors exhibit the lowest dropout voltage?
2. Which type(s) of power transistors lose ground current?
3. Which type(s) of power transistors exhibit the lowest output resistance?
4. What does "base drive" refer to in the context of power transistors?
5. What does "gate drive" refer to in the context of power transistors?
6. Which type(s) of power transistors dissipate the least power?
7. What parasitic components do vertical BJTs normally incorporate?
8. What parasitic components do lateral BJTs normally incorporate?
9. What parasitic components do substrate MOSFETs normally incorporate?
10. What parasitic components do welled MOSFETs normally incorporate?
11. Why is parasitic well resistance a concern in power PMOSFETs?
12. What layout practice(s) alleviate this effect?
13. What is the general shape of space-efficient metallic power planes in an IC?
14. What parasitic resistances exist between the input supply and the output?
15. Generally, what transistor configuration is the most suitable for the buffer?

16. Why are N-type followers not ideal for driving power NPN BJTs?

17. Which type(s) of transistors are best suited for driving power PMOSFETs?

18. What type of buffer is best suited for driving a power PNP BJT?

19. What are the features of a regenerative loop in a buffer?

20. What is the drawback of a regenerative loop?

21. Why is a load-tracking pole appealing in a buffer that drives a P-type power transistor?

22. Why are P-type differential pairs less popular than their N-type counterparts?

23. How can the error amplifier minimize the effect of load mirrors on headroom?

24. Which type(s) of load mirror(s) in the error amplifier minimize power-supply noise in the output of a regulator with an NPN power switch?

25. Which type(s) of load mirror(s) in the error amplifier minimize power-supply noise in the output of a regulator with a PMOS power switch?

26. Which transistors contribute the most to the random component of input-referred offsets?

27. What produces systemic input-referred offset?

28. How does load regulation manifest in the error amplifier?

CHAPTER 10

Linear Regulators

The purpose of this chapter is to use the compensation strategies and circuit blocks developed in Chaps. 8 and 9 to build linear-regulator systems. For that, the discussion starts with *low-dropout* (LDO) regulators because, of the possible alternatives, they usually dissipate the least power. And low power dissipation not only extends battery life but also frees the space that a heat sink would otherwise occupy. Surviving severe load dumps, however, is sometimes more critical than reducing power consumption, so high-bandwidth solutions are next. But since stand-alone circuits are also useful in many respects, the chapter continues with self-referenced regulators. Performance enhancers follow because they can, if nothing else, pacify the space, load-dump, accuracy, and power-supply requirements that emerging *system-on-chip* (SoCs) solutions demand. Some applications even call for current regulators, so the chapter also explores them.

10.1 Low-Dropout Regulators

10.1.1 Output-Compensated PMOS

The number and diversity of features that consumers enjoy, demand, and in some cases expect from portable, battery-powered devices often outpace technology- and design-driven reductions in energy and power. In other words, with respect to linear regulators targeted for the higher power portable market, the sum of smaller but far more numerous feature-specific currents remains relatively high at several tens and even hundreds of milliamps. As a result, because many of these features engage and disengage simultaneously at rates faster than their current-starved supply circuit can respond, high output capacitance C_O is imperative, and output compensation is not only merely justifiable but often also a necessity.

Combining low quiescent-current and low input-voltage requirements, which portable devices demand for high efficiency and extended battery life, with high output capacitance for fast load-dump response normally calls for *output-compensated P-channel metal–oxide–semiconductor (PMOS)* solutions, like the one Fig. 10.1 shows. Here,

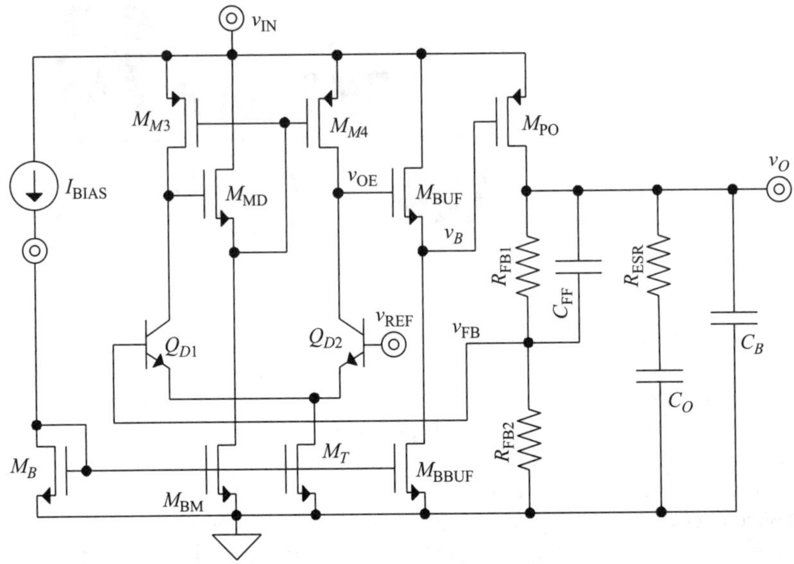

Figure 10.1 Output-compensated low-dropout PMOS regulator.

a P-type MOS mirror M_{M3}–M_{M4} loads an NPN differential pair Q_{D1}–Q_{D2} to drive an N-channel MOS buffer follower M_{BUF}, and with the buffer, control a power PMOS transistor M_{PO}. N-channel follower M_{MD} diode-connects M_{M3} to match to first order the combined effects of M_{PO}'s source–gate voltage v_{SGO} and M_{BUF}'s gate–source voltage v_{GSBUF} on M_{M4}'s source–drain voltage v_{SD4} to M_{M3}'s v_{SG3} and M_{MD}'s v_{GSMD} on M_{M3}'s v_{SD3}. This way, the systemic offset that mismatched voltages across the mirror would otherwise produce is minimal. The purpose of capacitor C_{FF} is to feed forward small signals from v_O to feedback node v_{FB}. This introduces an in-phase zero z_{FF} that helps save phase near the *unity-gain frequency* of the system. The input-pair transistors are bipolar-junction transistors (BJTs) because they generally match better than metal–oxide–semiconductor (MOS) field-effect transistors (FETs), so they produce lower input-referred offsets.

Compensation

The general strategy for compensating an output-compensated PMOS regulator is to place the output pole p_O at low frequency so p_O dominates over all other poles and zeros in the circuit, as Fig. 10.2 illustrates, and to keep the unity-gain frequency f_{0dB} well below parasitic poles in the system. In practice, when other poles and zeros remain at higher frequencies across temperature and fabrication corners and operating conditions, loop gain drops linearly with frequency past p_O in a single-pole roll-off manner. This way, f_{0dB}'s minimum point is when low-frequency loop gain A_{LG0} reaches 1 after

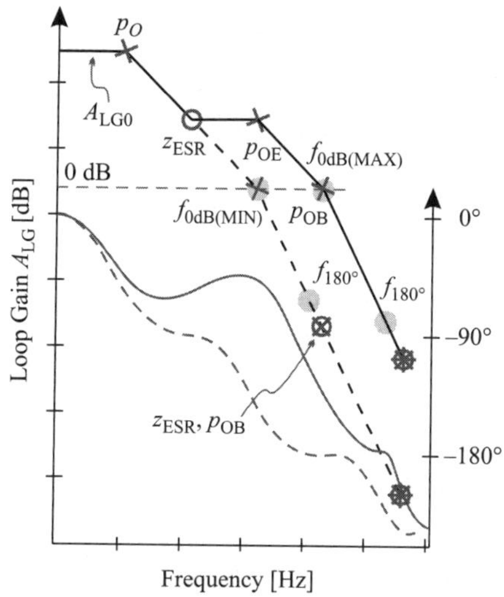

FIGURE 10.2 Frequency response of an output-compensated PMOS regulator.

traversing a gain–bandwidth product GBW, or $A_{LG0}p_O$, expanse of frequency:

$$\frac{A_{LG0}}{1+\frac{s}{2\pi p_O}} \approx \frac{A_{LG0}}{\left(\frac{s}{2\pi p_O}\right)}\Bigg|_{f_{0dB(MIN)} \approx A_{LG0}p_O} \equiv 1 \tag{10.1}$$

Note that $f_{0dB(MIN)}$ is often a specification target.

Building gain in the error amplifier A_E requires a high-resistance node v_{OE}, so the pole associated with that node p_{OE} is normally the second dominant pole in the circuit. Although the equivalent series resistance (ESR) of the output capacitor C_O introduces what could be a phase-saving zero z_{ESR} to the response near p_{OE}, R_{ESR} and, as a result, z_{ESR}'s location vary substantially across temperature and fabrication corners:

$$\frac{1}{2\pi R_{ESR(MAX)}C_O} \le z_{ESR} \le \frac{1}{2\pi R_{ESR(MIN)}C_O} \tag{10.2}$$

In consequence, placing p_{OE} well below $f_{0dB(MIN)}$ and expecting z_{ESR} to cancel p_{OE}'s effects is risky, but restraining $f_{0dB(MAX)}$ below $z_{ESR(MIN)}$ in all cases, on the other hand, is unfavorably pessimistic. The aim is to therefore place the second dominant pole p_{OE} near $f_{0dB(MIN)}$, well below $f_{0dB(MAX)}$:

$$p_{OE} \approx \frac{1}{2\pi(r_{sd4} \parallel r_{o2})C_{D4}} \approx f_{0dB(MIN)} \tag{10.3}$$

where C_{D4} is the total parasitic capacitance at M_{M4}'s drain. To keep p_{OE} near $f_{0dB(MIN)}$, however, the resistance at v_{OE} must be moderately low, which is why the channel lengths of the mirror devices M_{M3} and M_{M4} are relatively short, because channel-length modulation across short channels produces lower output resistances. Notice, by the way, that $z_{ESR(MIN)}$ extends $f_{0dB(MIN)}$ by how far $z_{ESR(MIN)}$ precedes p_{OE}:

$$f_{0dB(MAX)} \approx f_{0dB(MIN)}\left(\frac{p_{OE}}{z_{ESR(MIN)}}\right) \tag{10.4}$$

As operating frequency climbs, the next pole to appear is the buffer pole p_{OB} at M_{PO}'s gate v_B, and since p_O and p_{OE} already shift considerable phase in the absence of z_{ESR}, p_{OB} must remain well above $f_{0dB(MIN)}$, but not necessarily higher than $f_{0dB(MAX)}$. In fact, since z_{ESR} is the reason f_{0dB} shifts to $f_{0dB(MAX)}$ in the first place, z_{ESR} offsets the effects of p_O and adds margin for p_{OB} to reside near $f_{0dB(MAX)}$:

$$p_{OB} \approx \frac{g_{mBUF}}{2\pi C_{SGO}} \approx f_{0dB(MAX)} \tag{10.5}$$

After p_{OB}, though, there is no margin left, and all remaining parasitic poles in the system, which include the feedback pole p_{FB} at v_{FB}, C_B's bypass pole p_B, and the mirror pole p_M, must remain well above $f_{0dB(MAX)}$.

Keeping parasitic poles at high frequencies, however, demands quiescent current, which is why shifting them just high enough beyond $f_{0dB(MAX)}$ for stable conditions to remain is the general design approach. Although the effect of each pole a decade past $f_{0dB(MAX)}$ on phase margin is a few degrees, the relatively small, but compounded contributions of several poles can still have an avalanching effect on phase shift across the feedback loop near f_{0dB}. And a rapidly falling phase near f_{0dB} means that the circuit's proneness to unstable conditions is more sensitive to temperature and process variations, because a relatively small shift in f_{0dB} amounts to a substantial change in phase margin. In other words, the circuit is conditionally stable and may oscillate during startup or in response to fast load dumps.

Fortunately, nearby in-phase zeros recover some of the phase lost to dampen the avalanching effects that the remaining poles have on phase shift. The general design approach is therefore to keep parasitic poles barely a decade above $f_{0dB(MAX)}$ and introduce, when possible, just as many in-phase zeros to offset them. Feed-forward capacitor C_{FF} in Fig. 10.1 presents such a phase-saving zero z_{FF} when C_{FF} shunts feedback resistor R_{FB1} past $1/2\pi R_{FB1}C_{FF}$:

$$\left.\frac{1}{sC_{FF}}\right|_{z_{FF}=\frac{1}{2\pi R_{FB1}C_{FF}}\approx 5 f_{0dB(MAX)}} \equiv R_{FB1} \tag{10.6}$$

Introducing a zero below $f_{0dB(MAX)}$ in the presence of z_{ESR}, however, extends $f_{0dB(MAX)}$ closer to parasitic poles, where parasitic poles would have a more pronounced impact on phase margin. So to prevent temperature and process variations from pulling z_{FF} below $f_{0dB(MAX)}$, z_{FF} is, by design, slightly above $f_{0dB(MAX)}$.

Now that z_{FF} adds some margin to the system, parasitic bypass and feedback poles p_B and p_{FB} can comfortably reside barely a decade above $f_{0dB(MAX)}$. Recalling from Chap. 8 that C_B shunts C_O's R_{ESR} to produce p_B:

$$p_B \approx \frac{1}{2\pi R_{ESR} C_{B(MAX)}} \approx 10 f_{0dB(MAX)} \tag{10.7}$$

keeping p_B at $10f_{0dB(MAX)}$ bounds bypass capacitance below $C_{B(MAX)}$, which in many cases is not a problem because C_B might only be parasitic. Similarly, the capacitance present at the feedback node v_{FB} confines R_{FB1} and R_{FB2}'s combined parallel resistance because that capacitance should not shunt $R_{FB1} \parallel R_{FB2}$ below $10f_{0dB(MAX)}$:

$$\left.\frac{1}{s(C_{\mu 2} + C_{\pi 2} + C_{FF})}\right|_{p_{FB} = \frac{1}{2\pi(R_{FB1} \parallel R_{FB2})(C_{\pi 2} + C_{FF})} \approx 10 f_{0dB(MAX)}} \equiv R_{FB1} \parallel R_{FB2} \tag{10.8}$$

where Q_{D2}'s multiplying Miller effect on $C_{\mu 2}$ is negligible because the parasitic capacitance present at Q_{D2}'s collector already shorted the resistance at v_{OE} at a lower frequency, which means the gain across Q_{D2} is low. Unlike p_B, though, which often stays at high frequency without much effort, p_{FB} tends to creep toward lower frequency because feedback resistances R_{FB1} and R_{FB2} are usually high to limit their current to less than 2–5 μA.

Load mirror M_{M3}–M_{M4} generally introduces a closely spaced pole–zero pair, so its collective impact on phase is not as profound. Level-shifting follower M_{MD} and mirroring transistor M_{M3}, however, comprise an inverting feedback loop that must itself be stable. This inner loop has two poles: a high-frequency pole at M_{MD}'s source p_{G3} and a bandwidth-setting low-frequency pole at M_{M3}'s drain p_{D3}. p_{D3} must therefore be sufficiently high to ensure that the closed-loop bandwidth p_M of the mirror is slightly above $f_{0dB(MAX)}$. Since p_M is the loop's unity-gain frequency, p_M is the gain–bandwidth product $A_{V0}p_{D3}$ of M_{M3}'s diode connection:

$$p_M = A_{V0} p_{D3} \approx \frac{g_{m3}(r_{sd3} \parallel r_{o1})}{2\pi(r_{sd3} \parallel r_{o1})C_{D3}} \approx \frac{g_{m3}}{2\pi C_{D3}} \approx 5 f_{0dB(MAX)} \tag{10.9}$$

where p_M is the inverting feedback-loop translation of p_{D3} and C_{D3} is the total parasitic capacitance at the gate of M_{MD}.

To ensure that the loop in the mirror is stable, M_{MD}'s gate pole p_{D3} must be dominant and p_{G3} must reside near the closed-loop bandwidth p_M:

$$p_{G3} \approx \frac{g_{mMD}}{2\pi(C_{SG3} + C_{SG4} + C_{GSMD})} \approx p_M \approx 5 f_{0dB(MAX)} \tag{10.10}$$

where M_{M4}'s Miller-multiplying effect on C_{GD4} is negligible because, near p_{G3}, p_{OE} already shunted the effects of v_{OE}'s resistance, forcibly reducing the gain across M_{M4}. Similarly, the capacitance at M_{M3}'s drain shunted its parallel resistance at p_{D3}, well below p_{G3}, so C_{GSMD} also loads M_{M3}'s gate. Collectively, the loop introduces the feedback translation of p_{D3} to the system in the form of a mirror pole p_M, and when small signals from the input device Q_{D2} overwhelm those sourced by the loading mirror M_{M4}, the mirror introduces an in-phase zero z_M at approximately $2p_M$, which as before, helps reduce the avalanching effects this and other poles in the vicinity have on phase.

The feed-forward capacitor C_{GSBUF} in the N-channel voltage-following buffer M_{BUF} also introduces an in-phase zero z_{BUF}. The location of this zero is normally high because M_{BUF} is small, which means C_{GSBUF} is low:

$$i_C = \frac{v_{gs}}{sC_{GSBUF}}\Bigg|_{z_{BUF} \approx \frac{g_{mBUF}}{2\pi C_{GSBUF}} \geq 5 f_{0dB(MAX)}} \equiv i_{gm} = v_{gs} g_{mBUF} \tag{10.11}$$

Similarly, the feed-forward capacitor C_{GDO} in the power PMOS M_{PO} also introduces a zero z_{PO}, but this time the zero is out of phase with g_{mO}'s current. As a result, z_{PO} adds gain but subtracts phase, which is the worst possible combination, so z_{PO} must remain at or above $10f_{0dB(MAX)}$. Fortunately, z_{PO}'s location often resides at high frequency because g_{mO} is high. In other words, C_{GDO}'s current does not overwhelm g_{mO} until the operating frequency is high, past z_{PO}:

$$i_C = \frac{v_{sgO} - v_{sdO}}{sC_{GDO}} = \frac{v_{sgO}}{sC_{GDO}}\Bigg|_{z_{PO} \approx \frac{g_{mO}}{2\pi C_{GDO}} \geq 10 f_{0dB(MAX)}} \equiv i_{gm} = v_{sg} g_{mO} \tag{10.12}$$

Differential input-pair transistor Q_{D2}'s Miller capacitor $C_{\mu 2}$ also feeds forward out-of-phase signals across Q_{D2}, so $C_{\mu 2}$ establishes an out-of-phase zero z_{D2}. Like z_{PO}, z_{D2} must stay at or above $10f_{0dB(MAX)}$, but like z_{BUF}, z_{D2} naturally gravitates toward high frequency because Q_{D2} is small, which means $C_{\mu 2}$ is low. Plus, Q_{D2}'s g_{m2} is relatively high because Q_{D2}'s collector current varies exponentially with respect to Q_{D2}'s base–emitter voltage:

$$i_C = \frac{v_{be2} - v_{ce2}}{sC_{\mu 2}} = \frac{v_{be2}}{sC_{\mu 2}}\Bigg|_{z_{D2} \approx \frac{g_{m2}}{2\pi C_{\mu 2}} \geq 10 f_{0dB(MAX)}} \equiv i_{gm} = v_{be2} g_{m2} \tag{10.13}$$

10.1.2 Miller-Compensated PMOS

The need for higher power levels and quicker response times is as widespread as the demand for higher integration, which is where internal compensation schemes thrive. The fact is, packing more features into a single integrated circuit (IC) introduces systemic uncorrelated noise into the power supply. This noise reduces the *signal-to-noise ratio* (SNR) of sensitive analog electronics attached to that supply. Dedicated on-chip regulators help in this respect because they decouple the otherwise common noise from sensitive loads. Having no external pad or pin to rely on, though, the challenge in designing these *point-of-load* (PoL) regulators is the unavailability of off-chip capacitors. Luckily, the lower power levels that these targeted loads demand suffer correspondingly lower load dumps, so lower on-chip capacitors can often satisfy the modest needs of a lighter load.

A low-dropout (LDO) topology uses an MOS power transistor to forego the base-current requirements of a BJT, and a P-type device to reduce the circuit's dropout voltage. The *internally compensated* design shown in Fig. 10.3 employs an N-channel differential pair M_{D1}–M_{D2} loaded with a folding cascoded mirror M_{M3}–M_{M4} and corresponding biasing transistors M_{B7}–M_{B8}. Compensating Miller capacitor C_C pulls the error-amplifier pole p_{OE} at v_{OE} to lower frequency, and nulling resistor R_N limits C_C's feed-forward current to such an extent that R_N converts the out-of-phase zero of the power transistor M_{PO} into an in-phase zero z_N. As before, C_O helps supply and sink fast load dumps while the circuit reacts and adjusts M_{PO}.

Compensation

As with most internally compensated circuits, the error amplifier's pole p_{OE} is dominant. And here, the inverting gain A_{PO} across the power PMOS transistor M_{PO} Miller-multiplies capacitor C_C to shunt M_{B8}'s resistance r_{ds8}, which shunts and dominates over M_{C6}'s source-degenerated counterpart, at p_{OE}. Low-frequency loop gain A_{LG0} therefore drops

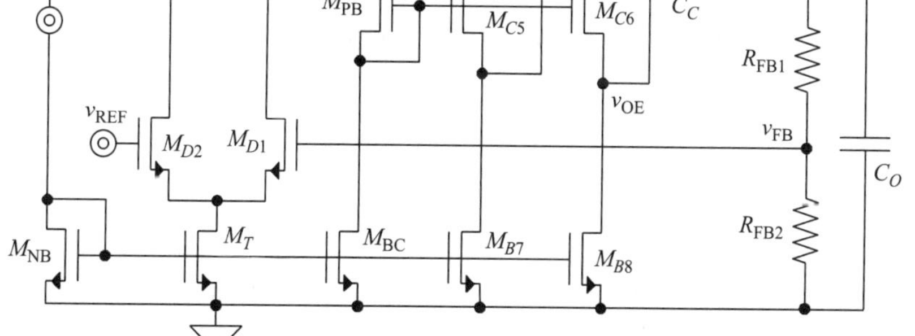

Figure 10.3 Miller-compensated low-dropout PMOS regulator.

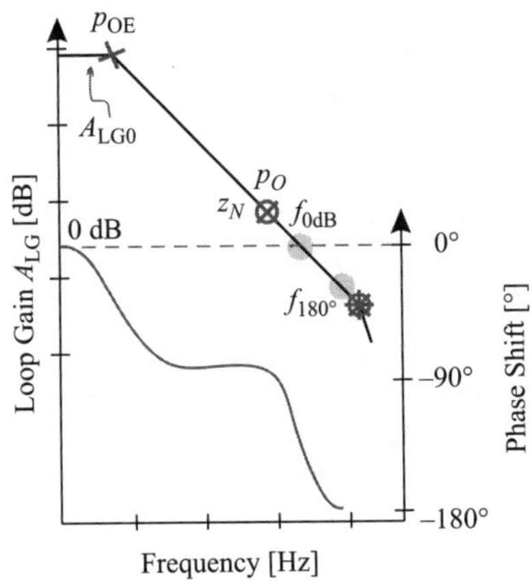

FIGURE 10.4 Frequency response of a Miller-compensated PMOS regulator.

linearly past p_{OE} and reaches 1 in the absence of other poles and zeros at f_{0dB}, as Fig. 10.4 demonstrates, at the gain–bandwidth-product frequency $A_{LG0}p_{OE}$, which is roughly $g_{m2}/2\pi(C_C + C_{GDO})$:

$$\frac{A_{LG0}}{1+\frac{s}{2\pi p_{OE}}} \approx \frac{A_{LG0}}{\left(\frac{s}{2\pi p_{OE}}\right)} \approx \frac{g_{m2} r_{ds8} A_{PO}}{r_{ds8}[A_{PO}(C_C + C_{DGO})]s}\Bigg|_{f_{0dB} \approx \frac{g_{m2}}{2\pi(C_C + C_{DGO})}} \equiv 1 \quad (10.14)$$

The output pole p_O then follows because output capacitance C_O is as high as possible to supply fast load dumps without drooping excessively. This time, however, p_O occurs when C_O and M_{PO}'s C_{DBO} shunt the $1/g_{mO}$ diode-connected resistance that results when C_C shorts M_{PO}'s gate–drain terminals. So given that C_C and C_{DGO} shunt their parallel resistance at a lower frequency, at p_{OE}, C_O and C_{DBO} shunt $1/g_{mo}$ at $g_{mo}/2\pi(C_O + C_{DBO})$:

$$\frac{1}{s(C_O + C_{DBO})}\Bigg|_{p_O \approx \frac{g_{mO}}{2\pi(C_O + C_{DBO})} \leq \frac{f_{0dB}}{10}} \equiv \frac{1}{g_{mO}} \quad (10.15)$$

Nulling resistance R_N is high enough to not only block out-of-phase feed-forward currents through C_C but also shift the feed-forward zero to the left half of the s plane. This way, the zero z_N that R_N establishes helps offset the effects of p_O. z_N asserts its influence when nulling-path current i_R matches and surpasses M_{PO}'s i_{gm}:

$$i_R = \frac{v_{sgO} - v_{sdO}}{R_N + \frac{1}{sC_C}} = \frac{v_{sgO}}{R_N + \frac{1}{sC_C}}\Bigg|_{z_N \approx \frac{1}{2\pi\left(R_N - \frac{1}{g_{mO}}\right)C_C} \approx p_O} \equiv i_{gm} = v_{sgO} g_{mO} \quad (10.16)$$

Unfortunately, p_O shifts considerably with output current i_O and z_N does not, and to make matters worse, z_N does not even track p_O with temperature or process. As a result, p_O and z_N should not precede f_{0dB} by more than a decade to avoid phase shift from dipping to 180° at any frequency below f_{0dB}.

The parasitic poles and zeros that appear past p_{OE}, p_O, and z_N reside only within one decade of f_{0dB}, and in-phase zeros hover nearby to prevent the phase shift across the loop from avalanching. The first pole to consider past p_O and z_N is the feedback pole p_{FB} at v_{FB} because $R_{FB1} \parallel R_{FB2}$ is moderately high to keep quiescent-current flow through R_{FB1} and R_{FB2} low:

$$\left.\frac{1}{s(C_{GD1}+C_{GS1})}\right|_{p_{FB}\approx\frac{1}{2\pi(R_{FB1}\parallel R_{FB2})C_{GS1}}\approx 10 f_{0dB(MAX)}} \equiv R_{FB1}\parallel R_{FB2} \tag{10.17}$$

where the Miller effects of C_{GD1} are negligibly small because the gain across M_{D1} is low, and R_{FB1} is in parallel with R_{FB2} because C_O already shunted the resistance at v_O past p_O.

The mirror and folding poles at the sources of M_{C5} and M_{C6} are next because the remaining resistances in the circuit are roughly $1/g_m$ and these nodes include C_{GS} capacitances, which are moderate. The mirror pole p_M occurs when C_{GS3}, C_{GS4}, C_{GD3}, C_{GD4}, C_{GD5}, and C_{GD7} shunt M_{M3}'s $1/g_{m3}$ resistance:

$$\left.\frac{1}{s(C_{GS3}+C_{GS4}+C_{GD3}+C_{GD4}+C_{GD5}+C_{GD7})}\right|_{p_M\approx\frac{g_{m3}}{2\pi(C_{GS3}+C_{GS4})}\approx 5f_{0dB}} \equiv \frac{1}{g_{m3}} \tag{10.18}$$

Luckily, the mirror zero z_M offsets p_M's effects soon after at $2p_M$, which is why p_M's placement can remain slightly above f_{0dB}. Like M_{MD} in the output-compensated PMOS regulator, cascode current buffer M_{C5} diode-connects M_{M3}, and the location of the pole at M_{C5}'s source p_{S5} should therefore match or exceed p_M to avoid exposing this loop to unstable feedback conditions. This requirement is automatically met when considering that the folding pole at M_{C6}'s source p_{S6} must remain a decade above f_{0dB} because no in-phase zeros exist to offset its effects. The influence of $p_{S5,6}$ appears when M_{C5}'s and M_{C6}'s C_{GSC}, M_{D1}'s and M_{D2}'s C_{GDD}, and M_{M3}'s and M_{M4}'s C_{GDM} shunt M_{C5}'s and M_{C6}'s $1/g_{mC}$ source resistances:

$$\left.\frac{1}{s(C_{GSC}+C_{GDM}+C_{GDD})}\right|_{p_{S5,6}\approx\frac{g_{mC}}{2\pi(C_{GSC}+C_{GDM}+C_{GDD})}\approx 10f_{0dB}} \equiv \frac{1}{g_{mC}} \tag{10.19}$$

where the loading influence of r_{ds8} on g_{m6} is negligible because the Miller capacitance present at v_{OE} shunts r_{ds8} at lower frequency, at p_{OE}. Incidentally, p_{S5} and p_{S6} affect each half of the differential signal equally, so their combined effect on the full differential response is one pole at $p_{S5,6}$.

The last zero to consider is the one that M_{D1}'s out-of-phase feed-forward capacitor C_{GD1} produces in the form of z_{D1}, which happens when C_{GD1}'s current overwhelms M_{D1}'s g_{m1} current i_{gm}:

$$i_C = \frac{v_{gs1} - v_{ds1}}{sC_{GD1}} = \frac{v_{gs1}}{sC_{GD1}}\Bigg|_{z_{D1} \approx \frac{g_{m1}}{2\pi C_{GD1}} \geq 10 f_{0dB(MAX)}} \equiv i_{gm} = v_{gs1} g_{m1} \tag{10.20}$$

z_{D1} should exceed f_{0dB} by at least a decade, which may not be difficult to accomplish because C_{GD1} is low. Notice that M_{PO}'s out-of-phase feed-forward parasitic capacitor C_{GDO} does not include a current-limiting resistor, so C_{GDO} introduces another right-half-plane zero z_{PO} to the mix. Fortunately, z_{PO} is at considerably higher frequency because the transconductor current that C_{GDO} must overwhelm to establish z_{PO} is very high. Like z_{FF} in the output-compensated PMOS case, a feed-forward capacitor across R_{FB1} can introduce another in-phase zero near f_{0dB} to help save phase and extend f_{0dB} slightly, except that doing so requires additional silicon area.

What is interesting and perhaps useful in the Miller-compensated LDO is the circuit's ability to yield stable operating conditions across output capacitance C_O. The case thus far presented assumes that C_O is relatively low and bounded to keep p_O within a decade of f_{0dB}. Consider, however, that raising C_O pulls p_O to lower frequency while simultaneously pushing p_{OE} to higher frequency. Pole p_{OE} shifts because shunting the resistance at v_O reduces the gain across M_{PO}, which reduces the Miller impact that M_{PO} has on C_C and C_{DGO}. In fact, if C_O is sufficiently high, p_O is dominant because C_O and C_{DBO} shunt r_{sdO}, R_{FB1}, and R_{FB2} at considerably lower frequency than the equivalent Miller pole would:

$$\frac{1}{s(C_O + C_{DBO})}\Bigg|_{p_O \approx \frac{1}{2\pi[r_{dsO} \| (R_{FB1}+R_{FB2})](C_O + C_{DBO})}} \equiv r_{dsO} \| (R_{FB1} + R_{FB2}) \tag{10.21}$$

Miller pole p_{OE} would then result when C_C and C_{GDO}, now that v_o is nearly at ground, shunt r_{ds8}:

$$\frac{1}{s(C_C + C_{GDO})}\Bigg|_{p_{OE} \approx \frac{1}{2\pi r_{ds8}(C_C + C_{GDO})} \geq \frac{f_{0dB}}{10}} \equiv r_{ds8} \tag{10.22}$$

Resistor R_N yields the effects of an in-phase zero z_N when R_N current-limits C_C:

$$\frac{1}{sC_C}\Bigg|_{z_N \approx \frac{1}{2\pi R_N C_C} \approx p_{OE}} \equiv R_N \tag{10.23}$$

past which point the total impedance across the C_C–R_N path reduces to R_N. Under these new operating conditions, which amount to swapping p_{OE} and p_O and using z_N to cancel p_{OE} in the frequency response of Fig. 10.4, the circuit is again stable.

Worst-case stability conditions occur when C_O is at an intermediate point, between maybe 500 nF and 2 μF, when p_{OE} and p_O are close. However, because z_N saves phase and load-induced variations in p_O cause the loop gain to be low, maintaining stable conditions across all values of C_O is possible, albeit with varying degrees of phase margin. This role reversal between p_{OE} and p_O amounts to shifting pole dominance between the internal and output nodes. Unfortunately, Miller capacitors C_C and C_{DGO} feed forward ripples in v_{IN} to v_O because the diode connection that they close around M_{PO} lowers the coupling impedance between v_{IN} and v_O to roughly $1/g_{mO}$. This means that power-supply rejection is lower here than in the output-compensated case of Fig. 10.1.

Miller Amplification

The silicon area that a large Miller capacitor demands from a dense SoC solution is costly and sometimes difficult to justify. Reducing the area that on-chip capacitors occupy by actively multiplying their capacitances presents some risk, but the area it saves can nonetheless offer appealing alternatives. Figure 10.5 illustrates and highlights how to leverage and modify the low-dropout PMOS regulator in Fig. 10.3 to multiply the effects of capacitors C_M and C_X on the dominant low-frequency pole at v_{OE}. The basic idea is to channel the capacitors' displacement current into amplifying current mirrors M_{B7A}–M_{B7} and M_{B8A}–M_{B8} and steer the multiplied currents back to v_{OE}. Mirror M_{B8A}–M_{B8}, for example, multiplies C_X's current i_{CX} and feeds the multiplied translation $A_M i_{CX}$ back to v_{OE}. This way, C_X's total influence on v_{OE} is i_{CX},

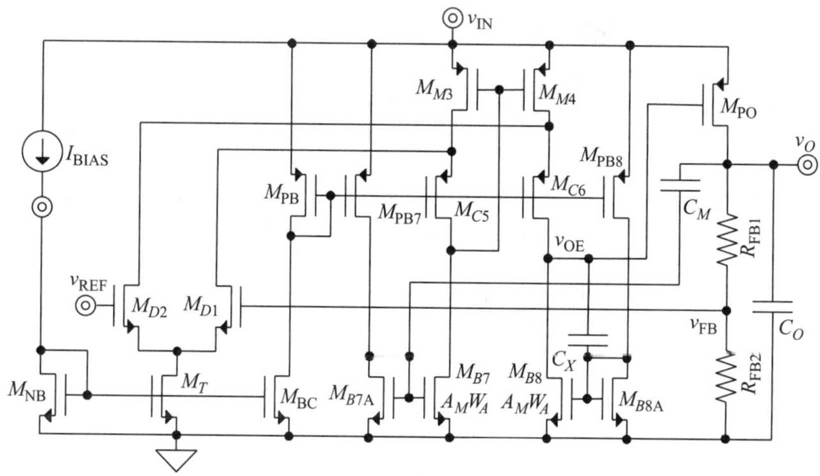

Figure 10.5 Miller-multiplied low-dropout PMOS regulator.

or $v_{oe}sC_X$, plus $A_M i_{CX}$, so C_X requires 1 + A_M times more current to droop, which means C_X presents a capacitance C_{EQX} that is equivalent to 1 + A_M times that of C_X:

$$i_{EQ} = i_{CX} + i_{B8} = i_{CX}(1 + A_M) \approx v_{oe}sC_X(1 + A_M) \equiv v_{oe}sC_{EQX} \quad (10.24)$$

Similarly, but now applied to Miller capacitor C_M, mirror M_{B7A}–M_{B7} and C_M together displace current from v_{OE} at the same rate that (1 + A_M)C_M would. In this case, C_M feeds current to v_{OE} through M_{B7A}–M_{B7} and M_{M3}–M_{M4}, instead of M_{B8A}–M_{B8}, to ensure that C_M's connectivity around M_{PO} remains noninverting, because a conventional Miller capacitor offers no feedback inversion from v_O to v_{OE}.

By multiplying C_C this way, 2 pF and a current-mirror gain of 10 achieve the same Miller objectives that 20 pF would without multiplication. Note, however, that the current multiplier has energy-shunting poles in its path that a conventional Miller capacitor does not. Plus, the diode-connected transistor in the mirror presents a series $1/g_m$ resistance that current-limits the capacitor that feeds it. These nonidealities produce a peaking effect near the unity-gain frequency that C_X at 2 pF, which is not a Miller capacitor and has a shorter and higher-bandwidth path, dampens. Lowering M_{B7A}'s $1/g_m$ resistance by, for instance, raising the shunt-feedback loop gain across M_{B7}'s drain–gate terminals also dampens this effect. Inserting a high-bandwidth, noninverting amplifier between M_{B7A}'s drain and gate terminals can increase this loop gain. Perhaps the most significant drawback here is the input-referred offset that mirrors M_{PB7}–M_{PB8}, M_{B7A}–M_{B7}, and M_{B8A}–M_{B8} add to the error amplifier A_E. Unfortunately, compounded mismatches in these mirrors produce a considerable offset current between M_{B7} and M_{B8}, so the overall accuracy of the regulator ultimately suffers.

10.2 High-Bandwidth Regulators

Low-dropout operation in dedicated on-chip supplies is not always a requirement because the power dissipated by the rest of the system, on average, can overwhelm that of the particular load in question. If other considerations allow, using this argument to ease dropout requirements is important because N-type power transistors, which suffer from higher dropout voltages, outperform their lower-dropout P-type counterparts in speed, and as a result, in on-chip capacitor area. Headroom may still limit dropout voltages to low values, however, like when a switching regulator supplies the entire IC and its output, which is the linear regulator's input supply, is already low to mitigate the losses associated with the high-power sectors of the system.

10.2.1 Internally Compensated NMOS

Considering that low quiescent current is always an overriding concern, the *high-dropout* (HDO) regulator illustrated in Fig. 10.6 employs

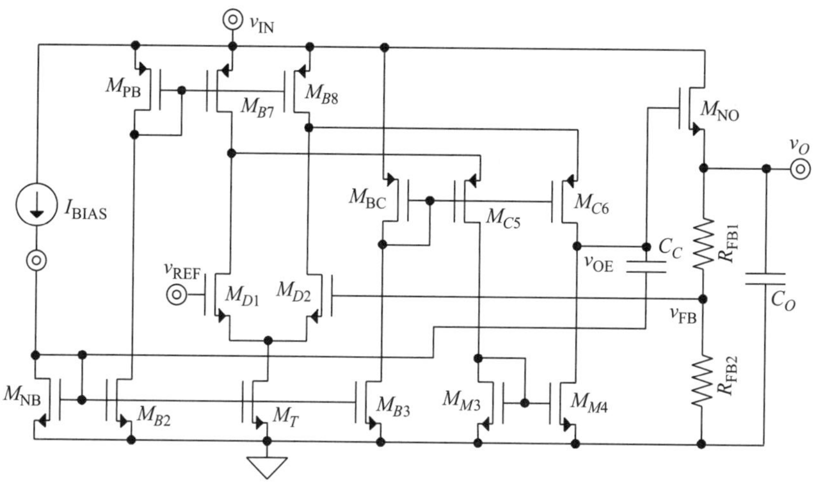

Figure 10.6 Internally compensated high-bandwidth NMOS regulator.

an NMOS power transistor M_{NO} because the gate draws no steady-state current. The design also features an N-channel differential MOS pair M_{D1}–M_{D2} that folds into a cascode pair M_{C5}–M_{C6} that feeds an N-channel mirror M_{M3}–M_{M4}. Compensation capacitor C_C ensures that the error amplifier pole p_{OE} at v_{OE} is dominant at lower frequency. M_{NB}'s $1/g_{mNB}$ diode-connected resistance current-limits C_C to establish an in-phase zero z_C that offsets the impact of the output pole p_O. Using M_{NB} to introduce a zero is more compelling than placing an additional in-phase feed-forward capacitor C_{FF} across R_{FB1} because C_{FF} requires more silicon area and components in $1/g_{mNB}$ track corresponding components in M_{NO}'s $1/g_{mNO}$ source resistance from p_O across temperature and fabrication corners. C_C connects to ground via M_{NB} instead of the input supply v_{IN} to avoid coupling input power-supply noise to v_O through the voltage-following action of M_{NO}. Output capacitor C_O, even if only at picofarads, helps source and sink fast load dumps while M_{NO} and the outer feedback loop that supports it react.

Compensation

The general compensation strategy in an *internally compensated regulator* is to first set a dominant low-frequency pole p_{OE} at the output of the error amplifier A_E, as Fig. 10.7 illustrates. p_{OE} is dominant because the resistance of the gain-setting node in A_E at v_{OE} is high at r_{ds} or r_o, or higher, and all other nodes, including the output v_O, exhibit lower resistances at roughly $1/g_m$. v_O's pole p_O is the second most dominant pole in the system because output capacitance C_O is high. So placing p_O within a decade of the system's unity-gain frequency f_{0dB} is next.

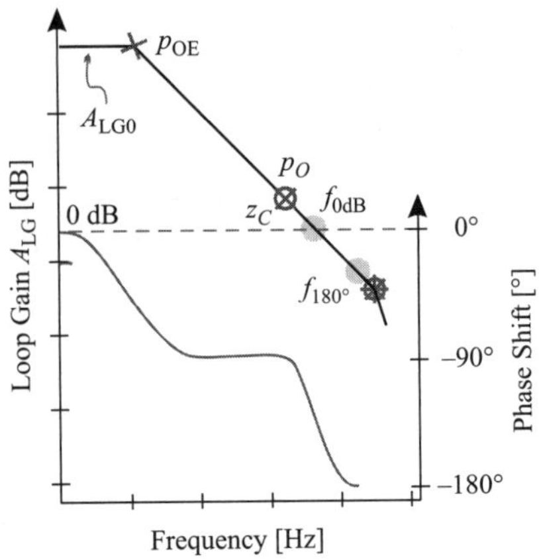

FIGURE 10.7 Frequency response of an internally compensated NMOS regulator.

The third step is to add an in-phase zero at or slightly above p_O to offset the effects of p_O. Pushing all parasitic poles within a decade above f_{0dB} and keeping inherent in-phase zeros in their vicinity to prevent phase shift across the loop from avalanching stabilizes the system. This way, loop gain drops linearly with frequency from p_{OE} to 1 with a constant gain–bandwidth product $A_{LG0}p_{OE}$, so the unity-gain frequency f_{0dB} is at the gain–bandwidth frequency, at $g_{m1}/2\pi(C_C + C_{GDNO})$:

$$\frac{A_{LG0}}{1+\dfrac{s}{2\pi p_{OE}}} \approx \frac{A_{LG0}}{\left(\dfrac{s}{2\pi p_{OE}}\right)} \approx \frac{g_{m1}r_{ds4}}{r_{ds4}(C_C + C_{GDNO})s}\Bigg|_{f_{0dB}\approx\frac{g_{m1}}{2\pi(C_C+C_{GDNO})}} \equiv 1 \quad (10.25)$$

where the gain across M_{NO} is roughly 1, M_{M4}'s r_{ds4} is considerably lower than M_{C6}'s drain resistance, and f_{0dB} is often a specification target.

C_O is intentionally as high as silicon and board space allow because C_O momentarily sinks and supplies all the load current that power transistor M_{NO} is not fast enough to satisfy during fast load dumps. As a result, the pole that C_O establishes tends to dominate over all remaining parasitic poles in the circuit. This output pole p_O happens when C_O and M_{NO}'s C_{GSNO} shunt M_{NO}'s $1/g_{mNO}$ output resistance:

$$\frac{1}{s(C_O + C_{GSNO})}\Bigg|_{p_{O(MIN)}\approx\frac{g_{mNO}}{2\pi(C_{O(MAX)}+C_{GSNO})}\geq\frac{f_{0dB}}{10}} \equiv \frac{1}{g_{mNO}} \quad (10.26)$$

where C_{GSNO} shunts $1/g_{mNO}$ to ground through the short that C_C established past p_{OE}, which is considerably below p_O. M_{NB}'s $1/g_{mNB}$

diode-connected resistance current-limits C_C to introduce an in-phase zero z_C when C_C's impedance falls below M_{NB}'s $1/g_{mNB}$:

$$\left.\frac{1}{sC_C}\right|_{z_C=\frac{g_{mNB}}{2\pi C_C}\equiv p_O} \equiv \frac{1}{g_{mNB}} \tag{10.27}$$

whose purpose is to offset the effects of p_O. With z_C nearby, p_O can precede f_{0dB} slightly.

Matching and correlating p_O and z_C across fabrication corners, and worst of all, across load is challenging because p_O, through M_{NO}'s g_{mNO}, shifts with output current i_O. For example, a two-decade variation in i_O, such as from 10 μA to 1 mA, produces a one-decade shift in g_{mNO} and p_O because g_{mNO} climbs with the square root of i_O. Mismatches between p_O and z_C exacerbate this variation. As a result, stability requirements when p_O precedes z_C by almost a decade constrain p_O and, by translation, C_O to remain roughly within a decade of f_{0dB}, which is why $p_{O(MIN)}$ must surpass $f_{0dB}/10$. Although not immediately obvious, fully dipping the phase to 180° below f_{0dB} before z_C can recover it, even if only briefly, increases the likelihood that f_{0dB} crosses the 180° phase-shift frequency $f_{180°}$ during startup conditions, when poles, zeros, and f_{0dB} transition across frequency to their steady-state locations.

Past p_{OE}, p_O, and z_C, less than 90° of phase margin is available at f_{0dB}. So to keep f_{0dB} as high as possible without requiring considerable quiescent current, parasitic poles may reside within one decade of f_{0dB}, but only if in-phase zeros hover nearby to prevent the phase response from avalanching. One of the first parasitic poles to appear after p_O is at the feedback node v_{FB} because, to reduce quiescent current flow, the resistance there, at $R_{FB1} \| R_{FB2}$, is normally moderate to high, and the capacitance present is moderate with at least one C_{GS}. This feedback pole p_{FB} asserts its influence when C_{GS2} and C_{GD2} shunt $R_{FB1} \| R_{FB2}$:

$$\left.\frac{1}{s(C_{GS2}+C_{GD2})}\right|_{p_{FB}=\frac{1}{2\pi(R_{FB1}\|R_{FB2})C_{GS2}}\approx 10 f_{0dB(MAX)}} \equiv R_{FB1} \| R_{FB2} \tag{10.28}$$

where M_{D2}'s Miller effect on C_{GD2} is negligible because the gain across M_{D2} is low and R_{FB1} is in parallel with R_{FB2} because p_O, which is at lower frequency, already shorted the resistance at the output v_O to ground.

All other resistances in the circuit are near $1/g_m$, so the next poles to consider are the mirror pole p_M and the folding poles p_{S5} and p_{S6} at the sources of cascode transistors M_{C5} and M_{C6} because their respective capacitances include at least one C_{GS}. The mirror pole p_M occurs when C_{GS3}, C_{GS4}, C_{GD4}, and C_{GD5} shunt the $1/g_{m3}$ resistance that M_{M3} establishes at approximately $g_{m3}/2\pi(C_{GS3} + C_{GS4})$:

$$\left.\frac{1}{s(C_{GS3}+C_{GS4}+C_{GD4}+C_{GD5})}\right|_{p_M\approx\frac{g_{m3}}{2\pi(C_{GS3}+C_{GS4})}\approx 5f_{0dB}} \equiv \frac{1}{g_{m3}} \tag{10.29}$$

and the mirror zero z_M offsets p_M's effects soon after at $2p_M$, which is why p_M's placement can remain slightly above f_{0dB}. Folding poles p_{S5} and p_{S6} occur when M_{C5}'s and M_{C6}'s C_{GSC}, M_{B7}'s and M_{B8}'s C_{GDB}, and M_{D1}'s and M_{D2}'s C_{GDD} shunt M_{C5}'s and M_{C6}'s $1/g_{mC}$ source resistances:

$$\left.\frac{1}{s(C_{GSC}+C_{GDB}+C_{GDD})}\right|_{p_{GS5,6}\approx\frac{g_{mC}}{2\pi(C_{GSC}+C_{GDB}+C_{GDD})}\approx 10f_{0dB}} \equiv \frac{1}{g_{mC}} \tag{10.30}$$

where the loading influence of M_{M4}'s r_{ds4} on M_{C6}'s $1/g_{mC}$ source resistance is minimal because, at these frequencies, p_{OE} already shunted r_{ds4}. Although these poles only include one C_{GS}, unlike p_M, which has two, they have no zero to help offset their effects, so they should reside a decade above f_{0dB}. Note that p_{S5} and p_{S6} affect each half of the differential signal equally, so their combined effect on the full differential response is one pole at $p_{S5,6}$.

Although the capacitance of the in-phase feed-forward capacitor C_{GSNO} in the power transistor M_{NO} is relatively high, the in-phase zero z_{NO} that C_{GSNO} introduces tends to reside at high frequency. This is because C_{GSNO}'s current must overwhelm the relatively high transconductor current that M_{NO} sources to exert any impact on the overall response:

$$\left. i_C = \frac{v_{gsNO}}{sC_{GSNO}}\right|_{z_{NO}=\frac{g_{mNO}}{2\pi C_{GSNO}}\geq 5 f_{0dB}} \equiv i_{gm} = v_{gsNO}g_{mNO} \tag{10.31}$$

Still, ensuring that z_{NO} remains slightly above f_{0dB} is important to avoid extending f_{0dB} to where several parasitic poles reside. However, if C_O is negligibly low, which is admittedly unusual, z_{NO} cancels p_O, so z_C should climb slightly above f_{0dB}. Irrespective of all this, the out-of-phase zero z_{D2} that M_{D2}'s feed-forward capacitance C_{GD2} produces when C_{GD2}'s current surpasses g_{m2}'s current must remain at least one decade above f_{0dB} because z_{D2} inverts the feedback signal:

$$\left. i_C = \frac{v_{gs2}-v_{ds2}}{sC_{GD2}} = \frac{v_{gs2}}{sC_{GD2}}\right|_{z_{D2}\approx\frac{g_{m2}}{2\pi C_{GD2}}\geq 10 f_{0dB(MAX)}} \equiv i_{gm} = v_{gs2}g_{m2} \tag{10.32}$$

Fortunately, C_{GD2} is typically low, so ensuring z_{D2} is well above f_{0dB} is not difficult.

10.3 Self-Referenced Regulators

Accuracy across temperature is paramount, and although the regulator's input-referred offset plays a key role in this respect, the reference voltage v_{REF} carries the bulk of the burden. Delegating the task to another circuit is usually prudent because other circuits in the system also rely on an accurate reference. Plus, sharing the resource reduces overhead in silicon area, power, and possibly test time. The value of a *very* accurate reference, however, is lost in a regulator because the temperature drift of the regulator's input-referred offset counters the efforts of the reference. Dedicating a reference to the regulator to correct this error is often too costly to consider, which is why reducing the offset in the first place is *a priori*.

Some applications call for self-referencing features in regulators because a global reference is not always available. Power-moding for the sake of extending battery life, for example, often involves powering down several functional units in a system at a time, including the reference, but not necessarily the power-supply circuit. Other applications demand a self-referenced regulator simply because the only available reference is off chip, and the cost associated with allocating a pin and board space to import that reference into the IC is prohibitively high.

10.3.1 Zero-Order Temperature Independence

Temperature independence in a regulator, like dropout performance, is not always a requirement, and overdesigning a circuit for sheer accomplishment raises silicon area, power, risk, and effort unnecessarily. Irrespective of what the application demands, understanding the basics of temperature dependence is important when designing circuits to survive standard and extended commercial temperature ranges of 0 to 85°C and –40 to 125°C. To this end, a Taylor-series expansion is useful because the series describes a parameter's dependence to temperature with respect to its zero, linear, quadratic, and higher-order terms:

$$p_X = \sum_{A=0}^{N} K_A T^A = K_0 + K_1 T^1 + K_2 T^2 + \cdots + K_N T^N \tag{10.33}$$

where p_X is any parameter, T is temperature, and K_0, K_1, K_2, and so on are its zero-, first-, second-, and higher-order coefficients. For reference, consider that the first- and higher-order coefficients of a perfectly temperature-independent reference are zero. By convention, as a result, when only the first- and second-order coefficients are zero, the parameter is more dependent on temperature and is said to be independent to second order. Similarly, when the first-order coefficient is nonzero and not negligible, irrespective of higher-order terms, the parameter is temperature independent to zero order, which is where self-referenced regulators without temperature-independent features fall.

Diode or base–emitter, gate–source, and zener voltages are valuable zero-order references because they produce reasonably predictable voltages across fabrication corners. In modern regulator

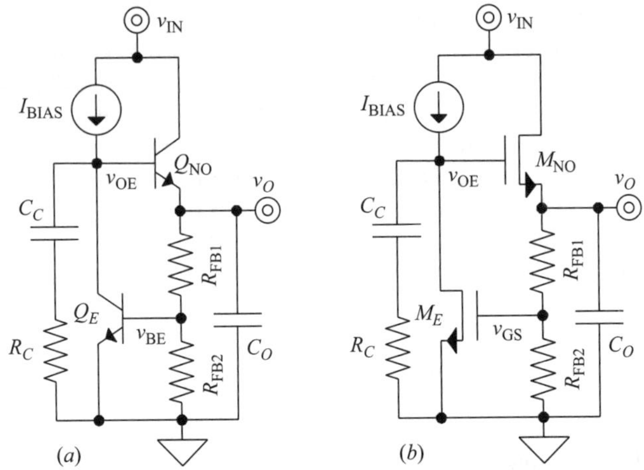

FIGURE 10.8 Self-referenced zero-order (a) NPN and (b) NMOS regulators.

applications, however, zener voltages at 5–8 V are not as practical as their base–emitter or gate–source counterparts because zener voltages often exceed the voltage-breakdown levels of driving process technologies, which nowadays often hover between 1.8 and 5 V. As a result, deriving a regulator output v_O from a base–emitter or a gate–source voltage, like Q_E's v_{BE} and M_E's v_{GS} in Fig. 10.8, and regulating its resistor-divided translation at $v_{BE}(R_{FB1} + R_{FB2})/R_{FB2}$ or $v_{GS}(R_{FB1} + R_{FB2})/R_{FB2}$ with a power transistor Q_{NO} or M_{NO} in shunt feedback is a popular means of self-referencing a regulator with zero-order temperature independence.

In the circuits shown, Q_{NO} and M_{NO} constitute high-dropout power devices with low output-resistance characteristics. Because v_O exhibits low resistance, the output pole p_O is nondominant, but since output capacitance C_O is nevertheless important and moderate to high to supply fast load dumps, compensation capacitor C_C pulls Q_{NO}'s base and M_{NO}'s gate pole p_{OE} at v_{OE} to lower frequency to ensure that p_{OE} is dominant across temperature and fabrication corners. The feedback pole p_{FB} at v_{BE} and v_{GS} is not negligibly high because Q_E's C_π and M_E's C_{GS} and feedback resistances R_{FB1} and R_{FB2} are moderate to high to keep quiescent current low. This is why R_C is there, to current-limit C_C, and in consequence, to introduce an in-phase zero that offsets the combined effects of p_O and p_{FB}. Note that reference transistors Q_E and M_E double as the error amplifier A_E in this configuration.

10.3.2 Temperature Compensation

Gate–Source Voltage

Including temperature independence raises the complexity of the circuit, although not necessarily to an extreme. Perhaps the simplest means of doing so is to find a naturally existing temperature-independent voltage or current, except such a voltage or current rarely exists in practice. The gate–source voltage v_{GS}, however, decomposes into threshold voltage v_T and drain–source saturation voltage $V_{DS(SAT)}$, where the latter includes the effects of transconductance parameter K' in its denominator:

$$v_{GS} = v_T + V_{DS(SAT)} = v_T + \sqrt{\frac{2i_D}{K'\left(\frac{W}{L}\right)}} \tag{10.34}$$

What is interesting about this decomposition is that v_T's zero-bias threshold V_{T0} and K' both fall with climbing temperatures, which means that the former has a tendency to lower v_{GS} and the latter to raise it. So since drain current i_D and width–length ratio W/L affect how much the temperature drift in K' alters v_{GS}, an optimum current-density setting $i_D/(W/L)$ exists that matches and opposes the effects of V_{T0}'s drift in v_{GS}. As a result, Fig. 10.8*b* also illustrates what could be a temperature-independent output, but only when $I_D/(W/L)$ is at its optimum point. In practice, process variations in V_{T0} and K' are appreciable at maybe ±100–±150 mV and ±15–±25 percent and uncorrelated, so v_{GS}'s drift, although reduced, remains far from zero. Still, the concept and circuit are sufficiently simple and robust to win the favor of many designers.

Bandgap Approach

Increasing temperature independence in a circuit, in a more general sense, ultimately reduces to combining temperature-dependent components whose complementary behaviors across temperature cancel. The most practical approach is to find two voltages or currents whose temperature-induced variations generally oppose one another. A *proportional-to-absolute-temperature* (PTAT) voltage v_{PTAT}, for example, which rises with temperature, can cancel the dependence of a *complementary-to-absolute-temperature* (CTAT) voltage v_{CTAT}, so their net sum v_{REF} is nearly constant across temperature, like Fig. 10.9 illustrates. Because a base–emitter's diode voltage v_{BE} mostly falls with rising temperatures, v_{BE} is usually the CTAT component in v_{REF}.

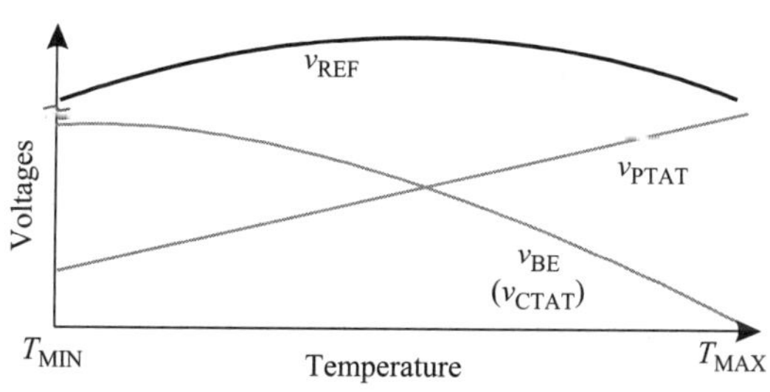

FIGURE 10.9 Complementary voltages and their temperature-independent sum.

The temperature dependence of the two components that comprise the reference need not be linear to produce a nearly constant response, just as long as they cancel each other's effects. In searching for the best candidates, looking for voltages or currents that do not vary significantly across fabrication corners, from die to die, wafer to wafer, and lot to lot is important because wide tolerances ultimately defeat the temperature-independent qualities of a supposedly accurate and predictable reference. Interestingly, the variation of diode and base–emitter voltages v_D and v_{BE} across fabrication corners is lower than their gate–source counterparts and the threshold voltages that partially define them.

The diode voltage across a PN junction generally falls with climbing temperatures at a rate that is equivalent to approximately –2.2 mV per degree. Because its linear component overwhelms higher-order terms, diode voltages often complement PTAT voltages, which is why they carry CTAT features. Note, however, that even though the compensated reference v_{REF} is mostly flat across temperature, higher-order components in v_{BE} add the characteristic curvature that v_{REF} in Fig. 10.9 exhibits, albeit at a lower scale.

Canceling the first-order component that v_{PTAT} and v_{CTAT} realize in Fig. 10.9 constitutes what is coined as *first-order compensation*. And reducing the effects of higher-order terms is *curvature correction*, where temperature drift can be less than the 20–100 ppm/°C that first-order references normally achieve when trimmed. While curvature correction is theoretically ideal, the fillers used in the plastic package to reduce the *thermal coefficient of expansion* of the plastic introduce uncompensated and uncorrelated offsets in the circuit that often render curvature-compensation efforts ineffectual. Ultimately, the additional circuit and trimming complexity required to improve the performance of a first-order reference by a small margin is oftentimes difficult to justify.

The *bandgap* reference, as engineers call it, is a popular circuit that uses, at its core, a PTAT voltage v_{PTAT} to compensate the CTAT behavior inherent in a diode or base–emitter voltage v_D or v_{BE}. The name bandgap arises because the zero-order temperature term of the diode voltage is the bandgap voltage of silicon, which is why the diode voltage is 1.2 V at 0 Kelvin. Additionally, and perhaps more importantly, because the PTAT voltage used to compensate v_D or v_{BE} has no zero-order term, which is to say v_{PTAT} is zero at 0 Kelvin, the reference voltage v_{REF} of a bandgap circuit also reduces to roughly 1.2 V, since canceling first- and higher-order terms leaves v_D's zero-order component intact. Generally, the same applies to diode- and PTAT-derived currents, except for the transresistance translation of the bandgap voltage into a resistor-defined current. Similarly, diode and PTAT fractions combine to yield references whose values fall below the bandgap mark of 1.2 V, realizing sub-bandgap values in the 0.2–1-V range. Irrespective of what ratio or derivative the circuit ultimately combines, using diode and PTAT components in any form to produce references follows the bandgap approach, even when applied to the core of a regulator.

Although the diode voltage v_D without its PTAT counterpart v_{PTAT} may be of little consequence, v_D itself can generate its own compensating PTAT voltage. This trait results because the exponential expression for diode current i_D includes the thermal voltage V_t, which climbs linearly with temperature:

$$i_D = I_S\left[\exp\left(\frac{v_D}{V_t}\right) - 1\right] \approx I_S \exp\left(\frac{v_D}{V_t}\right) \tag{10.35}$$

which solving for v_D yields

$$v_D \approx V_t \ln\left(\frac{i_D}{I_S}\right) \tag{10.36}$$

where reverse-saturation current I_S is proportional to the cross-sectional area of the PN junction, or the emitter–base junction in the case of BJTs. As a result, the difference of two diode voltages derived from matched but ratioed areas carrying matched and maybe ratioed currents is predictably, reliably, and consistently PTAT and linear across temperature:

$$v_{\text{PTAT}} = \Delta v_D = v_{D1} - v_{D2} \approx V_t \ln\left(\frac{i_{D1} I_{S2}}{I_{S1} i_{D2}}\right) = V_t \ln(C_I C_A) \propto T^1 \tag{10.37}$$

where i_{D1} is C_I times greater than i_{D2} and I_{S2}'s area A_2 is C_A times larger than I_{S1}'s A_1.

The important conclusion to draw from this discussion is that a difference in diode or base–emitter voltages Δv_D or Δv_{BE} with proportionately ratioed current densities i_D/A_D produces a PTAT voltage v_{PTAT}. The circuit core shown in Fig. 10.10, for instance,

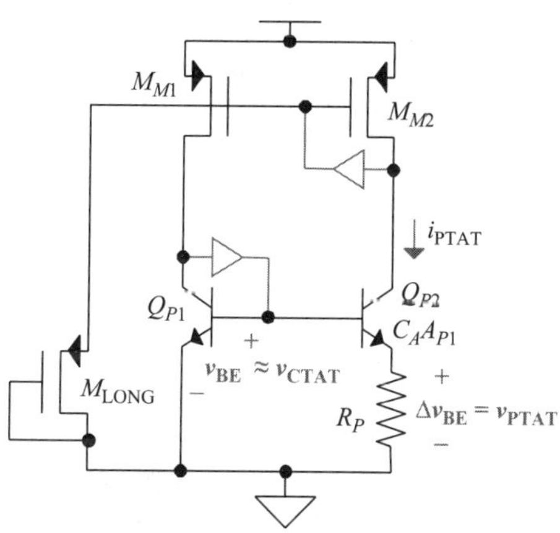

FIGURE 10.10 PTAT-generator core.

exploits this basic theorem by inserting a resistor R_P in the path of what would have otherwise been two equal base–emitter voltages. This way, the voltage across R_P is the difference between Q_{P1}'s and Q_{P2}'s base–emitter voltages. The resulting Δv_{BE} is PTAT only because the current mirror M_{M1}–M_{M2} attached to Q_{P1} and Q_{P2} forces their current densities to remain constant ratios of one another. This same PTAT core is the basis for most bias-current generators today.

Although not always the case, diode-connecting one of the bipolar transistors and one of the mirroring transistors is typical. A more general implementation of the core diode-connects them with other circuits whose purpose can be to level-shift the voltage or raise the gain of the regulating loop. For improved matching performance, the current-mirror gain ratio C_I is normally 1 because lower spreads in gain lead to reduced mismatches in currents. For similar reasons, engineers use equally sized BJT segments to define the emitter-area ratio C_A. A ratio of 8 between the matching BJTs is popular because surrounding one device with eight others produces a modular and common-centroid array that averages and matches two-dimensional gradients across the die. A ratio of 2 is also popular when accuracy is not as important because the resulting layout is still efficient with a common center of mass, although some of its modularity and cancellation of peripheral effects suffer.

Interestingly, the PTAT generator is bi-stable because the circuit is just as content to carry a PTAT current i_{PTAT} that is equal to v_{PTAT}/R_P as it is conducting no current at all. In other words, the mirror sources ratioed currents and the Δv_{BE} circuit ensures that the currents are PTAT or zero, if the mirror is off, in which case the Δv_{BE} circuit does not operate as intended. To avoid this zero-current state, a startup circuit typically samples this current or an equally telling voltage, compares it against a nominal threshold, and ensures that it stays above an acceptable threshold by sourcing or sinking additional current into the circuit. Sometimes this startup circuit is nothing more than a small and constantly flowing current into a diode-connected transistor, which is the purpose of the long-channel MOSFET M_{LONG} in Fig. 10.10. This way, the mirror current cannot be zero, and the positive-feedback action of the mirror and the BJT pair to which the mirror connects latches the current to the desired state.

Bandgap Conversion

The only difference between a zero-order diode-derived regulator and its first-order counterpart is the inclusion of a PTAT term in the output v_O. Consider, for example, how the self-referenced zero-order regulator in Fig. 10.8*a* and the PTAT current-generator core in Fig. 10.10 combine to produce the first-order circuit in Fig. 10.11. The basic idea is to include a PTAT voltage $\Delta v_{BE}R_{FB1}/R_P$ in the output v_O, which is what Q_{P3} achieves by sinking a mirrored version of i_{PTAT} from Q_E's base node. This way, R_{FB2} drops v_{BE} and R_{FB1} conducts both

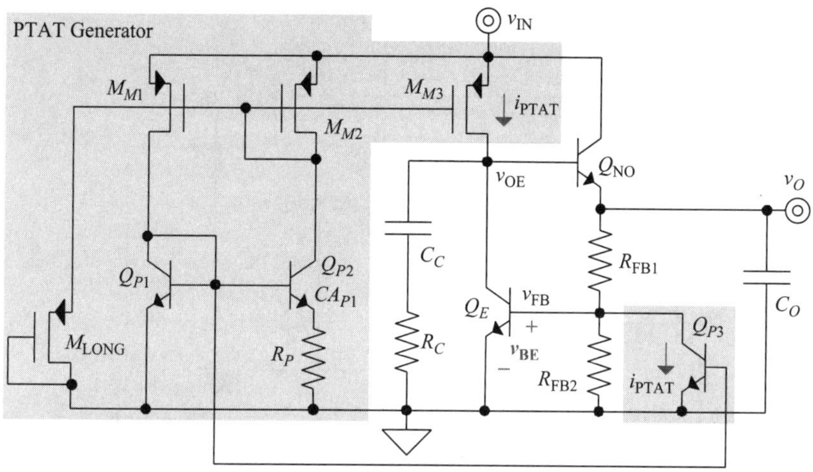

FIGURE 10.11 Bandgap-converted NPN regulator.

R_{FB2}'s v_{BE}/R_{FB2} current and Q_{P3}'s i_{PTAT} to establish an output voltage v_O that incorporates v_{BE}'s CTAT and Δv_{BE}'s PTAT components:

$$\begin{aligned} v_O &= v_{BE} + \left(\frac{v_{BE}}{R_{FB2}} + i_{PTAT}\right)R_{FB1} = v_{BE} + \left(\frac{v_{BE}}{R_{FB2}} + \frac{\Delta v_{BE}}{R_P}\right)R_{FB1} \\ &= v_{BE}\left(1 + \frac{R_{FB1}}{R_{FB2}}\right) + \Delta v_{BE}\left(\frac{R_{FB1}}{R_P}\right) \end{aligned} \quad (10.38)$$

In this case, the PTAT-generator core also mirrors and biases the error-amplifier and reference-transistor Q_E to conduct i_{PTAT} via mirroring device M_{M3}.

Bandgap Integration

Another variation of the same theme is to integrate the PTAT core into the feedback network of the self-referenced zero-order NMOS regulator in Fig. 10.8*b*, like Fig. 10.12 shows. As before, the core pulls a PTAT current from Q_E's base node that R_{P2} carries to establish a PTAT voltage v_{PTAT} above Q_E's base–emitter voltage v_{BE}. This way, v_{BE} and v_{PTAT} in v_O add and compensate each other's drifts across temperature:

$$v_O = v_{BE} + i_{PTAT}R_{P2} = v_{BE} + \Delta v_{BE}\left(\frac{R_{P2}}{R_P}\right) \quad (10.39)$$

In this case, resistors R_{P1} and R_{P2} in the core match and constitute a current mirror because the voltages across them match at $v_O - v_{BE}$, where Q_{P1}'s and Q_E's base–emitter voltages match because Q_{P3}, M_{PB}, and M_{BE} mirror, fold, and force Q_E and Q_{P1} to carry the same current densities.

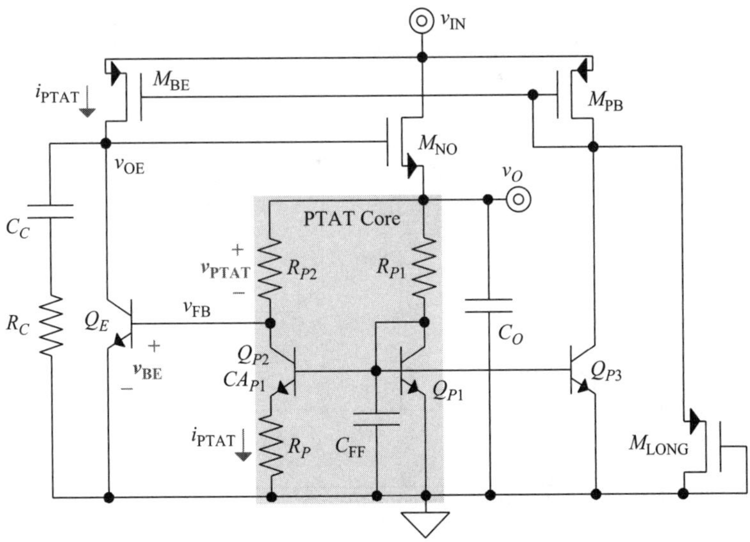

FIGURE 10.12 Bandgap-integrated NMOS regulator.

Long-channel startup PFET M_{LONG} pulls current from M_{PB}'s diode connection to ensure that M_{BE}, and as a result, M_{NO} never shut.

With respect to stability, C_C pulls v_{OE}'s pole p_{OE} toward lower frequency to ensure that p_{OE} is dominant across the feedback loop. R_C then current-limits C_C to establish an in-phase zero z_C that mitigates the avalanching effects that the output pole p_O and the feedback pole p_{FB} at Q_E's base v_{FB} have on phase response near the system's unity-gain frequency. C_O, like before, sources the load current that the regulator is not fast enough to supply during fast load dumps. PTAT-core components R_{P1}, Q_{P1}, and Q_{P2} feed forward out-of-phase signals from v_O to Q_E that produce a right-half-plane zero z_{FF} when Q_{P2}'s g_{m2} current exceeds R_{P2}'s counterpart. To push z_{FF} to higher frequency, R_P is below Q_{P2}, to degenerate Q_{P2}'s current. Plus, C_{FF} shunts some of this feed-forward energy to ground, further pushing z_{FF} toward higher frequency under the guise of a feed-forward filter pole p_{FF}.

The dual feed-through path in the core actually adds design flexibility because reversing the polarity of the main path amounts to moving R_P to the side of Q_{P1}, below Q_{P1}. This polarity inversion is exactly what a low-dropout PMOS transformation requires, as Fig. 10.13 demonstrates, because a P-type power transistor introduces an inversion that its N-type counterpart does not. The only other variation to the circuit relates to compensation, as C_C now enjoys the benefits of Miller multiplication, and R_C shifts the out-of-phase feed-forward zero across M_{PO} to the left half of the s plane. Q_{P1} and Q_{P2}'s base feedback pole p_{FB} is now part of the main loop, except p_{FB} resides

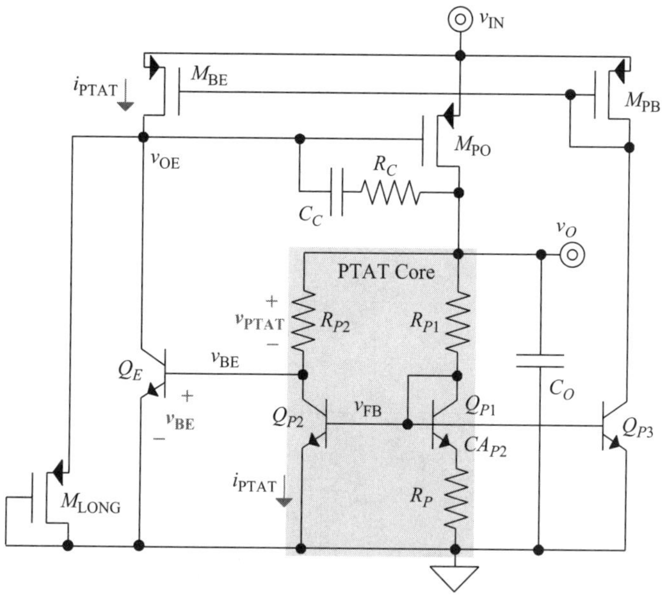

FIGURE 10.13 Bandgap-integrated PMOS regulator.

at high frequency because Q_{P1}'s $1/g_m$ diode-connected resistance is low and R_p is low to moderate at 10–60 kΩ. Q_{P2}'s C_μ also introduces an out-of-phase feed-forward path, but its resulting right-half-plane zero is also at high frequency because C_μ is low and Q_{P2}'s exponential behavior produces a high transconductance.

The basic idea of the PTAT core in Fig. 10.10 is to insert a resistor R_p in a voltage loop that includes the base–emitter voltages of two BJTs, so the PTAT-voltage difference Δv_{BE} appears across R_p. Where in the loop R_p lies does not change Δv_{BE}, as long as current densities through the transistors remain ratioed. Figure 10.14 achieves the same result by shifting R_p out of the emitter-degenerating position into the base junction of the two BJTs, which keeps R_p in the loop, but now between two bases. R_{P2} therefore carries R_p's PTAT current and drops an amplified version of Δv_{BE}'s v_{PTAT} across R_{P2} to offset to first order the CTAT tendencies of Q_{B1}'s base–emitter voltage v_{BE} in v_O and produce a bandgap response:

$$v_O = v_{BE} + i_{PTAT}(R_P + R_{P2}) = v_{BE} + \Delta v_{BE}\left(1 + \frac{R_{P2}}{R_P}\right) \tag{10.40}$$

Here, M_{M1}–M_{M2}, M_{PO}, and R_{B1} comprise the noninverting buffer that diode-connects Q_{P1}. Mirror Q_{B1}–Q_{B2} outputs the tail current that ensures that the total current flowing through the differential pair Q_{P1}–Q_{P2} is PTAT and constant, and mirror M_{M1}–M_{M2} forces that

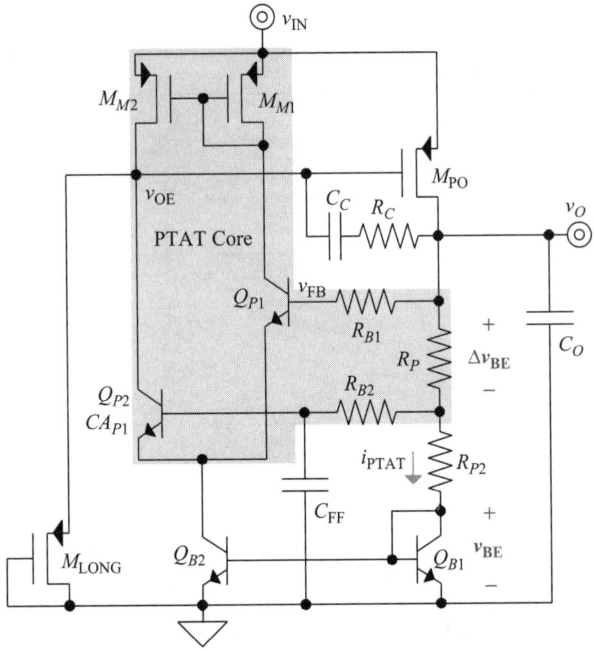

FIGURE 10.14 PTAT-generating differential pair in a PMOS regulator.

current to split equally between the two BJTs. Long-channel PFET M_{LONG}, as before, keeps the circuit from shutting. To do this, M_{LONG} pulls current from v_{OE} to induce M_{PO} to conduct current and start the circuit in the event that nothing else is on.

The error amplifier in the feedback loop now reduces to the differential pair Q_{P1}–Q_{P2} and the P-type mirror load M_{M1}–M_{M2}, except only Q_{P1} conducts the desired inverting feedback signal back to the power transistor M_{PO}. Fortunately, R_P drops some of the small signals present at v_O to reduce the out-of-phase feed-forward signal that reaches Q_{P2}. This effectively pushes the right-half-plane zero z_{FF} that Q_{P2} establishes to higher frequency. R_{B2}–C_{FF}'s pole p_{FF} further filters the feed-forward path to reduce its overall impact on the feedback system. Since base current produces a voltage across R_{B2}, R_{B1} is at the base of to Q_{P1} to match R_{B2}'s voltage and cancel any effect that R_{B2} would have otherwise had on the modified PTAT core. Besides amplifying the PTAT contribution in v_O, R_{P2} also sets the voltage across the collector–emitter terminals of the tail-current transistor Q_{B2} because Q_{P2}'s and Q_{B1}'s base–emitter voltages are similar. R_{P2}'s voltage $i_{PTAT}R_{P2}$ should therefore surpass Q_{B2}'s minimum collector–emitter voltage $V_{CE(MIN)}$ across the entire temperature range. C_C, R_C, and C_O, as before, pull v_{OE}'s pole p_{OE} to lower frequency, pull C_C's feed-forward zero z_C to the left half of the s plane, and supply fast load-dump currents, respectively.

10.4 Performance Enhancers

Extended operational life is undoubtedly a driving demand for a growing number of portable, battery-powered devices, such as self-powered wireless microsensors, smart cellular phones, tablets, biomedical implants, and so on. Aside from lowering dropout voltages, extending operational life also amounts to reducing quiescent-current flow, which unfortunately lowers load range, lengthens response time, degrades accuracy, and deteriorates power-supply rejection. Not surprisingly, improving these basic performance parameters is the subject of much attention in the industry, especially within the context of total on-chip integration. The following subsections therefore explore ways of enhancing the regulator to better address these areas.

10.4.1 Power Transistor

Many of the needs of the regulator emanate from the power transistor because it not only sets the driving load range and corresponding dropout voltage of the circuit but also presents a high parasitic capacitance that slows the response of the feedback loop to rapidly changing loads. Raising the driving range of the regulator without slowing the loop amounts to boosting the gate drive of the power switch without increasing its size. Two ways of achieving this objective under similar supply constraints are: (1) to reduce the effective threshold voltage of the transistor when supplying substantial load currents and (2) to add a supplementary, out-of-the-loop slave transistor that sources a significant fraction of the total load.

Bulk Boost

Although a low-threshold-voltage device can source higher currents, it also leaks more current during standby conditions, especially at high temperatures, which drains the battery and shortens the product's single-charge life. Alternatively, forward-biasing the PN junction across a PMOS transistor's source–bulk terminals *on demand* lowers the threshold voltage of the PMOS device $|v_{TP}|$ only when needed. This way, the lower $|v_{TP}|$ raises the transistor's *gate drive* $v_{SG} - |v_{TP}|$, or v_{SGT} for short, when the load is heavy while simultaneously retaining the low-leakage characteristics of a higher $|v_{TP}|$ when the load is light:

$$v_{SGT} = v_{SG} - |v_{TP}(v_{SB})| \propto -\sqrt{v_{SB}} = -\sqrt{i_{DS}R_{B1}} \propto -\sqrt{i_{DO}} \qquad (10.41)$$

To realize this on-demand feature, M_{PS} in Fig. 10.15 senses and mirrors a diminutive fraction of M_{PO}'s drain current i_{DO} into Q_{S2}, which buffers the current into the mirror that M_{M1} and M_{M2} realize. M_{M1}–M_{M2}'s output then pulls M_{PS}'s current i_{DS} from R_{B1} and R_{B2} to establish a source–bulk voltage v_{SB} across R_{B1} that rises linearly with

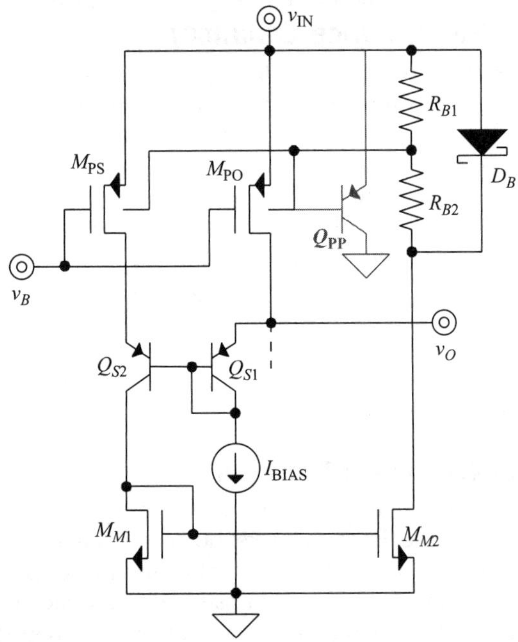

FIGURE 10.15 Threshold-reducing bulk-driver circuit for PMOS regulators.

M_{PO}'s i_{DO} and is nearly zero when i_{DO} is low. The purpose of the base-coupled pair that Q_{S1} and Q_{S2} implement is to ensure that M_{PS}'s and M_{PO}'s source–drain voltages are similar. This way, they both operate in the same region—both in saturation or both in triode.

Driving the bulk with a gate-derived signal v_B also constitutes an in-phase, feed-forward path across M_{PO} in the feedback loop of the regulator. The feed-forward bulk gain, however, normally remains well below its gate-driven counterpart because source–bulk transconductance g_{mb} is inherently lower than its source–gate counterpart g_m. Plus, Schottky diode D_B in the bulk-driver circuit loads and reduces the feed-forward gain, which means the in-phase zero does not typically surface. Interestingly, sensor Q_{S1}–Q_{S2}, mirror M_{M1}–M_{M2}, and M_{PO} establish an inverting feedback loop about v_O. Here, however, M_{PS} emitter-degenerates Q_{S2}, so Q_{S2} together with M_{M1}–M_{M2} and M_{PO}'s g_{mb} do not amplify small signals enough to influence in any significant way the feedback dynamics that regulate v_O about the reference.

Functionally, the current sensor Q_{S1}–Q_{S2} ensures that M_{PS}'s and M_{PO}'s drain–source voltages match to first order, so their mirroring ratio remains unchanged through dropout conditions, when M_{PO} enters the triode region. Mirror M_{M1}–M_{M2} reproduces M_{PS}'s sensed current i_{DS} and pulls it from the bulk-driver circuit. R_{B1} and R_{B2} then voltage-divide D_B's Schottky diode voltage to limit the extent to which R_{B1} forward-biases M_{PO}'s source–bulk terminals. Limiting v_{SB}

well below a diode voltage is important to keep the parasitic vertical PNP BJT Q_{PP} present in M_{PO} from activating and sinking current from the input supply v_{IN} into the substrate. Raising M_{PO}'s source–bulk voltage in this way can extend the reach of a 2000/1 PMOSFET with a 0.9-V threshold voltage that derives power from a 1.2-V input supply from 20 μA to about 500 μA. Likewise, applying 500 mV across this same transistor's source–bulk terminals can reduce the dropout voltage at 200 μA from 60 to 20 mV.

Master–Slave Approach

Another way of augmenting the driving capabilities of the low-dropout PMOS regulator without increasing the size of the power transistor M_{PO} in the control loop is by adding an auxiliary power device M_{PX} outside the loop. The basic idea is to introduce a *slave* transistor M_{PX} that derives its command unobtrusively from M_{PO}. This way, M_{PX}'s gate or base is outside the main loop, and the current M_{PX} supplies supplements that of M_{PO} to drive the load. The slave transistor therefore raises the total output current of the regulator without presenting additional capacitance to M_{PO}'s driving buffer—in other words, without slowing the response of the system to fast load dumps.

Linear Slave Figure 10.16 illustrates a linear embodiment of the foregoing master–slave approach. In this case, base-coupled differential pair Q_{P1} and Q_{P2} series-mixes M_{PO}'s source voltage with that of M_{PX} to impress across R_X the voltage dropped across R_S. Because R_X is A_I times lower than R_S, R_X's current i_{DX} is A_I times greater than M_{PO} and R_S's i_{DO}:

$$i_{DX} = \frac{v_{RX}}{R_X} \approx \frac{v_{RS}}{R_X} = \frac{i_{DO}R_S}{R_X} = A_I i_{DO} \tag{10.42}$$

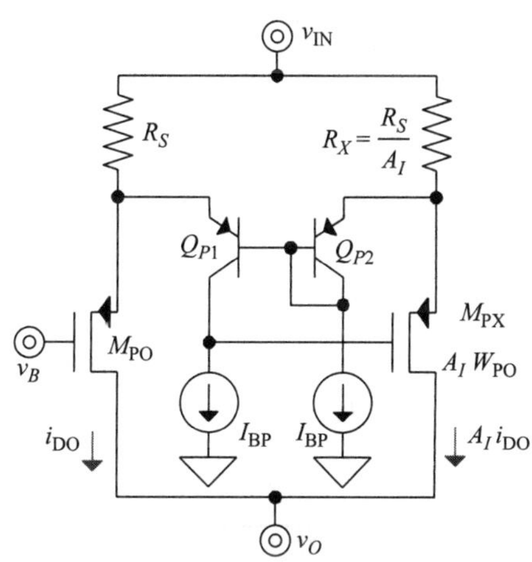

Figure 10.16 Linear master–slave power PMOS-transistor circuit.

R_X's v_{RX} mirrors R_S's v_{RS} because Q_{P1} and Q_{P2} feed the error difference between the two voltages to M_{PX}'s gate, whose feedback action reduces this difference to the extent that the gain across the inverting feedback loop allows. In other words, a separate feedback loop from the one that regulates v_O senses M_{PO}'s i_{DO} to output a proportionately higher linear translation of i_{DO} into the load.

Although the voltages across series resistors R_S and R_X raise the effective dropout voltage of the regulator, their impact need not be problematic, especially when using metallic resistors whose resistances are low. This nominal loss in silicon efficiency does not negate the benefits of higher bandwidth because M_{PO} and M_{PX} together supply the current of a larger device while only loading the loop with the capacitance of a smaller transistor M_{PO}. The elegance of the circuit, in fact, outside of its simplicity, rests on how unobtrusive the circuit is to the main feedback path of the regulator. The slave feedback loop is even self-compensating because only one low-frequency pole exists, and that is at the gate of M_{PX}.

Nonlinear Slave The relationship between auxiliary current i_{DX} and loop current i_{DO} need not be linear for the master–slave approach to work, as long as the main loop continues to regulate i_{DO} about whatever point i_{DX} establishes. In fact, i_{DX} need not even have a regulated component. With this in mind, the supplementary power transistor M_{PX} in Fig. 10.17 shifts the biasing current that flows through the main power transistor M_{PO} by sourcing a relatively large fraction of the total steady-state load. Here, M_{PX} engages when load current i_L rises to the extent that v_O falls past a window limit of v_{REF}. Once on, M_{PX} sources a steady-state current I_{DX} into v_O about which M_{PO} and its

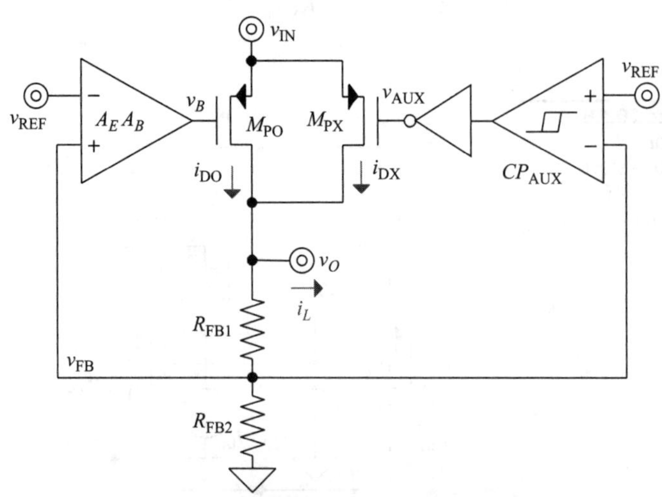

FIGURE 10.17 Nonlinear master–slave power PMOS-transistor circuit.

feedback loop can now regulate v_O to supply what remains of the load. In other words, M_{PO} supplies the difference between i_L and i_{DX}.

As in the linear case, the advantage of this circuit is that the regulator can now source the current of two power transistors M_{PO} and M_{PX} at the speed and bandwidth of one, of M_{PO}. The auxiliary path does present another inverting feedback path to v_O, but its small-signal gain is close to zero because hysteretic comparator CP_{AUX} latches M_{PX} on or off, not allowing M_{PX} to change with small variations in v_O. M_{PX} latches to its other state only when the load experiences a large-signal variation and v_O manages to reach CP_{AUX}'s other window limit. As a result, the circuit responds quickly at the speed of M_{PO} in the presence of load dumps that do not require M_{PX} to switch state. Under extreme conditions, however, when load dumps exceed M_{PO}'s working range, M_{PX}'s response and transition time slow the circuit, which is why CP_{AUX} and its driving buffer must be fast. Accelerating M_{PX}'s transition is especially important in portable applications where, in an effort to save energy and extend battery life, large sections of the system can engage and shut down quickly, and in the process present severe load dumps to the linear regulator.

10.4.2 Buffer

Bulk Feedback

Driving the base or gate of a large power transistor quickly with little to no current is one of the most challenging aspects of designing a regulator. To this end, the basic aim of the feedback loops presented in the buffer section of Chap. 9 is to reduce the driving impedance of emitter- and source-following buffers while dynamically adjusting their bias point on demand to optimally accommodate various load levels. Applying the feedback signal through the bulk, in the case of bulk-isolated MOS source followers, like Fig. 10.18 shows with PMOS buffer M_{BUF}, is perhaps an efficient means of lowering the output impedance without degrading the swinging limits of the buffer or considerably raising the number of transistors in the circuit. In fact, allowing the PN junction across the source–bulk terminals of M_{BUF} to

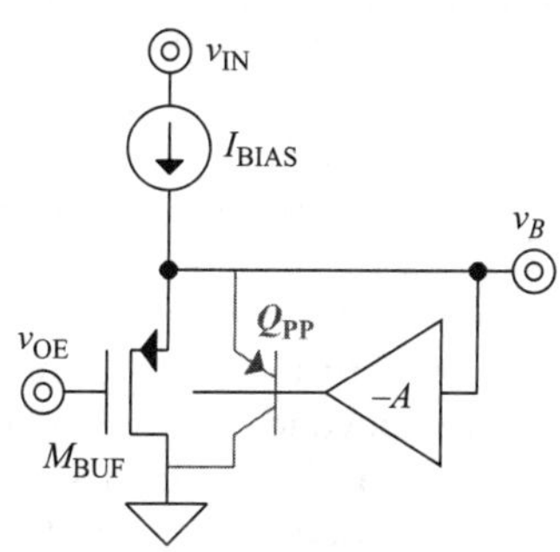

FIGURE 10.18 Bulk feedback in source-following buffers.

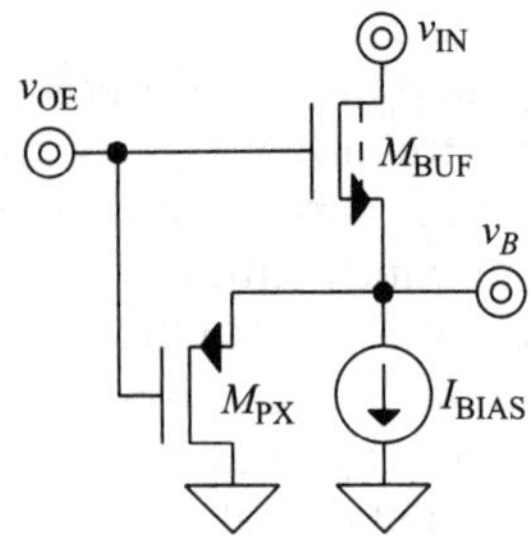

FIGURE 10.19 Voltage-following NMOS buffer with a complementary P-channel speed-up transistor.

forward-bias reduces M_{BUF}'s threshold voltage $|v_{TP}|$ to extend v_B's lower swing limit closer to ground. A positive voltage across the junction also activates the parasitic vertical BJT Q_{PP} attached to M_{BUF}'s source to help M_{BUF} sink even more current from the gate or base of the power transistor that M_{BUF} drives. Note, however, that Q_{PP} steers current into the substrate. Surrounding M_{BUF} with substrate contacts is therefore important here, to collect as much of this current as possible before it spreads through the substrate.

Speed-Up Transistor

Perhaps a more efficient means of improving the transient response of the buffer is to include a speed-up transistor that conducts and channels current to the base or gate of a large power transistor only during slewing conditions, on demand, when the buffer needs it most. Speed-up transistor M_{PX} in Fig. 10.19 does just this. Under steady-state conditions, v_B is a gate–source voltage below v_{OE}, so PMOS transistor M_{PX} is off. Although v_{OE} falls and source-following buffer M_{BUF} shuts in response to a load dump at the bandwidth of the loop, the large parasitic capacitance present at v_B slews with bias current I_{BIAS}, undesirably extending the response time of the power transistor. During severe slew-rate conditions, however, v_B lags v_{OE} so much that v_{OE} falls below v_B enough to engage M_{PX}, whose current helps pull v_B down faster. In other words, M_{PX} accelerates the slew-rate limited transition, and as a result, the overall response time of the circuit.

The only limitation of the speed-up transistor is the same feature that keeps it off in steady state: its threshold voltage $|v_{TP}|$ and that of the NMOS buffer v_{TN}. The problem is that v_{OE}, which is a gate–source voltage away from v_B in steady state, must traverse through at least two threshold voltages $|v_{TP}|$ and v_{TN} to engage M_{PX}. This means that, for M_{PX} to help, the load dump must be large and fast enough to induce the error amplifier A_E to swing v_{OE} more than two threshold voltages. Replacing the buffer transistor with a native NFET, whose threshold v_{TN} nears zero, reduces this burden. Still, v_{OE} must nonetheless swing approximately 200–400 mV beyond one $|v_{TP}|$ to offset M_{BUF}'s saturation voltage $V_{DS(SAT)}$ and possible bulk effects, if source–bulk terminals do not share a common connection. Further reductions in this voltage

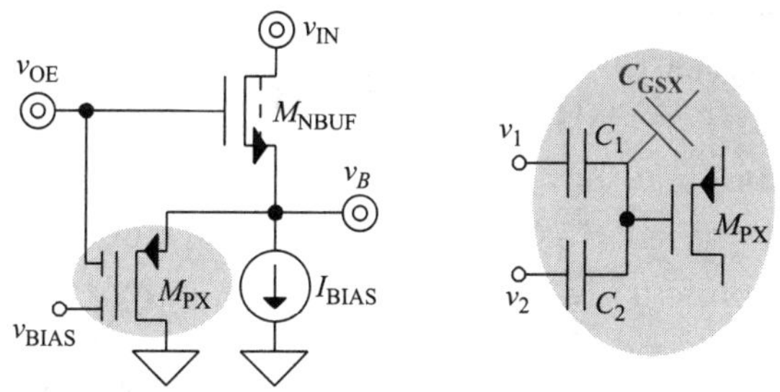

FIGURE 10.20 Voltage-following NMOS buffer with a capacitively coupled speed-up transistor.

excursion call for lower $|v_{TP}|$ values, which is where more expensive process technologies that can afford the mask-set composition that avails such devices claim superiority.

One way to lower $|v_{TP}|$ is through the bulk effect, by forward-biasing M_{PX}'s source–bulk terminals. But to keep the parasitic vertical BJT attached to M_{PX}'s source from conducting ground current in steady state, the forward-bias voltage should be either low or on demand. Pre-biasing M_{PX}'s source–gate voltage v_{SGX} to half its threshold-voltage point and subsequently coupling v_{OE} capacitively into M_{PX}'s gate, as Fig. 10.20 shows, achieves a similar result without the danger of inadvertently activating M_{PX}'s parasitic BJT. The idea is to precharge M_{PX}'s gate–source capacitance C_{GSX} with either a switched-capacitor network or a floating-gate input, where the gate voltage in the latter is a divider combination of its inputs:

$$v_{GX} = \frac{v_1 Z_{C2}}{Z_{C1} + Z_{C2}} + \frac{v_2 Z_{C1}}{Z_{C1} + Z_{C2}} = v_1\left(\frac{C_1}{C_1 + C_2}\right) + v_2\left(\frac{C_2}{C_1 + C_2}\right) \tag{10.43}$$

and capacitors C_1 and C_2 are considerably higher than M_{PX}'s C_{GSX}. The challenges with the floating-gate tactic are: (1) generating a bias voltage v_{BIAS} that consistently keeps v_{SGX} at roughly $0.5|v_{TP}|$ in steady state and (2) ensuring that the initial charge on the gate, before applying v_{OE} and v_{BIAS}, is zero.

10.4.3 Loop Gain

Another challenge in a linear regulator is the impact that limited loop gain A_{LG} has on load and line regulation and power-supply rejection. The problem is that low quiescent currents restrain parasitic poles to relatively low frequencies because the $1/g_m$ resistances that low currents define rarely fall below 20–30 kΩ. For reference, consider that $1/g_m$ for a 5-μm/1-μm transistor running 0.5 μA with a transconductance parameter of 100 μA/V² is roughly 45 kΩ, and the pole it produces when confronted with 0.5 pF is $g_m/2\pi C_C$, or approximately 7 MHz. As a result, A_{LG} in the presence of an in-band

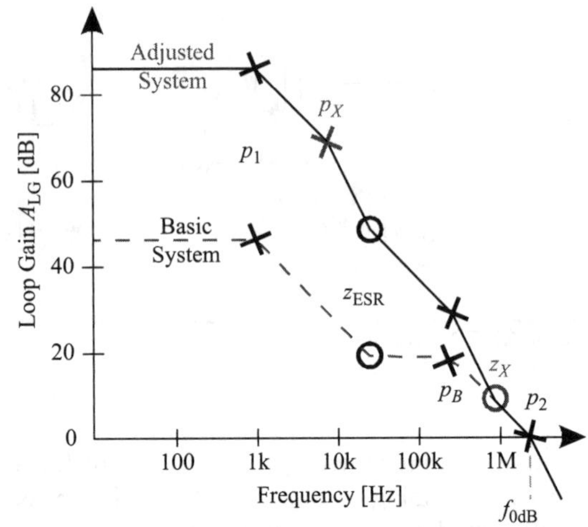

FIGURE 10.21 Loop gain with and without an additional pole–zero pair.

equivalent-series-resistor (ESR) zero z_{ESR} and its accompanying bypass pole p_B, whose combined effect is to extend the system's unity-gain frequency f_{0dB} like Fig. 10.21 illustrates, should not exceed 40–50 dB. Exceeding this boundary would otherwise extend f_{0dB} into the parasitic-pole region, which for many systems is between 5 and 10 MHz.

Another way to view this problem is through the impact that z_{ESR} and p_B have on the falling rate of the loop gain. Because z_{ESR} precedes p_B, the pair decelerates A_{LG}'s overall fall across frequency, requiring a wider band of frequencies for the gain to reach 1. Interestingly, accelerating this falling rate allows A_{LG} to start falling from a higher gain and still reach 1 at the same f_{0dB}. Introducing a pole–zero pair well below f_{0dB}, as p_X and z_X demonstrate in Fig. 10.21, achieves this objective without compromising the stability of the system. The idea is for p_X to accelerate the fall and z_X to recover the phase lost with p_X before reaching f_{0dB}. The main challenge with this tactic is that z_{ESR} and p_B are, for all practical purposes, unpredictable. This means f_{0dB} varies significantly across temperature, operating conditions, and fabrication corners, and sufficient design margin must therefore exist to accommodate all possible values of f_{0dB}.

Considering that power is a precious commodity, leveraging the circuit that already exists to introduce the aforementioned pole–zero combination is ideal. In an output-compensated low-dropout regulator, for instance, A_E's output v_{OE} normally constitutes a parasitic pole p_E in the system that could easily pose as the pole sought in the pole–zero pair because the resistance at v_{OE} is normally high. Trailing a zero past p_E amounts to feeding forward in-phase signals to v_{OE}, or around v_{OE} to v_O, or current-limiting the capacitor that shunts v_{OE}.

Figure 10.22 illustrates two embodiments of what might deceptively look like feed-forward zeros, but in practice are current-limiting zeros. Although C_Z feeds forward in-phase signals from the

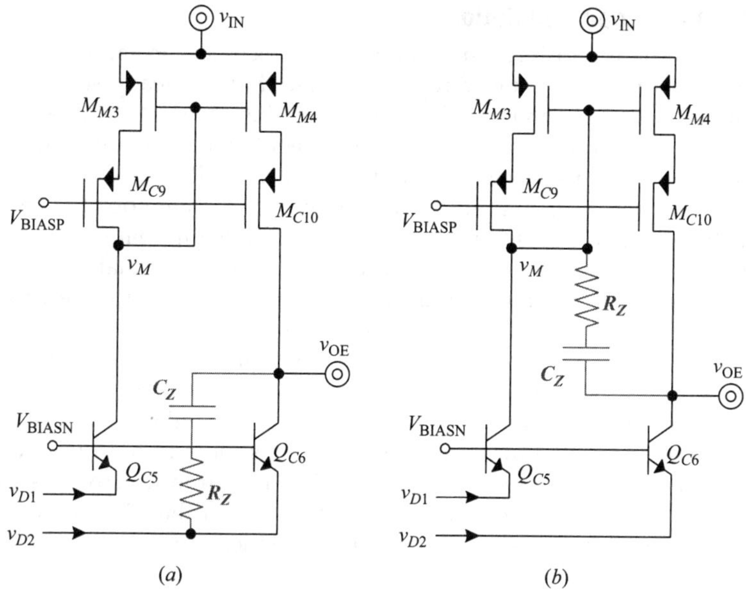

FIGURE 10.22 Integrating a pole–zero pair into the error amplifier.

differential pair to v_{OE}, C_Z's current does not usually surpass that of its competing current buffer Q_{C6} in Fig. 10.22*a*, or that of the mirroring transistor and current buffer M_{M4} and M_{C10} in Fig. 10.22*b*, so no feed-forward zero results. The reason for this is that Q_{C6}'s and M_{M4}'s currents do not drop until their base–emitter and source–gate capacitors shunt their respective terminals at high frequency. In fact, what often happens here is that C_Z first shunts v_{OE}'s output resistance R_{OE}, R_Z, and Q_{C6}'s $1/g_m$ or M_{M3}'s $1/g_m$ past p_X:

$$\left.\frac{1}{sC_Z}\right|_{p_X \approx \frac{1}{2\pi R_{OE} C_Z}} = R_{OE} + R_Z + \frac{1}{g_m} \approx R_{OE} \tag{10.44}$$

R_Z and $1/g_m$ then limit C_Z's current to arrest the effects of p_X with what amounts to an in-phase zero z_X:

$$\left.\frac{1}{sC_Z}\right|_{z_X \approx \frac{1}{2\pi\left(R_Z + \frac{1}{g_m}\right)C_Z}} = R_Z + \frac{1}{g_m} \tag{10.45}$$

The purpose of R_Z in this configuration is to decouple the design requirements of g_m from z_X. This way, $1/g_m$ need not be high to establish the desired z_X, which would otherwise pull the parasitic pole at Q_{C6}'s emitter in Fig. 10.22*a* or M_{M3}'s gate in Fig. 10.22*b* to lower frequency.

10.4.4 Load Regulation

An alternative to increasing loop gain, which compromises phase margin for the sake of load regulation, is to shift the steady-state output v_O with just enough voltage to cancel the effects of steady-state changes in output current i_O. Since load regulation is a manifestation of the regulator's closed-loop output resistance $R_{O.CL}$ and the loop gain that sets $R_{O.CL}$, i_O pulls v_O by roughly $i_O R_{O.CL}$. Therefore, the goal with this approach is to raise v_O by the same amount that i_O reduces it. One way to achieve this shift is to sense and feed a small fraction of the current that the output transistor M_{PO} sources into a resistor whose voltage shifts the reference v_{REF} of the system.

In Fig. 10.23, for example, M_{PS} mirrors a diminutive part of M_{PO}'s i_O into a 50–300-Ω resistor R_{OS} that is in series with v_{REF}. So when load current i_L and, as a result, i_O are low, M_{PS}'s i_S is negligibly low, and since the circuit's quiescent current i_Q is also low at 5–10 μA, R_{OS} hardly drops a few millivolts. When i_L climbs from microamps to milliamps, however, i_S can rise to such an extent that R_{OS} can drop and raise v_{REF} by the same amount that i_O reduces v_O:

$$v_O = v_{REF} - i_O R_{O.CL} + v_{OS} = v_{REF} - i_O R_{O.CL} + (i_Q + i_S) R_{OS}$$

$$= v_{REF} - i_O R_{O.CL} + (i_Q + i_O A_I) R_{OS} \approx v_{REF} - i_O R_{O.CL} + i_O A_I R_{OS} \qquad (10.46)$$

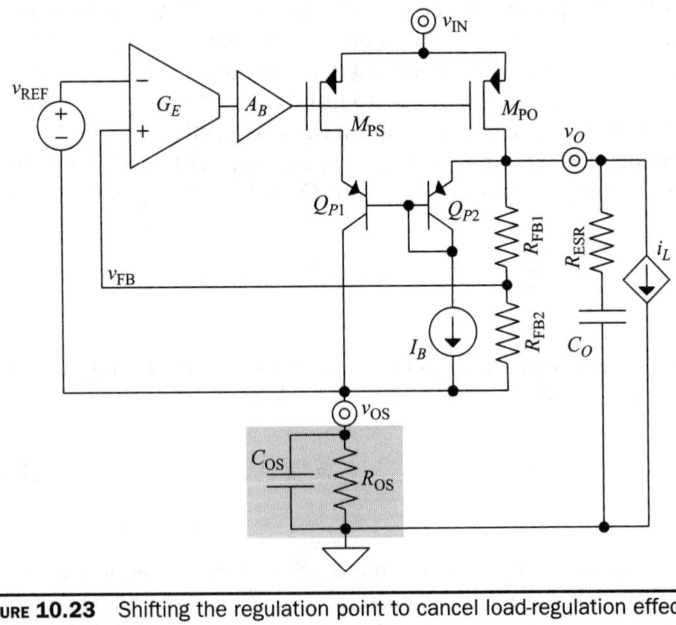

FIGURE 10.23 Shifting the regulation point to cancel load-regulation effects in a linear regulator.

where A_I is the attenuation factor from i_O to i_S. This means that load-regulation effects disappear in v_O when R_{OS} is the ratio of $R_{O.CL}$ and M_{PS}–M_{PO}'s mirror translation A_I, when R_{OS} is $R_{O.CL}/A_I$. In practice, a 60-mV variation in v_O can drop to less than 5 mV.

The purpose of sensor Q_{P1}–Q_{P2} in this circuit is to match the source–drain voltages of M_{PS} and M_{PO} to first order. This way, M_{PS} mirrors M_{PO}'s i_O across M_{PO}'s operating regions, even when M_{PO} enters triode and the regulator drops out of regulation. C_{OS}, on the other hand, shunts noise and the effects of displacement currents on R_{OS}'s voltage when responding to fast load dumps.

One key advantage of this technique, outside of improving accuracy, is low impact. First, R_{OS} requires a negligible amount of silicon area. Second, and arguably more important, R_{OS} exerts little to no effect on power efficiency because quiescent current i_Q in modern systems is low when the load is light, and although i_S rises with i_O, i_S nevertheless remains a miniscule portion of i_O through i_O's entire range. Plus, R_{OS} has no noticeable effects on frequency response or phase margin because the small signals that R_{OS} injects are common mode with respect to the entire circuit. In other words, as v_{REF} shifts, so do all signals in the error amplifier G_E and in the buffer A_B.

From a practical standpoint, matching and tracking R_{OS} to $R_{O.CL}/A_I$ is not trivial. The least costly approach with respect to power and test time is to match them nominally. In other words, size R_{OS} to cancel the systemic effects of $R_{O.CL}$ and accept the tolerance and random variations that result, reducing, for example, a 50–60-mV drop down to ±10–±15 mV. Increasing test time and silicon area to include two or three bits of trim for R_{OS} is another alternative, reducing the load-regulation variation to maybe less than ±5 mV.

The most significant drawback to this approach is that the circuit requires a dedicated reference v_{REF}, since shifting v_{REF} does not serve the needs of other analog circuits in the system that require an accurate reference. Quiescent current i_Q also produces a systemic offset across R_{OS} and in v_O that may be on the order of 0.5–2 mV. Floating the ground of the reference-setting components alone, however, as Fig. 10.24 shows, isolates R_{OS} from i_Q to remove this offset. The only shortcoming here is that variations in v_{OS} are no longer common mode to the circuit, so G_E, A_B, M_{PS}, Q_{P1}, R_{OS}–C_{OS}, and v_{REF} now constitute a noninverting feedback path. Fortunately, the feedback gain of this loop is considerably lower than that of its inverting counterpart, so its impact on the regulation performance of the system is normally manageable.

10.4.5 Load-Dump Response

In light of all the stability complexities and issues surrounding the linear regulator, looking outside the loop to improve load-dump response offers inherent value. Appending a high-bandwidth

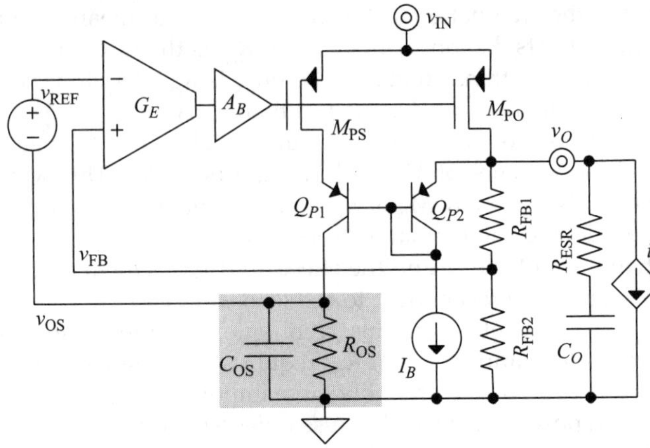

FIGURE 10.24 Shifting the reference to cancel load-regulation effects in a linear regulator.

shunt-feedback loop to the output v_O, for instance, as Fig. 10.25 shows, shortens response time because a faster local feedback network responds more quickly to changes in the load than the more complicated regulator loop. To ensure that the additional loop has negligible effects on steady-state accuracy, however, low-frequency gain must remain well below that of the regulator. This way, the main loop controls v_O below the system's bandwidth f_{0dB}, at low-to-moderate frequencies. In the case of portable systems, the additional loop should also draw little to no current to avoid shortening battery life.

Generally, under similar current-density constraints, more devices in a feedback loop slow the response of the circuit because each device adds energy-shunting nodes to the small-signal path, which is why compactness and speed go hand in hand. As a result, the fastest loop possible, from an architectural point of view, is the one already present in a transistor when its source or emitter terminal carries small signals, as in the case of source/emitter-degenerated transistors. Figure 10.25 illustrates how to bias and append two such single-transistor loops to v_O. Here, C_{NB} holds Q_{NO}'s base voltage V_{NB+}, so Q_{NO}'s base–emitter voltage v_{BENO} climbs immediately with reductions in v_O. This way, Q_{NO}'s current i_+ rises automatically when v_O falls in response to rising load dumps. Similarly, C_{PB} holds Q_{PO}'s base voltage V_{PB-}, so Q_{PO}'s emitter–base voltage v_{EBPO} similarly rises when v_O climbs to sink the current that the load no longer needs after a falling load dump. The purpose of I_{NB} and Q_{NB} and I_{PB} and Q_{PB} is to establish the voltages that R_{NB}–C_{NB} and R_{PB}–C_{PB} filter to bias Q_{NO} and Q_{PO}. C_{NB} and C_{PB} connect to ground instead of v_{IN} to avoid injecting supply ripple to v_O through the voltage followers that Q_{NO} and Q_{PO} realize.

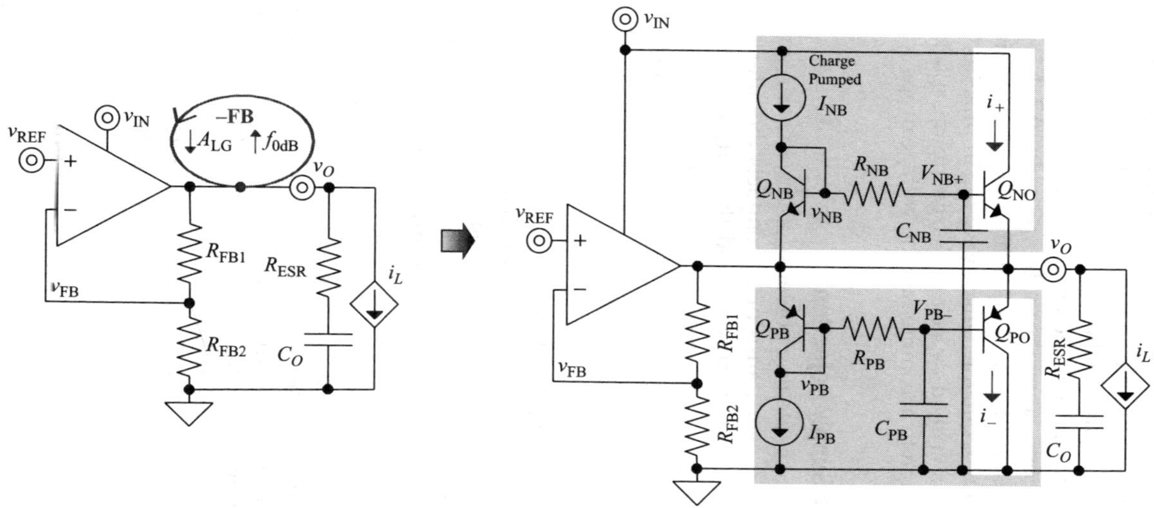

FIGURE 10.25 Bypass shunt-feedback loops about v_O.

As is often the case in design, the circuit in Fig. 10.25 also poses challenges. Raising Q_{NO}'s sourcing capability, for one, amounts to increasing Q_{NO}'s emitter area A_{NO}. Raising A_{NO} alone increases the mirror gain between Q_{NB} and Q_{NO}, so Q_{NO} sources more bias current into v_O, which means the circuit consumes more power. Raising both Q_{NB} and Q_{NO} while maintaining a low mirror gain, on the other hand, requires considerably more silicon area, which translates to higher cost. What is more, the voltage across I_{NB} and Q_{NB} when using a conventional current source can be roughly 0.7–1.1 V across temperature and fabrication corners, which negates the low-dropout features of a PMOS regulator. This is why I_{NB}, in practice, is usually the output of a switched-capacitor charge pump. Luckily, v_O is usually high enough to accommodate the voltage drops across Q_{PO}'s Q_{PB} and I_{PB}.

10.4.6 Power-Supply Rejection

When compared to switching power supplies, linear regulators are lossy because the power transistor S_O conducts the full steady-state load i_L across the input supply v_{IN} and the output v_O to dissipate $(v_{IN} - v_O)i_L$. Linear regulators nevertheless continue to be essential in, among other applications, battery-powered environments where power losses are critical. This is because, far from injecting switching noise, they suppress noise already present in the input supply. As a result, designers often use a switching regulator to strategically drop or boost the input supply efficiently and subsequently cascade a low-dropout regulator to reject the noise that the switcher generates.

Unfortunately, the regulator's feedback loop is less able to oppose the effects of noise near and past its unity-gain frequency f_{0dB} of maybe 0.1–1 MHz, which is near or below where switching noise is more than likely prevalent, at 0.1–10 MHz. The reason for this unfortunate mismatch in frequency is load-dump performance and power, because output capacitance is high and battery-supplied systems cannot afford to dissipate more power, so f_{0dB} cannot be high. The problem is worse in internally compensated systems because the dominant pole is inside the loop, so the transconductance gain across the loop starts dropping at low frequency. Output-compensated systems are in better shape because the transconductance gain across the loop does not attenuate until higher frequencies.

Passive Supply Filter

One way to suppress the noise that linear regulators cannot is to filter the regulator's input supply v_{IN} with a passive network. R_F and C_F in Fig. 10.26, for example, realize a *low-pass filter* in series with v_{IN} that can attenuate supply noise at and past the regulator's bandwidth. Unfortunately, R_F is in the power path, so R_F carries the load i_L and dissipates considerable ohmic power as a result. And raising C_F to accommodate a lower R_F is often impracticable with on-chip capacitors.

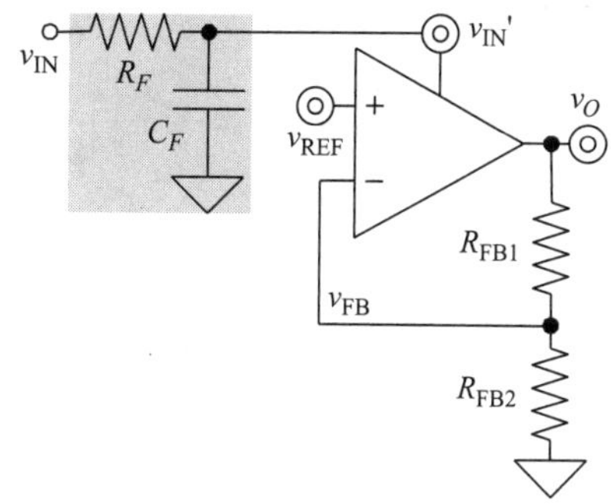

FIGURE 10.26 Passive supply filter.

LDO Supply Filter

Another way to boost the noise-suppression capabilities of a linear regulator is to cascade another regulator in series, like Fig. 10.27 shows. Because the preregulator is in the power path, it must drop a low voltage to minimize how much power it consumes, which means the preregulator must be a low-dropout circuit. Unfortunately, most portable applications cannot afford to dissipate the power that two series regulators consume, even if both are low-dropout regulators. Perhaps a more fundamental drawback here is that the preregulator suffers from similar bandwidth limitations as the main regulator does, so suppressing noise at and above 0.1–1 MHz remains a challenge with this approach.

Charge-Pumped Cascode

Conceptually, suppressing supply noise amounts to reducing the impedance from the output v_O to ground, which is what shunt feedback achieves in the linear regulator. Raising the impedance between the input supply v_{IN} and v_O also reduces supply noise, which is what R_F in the passive filter of Fig. 10.26 realizes. Another way of raising

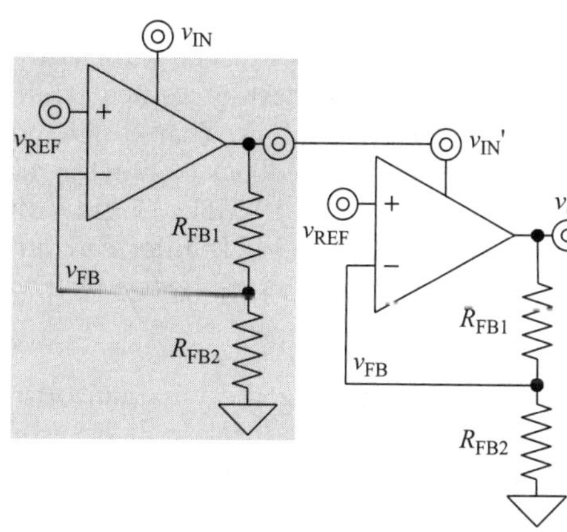

FIGURE 10.27 Low-dropout supply filter.

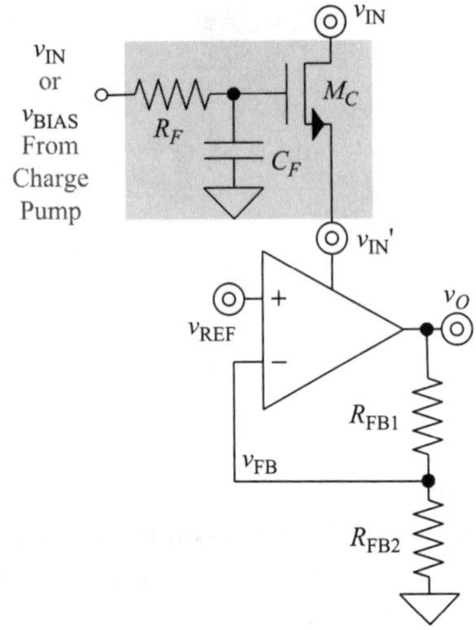

FIGURE 10.28 Charge-pumped cascode.

the top impedance is by inserting a cascode transistor M_C between v_{IN} and v_O, as Fig. 10.28 shows, whose series small-signal resistance is r_{ds}. Unfortunately, M_C is in the power path, so dropout voltage and ohmic losses still rise. To reduce this loss, either M_C's threshold voltage v_{TN} with its bulk effect, if any, must be low or M_C's gate voltage must rise above v_{IN}, which is why a *charge pump* drives M_C in the figure. The driving design objective with this circuit is to reduce M_C's drain–source voltage v_{DS} without pushing M_C into triode, which would otherwise reduce M_C's resistance and its impact on power-supply rejection. v_{DS} must therefore remain above M_C's saturation limit $V_{DS(SAT)}$.

As a voltage follower, M_C's source reproduces and injects the noise it receives at its gate, so filtering M_C's gate remains a necessity. Luckily, the filter's decoupling resistance R_F is no longer in the power path, so R_F can be high at 100–500 kΩ and C_F correspondingly low at 10–30 pF. The charge pump, however, generates additional noise, so the filter's corner frequency must be sufficiently low to suppress most of this noise and related harmonics near and above the regulator's bandwidth f_{0dB} of 0.1–1 MHz. Thankfully, noise below f_{0dB} is less problematic because the feedback loop in the regulator can suppress it well. With this approach, a 10-MHz 0.5-μm CMOS charge pump with a 100-mV$_{PP}$ output and a 3-kHz filter can further suppress supply noise by another 30 dB across frequency, well below and well past f_{0dB}.

Series Feedback

Another way of increasing the impedance to the supply is with a series current-sampling feedback loop. Inverting transconductor $-G_I$ in Fig. 10.29 does this, senses M_{PO}'s output current i_O and feeds this sampled impression back to M_{PO} to raise the impedance from the

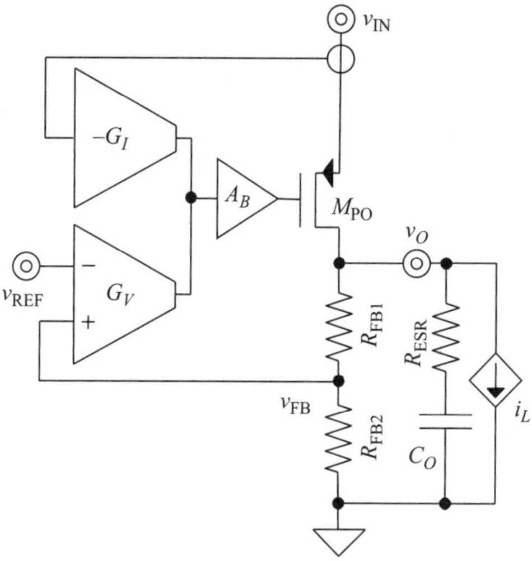

FIGURE 10.29 Current-sampling feedback network.

input supply v_{IN} to the output v_O. Note, however, that this current loop is in parallel with the voltage loop that regulates v_O. So to regulate v_O about the reference v_{REF}, the gain $A_{LG(I)}$ across the current loop must be lower than that of the voltage loop $A_{LG(V)}$ at low frequencies by a comfortable margin. But to decouple moderate- to high-frequency supply noise from the output, $A_{LG(I)}$ should rise above $A_{LG(V)}$ at high frequencies. To meet both of these requirements, a resistance can degenerate G_I and a capacitor can shunt that resistance to raise G_I at higher frequencies. This way, power-supply rejection can be 10 times higher near the regulator's bandwidth, when $A_{LG(V)}$ is low and therefore ineffectual. This improvement, however, does not extend to higher frequencies because $A_{LG(I)}$'s bandwidth ultimately limits G_I's ability to decouple v_{IN} from v_O. In fact, the main challenge with this approach is designing a current loop with a unity-gain bandwidth that exceeds that of the voltage loop with little quiescent power. Note, by the way, that the mirroring PMOS transistor M_{PS}, the base-coupled pair Q_{P1}–Q_{P2} that matches the drain voltages of M_{PO} and M_{PS}, and the bias current I_B in Figs. 10.23 and 10.24 and similar circuits in Chap. 9 can sense M_{PO}'s output current i_O without affecting dropout performance.

10.5 Current Regulation

In power supplies, current regulation is not really a driving objective, but rather a means to an end. Although currents used to bias the analog or mixed-signal building blocks of a system could benefit from current regulation, just to cite an example, they rarely require it. On the other hand, switching power supplies, especially boosting switched-inductor regulators, often regulate current to transform an inductor into an equivalent resistance. That way, the system reduces an LC feedback response that is difficult to stabilize to that of an RC, which is less challenging to stabilize. In these cases, regulating voltage loops embed these current loops, so they do not stand on their own. But not to digress from the general subject, the most appealing feature of current regulation in analog design is high output resistance. When testing power supplies, for example, subjecting the system to well-controlled variations in load current is important, so emulating the load with a regulated current can be useful. For these reasons and for the sake of completeness, the following discussion briefly addresses some important design considerations when regulating currents.

In the context of current regulation, negative feedback opposes the effects of fluctuations in the output voltage and the input supply on that current. And to regulate a current, the feedback loop must series-sample the current. This way, because variations in output voltage produce little changes in current, output resistance is high. This feature of high output resistance is especially desirable in current sources and current mirrors, where variations in current negate their functional benefits. Extending the operating range for which this feature remains true is also important, so surviving low output voltages is another important metric to evaluate in circuits of this sort.

10.5.1 Current Sources

A popular way of defining and regulating an output current i_O is to derive i_O from a well-defined voltage reference v_{REF}. In the case of Fig. 10.30, for example, a differential transconductor G_E series-mixes v_{REF} with M_{NO}'s source v_{FB} to reproduce v_{REF} in v_{FB}. The resistor R_I that drops v_{FB} therefore conducts a well-defined translation of v_{REF} that is nearly v_{REF}/R_I, and M_{NO} series-samples this current to output it in the form of i_O. Here, the loop regulates a cascode transistor M_{NO}, which is why engineers call these types of networks regulated-cascode circuits.

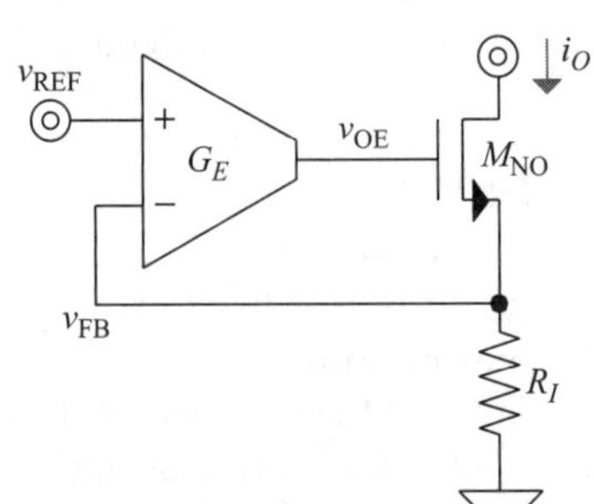

FIGURE 10.30 Voltage-derived regulated-cascode current source.

Here, G_E is the error amplifier and M_{NO} is the output transistor of the regulator. And the feedback loop that G_E and M_{NO} comprise only has two poles, at G_E's output v_{OE} and at G_E's feedback point v_{FB}. Of these, v_{FB}'s pole p_{FB} is at considerably higher frequency than v_{OE}'s p_{OE} is because M_{NO}'s source resistance is usually much lower than G_E's output resistance R_{OE}, which is normally on the order of r_o or r_{ds}, or higher. This is why this loop is often stable without additional capacitance. Plus, M_{NO}'s feed-forward capacitor C_{GS} helps the cause by introducing an in-phase zero z_{NO} soon after p_{FB}, rendering the circuit very stable.

The output resistance of the regulated cascode transistor M_{NO} is the amplified extrapolation of its nonregulated, but still source-degenerated counterpart. With respect to feedback, the feedback factor β_{FB}, or v_{FB}/i_O, is just R_I, and the forward open-loop transconductance gain $A_{G.OL}$, or $i_O/(v_{REF} - v_{FB})$, is G_E's gain G_ER_{OE} and M_{NO}'s degenerated transconductance G_{NO}, which reduces to G_ER_{OE}/R_I because $g_{mNO}R_I$ is normally much greater than 1 in G_{NO}:

$$A_{G.OL} \equiv \frac{i_O}{v_{REF} - v_{FB}} = G_ER_{OE}G_{NO} = G_ER_{OE}\left(\frac{g_{mNO}}{1+g_{mNO}R_I}\right) \approx \frac{G_ER_{OE}}{R_I} \tag{10.47}$$

where g_{mNO} is M_{NO}'s nondegenerated transconductance. Without G_E, source-degenerating M_{NO} with R_I raises M_{NO}'s output resistance r_{dsNO} to establish an open-loop output resistance $R_{O.OL}$ that is nearly equivalent to $r_{dsNO}g_{mNO}R_I$:

$$R_{O.OL} = r_{dsNO} + r_{dsNO}g_{mNO}R_I + R_I \approx r_{dsNO}g_{mNO}R_I \tag{10.48}$$

where $r_{dsNO}g_{mNO}R_I$ is normally much greater than r_{dsNO} and R_I combined. Using G_E to enclose M_{NO} in another series-sampling loop raises R_{OL} further by the loop gain $A_{G.OL}\beta_{FB}$ that G_E helps establish. This is why closed-loop output resistance $R_{O.CL}$ is G_ER_{OE} times greater than $R_{O.OL}$:

$$\begin{aligned} R_{O.CL} &= (1+A_{G.OL}\beta_{FB})R_{O.OL} \approx A_{G.OL}\beta_{FB}R_{O.OL} \\ &\approx \left(\frac{G_ER_{OE}}{R_I}\right)R_I(r_{dsNO}g_{mNO}R_I) = G_ER_{OE}(r_{dsNO}g_{mNO}R_I) \end{aligned} \tag{10.49}$$

Note that the minimum output voltage $v_{O(MIN)}$ that M_{NO} can sustain without succumbing to triode effects is R_I's v_{REF} and M_{NO}'s saturation voltage $V_{DS(SAT)}$, so low v_{REF} values are ideal.

Figure 10.31 illustrates a self-referenced transistor-level embodiment of the series-sampling circuit in Fig. 10.30. In this case, transistor M_E is G_E and the feedback node v_{FB} doubles as the reference. This way, the voltage across R_I is M_E's gate–source voltage v_{GS} and R_I's and M_{NO}'s current i_O is v_{GS}/R_I. v_{GS}, of course, varies with temperature and

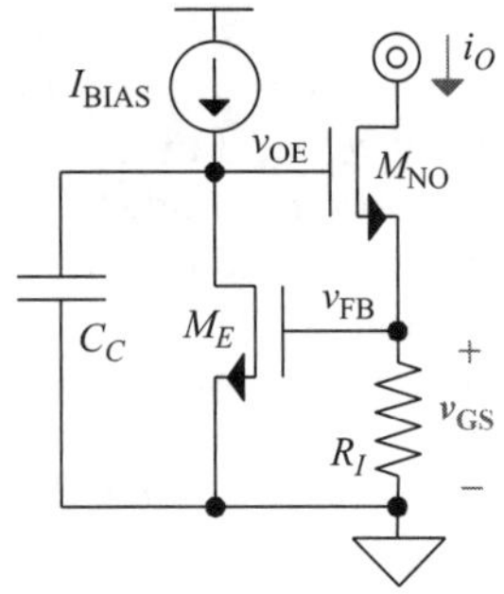

FIGURE 10.31 Self-referenced regulated-cascode current source.

fabrication corners, so i_O is not accurate. Still, M_E and M_{NO} regulate i_O to oppose variations in the output voltage, so closed-loop output resistance $R_{O.CL}$ is high at $G_E R_{OE} r_{dsNO} g_{mNO} R_I$, where G_E and R_{OE} are M_E's g_{mE} and r_{dsE}. Since v_{REF} is now v_{GS}, v_{GS} limits $v_{O(MIN)}$ to the sum of M_{NO}'s $V_{DS(SAT)}$ and M_E's v_{GS}.

10.5.2 Current Mirrors

Current regulators often rely on existing current references to derive their outputs. In other words, they realize the function of a current mirror, but with regulating features. To this end, adding a regulated-cascode transistor M_{NO} to the output of a basic current mirror like Fig. 10.32 shows achieves the objectives sought in a current regulator. Here, the output current i_O is a linear translation of the input current i_{IN} with little to no channel-length modulation error because both drain–source voltages are a gate–source voltage above ground:

$$i_O = i_{IN}\left[\frac{(W/L)_2(1+v_{DS1})}{(W/L)_1(1+v_{DS2})}\right] = i_{IN}\left[\frac{(W/L)_2(1+v_{GSE})}{(W/L)_1(1+v_{GS1})}\right] \approx i_{IN}\left[\frac{(W/L)_2}{(W/L)_1}\right] \quad (10.50)$$

Note that M_{NO} only regulates i_O against variations in the output voltage, not against mismatches in the mirror's width–length ratios.

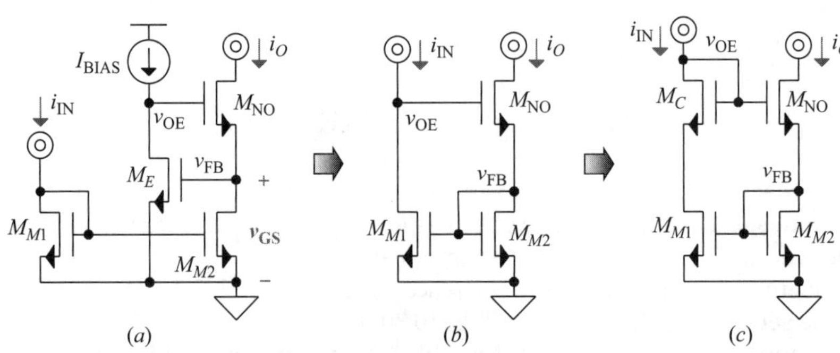

FIGURE 10.32 Regulated-cascode current-mirror evolution.

Also notice that M_{NO}, M_E, and M_{M2}'s r_{ds2} in Fig. 10.32*a* implement the same current-sampling feedback loop that their corresponding transistors and R_I do in the regulated-cascode circuit of Fig. 10.31. This means that $R_{O.CL}$ is just as high, where R_I is r_{ds2} and M_E's v_{GS} limits $v_{O(MIN)}$.

Interestingly, using the mirror's diode-connected transistor M_{M1} as the error amplifier M_E, as Fig. 10.32*b* illustrates, integrates the mirror into the feedback loop. This way, the T connection at M_{NO}'s gate shunt-mixes input current i_{IN} with M_{M1}'s drain current i_1, M_{NO}'s drain series-samples output current i_O, and M_{M1} and M_{M2} feed back a mirrored version of i_O to the mixing point. So with enough loop gain, i_1 mirrors i_{IN} and M_{M2}–M_{M1} translates i_1's impression of i_{IN} to i_O.

The open-loop current gain $A_{I.OL}$, or $i_O/(i_{IN} - i_1)$, is the gain into M_{M1}'s output resistance r_{ds1} and across M_{NO} and M_{M2}–M_{M1} to i_O. And since M_{M2}'s $1/g_{m2}$ diode-connected resistance hardly degenerates M_{NO}, the difference between i_{IN} and i_1 drops a voltage across r_{ds1} that M_{NO}'s degenerated transconductance G_{NO} outputs as i_O to reduce $A_{I.OL}$ to roughly $0.5r_{ds1}g_{mNO}$:

$$A_{I.OL} \equiv \frac{i_O}{i_{IN} - i_1} = r_{ds1}G_{NO} = \frac{r_{ds1}g_{mNO}}{1+\dfrac{g_{mNO}}{g_{m2}}} \approx 0.5r_{ds1}g_{mNO} \tag{10.51}$$

M_{M2}–M_{M1} then mirrors i_O to sink the feedback current i_{FB}, or i_1, from M_{NO}'s gate, so the feedback factor β_{FB}, or i_O/i_1, is $(W/L)_1/(W/L)_2$. With enough loop gain, the loop raises M_{NO}'s hardly degenerated output resistance of $r_{dsNO} + r_{dsNO}\,g_{mNO}/g_{mM2}$, or nearly $2r_{dsNO}$, to

$$\begin{aligned} R_{O.CL} &= (1 + A_{I.OL}\beta_{FB})R_{O.OL} \approx A_{I.OL}\beta_{FB}R_{O.OL} \\ &\approx (0.5r_{ds1}g_{mNO})\left[\frac{(W/L)_1}{(W/L)_2}\right](2r_{dsNO}) \end{aligned} \tag{10.52}$$

where

$$R_{O.OL} = r_{dsNO} + r_{dsNO}g_{mNO}\left(\frac{1}{g_{m2}}\right) + \frac{1}{g_{m2}} \approx 2r_{dsNO} \tag{10.53}$$

$R_{O.CL}$ here is lower than in the previous circuit because R_I is M_{M2}'s $1/g_{m2}$ instead of r_{ds2}. Ultimately, the drawback to this circuit is channel-length modulation error because M_{M1}'s and M_{M2}'s drain–source voltages do not match. Luckily, connecting a transistor M_C to M_{NO}'s gate like Fig. 10.32*c* shows that matches M_{NO} level-shifts M_{NO}'s gate voltage down by a matching gate–source voltage. This way, the drain voltage across M_{M1} matches that of M_{M2}. Note, by the way, that $v_{O(MIN)}$ for all three circuits is M_{NO}'s $V_{DS(SAT)}$ and M_{M2}'s or M_E's v_{GS}.

Since the loop mixes i_{IN}, shunt mixing reduces M_{M1}'s drain resistance r_{ds1} by the gain across the loop to roughly $2/g_{mNO}$ when the mirror gain is 1:

$$R_{I.CL} = \frac{r_{ds1}}{1 + A_{I.OL}\beta_{FB}} \approx \frac{r_{ds1}}{A_{I.OL}\beta_{FB}} = \frac{r_{ds1}}{(0.5 r_{ds1} g_{mNO})}\left[\frac{(W/L)_2}{(W/L)_1}\right] \approx \frac{2}{g_{mNO}}\left[\frac{(W/L)_2}{(W/L)_1}\right] \quad (10.54)$$

This closed-loop input resistance is nearly the same as that of its nonmixed predecessor from Fig. 10.32*a*, whose input resistance is M_{M1}'s diode-connected resistance $1/g_{m1}$. One disadvantage here is that the input voltage must now surpass M_{NO}'s and M_{M2}'s gate–source voltages, whereas its nonmixed counterpart must exceed only M_{M1}'s gate–source voltage.

Context In view of the foregoing current-mirror discussion, comparing the performance of the regulated-cascode current mirrors presented here against the more basic forefathers from Chap. 5 is prudent. Like Table 10.1 indicates, the fundamental benefit of the regulated-cascode circuit is to amplify the effects of its nonregulated counterpart on output resistance, except integrating the mirror into the loop lowers the gain back to its nonregulated levels. Including a loop that regulates the cascode transistor also tends to raise $v_{O(MIN)}$. Unfortunately, reducing $v_{O(MIN)}$ below its natural point complicates the circuit to the extent that power and accuracy suffer. In other words, regulating the cascode transistor raises power consumption, $v_{O(MIN)}$, and when keeping the mirror

	Basic Mirror	Low-Voltage Cascode Mirror	Regulated Cascode with the Mirror Outside the Loop	Regulated Cascode with the Mirror in the Loop and Level Shifted
$v_{IN(MIN)}$	v_{GS}	v_{GS}	v_{GS}	$2v_{GS}$
$v_{O(MIN)}$	$V_{DS(SAT)}$	$2V_{DS(SAT)}$	$V_{DS(SAT)} + v_{GS}$	$V_{DS(SAT)} + v_{GS}$
Accuracy	λ Error	–	–	–
R_{IN}	$1/g_m$	$1/g_m$	$1/g_m$	$2/g_m$
R_O	r_{ds}	$r_{ds}{}^2 g_m$	$r_{ds}{}^3 g_m{}^2$	$0.5r_{ds}{}^2 g_m$
Notes	Moderate R_O Low accuracy	High R_O Requires biasing	Highest R_O Requires biasing High $v_{O(MIN)}$	High R_O High $v_{O(MIN)}$

Table 10.1 Current-Mirror Comparison

outside the regulation loop, also output resistance. This is why the regulated-cascode circuit is only worthwhile when substantially higher output resistance is absolutely necessary, and when power and headroom are acceptable sacrifices. The basic cascode circuit is often more practical in this respect because it offers a more balanced cost–performance tradeoff. And even then, the demand for high output resistance is the only justifiable reason for its use over its more basic noncascoded predecessor.

10.6 Summary

The underlying aim of this chapter is to illustrate how to assemble and combine the ICs discussed in Chap. 9 into a linear-regulator system that addresses the overall design objectives described in Chaps. 1, 2, and 8. From this perspective, developing and applying a robust and efficient compensation strategy to ensure that the circuit is stable across all operating conditions is one of the first and most important steps in this part of the design process. The basic approach in this regard is to ensure that only one low-frequency pole dominates over all others, only one secondary pole resides near the system's unity-gain frequency f_{0dB}, and all other parasitic poles and out-of-phase zeros land a decade past f_{0dB}. In-phase zeros help extend f_{0dB} slightly by mitigating the avalanching effects that several poles exert on phase margin. Ultimately, however, even after optimization, low-power constraints (through their impact on $1/g_m$ resistances) limit bandwidth f_{0dB}, which means off- and on-chip output capacitors must source and sink load dumps whose rising and falling rates outpace the speed of the regulating loop.

The reference is another consideration, and although regulators often rely on external reference circuits, incorporating the workings of the reference into the feedback loop is not uncommon, especially in stand-alone applications and segmented, power-aware systems. The first step in this process is to integrate the PTAT current generator into the feedback loop, and to ensure that the noninverting feedback path it presents has lower gain than that of its inverting counterpart. A startup circuit must then guarantee that the reference reaches its desired state, even when initially latched otherwise. And achieving temperature independence amounts to introducing a CTAT component, which normally reduces to a diode voltage.

The importance of dropout, power, speed, accuracy, and power-supply rejection in linear regulators cannot be overstated, which is why developing strategies aimed at optimizing those specifications is so important. Complicating a circuit, however, normally slows its response, requires more power, occupies more silicon area, and raises risk. The key, then, is to find efficient means of exploiting and leveraging the transistors and devices that already exist in the circuit, which is not to say that risk and tradeoffs should disappear completely.

Driving the bulk of a power MOS transistor, for instance, raises not only gate drive but also the propensity for parasitic BJTs to activate. Any attempt at improving power-supply rejection, just to cite another example, also tends to raise dropout voltage, so power performance suffers. Still, finding a sensible and optimum balance between conflicting specifications, design objectives, and complexity is possible, especially when the designer understands the application and how the regulator affects the system it supports.

Generally speaking, designing linear regulator ICs is the art of assembling and combining the semiconductor components from Chap. 3 into the analog building blocks in Chaps. 4 and 5, and at a higher level, into systems that regulate their outputs according to the application and feedback requirements described in Chaps. 1, 2, 6, and 8. Understanding the application and the technologies that drive the design is imperative in this process, especially when attempting to extend battery life and improve performance. In light of these considerations, this chapter in many ways concludes the discussion on linear regulator ICs, because Chaps. 1 through 10 present almost all the technology and expertise necessary to design one. The only aspects that remain are circuit protection and characterization, which Chap. 11 addresses.

10.7 Review Questions

1. What is(are) the fundamental advantage(s) of output-compensated PMOS regulators?

2. What is(are) the disadvantage(s) of output-compensated PMOS regulators?

3. What is(are) the fundamental advantage(s) of Miller-compensated PMOS regulators?

4. What is(are) the disadvantage(s) of Miller-compensated PMOS regulators?

5. What is(are) the fundamental advantage(s) of N-type regulators?

6. What is(are) the disadvantage(s) of N-type regulators?

7. What is the role of a PTAT current in a self-referenced regulator?

8. What is the difference between "bandgap conversion" and "bandgap integration?"

9. How can the current-driving capabilities of a PMOS regulator be raised without adding capacitance to the regulating loop?

10. How can loop gain be elevated for better regulation performance without raising quiescent current flow?

11. How can load-regulation effects in the output be cancelled without raising loop gain?

12. What is(are) the fundamental advantage(s) of a passive filter in series with the input power supply?

13. What is(are) the disadvantage(s) of a passive filter in series with the input power supply?

14. What is(are) the disadvantage(s) of a preregulator in series with the input power supply?

15. What is(are) the fundamental advantage(s) of a charge-pumped cascode transistor in series with the input power supply?

16. What is(are) the disadvantage(s) of a charge-pumped cascode transistor in series with the input power supply?

17. Why is a regulated-cascode circuit used in current sources and current mirrors?

18. Is a regulated-cascode circuit always better than its nonregulated cascode counterpart? Why?

CHAPTER 11

Protection and Characterization

Understanding the performance objectives of a linear regulator and designing an integrated circuit (IC) that fulfills those goals are, of course, first and foremost in the design process. However, neglecting to protect the regulator against damaging and often catastrophic, though unintended, but nonetheless realistic events renders the IC ineffectual and the system it supports inoperable, if not worse. The first line of defense against these harmful conditions is to protect the IC at the circuit level, especially when power levels can surpass the operating and breakdown limits of the regulator.

Characterization is next, because comprehending and reporting the operating and performance limits of the fabricated IC keeps users from overstressing it. Determining these constraints from measurements is imperative because the *three-sigma* (3σ) semiconductor models used to predict the performance of the IC are imperfect. For one, simulations do not account for a number of parasitic devices in the layout and effects in the system, such as noise injection into the substrate, noise coupling, and substrate de-biasing. Plus, models do not predict performance variations well across dies, fabrication lots, or time, such as aging effects. So in harmony with the design sequence thus far presented in the book and the product- and prototype-development cycle it seeks to support, this chapter discusses circuit protection and characterization.

11.1 Protection

The purpose of on-chip circuit protection is to avoid exposing any part of the IC to conditions that surpass its breakdown limits. Bearing in mind that the power transistor S_O conducts all the load current, S_O is liable to exceed its *safe-operating area* (SOA), so the IC must prevent S_O from reaching its maximum rated limit $P_{SW(MAX)}$. Commercial products therefore incorporate *overcurrent* and *thermal-shutdown* features into the IC. And for similar reasons, but to address different concerns, they also include *reverse-battery* and *electrostatic-discharge protection*.

Protection, however, should not risk or interfere with the functionality or performance of the regulator. In other words, protection should engage only when needed, on demand, and be not only reliable but also transparent when the system is operating normally. Transparency, among other things, also implies drawing little to no quiescent current from the power supply, which is why trip-point accuracy is normally poor. Luckily, because regulation and efficiency sells more than protection, especially in the battery-powered segment, engineers often overdesign power capabilities for the sake of accuracy and dropout performance. As a result, the maximum power lost in the IC is usually well below its rated limit, so the operating threshold that prompts a protection circuit to engage need not be accurate. Margin between actual and rated limits also helps keep transient glitches in the system from inadvertently tripping protection circuits. Incorporating deglitch filters also helps prevent noise from initiating false alarms.

11.1.1 Overcurrent Protection

Fixed Limit

Overcurrent protection, whose aim is to protect S_O, is another term for *short-circuit* and *overload protection*. Its objective is to keep the power dissipated across S_O as P_{SW} below its maximum rated limit of $P_{SW(MAX)}$. To that end, fixed overcurrent protection limits S_O's output current i_O to I_{OCP}, as Fig. 11.1 illustrates. In practice, however, I_{OCP} is not accurate, so I_{OCP}'s minimum point $I_{OCP(MIN)}$ limits $i_{O(MAX)}$. This means that the IC can supply no more than the power that $I_{OCP(MIN)}$ and v_O's target $V_{O(TAR)}$ establish:

$$P_{O(MAX)} = v_O i_{O(MAX)} \approx V_{O(TAR)} I_{OCP(MIN)} \tag{11.1}$$

Since S_O drops v_{IN} to v_O when supplying i_O, the power that S_O dissipates as P_{SW} is

$$P_{SW} = v_{SW} i_{SW} = (v_{IN} - v_O) i_O \tag{11.2}$$

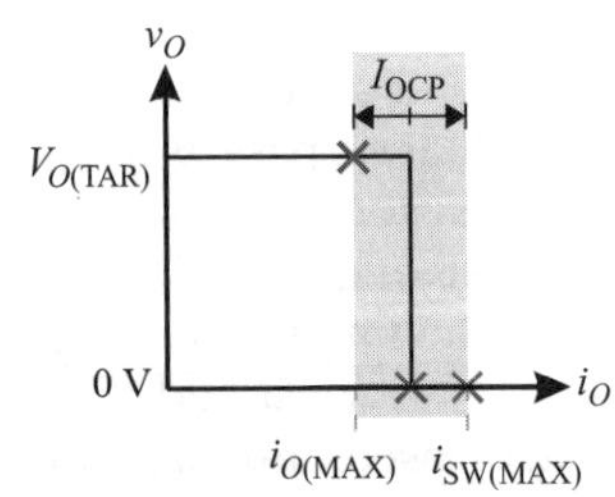

Figure 11.1 Fixed overcurrent protection.

This means that P_{SW} is highest when v_O short-circuits (SC) to ground and the system limits i_O to I_{OCP}'s maximum limit $I_{OCP(MAX)}$:

$$P_{SW(MAX)} = v_{SW} i_{SW}\Big|_{SC} = (v_{IN} - 0) i_{SW(MAX)} = v_{IN} I_{OCP(MAX)} \tag{11.3}$$

As a result, together with $V_{O(TAR)}$ and v_{IN}, $I_{OCP(MIN)}$ and $I_{OCP(MAX)}$ keep the maximum power that S_O can deliver to the load as $P_{O(MAX)}$ at less than a $V_{O(TAR)}/v_{IN}$ fraction of S_O's rated limit of $P_{SW(MAX)}$:

$$\frac{P_{O(MAX)}}{P_{SW(MAX)}} = \left(\frac{V_{O(TAR)}}{v_{IN}}\right)\left(\frac{I_{OCP(MIN)}}{I_{OCP(MAX)}}\right) < \frac{V_{O(TAR)}}{v_{IN}} \tag{11.4}$$

Note, by the way, that I_{OCP} must account for temperature extremes, so $P_{SW(MAX)}$ corresponds to the highest possible rating across temperature.

Figure 11.2 depicts an illustrative example of a circuit that protects power transistor M_{PO} according to the fixed-overcurrent profile in Fig. 11.1. Here, sensing transistor M_{PS}, current buffer Q_{S1}, common-source transconductor M_{S2}, and mirror M_{S3}–M_{S4} close an inverting feedback loop that engages when M_{PS}'s current i_{PS} nears threshold current I_{OCP}/A_I. Since M_{PS}'s i_{PS} is A_I times lower than M_{PO}'s current i_O, the loop develops gain only when i_O nears overcurrent limit I_{OCP}—otherwise, M_{S2}'s gate voltage v_{G2} is zero and M_{S3}–M_{S4} is off. Once engaged, the T connection at M_{S2}'s gate shunt-mixes I_{OCP}/A_I with i_{PS}, and M_{PS} shunt-samples M_{PO}'s gate voltage v_B. This way, the circuit reproduces I_{OCP}/A_I in i_{PS}, and by a mirror translation to the output, I_{OCP} in i_O, and the loop reduces v_B's impedance in shunt-feedback fashion.

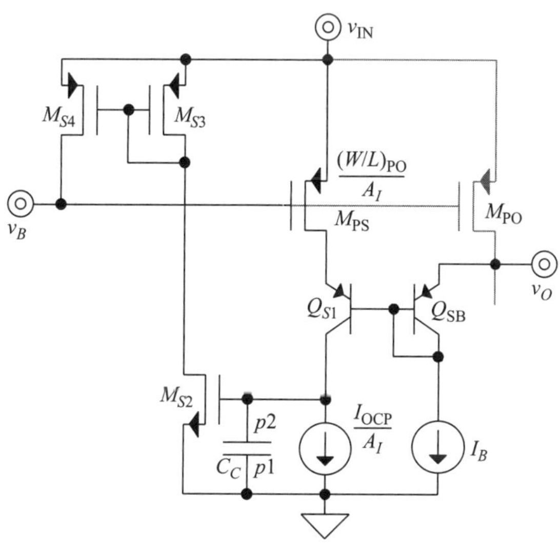

FIGURE 11.2 Fixed current limiter.

As a result, the loop is off when i_O is below I_{OCP}, and when i_O reaches and exceeds I_{OCP}, the loop overwhelms the error amplifier A_E and the buffer A_B to regulate i_O about I_{OCP}.

The purpose of base-coupled bipolar-junction transistors (BJTs) Q_{S1} and Q_{SB} is to impress M_{PO}'s drain–source voltage across M_{PS}. For this, bias current I_B drops an emitter–base voltage v_{EB} across Q_{SB} that Q_{S1}'s v_{EB} shifts up and cancels to ensure that M_{PS}'s drain voltage follows M_{PO}'s v_O. Therefore, M_{PS} mirrors M_{PO} across all possible states of v_O, which can only happen when both devices are in the same operating region. Compensating capacitor C_C is at M_{S2}'s gate v_{G2} to pull v_{G2}'s pole p_{G2} to low frequency, because higher capacitance shunts the resistance across its terminals at lower frequency. As a result, because the resistance at v_{G2} is high, p_{G2} dominates and the gain across M_{PS}, M_{S2}, M_{S3}, and M_{S4} falls to 1 before the buffer pole p_B at v_B asserts its influence. Note that M_{PO} and the feedback loop that controls v_O shunt-samples v_B, so p_B is normally at high frequency.

Fold-Back Limit

In the portable space, overdesigning the linear regulator to source less power than it can actually handle is often a natural byproduct of the dropout specification, which forces the size of the power device to be larger than its power rating requires. Many higher-power applications, however, do not enjoy this providence, which is where *fold-back current limiting* finds a niche. Recall that the drawback to a fixed overcurrent limit is that S_O conducts more current when shorted at $i_{SW(MAX)}$, or $I_{OCP(MAX)}$, than when maxed at $i_{O(MAX)}$, or $I_{OCP(MIN)}$, so $I_{OCP(MAX)}$ sets $P_{SW(MAX)}$. One way to ease this constraint on $P_{SW(MAX)}$ is to lower or "fold back" I_{OCP} to I_{OCP}' in Fig. 11.3 with reductions in v_O. This way, S_O conducts less current when shorted at $i_{SC(MAX)}$, or $I_{OCP(MAX)}'$, than when maxed at $i_{O(MAX)}$, or $I_{OCP(MIN)}$. In other words, for the same $P_{SW(MAX)}$ rating, I_{OCP} and, as a result, $i_{O(MAX)}$ can be higher in the fold-back scheme than in the fixed counterpart.

The advantage of folding back I_{OCP} is that S_O's P_{SW} when maxed at $I_{OCP(MAX)}$ can be as high as when shorted at $I_{OCP(MAX)}'$:

$$P_{SW(MAX)} = v_{IN} i_{SC(MAX)} = v_{IN} I_{OCP(MAX)}' \equiv (v_{IN} - V_{O(TAR)}) I_{OCP(MAX)} \tag{11.5}$$

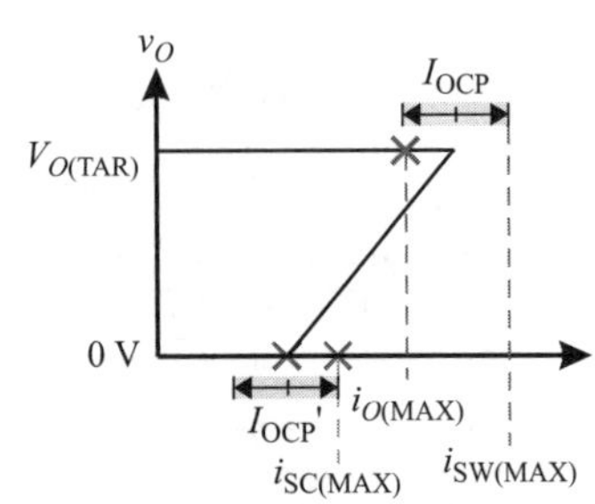

FIGURE 11.3 Fold-back overcurrent protection.

This way, $I_{OCP(MAX)}$ can be well above $i_{SC(MAX)}$:

$$\frac{I_{OCP(MAX)}}{i_{SC(MAX)}} = \frac{I_{OCP(MAX)}}{I_{OCP(MAX)}'} = \frac{v_{IN}}{v_{IN} - V_{O(TAR)}} > 1 \tag{11.6}$$

and $i_{O(MAX)}$ at $I_{OCP(MIN)}$ can therefore be greater than $i_{SC(MAX)}$ at $I_{OCP(MAX)}'$. This means that $P_{O(MAX)}$ can be higher than a $V_{O(TAR)}/v_{IN}$ fraction of $P_{SW(MAX)}$:

$$\frac{P_{O(MAX)}}{P_{SW(MAX)}} = \frac{V_{O(TAR)} i_{O(MAX)}}{v_{IN} i_{SC(MAX)}} = \frac{V_{O(TAR)} I_{OCP(MIN)}}{v_{IN} I_{OCP(MAX)}'} > \frac{V_{O(TAR)}}{v_{IN}} \tag{11.7}$$

Figure 11.4 shows a circuit that protects power transistor M_{PO} according to the fold-back profile in Fig. 11.3. Here, sensing transistor M_{PS} mirrors a diminutive fraction of M_{PO}'s i_O through R_S into v_O to drop a voltage v_R that rises with i_O and shifts with v_O. Voltage divider R_{T1}–R_{T2}, transconductor amplifier A_{OCP}, common-source transconductor M_{S1}, and M_{PS} then close an inverting feedback loop that engages when R_S's v_R rises above the threshold that voltage dividers R_{T1}–R_{T2} and R_{B1}–R_{B2} establish. For this, R_{T1}–R_{T2} drops more voltage than R_{B1}–R_{B2}, so A_{OCP}'s noninverting input v_P is higher than its inverting counterpart v_N when R_S's v_R is low, which means A_{OCP}'s output is high and M_{S1} is off as a result. A_{OCP} engages when i_{PS}'s impression of i_O drops enough voltage across R_S to lower v_P below v_N.

Once engaged, A_{OCP} series-mixes v_P with v_N and M_{PS} shunt-samples M_{PO}'s gate v_B to reproduce v_P in v_N and reduce v_B's impedance in shunt-feedback fashion. Since v_N mirrors v_P and R_{T1}–R_{T2}'s gain A_T is

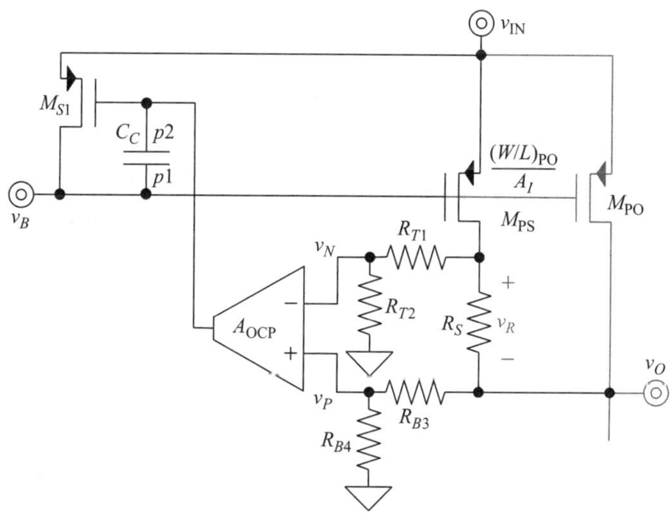

FIGURE 11.4 Fold-back current limiter.

lower than R_{B1}–R_{B2}'s counterpart A_B, R_S's top voltage falls more than R_S's bottom voltage when v_O drops. In other words, R_S's voltage v_R falls with v_O when A_{OCP}'s loop controls i_{PS} to lower i_{PS}, and by A_I's translation, lower i_O with v_O:

$$v_N = (v_O + i_{PS}R_S)A_T = \left[v_O + \left(\frac{i_O}{A_I}\right)R_S\right]A_T \approx v_P = v_O A_B \tag{11.8}$$

so

$$i_O \approx v_O A_I \left(\frac{A_B - A_T}{R_S A_T}\right) \tag{11.9}$$

As a result, the loop is off when the voltage that i_{PS}'s impression of i_O drops across R_S is low enough to keep v_N below v_P. And when i_O is high enough to raise v_N above v_P, A_{OCP} and M_{S1} overwhelm the error amplifier A_E and the buffer A_B to regulate i_{PS}, and by an A_I translation, regulate i_O about a level that falls with v_O, as Fig. 11.3 shows. Here, Miller-compensation capacitor C_C pulls the pole at M_{S1}'s gate to low frequency and pushes the one at M_{PO}'s counterpart v_B to high frequency to stabilize the system.

To gain more insight into the inner workings of this circuit, consider the case where a conventional amplifier with an intentional input-referred offset V_{OS} replaces A_{OCP} and M_{S1}, the gains across the voltage dividers match, and a bias voltage source drives R_S's bottom terminal. This way, the loop ensures that the voltage drop across R_S is constant at a voltage-divided translation of V_{OS}, and since i_O's mirrored impression in i_{PS} establishes v_R, i_O is also constant. Removing the conventional amplifier and its offset and replacing it with A_{OCP} and M_{S1} means the loop can only source current into v_B, so when v_R is negative, M_{S1} is off. Increasing the gain of the bottom voltage divider raises v_P, so A_{OCP} shuts M_{S1} even when v_R is above zero by a small amount. And removing the voltage source lets v_P fall with v_O to engage the loop when v_P drops below v_N. Once engaged, the gain difference between the voltage dividers ensures that v_R is positive when v_P and v_N are equal, so i_O is $i_{PS}A_I$ and therefore also $A_I\, v_R/R_S$. Since the bottom divider's gain is higher than that of the top, R_S's top voltage falls more than v_O, and as a result, v_R and by translation i_O fall with v_O.

11.1.2 Thermal Shutdown

In practice, *junction temperature* T_J is sometimes an indirect measure of output current i_O because, barring the existence of other power supplies in the IC, T_J climbs with the power dissipated in the regulator P_{REG}, almost all of which the power transistor S_O dissipates as P_{SW}. In fact, P_{SW} and the *thermal impedance* θ_P of the package determine how much higher T_J is over *ambient temperature* T_A:

$$T_{J(MAX)} = T_A + P_{REG(MAX)}\theta_P \approx T_A + P_{SW(MAX)}\theta_P = T_A + v_{IN} I_{OCP(MAX)}\theta_P \tag{11.10}$$

where S_O consumes the most power when shorted to ground and $I_{OCP(MAX)}$ becomes $I_{OCP(MAX)}'$ when folding back I_{OCP} to I_{OCP}' as the output v_O drops to zero. As a result, when considering the regulator on its own, thermal shutdown and the overcurrent function protect the IC from similar stressing conditions, so implementing both functions on a stand-alone linear regulator can be redundant. With multiple regulators and power devices on the same die, however, the situation changes because *hot spots* and *thermal time constants* decouple T_J from S_O's i_O.

Because low-dropout specifications usually raise S_O's power rating well above S_O's breakdown limits, the purpose of thermal protection is not always to protect S_O, but rather to keep the package from melting, as is typically the case in portable applications. Therefore, since the melting point of plastic packages T_P is near 170°C, the thermal trip point T_{SHUT} of the IC must fall below T_P. And because low quiescent current, limited silicon area, and little test time induce inaccuracies and variations in T_{SHUT}, T_{SHUT} should precede T_P by some margin to, say, 150°C.

In practice, temperature varies slowly, which is another way of saying thermal time constants are long. This means that temperature will more than likely hover about T_{SHUT} when the load stresses S_O to that extent. And since temperature rises when S_O conducts and falls otherwise, the circuit can shut and reengage uncontrollably about T_{SHUT}. *Thermal hysteresis* T_{HYS} eliminates this effect, because the circuit trips when temperature reaches T_{SHUT} and does not reengage until temperature falls below T_{SHUT} by T_{HYS}.

Perhaps the most practical and predictable means of establishing a temperature trip point, as Fig. 11.5 illustrates, is to sense when a *proportional-to-absolute-temperature* (PTAT) voltage v_{PTAT} falls past its rising counterpart, a *complementary-to-absolute-temperature* (CTAT) voltage v_{CTAT}. Luckily, the difference between two matched but ratioed diode or base–emitter voltages Δv_D or Δv_{BE} produces a linear and reliable PTAT voltage. And although not as linear, a diode or base–emitter voltage v_D or v_{BE} falls reliably with rising temperatures to establish the CTAT voltage needed.

The purpose of the circuit in Fig. 11.6 is to generate a shutdown signal v_{SHUT} that rises when temperature reaches T_{SHUT} and does not fall until temperature falls below T_{SHUT} by T_{HYS}, like the shutdown

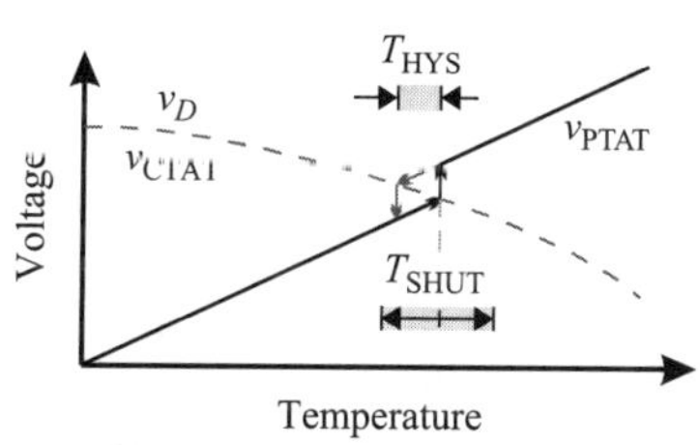

FIGURE 11.5 Thermal-shutdown graph.

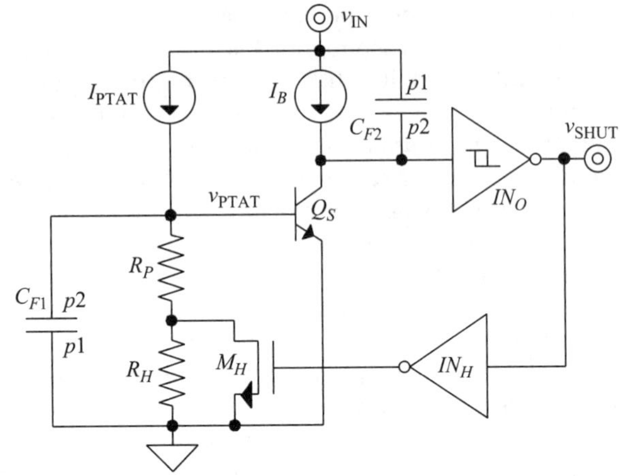

FIGURE 11.6 Thermal-shutdown circuit.

profile in Fig. 11.5 dictates. Here, when temperature is low, bias PTAT current I_{PTAT} establishes a PTAT voltage v_{PTAT} across R_P that is not sufficiently high to induce temperature-sensing transistor Q_S to sink bias current I_B. As a result, Q_S's collector voltage v_{CS} is high, inverter IN_O outputs a low v_{SHUT}, and inverter IN_H outputs a high to close M_H and short R_H. As temperature rises, so does Q_S's base voltage v_{PTAT}, and when v_{PTAT} is sufficiently high for Q_S to sink and overwhelm I_B, v_{CS} collapses and IN_O raises v_{SHUT} to shut the system.

Since v_{PTAT} rises slowly, Q_S's collector voltage v_{CS} also falls slowly, which causes uncertainty in IN_O's transition, in v_{SHUT}. Inverter IN_O includes hysteresis to keep v_{SHUT} from uncertain states when v_{CS} is halfway across the supply. This way, IN_O rails v_{SHUT} to the input supply v_{IN} as soon as Q_S's v_{CS} falls below IN_O's lower threshold, and it does not lower v_{SHUT} until v_{CS} rises above IN_O's higher threshold. Once high, IN_H opens M_H to raise v_{PTAT} by the voltage that I_{PTAT} drops across R_H, as v_{PTAT} near T_{SHUT} in Fig. 11.5 shows. As a result, v_{CS} does not rise again until temperature is low enough to lower the new v_{PTAT} below the base–emitter voltage that induces Q_S to sustain I_B, establishing in the process the desired T_{HYS}. Filter capacitors C_{F1} and C_{F2} keep ground and supply noise from inadvertently asserting shutdown events.

11.1.3 Reverse-Battery Protection

The purpose of reverse-battery protection, as the name implies, is to shield the regulator from the effects of sustained reverse-battery conditions. The most efficient method of achieving this objective when considering quiescent current, silicon area, and electrical interference in the form of noise and undesired voltage drops is to altogether avoid the condition. Luckily, "keying" the battery holder with plastic guards, tracks, or guides ensures that the battery fits only when properly oriented. This kind of protection is possible and popular in many consumer products, such as cameras. Coin cells, unfortunately, are difficult to key, so they often include protection in the electrical domain, which is usually more costly.

Electrical reverse-battery protection should remain off and transparent, but alert during normal operating conditions, and assert only when needed, on demand. Transparency, unfortunately, implies that quiescent current i_Q, series voltage drops in the supplies v_S, and reverse current i_R should all be zero, or at least, near zero. And to sustain these extreme conditions for any length of time, the electrical components exposed must be capable of sustaining high power and high electric fields. These devices must therefore be large, with deep and continuous guard rings, much like the power transistor, and with lightly doped junctions to survive high voltages.

Functionally, reverse-battery protection either shunts current away from the regulator circuit or blocks it. Diverting current with a shunt diode, for example, as Fig. 11.7 shows, protects the regulator, but it does so at the expense of considerable reverse-battery current i_R, which can quickly drain the battery. Blocking current, as a result, is better with respect to battery life, but not forward efficiency, because blocking often introduces a series voltage v_S in one of the supplies during normal operating conditions. Although not as much as with shunt protection, this series voltage drop dissipates power and reduces the effective input supply to the circuit, adversely affecting headroom and dropout performance.

A series diode is one way to block reverse-battery current. The drawback to the diode is that the 0.6–0.7 V dropped dissipates considerable power in steady state. Series transistors are better blockers in this respect because their series voltages are lower, especially when using N-type devices whose mobility is normally two to three times higher than that of their P-type counterparts. Metal–oxide–semiconductor (MOS) field-effect transistors (FETs) are better than BJTs because base currents i_Q lose quiescent power across the supplies. The drawback to

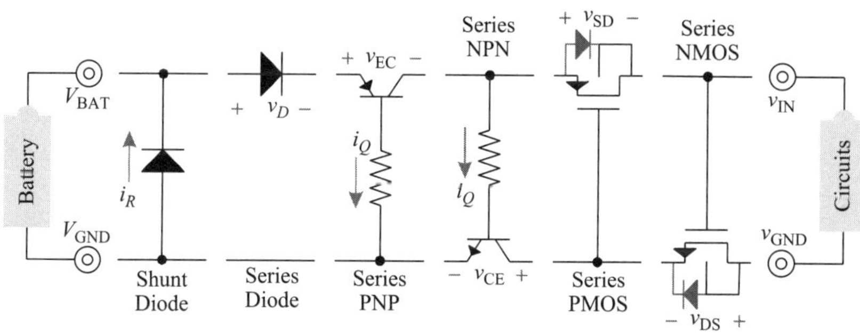

FIGURE 11.7 Reverse-battery circuits.

MOSFETs is silicon area because MOSFETs normally require more area than BJTs to conduct the same current. Still, irrespective of the means, unless mechanical restraints are possible, altogether avoiding reverse current i_R, series voltage v_S, or quiescent current i_Q in reverse-battery circuits is challenging. Note, by the way, that MOS bulks connect to drains to ensure that the parasitic body diode between the two terminals does not activate during reverse-battery conditions.

11.1.4 Electrostatic-Discharge Protection

Electrostatic discharges (ESDs) are not entirely unlike reverse-battery conditions, except ESDs are short-lived, occur in both directions, and the voltages involved are substantially higher. Most ESD strikes result from one of three basic mechanisms. The first and most obvious source is the human touch, after generating static charge by walking across a carpet, for example, which the *human-body model* (HBM) emulates. Similarly, although from a different source, the *machine model* (MM) mimics the charge transferred to ICs when machines handle packaged ICs during the bundling and shipping process. Lastly, the *charged-device model* (CDM) describes the electrical conditions that result after sliding microchips down their plastic container or tube, when precharged devices touch chassis ground. The common thread in all these cases is the sudden release of charge, which is why 100–300-pF capacitors precharged to 0.2–5 kV model these events.

The performance of ESD circuits depends strongly on physical layout and manufacturing process technology, and predicting the extent to which they protect the IC is largely empirical. ESD structures, as a result, vary significantly across process technologies, across the junctions and the polysilicon materials they protect, and even across semiconductor companies. What is worse, all input and output pins are susceptible to positive and negative ESD strikes, so they all need protection.

Most ESD circuits use forward-biased diodes, diodes in breakdown, or latching BJTs to clamp the voltage of the pin they protect. One common structure used in digital ICs is the pair of stacked forward-biasing diodes in Fig. 11.8. Here, a positive ESD strike

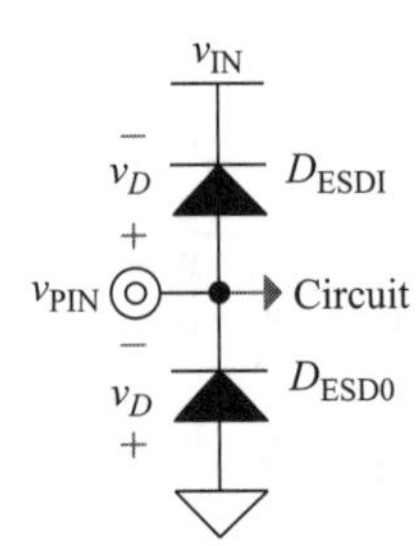

FIGURE 11.8 Stacked diode clamp.

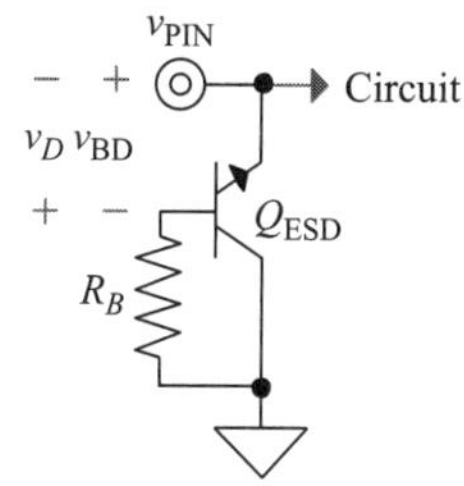

FIGURE 11.9 Dual-mode diode clamp.

forward-biases the diode between the pin v_{PIN} and the input supply v_{IN} to clamp v_{PIN} to a diode voltage v_D above v_{IN}. And a negative strike forward-biases the diode between v_{PIN} and ground to clamp v_{PIN} to a diode voltage v_D below ground. Although the circuit is simple and robust, and its clamp points are well defined about the supplies, positive strikes drain energy through v_{IN}, so current flows through v_{IN} before finally reaching ground, which is the lower-impedance plane.

Fortunately, reverse-biasing the bottom diode in Fig. 11.8 to the point that it "breaks down" also shunts energy away from the pin, so one diode can clamp both positive and negative ESD strikes. The emitter–base junction of the BJT in Fig. 11.9, for example, clamps positive strikes to the junction's breakdown voltage v_{BD} and negative strikes to a diode voltage v_D below ground. Note that, although the circuit no longer drains strikes through v_{IN}, v_{PIN} now rises to a voltage that is not as well defined about v_{IN}. Still, the breakdown voltage is predictable and reliable, so this structure enjoys popularity. As a side note, engineers normally use the base–emitter junction of vertical NPN BJTs to build *zener diodes* that break down when reversed to 6–7 V.

Since ESD events are sudden and transient in nature, coupling some of the energy through a capacitor is another way of triggering a transistor. The capacitor in Fig. 11.10, for example, derives energy from the strike to feed the base of an NPN BJT. When engaged this way, the BJT sinks ESD energy away from the circuit and into ground.

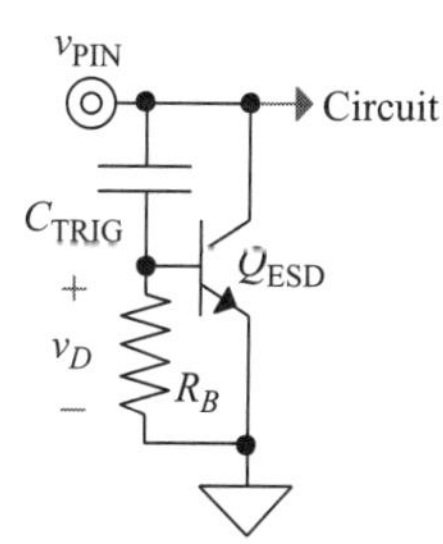

FIGURE 11.10 Capacitor-coupled BJT clamp.

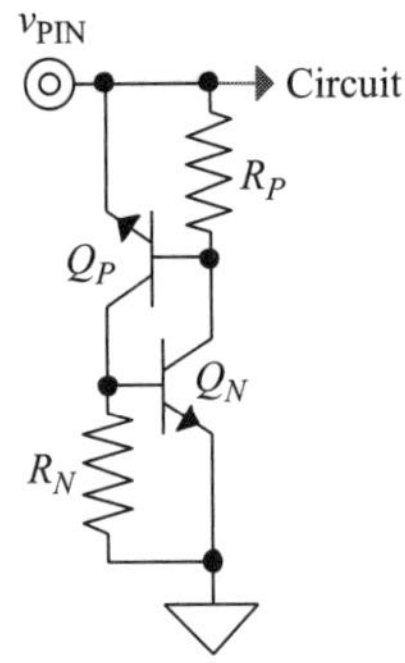

Figure 11.11 Latching BJT clamp.

For the circuit to trigger, though, the time constant of the resistor–capacitor filter should match the rise and fall times expected in the strikes, so tuning the filter is important.

Latching NPN and PNP BJTs built from consecutive N–P–N–P or P–N–P–N junctions are also popular in the industry. These so-called *silicon-controlled rectifiers* (SCRs) rely on positive feedback to latch and steer ESD energy away from a sensitive circuit. Operationally, the voltages across the base–emitter junctions of the BJTs are zero under normal conditions, so the circuit that the BJTs comprise in Fig. 11.11 is off. An ESD event, however, can couple sufficient transient energy into one of the bases to induce one of the BJTs to feed the other, and as a result, to start the positive-feedback action that latches the loop. In other words, as Q_N prompts Q_P to conduct, Q_P responds by forward-biasing Q_N further, which in turn causes Q_P's current to rise even more. This mounting current ultimately drains ESD strikes.

Unfortunately, polysilicon gates are especially sensitive to charged-device strikes. In these cases, inserting a resistor R_{ESD} in series with the clamp helps, like Fig. 11.12 illustrates. This way, the voltage dropped across the resistor limits the gate's exposure to a fraction of the strike voltage. For better results, these clamps should be near the

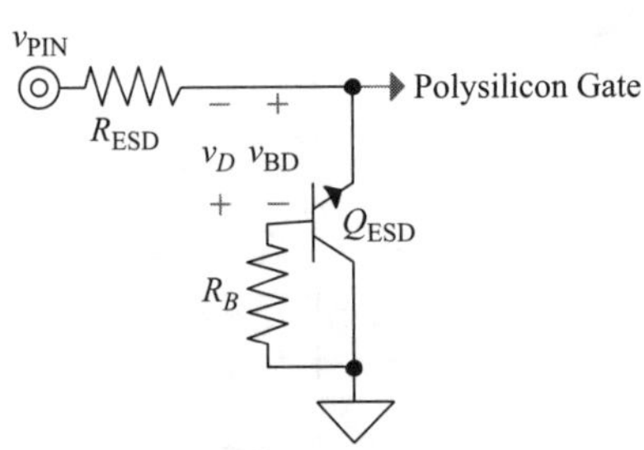

Figure 11.12 Resistor-limited dual-mode diode clamp.

gates they protect, and never in front of nothing but gates, because gates draw no steady-state current and other terminals do. Although not always the case, 1–10 kΩ is usually sufficient to protect CMOS gates.

Luckily, most power transistors already incorporate some form of protection. This is because the power device is already capable of drawing and supplying substantial power from the input supply v_{IN} to the output v_O. And, to reduce series resistance, the transistor is normally near the edge of the die, close to v_{IN} and ground. Plus, to keep diffusions from dropping voltages that could otherwise activate parasitic diodes and BJTs in the power transistor, heavily contacted N- and P-type diffusion and metallic guard rings surround the device. These rings can therefore collect and steer currents with ease to the supplies. This is why the large body diode of a power PMOSFET, for example, protects v_O against positive ESD strikes, because it can steer energy to v_{IN}, away from sensitive components attached to v_O. But if v_O also connects to a polysilicon gate, as in the case where the error feedback amplifier uses a MOS differential pair, adding a local resistor-limited dual-mode diode clamp near the gate exposed is good practice.

11.2 Characterization

Well-specified datasheets shield ICs from overstressing conditions as much as protection circuits do, if not more. The reality is that inaccurate or incomplete information on the statistical performance and operating limits of the IC tempts system designers and end users to overstress the IC. The worst byproduct of this kind of neglect is that the system may not fail immediately but later, after the product that incorporates the IC reaches the market. Unfortunately, recalling and fixing problems after the fact costs a considerable amount in the form of manufacturing and personnel expenses and reputation, which also translates to a reduction in future sales because users and consumers lose confidence in the goods the company sells. Datasheets must therefore specify and include carefully tested and statistically meaningful information on the performance and operating limits of the IC.

This section discusses the most common test methods used to determine the operating limits of a linear regulator. Since semiconductor profits hinge on large-scale production and volume sales, increasing the statistical value and confidence level of the data is extremely important. Testing many ICs from several different wafers and fabrication lots across worst-case operating conditions and temperature corners is therefore a necessary step in the manufacturing process. The underlying aim is to determine the operational and performance limits of the regulator. Although not often quoted in datasheets, the designer must also ensure that the IC starts and recovers properly from worst-case power-up and -down sequences.

11.2.1 Emulating the Load

Knowing the nature of the load is helpful when determining the effects of load range and *load dumps* on the feedback dynamics of the loop that regulates the output. Unfortunately, outside of the specified load-current range, the load is largely unknown and unpredictable, so understanding how the load could possibly affect the system is imperative. Thankfully, a Norton-equivalent model is sufficient, but only for a particular load setting, not the entire range of possibilities. Consider, for example, that while a variable current source can emulate variations in current, one parallel resistance cannot account for all possible load combinations.

Still, modeling the load with a variable resistor, like Fig. 11.13 shows, is practical when considering the physical limitations of a laboratory environment. The idea is to use the regulator's output voltage v_O to establish the load current i_L. This way, since v_O is the ohmic drop across R_L, i_L is v_O/R_L. The problem with this setup is that R_L may not match the small-signal resistance of the actual load when sinking a particular load current i_L. In the case of an operational amplifier that draws 0.5 mA from a 1.8-V regulator, for example, the resistance across the amplifier is probably not 1.8 V ÷ 0.5 mA, or 3.6 kΩ, especially considering that quiescent current changes little with variations in the regulator's v_O. In all fairness, though, the model of a digital IC, just to cite a different example, often reduces to an equivalent resistance because its current is, for the most part, proportional to its supply, to v_O in this case, so a purely resistive load model is sufficient under these circumstances.

The purpose of capacitors $C_{IN(T)}$ and $C_{IN(C)}$ is to shunt noise energy away from the regulator. This is important because the supply voltage v_{IN} can be the output of a switching supply circuit that can ripple 20–50 mV at anywhere from 50 kHz to 1 MHz. Since v_{IN} often derives power from an ac outlet, v_{IN} might also include 50–60-Hz ac noise. A *tantalum* or an *electrolytic capacitor* $C_{IN(T)}$ can therefore suppress low-frequency noise because capacitance and the time constant that its equivalent series resistance (ESR) sets are both, in relative terms, high.

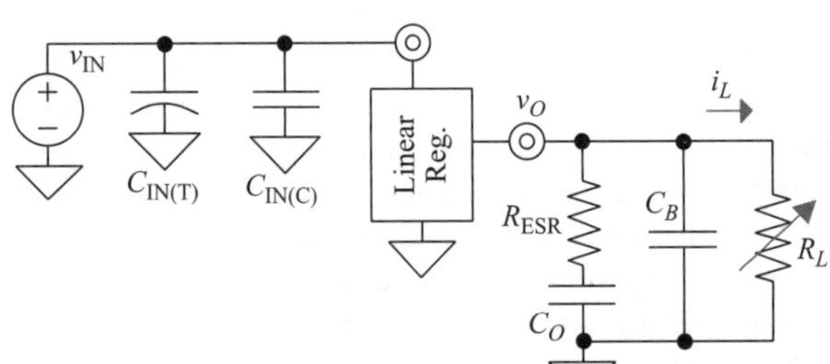

FIGURE 11.13 Emulating the load with a variable resistor.

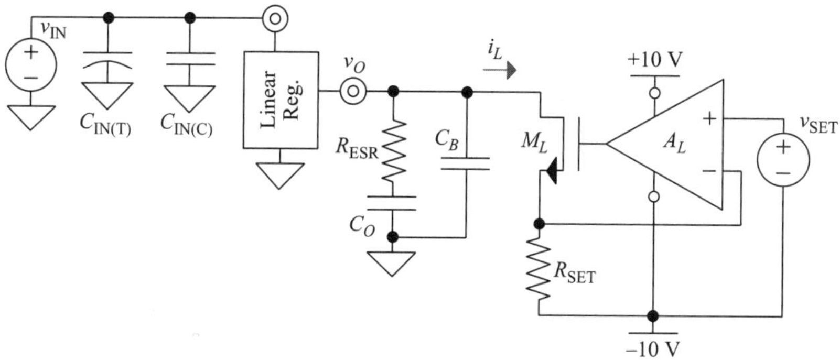

FIGURE 11.14 Emulating the load with a variable current source.

A *ceramic capacitor* $C_{IN(C)}$ usually accompanies $C_{IN(T)}$ because, although the capacitance of a ceramic capacitor is lower, so is the time constant that its ESR establishes. $C_{IN.C}$ can therefore shunt high-frequency noise away from the IC. Another way to eliminate all this noise is to use a battery to supply the IC. Since the output of the battery is neither switched nor connected to an ac outlet, its output is free of these types of systemic noise.

On the other extreme, as mentioned earlier, modeling the load with a current source does not account for variations in load resistance R_L across load current i_L. Still, examining this other extreme is nonetheless useful because the load is, as already stated, principally unknown. The circuit in Fig. 11.14 emulates what amounts to an all-active load. Here, the operational amplifier A_L series-mixes a programmable voltage source v_{SET} with the voltage across R_{SET} to set the current i_L that transistor M_L series-samples. A_L also amplifies the degenerating effect of R_{SET} on M_L's output resistance R_L. As a result, R_{SET}'s voltage is nearly v_{SET} and i_L is therefore v_{SET}/R_{SET}, and R_L is A_L times greater than M_L's g_m r_{ds} R_{SET}. Although modeling the load in this fashion does not include the effects of a purely resistive load, implementing both setups examines the effects of both extremes of R_L on the loop dynamics of the system. Note, by the way, that R_{SET} connects to a negative supply to ensure that the voltage across M_L is high enough to keep M_L operating in the saturation region—in other words, to ensure that M_L's r_{ds} remains high across i_L's operating range.

11.2.2 Regulation Performance

Load

Load regulation refers to how much the output v_O varies when subjected to full-scale, steady-state changes in load current i_L, as when v_O drops 15 mV after i_L climbs from nil to 5 mA. Measuring load regulation amounts to subjecting a linear regulator to the variable resistor load

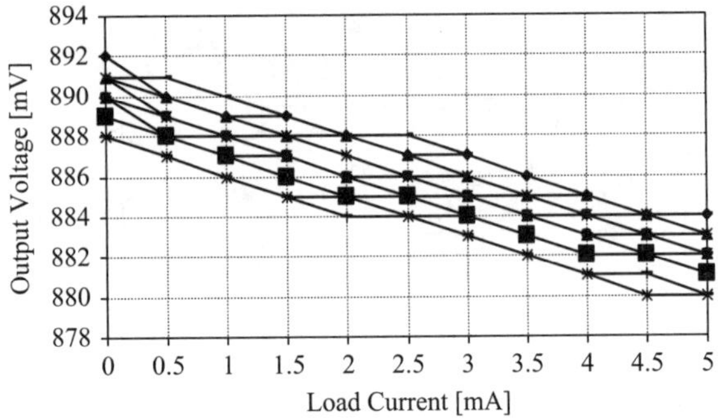

FIGURE 11.15 Load regulation.

in Fig. 11.13 or current-source load in Fig. 11.14, sweeping i_L, and monitoring v_O. To ensure that v_O settles to its steady-state value before recording a measurement, changes in i_L must be slower than the speed of the regulator. In other words, *soak time*, which ultimately translates to *test time* in production, should be long enough for the IC to respond and settle. With sufficient soak time, and at a nominal input-supply setting, product engineers measure load regulation across temperature, dies, wafers, and fabrication lots. When finished, a load-regulation graph of the output across all load levels, like the Fig. 11.15 illustrates, and an accompanying summary table report the three-sigma variation of the output across the specified load range, temperature, and fabrication corners.

Line

Line regulation refers to how much the output v_O varies when subjected to full-scale, steady-state changes in the input supply v_{IN}, as when v_O falls 5 mV after v_{IN} drops from 5 to 1.8 V. The setup and procedure for extracting line regulation closely resemble those of load regulation. The only difference is that v_{IN} varies while the load i_L remains constant. Since line regulation refers to the range of v_{IN} for which the IC remains linear, v_{IN} should not extend down to the headroom limit $v_{IN(MIN)}$ of the circuit. So to avoid mixing the temporary effects of a headroom crisis and startup, whose time constants can be long, v_{IN} should fall, rather than rise. v_{IN} should therefore drop from its maximum specified limit to its minimum counterpart with sufficient soak time before each recording of v_O to allow v_O to settle to its final steady-state value. When finished, a line-regulation graph of the output across all allowable supply levels, like the one in Fig. 11.16, and an accompanying summary table report the three-sigma variation of the output across the specified supply range, temperature, and fabrication corners.

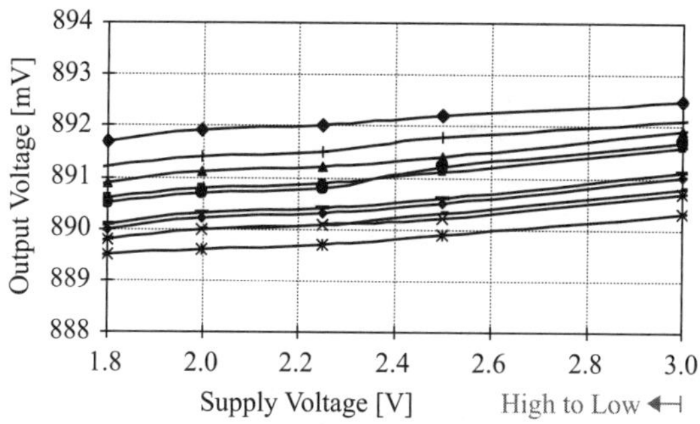

FIGURE 11.16 Line regulation.

Temperature

Temperature drift is another steady-state parameter, which means that testing drift requires a similar circuit and analogous measures to those of load and line regulation. The difference, of course, is that temperature varies while both the load i_L and the input supply v_{IN} are constant. Soak time in this measurement is normally longer because the *thermal time constant* of the IC in a plastic package is considerably longer than the circuit's bandwidth. If one is not careful, in fact, the output v_O may show false signs of *thermal hysteresis* when raising and lowering the temperature. In other words, the drift in one direction may differ from that in the other. If the hysteresis is indeed an artifact of the measurement and not inherent to the circuit, which is typically the case, it should disappear when soak time extends beyond the thermal time constant of the packaged IC.

Temperature drift in v_O across process corners for trimmed parts may not be monotonic or consistent, even when trimmed to its optimal setting. This means that quoting the slope in volts-per-degree or the corresponding *fractional temperature coefficient* TC in parts per million per degree is not entirely relevant. The best means of describing drift is, as with a reference voltage, the *box method*. The idea is to enclose all extreme values of v_O across temperature in a box, and quote the three-sigma variation of its vertices. When finished, a temperature-drift graph of the output across all allowable temperatures, like the one in Fig. 11.17, and an accompanying summary table report the worst-case three-sigma variation of the output across the specified temperature range and across fabrication corners.

Trimming

In the case of Fig. 11.17, notice that the tendency of the mean is to drop with temperature. This means, the reference v_{REF} that sets the output v_O is not at its optimal setting. Reversing this trend amounts to raising v_{REF}'s PTAT component slightly. Fortunately, most systems *trim* their references by adjusting PTAT voltages. As a result, elevating

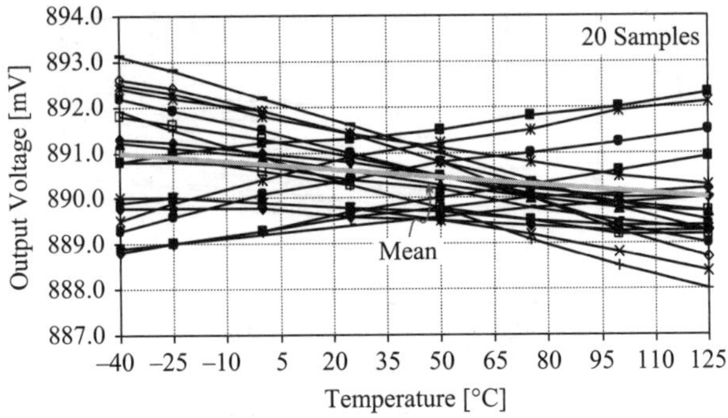

FIGURE 11.17 Temperature drift.

the trim target adds PTAT voltage to v_{REF}, which in turn, pivots the mean counterclockwise so that v_{REF}, and as a result, v_O fall less with temperature. Changing v_{REF} this way, however, means the feedback resistors that translate v_{REF} to v_O must also adjust to accommodate the new value of v_{REF}.

Knowing how to determine the trim target of a particular part is important in this process. In the case of bandgap-like references, a linear PTAT voltage v_{PTAT} counters the falling tendency of a diode voltage v_D to establish a reference v_{REF} that is largely independent of temperature. But because v_D falls more at higher temperatures, v_{REF} usually bends like Fig. 11.18 demonstrates. As a result, v_{REF} incurs the least variation when its slope and resulting fractional temperature coefficient are zero halfway across the temperature range, at T_{MID}.

Since the trimming process adjusts the PTAT component of v_{REF}, raising the trimming target increases the slope of v_{REF}. The lowest trim point of v_{REF} therefore produces the lowest slope and lowest temperature coefficient, and vice versa, as Fig. 11.19 shows. So because v_{PTAT} raises the slope linearly, the linear extrapolation of the trim range that produces a slope of zero corresponds to the optimal setting

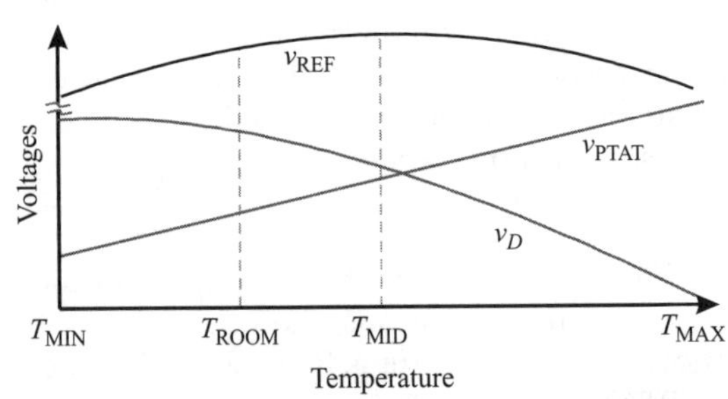

FIGURE 11.18 Trimmed reference.

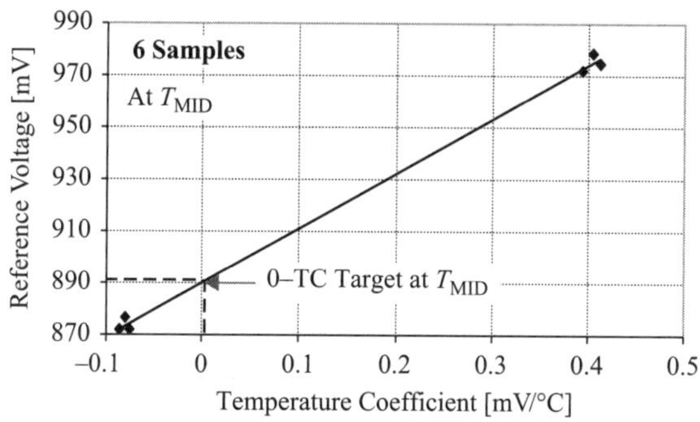

FIGURE 11.19 Temperature coefficient of the reference across trimming range at T_{MID}.

of v_{REF} at T_{MID}. In this case, 890 mV is the best T_{MID} target. This means that targeting a voltage at room temperature T_{ROOM} in Fig. 11.18 that corresponds to 890 mV at T_{MID} produces the least variation in v_{REF}. This also means that the temperature coefficient of v_{REF} at T_{ROOM} should not be zero for optimal results. Repeating this extraction process for several ICs from different wafers and fabrication lots reveals a statistical mean target that produces the least variation in v_{REF} and v_O.

Load Dumps

Specifying transient variations in the output v_O when subjected to sudden load changes is more difficult than its load, line, and drift siblings because predicting the nature of the load dumps is mostly guesswork. Even if the total transient variation in the load Δi_L were known, and notice that the value often differs from its full-scale range, rise and fall times t_R and t_F may not be, and the total variation in v_O depends on all three, on Δi_L, t_R, and t_F. This is why the nominal time-domain graph that datasheets typically show is unfortunately inconclusive, because t_R and t_F are neither apparent from the graph nor quoted. What is more, the response is normally optimistic because t_R and t_F are often in microseconds, near the bandwidth of the regulator, which means the regulator is fast enough to track changes in i_L. In all fairness, though, quoting difficult-to-achieve rise and fall times of 1–100 ns may not be realistic. Still, even if not quoted, system designers understand this inaccuracy, so they normally allocate 3 to 7 percent of margin in accuracy to variations in v_O when responding to rising and falling widespread load dumps.

Unfortunately, regulators targeted for *system-on-chip* (SoC) platforms do not enjoy the "specmanship" flexibility that stand-alone devices do. In fact, modern mixed-signal systems not only switch at high frequencies but also rely on low on-chip output capacitances. As a result, synchronized transitions are quick, and their effects on the output v_O can be substantial. Luckily, the load is usually more predictable and its magnitude is less severe in on-chip systems.

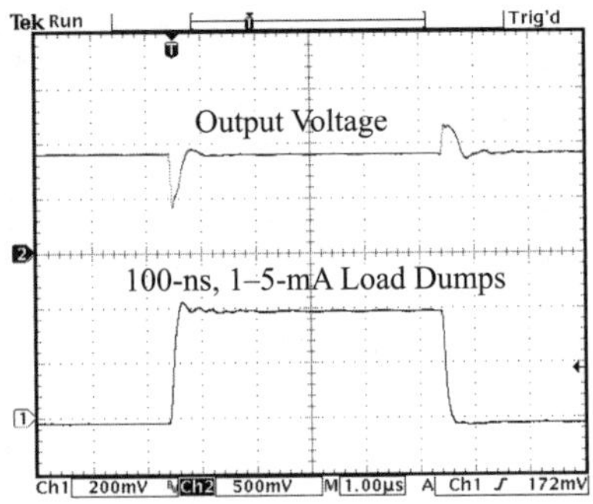

FIGURE 11.20 Load-dump response.

Figure 11.20 is an example that illustrates how a 100-ns, 1–5-mA load dump switching at 100 kHz produces 500-mV excursions in v_O.

Emulating worst-case load dumps with a resistor R_L is possible. The amplifier network in Fig. 11.21, for example, sets the voltage of R_L's bottom terminal v_B to pull a load current i_L from v_O that is proportional to the ohmic drop across R_L and inversely proportional to R_L. For this, the amplifier A_L series-mixes the output of a pulse generator v_{PLS} with v_B to impress v_{PLS} on v_B. This way, assuming the IC regulates v_O about its target $V_{O(TAR)}$, i_L is roughly:

$$i_L = \frac{v_O - v_B}{R_L} \approx \frac{V_{O(TAR)} - v_{PLS}}{R_L} \tag{11.11}$$

So if $V_{O(TAR)}$ is 2 V, R_L is 40 Ω, and v_{PLS} pulses between 1.96 and 0 V, i_L pulses between 1 and 50 mA.

The underlying purpose of the amplifier A_L is to sink the current that the generator that outputs v_{PLS} cannot. Similarly, the reason for buffering A_L's output with an N-channel power transistor M_L is that A_L might not be able to sustain i_L. In this case, M_L's source connects to

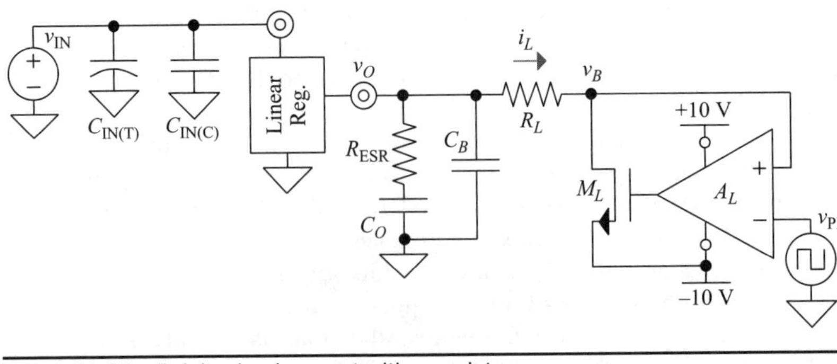

FIGURE 11.21 Pulsing load current with a resistor.

a negative supply that is well below ground to ensure that M_L operates in saturation even when M_L's drain terminal, which carries v_B, falls below ground. A_L, by the way, should be sufficiently fast and powerful to drive M_L's gate within the rise and fall times needed to test the IC.

This circuit, unfortunately, suffers from two drawbacks. One is that v_O is not always near $V_{O(TAR)}$ when fully loaded, so the maximum load current that the circuit can establish may not be as predictable as the minimum. The second disadvantage is that establishing milliamps from a few volts requires tens of ohms. This means R_L can only test load-dump responses when the regulator's load resistance is low, which is not always the case in practice, especially when loaded with on-chip amplifiers, bias circuits, and other analog functions whose currents do not change with the input supply. Luckily, the current-source circuit in Fig. 11.14 is better in both respects because v_{SET} and R_{SET} alone set i_L, R_{SET} degenerates the transistor M_L that sinks i_L, and the amplifier A_L magnifies the degenerating effect of R_{SET}. As a result, i_L is independent of v_O, and the output resistance of M_L is high and equivalent to that of a regulated-cascode current source.

Accuracy

The total accuracy performance specified in a datasheet typically includes the combined three-sigma effects of load, line, and temperature on the output v_O across several parts, wafers, and fabrication lots. Perhaps the most practical means of integrating and visualizing these effects into a single specification is to monitor, measure, and graph v_O across load current i_L, as though the measurement were for load regulation, but at the extremes of the input supply v_{IN} and temperature. In other words, the combined accuracy graph would mimic the load-regulation graph in Fig. 11.15, except with four different families of curves belonging to all possible extreme combinations of v_{IN} and temperature: at the highest v_{IN} and highest temperature, lowest v_{IN} and lowest temperature, highest v_{IN} and lowest temperature, and lowest v_{IN} and highest temperature. Then, using the box method, the specification would report the three-sigma variations of the worst-case corners of the box. Note, by the way, that the accuracy specification quoted on a datasheet normally excludes the effects of load dumps. So a ±3-percent error specification does not necessarily imply v_O will remain within 3 percent of its target when subjected to load dumps.

Power-Supply Rejection

One of the reasons why *power-supply ripple rejection*, as it is often called, also describes *power-supply rejection* (PSR) is because the output of a switching dc–dc converter normally feeds the linear regulator, so the setup used to measure PSR literally produces a "ripple" in the supply voltage v_{IN}. The measurement therefore amounts to introducing a sinusoid of a particular frequency f_O into v_{IN}, monitoring the resulting ripple in the output v_O, measuring the ratio of the peaks in v_{IN} and v_O,

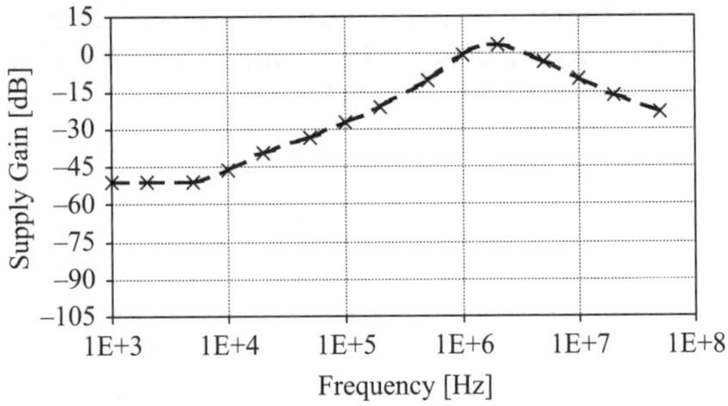

FIGURE 11.22 Power-supply rejection.

and repeating the procedure at other frequencies. Since the purpose of the circuit is to regulate v_O against variations in the load i_L and v_{IN}, the supply gain A_{IN} across the circuit $\Delta v_O/\Delta v_{IN}$ should be low between 0 and perhaps –60 dB, like Fig. 11.22 demonstrates. And because rejection is the complement of gain, PSR is the reciprocal of A_{IN}, so PSR refers to $\Delta v_{IN}/\Delta v_O$ and should therefore range between 0 and 60 dB. Datasheets often exclude a graph and instead list a few representative data points, like 50 dB at 1 kHz when loaded with 10 mA and supplied from 3 V.

In practice, a function generator alone is not a suitable replacement for v_{IN} because the generator cannot normally drive the currents that a linear regulator can supply. This is why the amplifier A_{IN} and the power transistor M_{IN} in Fig. 11.23 buffer what the generator outputs as v_{SIN} to establish a rippling v_{IN}. In this case, A_{IN} series-mixes the

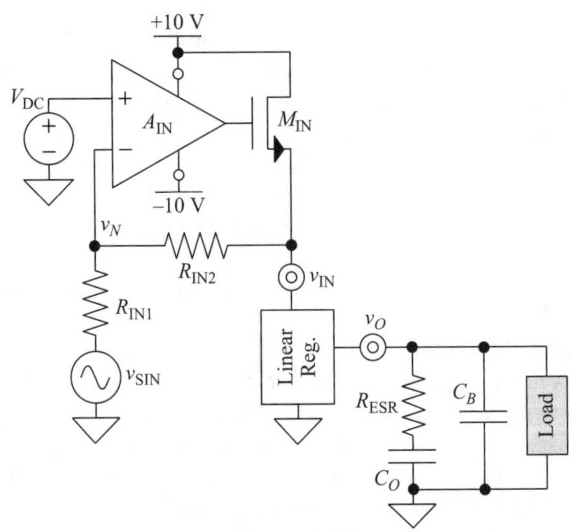

FIGURE 11.23 Injecting a ripple into the supply.

output of a dc voltage source V_{DC} with A_{IN}'s inverting input v_N, and the feedback resistors R_{IN1} and R_{IN2} translate v_N's impression of V_{DC} to establish a steady-state voltage at v_{IN}. The T connection at v_N shunt-mixes the current that the function generator v_{SIN} supplies through R_{IN1} into the loop to integrate a ripple into v_{IN}. This way, the circuit combines the noninverting network that amplifies V_{DC} with the inverting counterpart that amplifies v_{SIN} to establish a supply voltage that ripples according to v_{SIN} about a steady-state value that V_{DC} sets:

$$v_{IN} \approx V_{DC}\left(\frac{R_{IN1} + R_{IN2}}{R_{IN1}}\right) - v_{SIN}\left(\frac{R_{IN2}}{R_{IN1}}\right) \tag{11.12}$$

The peak-to-peak value of the input ripple is normally on the order of 20–100 mV to mimic the ripple that switching power supplies typically generate. When evaluating the results, consider that the measured results may not correlate exactly with the small-signal response obtained from ac simulations. This is because 100-mV variations in v_{IN} may not constitute the "small signals" that the simulator expects, especially when the regulator is close to dropout. Time-domain simulations that inject a ripple into v_{IN} and monitor the resulting ripple in v_O, as expected in real life and applied to the actual measurement, are therefore more accurate, especially when v_{IN} causes transistors to transition across operating regions from saturation to triode and vice versa.

11.2.3 Power Performance

Power-Conversion Efficiency

Determining the *power-conversion efficiency* η_C of a linear regulator amounts to measuring the input current i_{IN} that the input supply v_{IN} sources across the entire load-current range. This is because, outside of i_{IN} and load current i_L, η_C depends only on v_{IN} and the output v_O, both of which the user or the targeted application define:

$$\eta_C = \frac{P_O}{P_{IN}} = \frac{i_L v_O}{i_{IN} v_{IN}} = \eta_I\left(\frac{v_O}{v_{IN}}\right) \tag{11.13}$$

Therefore, because specification targets normally set v_{IN}, v_O, and i_L to prescribed levels, extracting and quoting ground current i_{GND}, or equivalently, quiescent current i_Q, which flows across v_{IN} and ground, is more meaningful. More meaningful yet is current efficiency η_I because, while i_Q is often negligible when i_L is high, i_Q can become a considerable fraction of i_L otherwise, when the system idles and i_L is low:

$$\eta_I \equiv \frac{i_L}{i_{IN}} = \frac{i_L}{i_{GND} + i_L} = \frac{i_L}{i_Q + i_L} \tag{11.14}$$

Extracting η_C and η_I amounts to subtracting i_L from i_{IN} when measuring load regulation at several v_{IN} and v_O settings. A summary table therefore specifies three-sigma variations in i_{GND}, or i_Q, at the extremes of the load range, or a nominal current at the halfway point, like for example, 20 μA when loaded with 10 mA. Note that i_{GND} is not always a function of i_L, but if i_{GND} were, a nominal graph of i_{GND} across i_L may also accompany the specification table.

Dropout

Measuring *dropout voltage* V_{DO} is not always straightforward. The difficulty usually arises in applications that demand low output voltages v_O because, when lowering the input supply v_{IN}, the regulator may exhaust its headroom $v_{IN(MIN)}$ before dropping out of regulation. In those cases, modifying the resistor divider so the feedback network raises v_O's target $V_{O(TAR)}$ raises the input onset voltage of the dropout region. This way, reducing v_{IN} pushes the regulator into dropout before headroom can run its course. But to ensure that this is the case, understanding how the regulator behaves across all of its regions of operation is important.

Consider, for example, the results graphed in Fig. 11.24, as obtained by expanding the line-regulation measurement in Fig. 11.16 to include lower v_{IN} values. The regulator tested here outputs roughly 0.9 V when v_{IN} is between 1.25 and 2.5 V. This v_{IN} range corresponds to the *linear region* because the feedback loop still has sufficient gain to regulate v_O about $V_{O(TAR)}$. Between 1.15 and 1.25 V, v_O rises slightly to 1.15 V. This happens because v_{IN} is low enough to push one or several transistors in the regulator closer to triode, which reduces the gain across the loop, and as a result, lowers the regulating accuracy of the circuit. Although not all regulators raise v_O in this region, v_O always deviates from $V_{O(TAR)}$ when gain drops.

Lowering v_{IN} below this point, below 1.15 V in Fig. 11.24, reduces the loop gain to zero, which is when the circuit enters dropout.

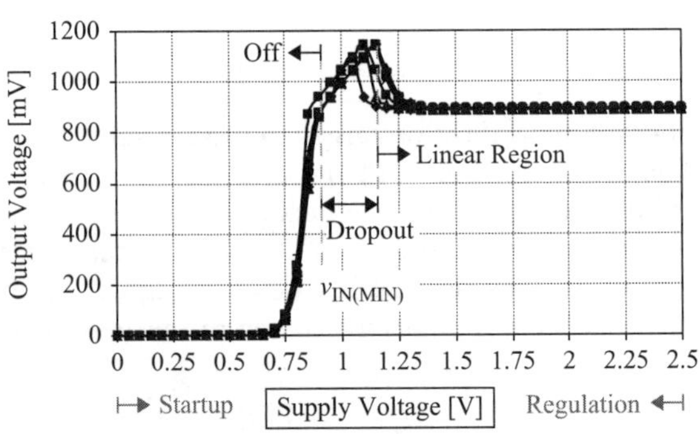

FIGURE 11.24 Dropout and headroom measurement.

In this mode of operation, v_O falls almost linearly with v_{IN}, and the dropout voltage V_{DO} is the voltage dropped across the power transistor: $v_{IN} - v_O$. In practice, V_{DO} drops with lower v_{IN} values because the base or gate drive of the power transistor also falls with v_{IN}. This is one of the reasons why product engineers often quote V_{DO} at its onset, at the edge of the linear region. More importantly, though, designers aim to keep the regulator just above the onset of dropout. This way, the regulator still regulates v_O while dropping the lowest possible voltage across the power transistor. In other words, V_{DO} at the onset of dropout is arguably the most relevant value to quote.

Datasheets normally report V_{DO} only when fully loaded, at the worst possible load. Also note that, as with line regulation, reducing v_{IN} from its maximum specified setting is better than raising v_{IN} from zero. This is because the latter suffers from the effects of startup, whose time constant can be so long that soak time may be insufficient to secure an accurate measurement.

11.2.4 Operating Requirements

Load Range

Although not always the case, the onset of dropout sometimes sets the regulator's maximum output current $i_{O(MAX)}$. In fact, setting $i_{O(MAX)}$ to the current required to induce a dropout voltage of 200 mV is not unusual in industry. For this, sweeping i_L when v_{IN} is about 100 mV above the targeted output $V_{O(TAR)}$ and monitoring when the output v_O drops 200 mV below v_{IN} indicates when i_O reaches $i_{O(MAX)}$.

This method of extracting $i_{O(MAX)}$, however, only applies to stand-alone regulator ICs, whose target market is not as clear as their application-specific counterparts. System-on-chip designs, just to cite an increasingly ubiquitous example, must supply specific load-current ranges under predetermined dropout conditions. In other words, extracting $i_{O(MAX)}$ is not as relevant or as meaningful as measuring load regulation across the actual or desired load range.

Headroom

The dropout measurement in Fig. 11.24, which is an expanded version of the line-regulation graph of Fig. 11.16, also includes the effects of *headroom*. $v_{IN(MIN)}$, which for Fig. 11.24 is 0.9 V, is the input-supply setting below which the circuit is no longer able to determine the state of the output v_O. From an operational perspective, transistors no longer have headroom to conduct current when v_{IN} is below $v_{IN(MIN)}$. This means that the loop effectively opens when v_{IN} falls below $v_{IN(MIN)}$, which is another way of saying negative feedback loses sense of v_O. As a result, the power transistor no longer tries to sustain i_L, so i_L pulls v_O toward ground.

Filter

Determining the acceptable filter range for which the regulator is stable is, perhaps, more involved than extracting other parameters. First consider that testing stability amounts to disturbing the system in the worst way possible and monitoring the output to see whether it settles or oscillates. In other words, a regulator is stable if it survives being subjected to the fastest and widest rising and falling load dumps possible. The main challenge here, as before, is the unpredictable nature of the load, because overstressing the regulator with excessively fast and wide load dumps is as pessimistic as understressing the system is unrealistic. Coupling this uncertainty with how involved the measurement is in practice explains the inconsistency of the methods used to extract this range in industry and the seemingly incomplete datasheets that result.

Ultimately, *stability*, unlike *phase margin*, which quantifies proneness to unstable conditions, is a binary parameter because the circuit under a given filter scenario is either stable or not stable. As a result, stability tests should include all possible worst-case conditions, such as full-scale load dumps from zero to $i_{O(MAX)}$ with the fastest possible rise and fall times at maybe 10–100 ns with 10 ms between them at the extreme limits of the temperature range. The problem with slower rise and fall times is that they do not expose the circuit to high-frequency perturbations, which spark oscillations.

One difficulty here is repeating the experiment for many combinations of output capacitance C_O and equivalent series resistance R_{ESR}. The easiest approach, although not always the most practical, is to find enough capacitors to populate the entire C_O–R_{ESR} space and use them to filter v_O when subjected to fast load-dump conditions. Then, if no oscillations persist, the circuit is stable. The problem with this approach is finding enough capacitors with different C_O–R_{ESR} combinations to fully characterize the regulator. Bear in mind that using transistors to switch resistors in the path of C_O is not always practical because switch resistance can be on the order of the R_{ESR} targeted, or higher, which means switch resistance can be both considerable and parasitic.

Soldering several low-ESR ceramic multilayer 1–10-nF capacitors in parallel, like Fig. 11.25 shows, and connecting the entire array in series with a bank of 1-Ω power resistors in parallel to build and emulate several C_O–R_{ESR} combinations is perhaps more practical. With this approach, the test can start with the highest possible C_O and lowest possible R_{ESR} combination, when all capacitors and resistors connect. Disconnecting one resistor after each low- and high-temperature measurement is next. Then, resoldering all resistors and repeating the process after disconnecting one capacitor follows. The procedure repeats this sequence until all but one capacitor remains in the filter, at which point the process finishes testing the entire C_O–R_{ESR} array.

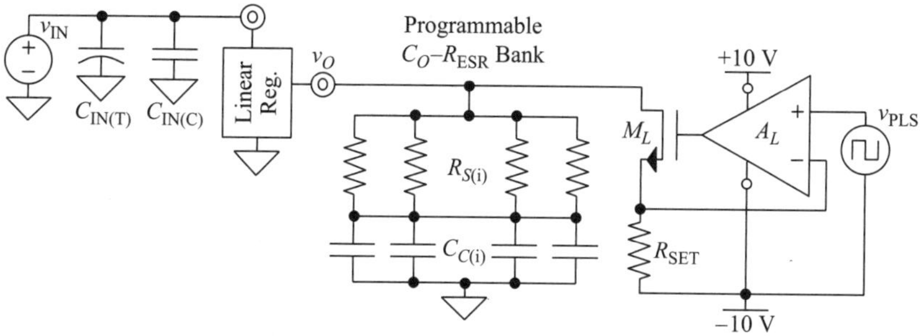

FIGURE 11.25 Testing stability across output resistance–capacitance combinations.

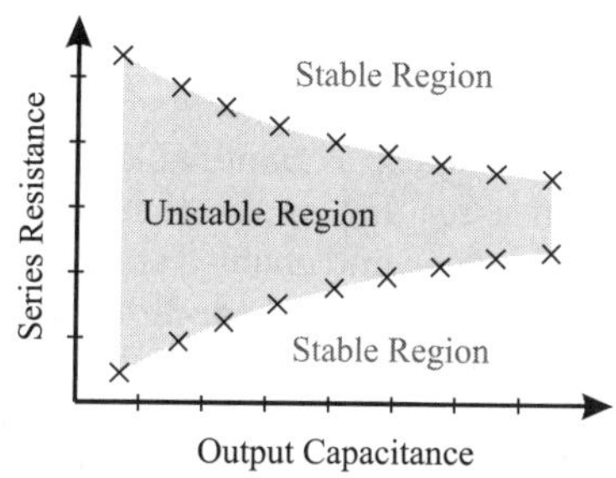

FIGURE 11.26 Stability across resistance–capacitance combinations.

Figure 11.26 demonstrates the results of a sample set of stability measurements for an output-compensated regulator. Each data point represents the edge beyond which a rise or fall in R_{ESR} causes sustained oscillations in v_O. Some designers call the C_O–R_{ESR} boundaries that result "the curves of death" because they map where the regulator is and is not stable, that is to say, the boundaries within which the circuit survives. As expected, the graph shown depicts a regulator that is generally more stable with higher output capacitance C_O, which corresponds to a regulator whose output pole p_O is dominant. For stable conditions to persist at low C_O, R_{ESR} must be low enough to keep the zero that R_{ESR} establishes from extending the unity-gain frequency f_{0dB} of the system into the parasitic-pole region. Or R_{ESR} must be sufficiently high for the ESR zero to cancel the effects of the system's second dominant pole, which is normally the pole at the output of the error amplifier p_E.

An internally compensated response generally complements the graph shown because the regulator would generally become more stable with lower values of C_O. With high output capacitances, an internally compensated regulator normally requires the ESR zero that R_{ESR} establishes to be low enough to offset the effects of p_O before reaching f_{0dB}. The system can also be stable with high output capacitances if another zero offsets p_O and R_{ESR} is low enough to keep the ESR zero from extending the system's f_{0dB}.

11.2.5 Startup

Startup or power-up sequences are sometimes absent in datasheets because the mere fact that the datasheet or the designer claims the circuit works is a guarantee that the circuit starts. Emerging portable applications, however, are starting to demand this specification, not because they wish to verify its functionality, but because they can only tolerate short startup times, within maybe 1 μs or less. The driving motivation for characterizing this parameter is that small, battery-powered devices cannot afford to dissipate any quiescent power indefinitely, and they must therefore engage and disengage circuits to preserve energy and extend operational life. So when enabling a block on demand, startup time is part of the system's response time. As a result, testing startup is not only a functional test but also a quantifiable parameter.

Startup tests, as in a real-life application, must consider supply ramps, supply and load pulses, and enabled starts. Since full-rail supply ramps can occur in both microseconds and milliseconds, tests should include both extremes. Tests should also include intermediate rise times to ensure that the corresponding frequency of the ramp does not prompt uncontrolled oscillations in the output v_O. Similarly, the regulator should recover from noise- and load-induced pulses in the input supply v_{IN} and v_O because the switching supply that feeds the regulator might rise and fall 1 V within 1 μs and v_O might momentarily short to ground. The circuit used to test power-supply rejection in Fig. 11.23 can also emulate short and steady variations in v_{IN}. The experimental setups used to test load regulation in Figs. 11.14 and 11.21 can similarly test momentary short-circuit events. As an example, Fig. 11.27 shows that 20 regulator ICs require roughly 0.5 ms to start and settle their outputs to their target when the supply ramps up in 100 ns.

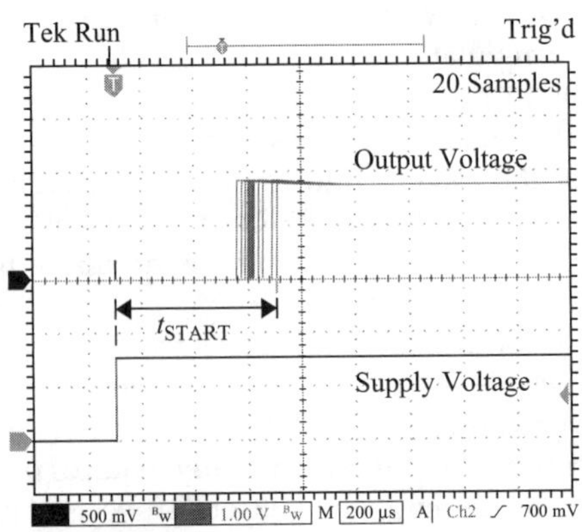

FIGURE 11.27 Startup response.

11.3 Summary

Protecting a linear regulator IC is as important as designing it in the first place. This is because, in practice, the load may short, just to cite an example, and stress the IC long enough to elevate the junction temperature beyond the melting point of the plastic package. Including overcurrent protection or thermal shutdown in the IC guards the regulator against this type of event. And more specifically, it protects the power transistor from such strenuous and damaging conditions.

The designer must also anticipate the human factor because people can not only reverse the battery but also expose the IC to sudden electrostatic discharges. Unfortunately, one of the most challenging aspects of incorporating protection that is always alert and ready to respond to all possible violations is maintaining transparency. In other words, protection circuits should not dissipate power, introduce voltage drops, or react to false alarms produced by noise.

Ignorance on the part of the user or system designer can be equally destructive, if not more so, because systematically overstressing the regulator with a load that the circuit cannot handle may cause it to fail not immediately but later, when a product using the IC is in full production. Characterization is by far the best means of increasing reliability and mitigating the risk of a *recall*. The data, however, is only as valid as the measurements, which is why emulating the load and carefully considering the experimental setup are so important in this process. While a resistor alone, for example, can fully model a particular load, it cannot represent all possible loads, so complementing the measurements with a current-source load is often a sensible precaution. Bear in mind that, although not always evident, the savings from avoiding short- and long-term product failures justifies—within reason—the test time, effort, and other related costs incurred when fully characterizing the regulation, power, and operating limits of the system.

11.4 Review Questions

1. What is meant by "transparency" of protection circuits?

2. What is the ultimate advantage of fold-back current limiting over a fixed limit?

3. When the power transistor is not the limiting agent, what phenomenon should determine the onset target temperature of a thermal-shutdown circuit?

4. What is the most benign or "transparent" form of reverse-battery protection?

5. What is the drawback to that form of protection?

6. What are the three most popular electrostatic-discharge models?

7. What range of voltages can electrostatic discharges impress on chips?

8. What are the most popular electrical components used to absorb electrostatic discharges?

9. What is the drawback to stacked forward-biasing diode clamps to the input supply and ground when absorbing electrostatic discharges?

10. How can the load of a regulator be emulated?

11. What does load regulation describe?

12. Should line regulation be tested when the input supply rises or falls? Why?

13. What is soak time?

14. What is(are) the effect(s) of lowering the trim target on the reference voltage?

15. At what temperature should the temperature coefficient of the reference be zero for optimal results?

16. Which variation in the output does total accuracy normally exclude in datasheets?

17. What is another name for power-supply rejection?

18. Why use an off-chip power transistor and an off-chip amplifier when testing power-supply rejection, load regulation, and load-dump response?

19. Which parameters describe power-conversion efficiency, and of those, which are normally not open design variables?

20. What happens to the circuit when the regulator enters the dropout region?

21. Why does dropout voltage normally rise with reductions in the input supply voltage?

22. At which point of the dropout region is dropout voltage normally quoted?

23. If the regulator skips the dropout region when transitioning from the linear region to cut off, how can dropout voltage be measured?

24. How is stability tested?

25. Why is startup time an increasingly important parameter in emerging applications?

Index

Note: Page numbers followed by *f* denote figures; page numbers followed by *t* denote tables.

D

E

F

J

K

L

M

S

Z